TRANSPORT PHENOMENA IN POROUS MEDIA

TRANSPORT PHENOMENA IN POROUS MEDIA

Edited by

DEREK B. INGHAM
Department of Applied Mathematical Studies
University of Leeds
Leeds LS2 9JT
United Kingdom

IAON POP
Faculty of Mathematics
University of Cluj
R-3400 Cluj, CP 253
Romania

PERGAMON
An Imprint of Elsevier Science

ELSEVIER SCIENCE Ltd
The Boulevard, Langford Lane
Kidlington, Oxford OX5 1GB, UK

First edition 1998

Library of Congress Cataloging in Publication Data
A catalog record from the Library of Congress has been applied for.

British Library of Cataloguing in Publication Data
A catalogue record from the British Library has been applied for.

ISBN: 0 08 042843 6

♾ The paper used in this publication meets the requirements of ANSI/NISO Z39.48-1992 (Permanence of Paper).
Printed in The Netherlands.

CONTENTS

PREFACE

Although this book reviews the state-of-the-art in transport phenomena in porous media, it will also be very useful to all those researchers working in fluid mechanics and heat transfer. In preparing it, we were motivated by the fact that the area of heat transfer, which is closely related to energy utilization and conservation, has received considerable attention, especially since the worldwide energy crisis arose. Further, the field of heat transfer in porous media is of great interest in many varied applications in contemporary technologies. Prominent among these applications being geothermal energy resources; nuclear energy systems; petroleum reservoirs; analysis of insulating systems; storage of grain, fruit and vegetables; flows through tobacco rods; pollutant dispersion in aquifers; post accident heat removal from nuclear reactor rubble beds; the processes of crystal manufacture, foam metals and fibrous media; to name but a few.

The book contains seventeen chapters which have been prepared by some of the outstanding researchers in the modern area of convective flow in porous media. Although much of the subject matter contained in this book is available in the research literature, in journals and conference proceedings, this book presents an overview of the most recent developments over a large variety of the most important topics in transport phenomena in porous media. This research area is increasingly being recognised as an extremely important factor in current practice of research and therefore the present initiative will meet an important need and serve to make the subject of porous media more easily accessible to researchers in heat transfer and fluid mechanics. In spite of the intention of this book to be used as a reference book for research in heat transfer in porous media, the chapters are written in a completely self-contained manner. Every chapter offers an immediate source of reference to the physical concepts, explanations, terminology and the methods employed in numerous topics in convective flow through porous media.

The collection of research topics included in the book is suitable for a wide range of interests, ranging from advanced students to researchers and practitioners in porous media.

The Editors would like to acknowledge the substantial contributions made by a large number of people who have been very helpful in the completion of this project. We especially would like to thank all the authors who have generously agreed to contribute a chapter and thus sharing with us their great knowledge and experience. Without their excellent contributions the Editors acknowledge that this project could not have been realised. Further, we would like to acknowledge some of the research workers who have assisted us in the reviewing of the chapters of the book, and all of them have taken a keen interest in this project.

In this regard, we especially would like to thank Professors A. Bejan, T. Na and D.A. Nield, Drs. L. Elliott and D. Lesnic and the Ph.D. student, Mr. R. Bradean. Additionally, we wish to express our most sincere appreciation to Professors A. Bejan, P. Cheng and D.A. Nield for the encouragement they have given us throughout all our research endeavours in fluid flows in porous media. Their example, through the quality and innovativeness of their work has been an inspiration to many researchers within the porous media community and this has certainly had much to do with the current healthy state of research into porous media.

Finally, we wish to express our gratitude to Mr. Keith Lambert, Senior Publishing Editor, for his enthusiastic handling of this project.

Leeds / Cluj
August, 1997

Derek B. Ingham / Ioan Pop

THE FUNDAMENTAL THEORY OF FLOW THROUGH PERMEABLE MEDIA FROM DARCY TO TURBULENCE[†]

J. L. LAGE

Mechanical Engineering Department, Southern Methodist University,
P. O. Box 750337, Dallas, TX 75275-0337, USA

INTRODUCTION

Professor Arthur Bergles began his keynote speech during the 1996 ASME National Heat Transfer Conference in Houston saying: "Old is good!". It is indeed the responsibility of the new generation to show interest and recognition for the *old*. But it is also our responsibility to challenge existing ideas, and to question and seek confirmation of old *dogmas*, i.e. myths hampering the progress of knowledge.

During the past years I have observed several *dogmas* in the area of flow through permeable media like the writing of *D'Arcy* as the last name of Henry Darcy, the pioneer of hydraulics in saturated porous media, or the writing of Forchheimer as *Forschheimer* (notice the 's'). Although irrelevant, these errors are nevertheless disturbing: if it can happen with names, it can very well happen with model equations, mathematical solutions, experimental data or with the interpretation of physical phenomena.

Writing this invited review chapter on the theory of incompressible flow through permeable media was the opportunity to seek answers to some more fundamental questions (dogmas). In my journey I travelled back to the 19th century, when Darcy reported his theory, and follow the historic events leading to the most recent (1997) ideas on modeling turbulent flow through a permeable medium.

The topic of flow through permeable media extends over several areas of science and engineering. The amount of knowledge available is huge. To select what to include and what to leave out was to trust my own instincts (scholarship). All of it was written in my home-office, strategically located in our *games* room where my two children and I learn that working (at what we like) and playing are not dissimilar – literally!

DARCY EQUATION

It was in 1856 that Henry Philibert Gaspard Darcy, then the *Inspecteur Général de Ponts et Chaussées*, or "*Dean* of the School of Bridges and Roadways," had his report

[†] Dedicated *in memory* of my dear friend and father-in-law, Mr. Fernando D'olne Soares de Barros, 1902-1997.

on the public fountains of Dijon, a middle size city in the central-eastern region of France, published [1]. After obtaining his education in Paris at the *École Polytechnique* and at the *École de Ponts et Chaussées*, Darcy became a member of the French Imperial Corps of Bridges, Roadways and Mines, returning to Dijon, his birthplace, where he worked on hydraulics until his premature death by a neurological disease in 1858 [2].

Darcy's major contribution is undoubtedly in the area of filter hydraulics with the discovery of an empirical law that bears his name. His report on the public fountains of Dijon [1] presented the principles to follow and the formulae to use in the design and construction of water distribution systems, water filtering, sewer systems, and the manufacturing of valves.

The French term *fontaines* in the main title of his report has induced some English-speaking researchers to erroneously relate Darcy's work to *water springs*. In his report, Darcy discussed in detail experiments on water flow in a conduit, water flow in the hydraulic network of Dijon, and the effect of trapped air on the flow rate of a hydraulic network. It is clear that the objective of his study was very broad, including administrative and judicial aspects of water distribution in a city as related to the well being of the people living in it. It is fascinating to read his witness to the seriousness with which water quality was treated in a small French city back in 1856.

The more technical aspects of his essay were detailed in seven appendices. This fact should not be interpreted, in any way, as an indication of reduced importance. In fact, Darcy stresses the significance of these appendices in the introduction of his report, acknowledging the original aspect (based on his own discernment) of the information included there.

One of the appendices, appendix *D* (pp. 559-603), contains a description of his studies on filtration leading to his cornerstone equation of flow through a porous medium. It is worth noting that appendix *D* takes up only 44 pages out of the 647 pages of his book.

In appendix *D*, Darcy brings to light some generic concepts of filtration. He describes the two main procedures used at the time to clear up water (i.e. to remove solid particles making the water *clear*), specifically, precipitation and filtering. He remarks on the practical difficulty in using precipitation, including the required extended time, the bacterial growth effect because of the long exposure time, the size of the tanks necessary to supply water to a city, etc.. He indicates that precipitation might be used as a preliminary technique for treating water to be supplied to a city, but should not be the only one.

He then explains a very simple filtration technique, i.e. to make liquid flow through passageways small enough to retain solid particulates to be eliminated from the water without severely obstructing the flow of liquid. This technique can be applied in two ways: artificially – water flows under the influence of variable pressure through layers of fine sand, gravel, and small stones; naturally – water flows through the alluvium accumulated naturally along a river bed, for instance.

Darcy mentions the existence of self-acting filters, i.e. filters through which water flows spontaneously by gravity through sand layers. He points out the experimental results from six filters in France and England. These filters were composed of sand supported by gravel, with a filtering flow of 3-13 m^3/m^2 per 24 hours. He notices the difficulty in obtaining a general law of filtering because the sand used in each experiment was of different origin, the loads (ratio of pressure to flow rate) were different, and because the water arrived at the filters with different degrees of cleanness.

Having decided to investigate the phenomenon of water filtering, Darcy describes in his report a simple and ingenious apparatus. He then presents the results of carefully performed experiments done by himself and a fellow engineer in Dijon. The same experiments were later on performed by a chief engineer, his acquaintance, for independent confirmation.

The experimental apparatus used by Darcy and his colleagues was a 3.5 m high vertical column – a circular duct of interior diameter 0.35 m – closed at the extremities with a screwed plate. Inside, at 0.20 m from the bottom, there was a horizontal separator to support the sand layer, dividing the column into two chambers. This separator was formed, from bottom to top, of an iron grid of 7 mm prismatic bars spaced by 7 mm, a grid of 5 mm cylindrical bars spaced by 5 mm (the prisms were placed perpendicular to the cylinders), and finally a metallic screen of 2 mm thickness.

Water was fed through a pipe extending from the hydraulic network of the hospital in Dijon, were his laboratory was located. The water flow rate was controlled by a valve placed along the feeding pipe connected to the top of the column.

The lower section of the column led to a one meter wide reservoir collecting the water for measuring the volumetric flow rate. The pressures above and below the sand layer were indicated via two U-shape mercury manometers equipped with diaphragms (piezometers). Finally, the top chamber had an air bleeding valve for charging the column with water.

The experiments were performed with silica (quartz) sand from the Saône river, with the following composition: 58 percent of sand with grains smaller than 0.77 mm diameter, 13 percent of sand with 1.1 mm diameter grains, 12 percent of sand with 2 mm grains, and 17 percent of gravel and shell fragments of various sizes. This sand presented an average porosity of 0.38.

Darcy's experimental apparatus is detailed in the Figure 3 of chart 24 (the figures of his book were published in a separated *Atlas*, as an addendum to the book), and shown here schematically in Figure 1.

The sand was placed and packed down in the column previously filled with water to avoid trapping the air, if done otherwise. The sand height was measured only at the end of a series of experiments after the water flow had conveniently compacted the

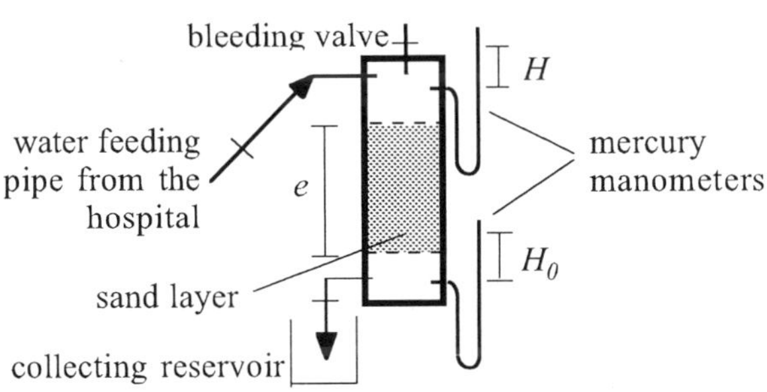

Figure 1 Simple sketch of Darcy's experimental apparatus

sand. Each experiment consisted of establishing determined pressures, read from the manometers, in the upper and lower chambers of the column.

Darcy measured the flow rate by collecting a certain volume of water during a certain period from the lower section of the column until two consecutive measurements were equal, thus establishing a steady flow. He related the height of water inside the column via the manometer height, i.e. without visual access to the column he could predict the height of water column above the sand layer. At high flow rates (or high pressures) he observed continuous oscillations in the manometer level, related by him to pressure variation in the hospital hydraulic system. Pressures were then reported with an estimated average value and a deviation. The maximum pressure deviation observed by Darcy was of five millimeters of mercury, and, at low flow rates, the pressure deviation was within one millimeter of mercury.

The experimental results were presented by Darcy in terms of volumetric flow rate (in liters per minute) and pressure difference across the sand (in meters of water-column). Defining a coefficient k – now called *hydraulic conductivity* – dependent on the permeability of the sand, and using the column cross-section area s ($s = 0.0962$ m^2), the height of the sand layer e, the pressure above the sand layer $P + H$ (atmospheric pressure P augmented by the manometer reading H), and $P + H_0$ as the pressure under the sand layer, Darcy proposed the following equation for estimating the volumetric flow rate Q through the column:

$$Q = \frac{ks}{e}(H + e - H_0) \tag{1}$$

Darcy reported results of five experiments. Four of them using different sand layer height ($e = 0.58$, 1.14, 1.71, and 1.70 m) and keeping the pressure under the sand layer equal to the atmospheric pressure ($H_0 = 0$). The fifth set of data, the most extensive of all with twelve points and the only one with nonzero H_0 values (H and H_0 can be positive or negative), was obtained with a 1.1 m high sand column.

Table 1 presents the results of the fifth experiment written in terms of average

Table 1 Modified experimental data reported by Darcy [1], p. 593

U (10^{-3} m/s)	Δp (kPa)	k (10^{-8} m^{3}s/kg)
3.26	128.3*	2.79
3.17	126.3*	2.76
3.12	123.4*	2.78
3.02	121.7	2.73
3.14	121.1	2.85
2.58	95.0	2.99
2.10	82.8*	2.79
1.70	65.8	2.84
1.37	56.7*	2.66
1.50	54.7	3.01
0.78	29.2	2.93
0.72	29.2*	2.71

fluid speed (Darcy velocity) $U = Q/s$ and total pressure difference across the sand $\Delta p = \rho g\ (H + e - H_0)$, assuming water at 10 °C (density ρ = 999.7 kg/m^3) and acceleration of gravity g = 9.81 m/s^2. Asterisks denote tests performed under strong pressure oscillation.

From Table 1, the average k value is equal to 2.82×10^{-8} m^{3}s/kg with a maximum deviation of 6.7 percent. Keep in mind that the values of parameter k listed in Table 1 were obtained, from the results reported by Darcy, via the *Darcy equation* (1) rewritten as:

$$k = U \frac{e}{\Delta p} \tag{2}$$

LIMITATIONS OF THE DARCY EQUATION

Darcy's Law, equation (2), is presently known to be limited in several aspects. It is valid, essentially, for incompressible and isothermal creeping flow (very low speed) of Newtonian fluid through a relatively long, uniform and isotropic porous medium of low hydraulic conductivity. Moreover, the hydraulic conductivity k, as defined by Darcy, is in fact fluid dependent besides being affected by the solid matrix of the porous medium.

We describe in the following subsections the most important physical implications of these limitations and the steps taken by scientists following Darcy to generalize equation (2).

Viscosity Effect: The Hazen-Darcy Equation

The empirical hydraulic equation for flow through porous media proposed by Darcy differs from the equation

$$U = \left(\frac{K}{\mu}\right)\frac{\Delta p}{e} \tag{3}$$

presented as *Darcy's Law* in some modern textbooks. Equation (3) has the hydraulic conductivity k of the original Darcy equation (2) substituted by (K/μ), where K is called the *specific permeability* and μ is the fluid dynamic viscosity. The specific permeability K is a hydraulic parameter supposedly independent of fluid properties.

In his report, Darcy does not make any reference to fluid viscosity. His experiments were performed with a single fluid, namely water. With no apparent means to control the water temperature, it was impossible for Darcy to investigate fluid viscosity effects with his experimental apparatus.

It is not very surprising that Darcy did not consider viscosity effects. Although the concept of fluid friction, or *lack of slipperiness*, was introduced by Newton in 1687, over 170 years before Darcy's report was published, viscous flow effects were seriously considered as an important parameter to determine fluid motion only much later, e.g. by Navier [3] in 1822 and in more detail – including the temperature effect on fluid viscosity – by Poiseuille [4] in 1844. These were then relatively *new ideas* at the time of Darcy's work.

Only years after the publication of the Darcy equation (2) was the effect of fluid viscosity observed indirectly and included in Darcy's law. Hazen [5], in 1893, seems to have been the first to notice the viscosity effect in connection to the original flow equation proposed by Darcy. By altering the temperature of the fluid prior to entering the filter, he observed the influence of water temperature on the hydraulic conductivity coefficient obtained experimentally following the Darcy equation. He proposed a modification to the original equation in the form

$$U = \left(\frac{T+10}{60}\right) k_{50} \frac{\Delta p}{e} \tag{4}$$

where T is the water temperature in degrees Fahrenheit, and k_{50} is the reference hydraulic conductivity value measured with water flowing isothermally through the permeable medium at a temperature equal to 50 °F.

A quick inspection reveals that the correction $(T+10)/60$, called the *Fahrenheit-factor*, mimics the water dynamic viscosity μ dependence on temperature, namely $\mu(50\ °F)/\mu(T)$, to less than one percent error for $40\ °F \leq T \leq 90\ °F$. One can then compare equation (4) with equation (3) to realize that $K = \mu(50\ °F)\ k_{50}$.

Keep in mind that Hazen's experiments were performed under isothermal conditions, i.e. invariant fluid temperature from inlet to outlet of the filter. We refer to equation (3) as the *Hazen-Darcy equation* to differentiate it from the original equation (2) proposed by Darcy.

Assuming that Darcy's experiments were performed with water at 10 °C (the experiments leading to the results shown in Table 1 were performed during February, 1856, in Dijon), the average permeability of his filter according to our calculations would be 3.7×10^{-11} m^2 (or, equivalently, equal to 37 Darcy) a reasonable value for a packed-sand column (see for instance the permeability values for sand reported by Scheidegger [6]).

Historically, then, the initial evolution of Darcy's original equation to a viscosity-dependent form was purely empirical and indirect, i.e. related to fluid temperature and not to viscosity directly. It was not until much later that the fluid viscosity parameter appeared as an individual component in the Darcy equation, e.g. in 1918 by Krüger [7], in 1920 by Zunker [8], and in 1927 by Kozeny [9]. Nutting [10], for instance, presented in 1930 a simple method to measure specific permeability already considering the fluid viscosity in the Darcy equation.

The previous paragraphs contradict the claim by Muskat [11], p. 70, cf. footnote 3: "In fact, the viscosity μ has not been expressly represented ... this omission has been probably due to the use by most of the investigators of a single fluid in their experiments ... . The first explicit separation of the term μ ... appears to be that of Wyckoff, Boset, Muskat, and Reed, Rev. Sci. Instr., 4, 394, 1933 ...". Even using a single fluid (e.g. water), it is evident that the viscosity effect could have been observed by simply varying the temperature of the fluid [12].

Darcy mentions in his report that he had already observed the interesting result revealed by equation (1) during his experiments with water flow through capillary conduits, where the water velocity does not go beyond 11 mm/s. He does not elaborate on the physical meaning of his equation. It is not clear if Darcy was aware of Hagen's [13] or Poiseuille's [14] work on flow through small-diameter pipes. Even if Darcy was aware of Hagen-Poiseuille's work, it is not surprising that he did not attempt to compare the two flow equations in view of the geometric differences between a sand column and small-diameter straight pipes.

Kozeny [9] seems to have been the pioneer in providing a physical explanation for the fluid-viscosity dependency of Darcy equation. By associating Poiseuille's theory of flow through capillaries with Darcy's empirical law of filtration, Kozeny demonstrated that by assuming uniform pressure drop and integrating Poiseuille's partial differential equation along a certain capillary length, an equation similar to the Darcy equation is obtained.

The Hagen-Poiseuille law, written as

$$\frac{s}{8\pi\mu} = U\frac{e}{\Delta p} \tag{5}$$

is remarkably similar to the Darcy equation: if one writes $k = s/(8\pi\mu)$ then equations (2) and (5) became identical. The analogy with Poiseuille flow allows the interpretation of the Darcy equation as a balance between pressure and global (macroscopic) viscous stress, and it also supports the fluid viscosity effect on k as observed empirically by Hazen [5].

Noteworthy is the almost universal use of constant fluid properties in theoretical and numerical studies of convection through porous media (see for instance Kaviany [15] and Nield and Bejan [16] for recent reviews of the subject).

In fact, there is a lack of experimental work validating the Hazen-Darcy equation (3) considering viscosity effects due to the temperature variation along the flow path, a situation typical of convection in a porous medium.

Szekely and Carr [17] studied nonisothermal gas flow through a permeable medium. The pressure variation along the medium, which is in excellent agreement with the experimental data, is obtained with an iterative scheme following an energy balance equation.

Ling and Dybbs [18] presented a numerical-theoretical investigation of the influence of temperature-dependent fluid viscosity on the steady forced convection over an isothermal flat plate bounding an infinite porous medium. Their results indicate a deviation of up to forty percent in the Nusselt number when compared with results obtained assuming uniform fluid viscosity.

Porneala and Lage [19] recently demonstrated the validity of equation (3) for the case of forced convection of a liquid with temperature-dependent viscosity between uniformly heated parallel flat plates (they used polyalphaolefin, a synthetic oil commonly employed for cooling avionics). Their experimental apparatus was designed to minimize the fluid temperature variation in the direction transverse to the flow, simplifying the theoretical analysis.

Subsequently, Nield and Lage [20] proposed a theoretical model to account for the fluid temperature variation along the direction transverse to the flow direction, and the effect of this temperature variation on the fluid viscosity, velocity, and total pressure drop.

Solid Form Effect: The Hazen-Dupuit-Darcy Equation

> *"It is necessary to be careful with the information presented by an experimentalist who lacks theoretical principles ... [he] gathers at random several facts and present them as proofs ... scientific knowledge without reasoning [theory] does not exist."* (by d'Alembert [21])

Essentially an experimentalist, Darcy investigated the phenomenon of water filtering using limited physical reasoning. It was Arsène Jules Emile Juvénal Dupuit, who coincidentally held the same title as Darcy (i.e. *Inspecteur Général des Ponts et Chaussés*), who was the first scientist to use theoretical principles to derive and physically explain Darcy's Law.

Dupuit, Darcy's contemporary, had a second edition of his report [22] on the *theoretical and practical studies of water flow in uncovered channels and through permeable terrain* published in 1863 (the first edition was published in 1848, before the publication of Darcy's work). The second edition of Dupuit's book, more than a simple revision of the first edition, includes a fundamental analysis of the Darcy equation.

The tremendous value of Dupuit's scientific contribution was recognized by a Commission appointed by the French Academy of Sciences. In a report presented to the Academy in 1861, the Commission recommended the addition of Dupuit's essay on *water movement through permeable terrain* – basically a summary of the most important additions to the second edition of his book – to the scientific collection of papers written by foreign scientists (although of French descent, Dupuit was born in Italy) and held by the Academy. A full transcript of the Commission's report is found in Dupuit's book following the Preface.

To fully appreciate Dupuit's contribution we have to recall the *scientific turmoil* in France during the first half of the 19th century. Within a span of a few decades, French scientists like Lagrange, Poisson, Laplace, Morin, Saint-Venant, Cauchy, Legendre, Clapeyron, Navier, Poiseuille, and others, advanced science at a rate never before seen, by challenging existing dogmas. The exchange and availability of information at the time were very precarious, making it very difficult for one to become rapidly abreast of all the new ideas.

Dupuit's familiarity with Prony's earlier work on flow of water in open channels [23] and with Navier's propositions on fluid motion [3], placed him in a very favorable position to explain the physics of Darcy's empirical discovery.

We trace back Dupuit's reasoning by initially considering the longitudinal cross section of a very wide open channel inclined to the horizontal at an angle γ, as seen in Figure 2. A fluid layer of thickness L flows steadily with average speed U.

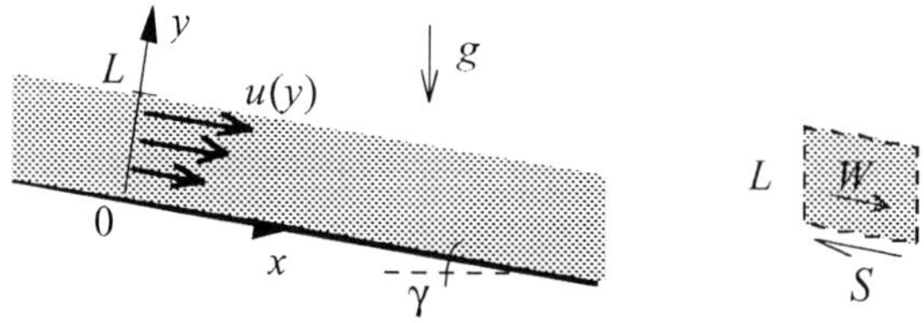

Figure 2 Diagrams of open channel flow (left) and of a fluid element (right)

For the steady flow considered in Figure 2, the linear momentum is constant, and therefore the resultant of forces along the x-direction must be zero, according to Newton's second law. Dupuit and his predecessors considered only two forces: weight W and shear resistance S.

The role of the shear resistance force was not clearly understood in the early part of the 19th century. Some scientists believed at the time that the total resistance was caused by two distinct phenomena, namely the shear resistance between two adjacent fluid layers, known as *fluid cohesion,* and the no slip resistance between fluid and solid surface, known as *fluid adherence*. Those following Prony considered the adherence effect to be much smaller than the cohesion effect, thus assuming a nonzero fluid velocity at the interface fluid and solid.

Navier, who proposed the proportionality between the cohesion force and the fluid velocity gradient, believed that both effects have the same physical origin. It was difficulty to reconcile these two concepts with the available experimental evidence.

Prony showed in his essay about *physico-mathematical researches on the theory of flowing water* [23] that the shear resistance should be a polynomial function of the fluid velocity at the solid surface, $u(0)$. When compared with experimental data, a quadratic polynomial seemed to provide a good fit. Assuming small velocity variation between $y = 0$ and $y = L$, the polynomial could be rewritten in terms of the average velocity U. The equation of equilibrium for the fluid element of Figure 2 is then

$$0 = L \sin\gamma - \alpha\, U - \beta\, U^2 \tag{6}$$

Equation (6), known as the *Prony equation*, was widely used in hydraulics of open channels. Dupuit showed in his report that equation (6) provides only a very crude (inaccurate) description of the flow behavior in an open channel, mainly because the real fluid velocity profile is quadratic in y and not *almost uniform* as assumed by Prony.

Dupuit then reasoned that the same equation (6) could be successfully applied for predicting steady flow through a homogeneous permeable medium because the large flow resistance imposed by each small channel (pore) being uniformly distributed would induce a uniform fluid velocity across the permeable layer in agreement with Prony's assumption. Substituting the weight term by the pressure gradient along a porous medium, equation (6) becomes:

$$0 = \frac{\Delta p}{\Delta x} - \alpha\, U - \beta\, U^2 \tag{7}$$

What Dupuit did, *de facto*, when proposing equation (7), was to *mentally visualize* the flow contribution of each pore (microscopic level) to the flow characteristic of the entire layer (global level).

It is surprising to see equation (7) named after Philipp Forchheimer in so many papers and textbooks. A careful reading of Forchheimer's original manuscript [24] indicates that he was not the first one to verify the breakdown of the Darcy equation. In fact, on page 1781 of his paper, Forchheimer stated Masoni's earlier experimental observation that the linear equation proposed by Darcy was inappropriate for correlating his (Masoni's) experimental data.

By the end of the 19th century, there was one group of scientists defending the linear Darcy equation and another defending the quadratic Dupuit equation. Forchheimer's paper was written mainly to state and defend his position regarding the most appropriate existing model for predicting the pressure drop through a permeable medium. Forchheimer sides with those in favor of the quadratic equation, supporting his position with a compilation of published experimental results indicating that the quadratic model fits the data more accurately than the linear model.

We notice that Forchheimer does not refer to Dupuit's report on flow through permeable media. He does, however, refer to Prony's quadratic model equation for open channel flow without justifying why the equation should be valid also for flow through a permeable medium. Those unaware of Dupuit's work could easily have taken Forchheimer as the one proposing the quadratic model. Nevertheless, naming equation (7) as the *Forchheimer equation* seems to be unwarranted and historically unfair to Dupuit.

We now observe that the original derivation of equation (6) is flawed because it considers the shear resistance as quadratic in velocity. Also, when going from equation (6) to equation (7), Dupuit modified the equation by accounting for the pressure effect and neglecting the weight effect (note that in the open channel flow sketched in Figure 2 the pressure gradient is zero).

Considering what we now know, specifically that: *adherence* and *adhesion* are of the same physical origin; the fluid velocity at the solid surface is zero; and the velocity profile $u(y)$ is a parabola, then the shear resistance effect in the Prony equation should have been written as linearly proportional to fluid velocity, and equation (7) would become physically inconsistent. However, equation (7) has been experimentally verified in several occasions. Hence, if shear resistance can not be responsible for the quadratic velocity term then a different phenomenon, not accounted for by Dupuit, must take place in the permeable medium to justify the quadratic term.

The physical phenomenon responsible for the quadratic term is the *form force* imposed to a fluid by any solid surface obstructing the flow path. This resistive (to unidirectional flow) force was proposed by Newton to be proportional to the average fluid velocity square. Using this concept, equation (7) is rewritten as

$$0 = \frac{\Delta p}{\Delta x} - \frac{\mu}{K} U - C \rho U^2 \tag{8}$$

where ρ is the fluid density and C is a form coefficient related to the geometry of the solid permeable medium.

Equation (8), the *Hazen-Dupuit-Darcy equation*, is identical to equation (7), except by the inclusion of: permeability, fluid viscosity and density, and form coefficient. Both equations are global in nature, valid for a permeable medium of finite length and cross section area. The last two terms of equation (8) represent the lumped viscous and the lumped form effects within the permeable medium, respectively.

Ward [25] proposed to replace the form parameter C in equation (8) by $c/K^{1/2}$, where the dimensionless parameter c was believed to be a universal constant. Ward's suggestion results from a dimensional analysis involving only four basic parameters, i.e. he assumed that the pressure-drop versus fluid-speed relationship (8) would depend only on U, K, ρ, and μ. Therefore, the form coefficient C should depend on the permeability K of the medium. This is obviously incorrect because the permeability, as linked to viscous drag, relates to the *effective surface length* of the solid porous matrix. The form coefficient C, on the other hand, does not depend on the length of the surface, but on the *form* of the solid matrix. Moreover, if Ward's suggestion was correct then c should be a universal constant, something which has been shown not to be the case (e.g. Beavers et al. [26] and Antohe et al. [27]). Therefore, the replacement of C by $c/K^{1/2}$ (another *dogma*) only masks the real physical meaning of the form parameter, with no practical benefit.

The derivation of equations (7) and (8) is presented here to demystify another very common dogma in fluid flow through porous media. The left side of equation (8) being zero reminds us that the fluid in and out of a permeable medium section retains its momentum. We see on the right side of equation (8) the terms representing all the considered forces acting upon the fluid. Of special interest is the rightmost term of equation (8), the *form force* term, often referred to as the *inertia force* or the *inertia term*. This terminology is obviously erroneous and confusing. It is erroneous because, with constant momentum, the flow has zero inertia (*inertia* being the time rate of momentum change). The confusion is created because the form force is proportional to U^2, and in certain special cases the flow inertia term is also written as velocity square. We know, however, that the rightmost term of equation (8) represents a force, that happens to vary with U^2. This force is *inertia related* as much as any other force term is, i.e. via Newton's second law.

The *inertia dogma* has spread out like a pest influencing many scientists and generating other misconceptions. It has led, for instance, many authors to mistakenly relate the quadratic velocity term in equation (8) to turbulence (e.g. Ward [25], Ergun [28] and Beavers and Sparrow [29]).

In several studies (e.g. Ward [25], Wyckoff et al. [30] and Fand et al. [31]) the permeability based Reynolds number,

$$Re_K = \frac{\rho U K^{1/2}}{\mu} \tag{9}$$

has been used to indicate the departure from the linear equation in fully developed steady flow through a porous medium, a direct consequence of the *inertia dogma*. Because steady fully developed flow has zero inertia, the Reynolds number of equation (9) has no physical meaning, i.e. it should not be confused with a ratio between inertia and viscous force. To use a Reynolds number to estimate the transition from viscous dominated flow regime (Hazen-Darcy equation) to form dominated regime (Hazen-Dupuit-Darcy equation), as done in several textbooks and technical papers, is nonsense.

What should be used to determine the departure from the linear flow regime, we suggest, is the physically meaningful ratio between the form force D_C and the viscous force D_μ. From equation (8)

$$\frac{D_C}{D_\mu} = \frac{C\rho U^2}{\left(\frac{\mu}{K}\right)U} = \frac{\rho C K}{\mu}U \tag{10}$$

The use of D_C/D_μ seems also to reduce the scattering of data observed when permeable media of very different form are used in hydraulic experiments (e.g. Macdonald et al. [32], Figure 1, p. 200 – notice that they used a Reynolds number similar to equation (10), with C replaced by $c/K^{1/2}$).

It is true, from many reports, that when Re_K exceeds a value of about 0.1, the relationship between fluid-velocity and pressure-drop departs from the linear equation (2) and follows the quadratic equation (8). The fact that Re_K is reasonably effective to determine the transition from *viscous* to *form* resistance dominated flow regime is coincidental, having to do with one limiting aspect: the majority of experimental data available is from flow through permeable media made up of particles with similar shape (spheres), and therefore the form coefficient C is relatively uniform and of the order of $O(K^{-1/2})$.

When the permeable medium is sparse, e.g. solid particles distant from each other, the viscous and form forces can be described as the *viscous and form drags* acting on each individual particle. An interesting recent numerical study by Zhang et al. [33] reported the individual contribution of viscous and form drags to the total flow resistance caused by in-line or staggered rectangular fins.

Equation (8) provides a global balance of forces acting on a finite fully saturated permeable volume with unidirectional steady flow of an incompressible fluid. The transformation of equation (8) into a differential formulation can be accomplished heuristically or formally though volume averaging (see, for instance, Bear and Bachmat [34] for more details).

A heuristic extension of the Hazen-Dupuit-Darcy model, in differential form, to multidimensional flow was presented by Stanek and Szekely [35]. Knupp and Lage [36] extended the model to the tensor permeability case (anisotropic medium) by building a convex variational principle that minimizes the flow work.

Liu et al. [37] indicated a decrease in the form coefficient as the flow velocity increases (the rate of increase of the ratio pressure-drop/fluid-speed is reduced as the particle based Reynolds number increases beyond 100) which could indicate a breakdown of the Hazen-Dupuit-Darcy model.

Lage, Antohe and Nield [38] have recently observed experimentally a transition from one quadratic regime to another quadratic regime, with different form coefficient, when increasing the speed of a fluid flowing through a porous medium. A cubic extension to the quadratic Hazen-Dupuit-Darcy model correlates well the experimental data though the transition. These two aspects can explain and justify the cubic extension suggested by Forchheimer [24]. This suggestion, like others presented in his paper, lacks physical reasoning and is purely empirical.

Considering Darcy's experimental results, Dupuit [22] justified the empirical equation derived by Darcy as a particular case of equation (7) obtained when the fluid average velocity is sufficiently small to hold the rightmost term negligible.

We now present in Table 2 another set of experimental results reported by Darcy [1]. This data set was obtained using a 0.58 m high sand column, and maintaining the pressure below the column equal to the atmospheric pressure. What differentiates the data presented in Table 2 from the data presented in Table 1 is the tendency of k to decrease with increasing the pressure drop Δp, notably within the high fluid speed range $U > 4 \times 10^{-3}$ m/s. This tendency is unnoticed in Table 1, where the maximum fluid speed equals 3.26×10^{-3} m/s.

Davis et al. [39] have applied a validation technique to the original Darcy equation considering only the data in Table 2. Their analysis indicated an increase in variance as the pressure drop increases, in other words, an apparent systematic relationship between the residuals. Based on their results Darcy original equation would not be validated by this experimental data set. One of the possible causes could be *systematic measurement errors*. A more likely explanation is simply the departure from the linear flow regime.

To prove our point, we first assume the data in Table 2 to follow the linear equation (2). We then curve fit the data and obtain $k = 2.909 \times 10^{-8}$ m^{3}s/kg from the best fit. With this k value and equation (2), we calculate a theoretical pressure drop Δp_t for every fluid speed and compare the results with the experimental Δp values in Table 2. In this case, the maximum pressure-drop deviation defined as $(\Delta p - \Delta p_t)/\Delta p$ is 14.8 percent. Repeating the same procedure, but now assuming the data to follow the quadratic equation (8) instead of equation (2), we obtain $k = K/\mu = 3.385 \times 10^{-8}$ m^{3}s/kg, $C = 1.1689 \times 10^6$ m, and a maximum deviation of 3.8 percent only.

It is evident that the quadratic model correlates better the experimental data of Table 2. Using equation (9), the maximum Reynolds number is $Re_K = 0.0235$, a value well below the usual value of 0.1 for transition to the quadratic regime. On the other hand, the maximum ratio of form to viscous forces, equation (10), is 0.2 indicating that the magnitude of these forces are comparable.

An equivalent analysis of the experimental data in Table 1 leads to $k = 2.809 \times 10^{-8}$ m^{3}s/kg from the linear equation, with 7.4 percent maximum pressure-drop deviation. For the quadratic equation we have: $k = K/\mu = 2.909 \times 10^{-8}$ m^{3}s/kg, $C = 0.4293 \times 10^6$ m, and a maximum deviation of 7.1 percent. The maximum Reynolds number in this case is: $Re_K = 0.0151$, and the maximum ratio of form to viscous

Table 2 Modified experimental data reported by Darcy [1], p.592 (asterisks denote strong pressure oscillation)

U (10^{-3} m/s)	Δp (kPa)	k (10^{-8} m^{3}s/kg)
5.09	106.8*	2.77
4.82	96.7*	2.89
4.24	84.1*	2.93
4.05	79.7*	2.95
3.78	74.8	2.93
2.62	49.2	3.10
2.48	48.1	2.99
2.08	39.2	3.07
1.33	23.1	3.32
0.62	10.9	3.32

forces is only 0.04.

Our previous and pioneer analysis of Darcy's data is also indicative of the following important concepts: (1) any experimental data can be curve fitted using the linear or the quadratic equations; (2) permeability and form parameter are representative quantities obtained from curve fitting, therefore they depend on the range of experimental data (see Antohe et al. [27] for details); (3) the ratio of form to viscous forces, equation (10), is the correct parameter to determine the flow regime, not the Reynolds number.

Flow Shear Effect: The Brinkman-Hazen-Dupuit-Darcy Equation

Brinkman [40] studied the limitation of the Hazen-Darcy equation regarding the viscous shear stresses caused by the surfaces of a permeable medium on an adjacent volume element of fluid. By drawing a parallel between the Hazen-Darcy equation, in multidimensional differential form

$$0 = -\nabla p - \frac{\mu}{K}\,\mathbf{u} \tag{11}$$

where $\mathbf{u}$ is the vector fluid velocity, and the Navier-Stokes equation for fully developed flow (also known as *Stokes flow*, or *creep flow*)

$$0 = -\nabla p' + \mu \nabla^2\,\mathbf{u}' \tag{12}$$

Brinkman suggested a modification to equation (11) by simply adding to it the Laplacian term of equation (12). By doing so, the new equation would hold for flow through high permeability media as well.

Brinkman recognized the different physical origin of each viscous term, i.e. fluid viscous shear stress, and solid viscous drag. He noticed that the fluid viscous shear stress effect should become negligible as compared to the viscous drag when the permeability is small.

He also pointed out that in a porous medium the fluid viscous shear stress caused by one surface can affect, by increasing or decreasing, the fluid shear stress caused by a neighboring surface. This fundamental phenomenon known as *interference* in aerodynamics, suggested to Brinkman the use of an effective fluid viscosity in the Laplacian term.

We must question Brinkman's reasoning in this respect. It is true that the fluid viscous shear effect is affected by neighboring particles, but this is so because the Laplacian of the fluid velocity distribution is affected. Equation (12), with the molecular fluid viscosity, is capable of capturing this effect. Therefore, the reason for using an effective fluid viscosity is in reality distinct from the one invoked by Brinkman. He failed to recognize that equations (11) and (12) are defined at different dimensions, although they are both written in differential form. Equation (12) refers to a *point* force – the fluid viscous shear – acting at any fluid location where the Laplacian of velocity is nonzero. Equation (11) refers to a *surface* force – the solid viscous drag acting along a fluid-solid interface within a permeable medium. Equation (11) requires a surface, and a surface has a higher dimension than a point. Including the laplacian term in equation (11) means transferring a pointwise quantity to a higher dimension. The laplacian term in this new dimension represents a lumped effect justifying then the use of an effective viscosity.

Another aspect overlooked by Brinkman is the difference between the two pressure terms of equations (11) and (12): the pressure p in equation (11) represents a force per unit of permeable area including solid and fluid, while the pressure p' in equation (12) is a force per unit of fluid area only. The same is true also for the fluid velocities **u** and **u**'.

Following the reasoning by Brinkman, one could very well replace equation (11) with the Hazen-Dupuit-Darcy equation in multidimensional and differential form, to obtain the *Brinkman-Hazen-Dupuit-Darcy equation*

$$0 = -\nabla(\phi p') + \mu_e \nabla^2 \mathbf{u}' - \frac{\mu}{K} \phi \, \mathbf{u}' + C \rho \phi^2 \left|\mathbf{u}'\right| \mathbf{u}' \tag{13}$$

where μ_e is the effective fluid viscosity, a function of the fluid viscosity *and* the geometry of the permeable medium. The surface porosity $\phi = A_f / A_m$, where A_f is the area occupied by the fluid and A_m is the area occupied by the solid, is introduced in equation (13) for consistency.

A fundamental contribution of Brinkman is the recognition of the fluid capability to transmit force by viscous shear as independent of the viscous drag.

Inertia Effect: General Equation

Vafai and Tien [41], and more recently Hsu and Cheng [42], presented a formal derivation of a *general equation* for fluid flow through an isotropic, rigid and homogeneous porous medium. The volume averaging technique leads to a closure problem, i.e. some of the forces acting on a representative permeable volume need to be modeled. Hsu and Cheng used the effective viscosity idea of Brinkman, and lumped the forces acting on the fluid phase of the permeable medium into effective viscous and form forces, modeled using the Hazen-Dupuit-Darcy model with Ward's suggestion. The equivalent final equation for incompressible fluid

$$\rho\left[\frac{\partial \mathbf{u}'}{\partial t} + (\mathbf{u}'\nabla)\mathbf{u}'\right] = -\nabla(\phi \, p') + \mu_e \nabla^2 \mathbf{u}' - \frac{\mu}{K} \phi \, \mathbf{u}' + C \rho \phi^2 \left|\mathbf{u}'\right| \mathbf{u}' \tag{14}$$

is an extension of equation (13) with the inertia term included.

Observe that equation (14) has six physical properties: fluid density ρ, fluid dynamic viscosity μ, effective viscosity μ_e, permeability K, porosity ϕ, and form coefficient C. Parameters ρ, μ, and ϕ can be measured independently but the other three, namely μ_e, K, and C, depend on the geometry of the permeable medium. They can not be measured directly, nor calculated analytically, because there is no model relating them to more basic (measurable) quantities valid for all porous media. In principle, these quantities should be obtained simultaneously by matching the results from solving equation (14) to experimental data.

An alternative to the simultaneous matching procedure was used by Givler and Altobelli [43] who estimated the effective viscosity following the determination of the permeability and the form coefficient from global measurements.

TRANSITION TO TURBULENCE: CONDUITS AND BLUFF BODIES

Hassanizadeh and Gray [44] have shown with scaling arguments that for *typical* porous media the *inertial force* (inertia or form drag – Hassanizadeh and Gray assumed them having the same *scales*) is three orders of magnitude smaller than the viscous force during the transition to quadratic regime, therefore, they can not be responsible for it. In some cases, then, the nonlinear evolution of the viscous force with fluid speed would be responsible for the nonlinear transition. Although their argument is physically plausible when considering the transition to turbulence in a straight-conduit-like media, Barak [45] refuted their argument by considering the experimental evidence for bluff-body-like media (e.g. circular disks placed perpendicular to the main flow direction) where vortices induce an *inertia* (form) dominated transition to a quadratic regime.

A comment made by Barak led Hassanizadeh and Gray [46] to incorrectly correct a *typographical error* by stating that "... macroscopic inertial forces have been shown to be small at the onset of nonlinear flow ..." – *macroscopic inertia forces* (convective inertia) are zero in steady flow through a homogeneous and isotropic porous medium. This particular instance exemplifies well the actual confusion between inertia (as change of momentum), viscous and form forces, and the concepts of *global* and *microscopic* domains.

It is instructive to discuss the different pressure loss behavior in transition to turbulence of flow through a straight conduit with regular cross section and flow around a bluff body (e.g. a disc). These two distinct geometries represent, independently, the two fundamental momentum transport processes within a porous medium. In the first one, a straight conduit, the *global* momentum balance for fully developed laminar flow is strictly between pressure drop and viscous diffusion represented by a linear velocity term. In the second case, of a bluff body, the balance during transition is mainly between pressure drop and form drag, proportional to the square of fluid velocity. Most porous media are a combination of these two particular geometries. Some can have a predominant conduit-like structure, some a predominant bluff-body-like structure.

By analogy with the pressure-drop versus flow speed relationship developed by Hagen-Poiseuille for flow through straight conduits, it became the norm to refer to the flow regime obeying Darcy's law as the *laminar* (or viscous dominated) regime. Any flow regime departing from Darcy's law would then be taken as *turbulent*, because a flow departing from the Hagen-Poiseuille flow was termed *turbulent flow*. Obviously this terminology is short-sighted. It implies that flow through any porous medium is similar to flow through a straight conduit.

A secondary consequence of this line-of-thought is the insistence by many investigators in using the particle diameter, or a representative pore diameter, as the length parameter when defining the Reynolds number to predict transition to turbulence of flow through porous media. Observe that although from a global point-of-view the inertia is zero in a fully developed flow through a homogeneous and isotropic porous medium, a Reynolds number defined with the mean fluid velocity may still represent the ratio between inertia and viscous force at the pore level when an appropriate microscopic length scale is used. Particle and pore diameter (even equivalent quantities) are not general enough, being appropriate only for very specific permeable media.

Consider, for instance, a medium of particles placed apart from each other in a lattice. The Reynolds number could be dependent on the characteristic distance between particles and on the particle diameter. That is because the microscopic inertia varies when the particles are moved apart from each other. And, depending on the shape of the particles, one can expect significant variation of the form force as compared with viscous forces. Therefore, an equivalent particle diameter is necessary to capture the transition due to the form effect.

Geertsma [47] made the important contribution of relating the length parameter in the Reynolds number to a combination of permeability and form coefficient, obtaining an expression similar to equation (10). The fact that one needs to obtain the permeability and the form coefficient of the porous medium prior to using the Reynolds number indicates the dependency on medium structure of the transition to turbulence by form and viscous forces. It becomes obvious that for the same fluid viscosity and mean fluid speed, different transition Reynolds numbers will be obtained by modifying the structure of the medium.

If one considers a porous medium made of spheres positioned in a lattice apart from each other, the pressure drop behavior with increased velocity should follow that of a single sphere. The transition to turbulence in flow around a bluff body is generally preceded by a pressure-drop deviation from the linear to quadratic (and still laminar) velocity dependency. Therefore, flow through a porous medium can depart from the linear equation to a quadratic equation before transition to turbulence. Although this flow sequence (linear-quadratic-transition) seems to be the norm rather than the exception, we believe that the structure of the porous media used frequently in previous experiments (packed spheres) is responsible for this sequence. If a medium resembling a maze of spaced conduits is used instead, one would expect a sharp transition from linear to turbulent flow more like the transition to turbulent flow through straight pipes.

Dudgeon [48] presented a detailed hydraulic study of the permeability of several heterogeneous porous media with irregular and nonuniform size packed particles. His experimental results support, although not conclusively, our hypothesis that turbulence in porous media can be of pipe-type (straight conduit) or of bluff-body-type. Table 3 reports some of his experimental findings, where the Reynolds number

Table 3 Particle-diameter based Reynolds number values for maximum flow speed (Re_{max}), for transition to quadratic flow regime (Re_q), and for the start of pressure oscillations (Re_{osc}) with water flow (from Dudgeon [48])

material	ϕ	d (mm)	Re_{max}	Re_q	Re_{osc}	Re_q/Re_{osc}
river sand	0.39	0.53	5.51	1.84	0.74	2.5
marble mix	0.38	24.9	7,650	60	14	4.2
stones	0.48	3.2	261	3.4	5.3	0.6

number is based on the equivalent average particle diameter and average (Darcy) fluid velocity. He reported the flow speed at which pressure fluctuations were detected by a piezo-electric probe placed inside the packed bed (he does not provide the location of the probe).

Except for the case of stones in Table 3, the beginning of pressure oscillation precedes the observed transition from the linear to quadratic regime. The results by Dudgeon verify that transition to turbulence prior to the flow reaching the quadratic regime is possible. In this case form drag is not predominant, therefore the pressure oscillation is more likely to be due to pipe-like transition effects. In the case of stones, the beginning of pressure oscillation occurs when form drag is effective, with transition more likely to be dominated by bluff-body effects (i.e. separation of the boundary-layer).

Bear [49], offered a brief discussion on turbulence, inertia forces and separation. We noticed with interest his assertion that "Most experiments indicate that actual turbulence occurs at Re values at least one order of magnitude higher than the Re at which deviation from Darcy's law is observed.", and that "... the deviation from Darcy's law [due to inertial forces] ... can not be attributed to turbulence.". Bear's comment invalidates the model hypothesis of Masuoka and Takatsu [50] who assumed the deviation from Darcy's law (or the form flow resistance) to be caused by turbulence (see also Nield [51]).

Another indication of the difference between a bluff-body like structure and a straight-conduit like structure is given indirectly by the study of turbulence intensity of flow exiting a porous medium. Hall and Hiat [52] observed experimentally that the spatially averaged turbulence intensity exiting a high porosity ceramic foam increases with the pore size. Because the limit of increasing the pore size is the unobstructed pipe flow configuration (smaller turbulence intensity), there must be an optimum pore size for maximum turbulence intensity. The evolution of turbulence intensity as the pore size increases is evidence again of the transition from conduit-like turbulence (when the pore size is small) – when increasing the pore size for the same flow velocity increases the turbulence intensity – to a bluff body like turbulence – when increasing further the pore size (the same as reducing the characteristic dimension of the bluff body) leads to a decrease in turbulence intensity.

EFFECT OF SOLID PARTICLES ON TURBULENCE

An important area of research is turbulent flow of fluids carrying solid particles. It is known that suspended particles can enhance or suppress turbulence, as well as modify the turbulence energy spectrum distribution, depending on the particle size and density, and concentration [53].

Generally speaking, small particles at small concentration tend to laminarize a turbulent flow and so reduce the drag. This is so because for small particles the viscous drag along the surfaces bounding the flow predominates and small particles *stiffen* the flow by enhancing the turbulence energy transfer to smaller eddies, in effect delaying turbulence and keeping the drag small. Large particles at high concentrations turn the viscous drag into form dominated drag, in which case delaying turbulence leads to a larger drag.

In the case of disperse solid particles (small particle concentration, generally smaller than two percent in volume) the interparticle effect is very small. The interaction between fluid and solid, in a reference frame attached to the solid particles (i.e. using the relative velocities between solid particles and fluid), shows characteristics similar to what would be expected of flow through a solid matrix composed of sparsely distributed solid particles.

Schreck and Kleis [54] showed that the turbulence intensity decreases with increasing particle concentration, i.e. the dissipation rate of the turbulence energy increases with particle concentration (decrease of the Kolmogorov microscale). They speculated that the suspended particles enhance the transfer of kinetic energy to smaller eddies and so extend the dissipation spectrum to smaller scales. Another important observation is that particles increase the isotropy of the flow (this information supports the use of isotropic turbulence relations, as is done in experimental studies of turbulent flow through packed beds described in the next section).

TURBULENT FLOW IN A POROUS MEDIUM: MICRO- VERSUS MACRO-SCALE

Several modern engineering processes can benefit from a better understanding of turbulence and its effect on the transport of momentum, energy, and species, in a porous medium.

Goeckner et al. [55] showed, for instance, that combustion within porous inert media leads to a reduction in gas temperature and NO_x emissions. Obviously, flame thickness, gas temperature, species concentration distribution, burning speed, etc., are all affected by turbulence within the porous structure.

Another important area of practical interest is one in materials science, namely in the casting and solidification of metal alloys where turbulence in the mushy (phase-change porous) zone can have an important effect on the quality of the final product. Although transport modeling of this particular problem has improved, the account for possible turbulence effect in the high porosity interface region between mushy zone and fluid is still omitted [56].

Short range pollution dispersion within urban regions [57] is a relatively new area of interest to which advances on turbulence in porous media can have a positive impact.

The study of turbulence in porous media is also fundamental for forestry and agriculture engineering, in chemistry for the design of chemical reactors and catalytic converters for instance, in bio-chemistry for the analysis of porous organs and bio-filters, and in petroleum exploration where gas can accelerate near the bore reaching speeds conducive of turbulent flow [58].

Evidence of Micro-turbulence in Porous Media

A fundamental characteristic of turbulent flow in porous media is the distinction between *microscopic* and *macroscopic* turbulence. The only experimental evidence of turbulence in a three-dimensional porous medium is of microscopic turbulence, that is, turbulence detected by point-wise probes placed within the medium.

There has been no attempt to volume-average the local signals within a representative volume to obtain the signature of the macroscopic turbulence. It is possible that when averaging a very large number of local (random) signals within a representative elementary volume, the microscopic turbulence would be smoothed out. Moreover, turbulence is not uniform along the tortuous and irregular paths inside a porous medium. Certain locations will be prone to reach turbulence at lower local fluid speeds than others. The macroscopic effect (volume averaged) might be negligible in certain cases.

It is obvious that increasing the fluid flow through a porous medium can lead to microscopic (pore level) turbulence. Separation and vortex shedding are the primary turbulence generation mechanisms for a porous medium resembling an agglomerate of bluff bodies. When the bodies are far apart (a high porosity medium) the vortices generated by each body individually will have space to develop and thus sustaining the energy cascade process so characteristic of turbulent flow. However, when the bodies are brought closer to each other (smaller porosity), the vortices generated by an upstream body will be obstructed by the neighboring bodies. The energy cascade process will be then less effective when the porosity is reduced and turbulence caused by form drag may not persist.

A porous medium resembling a maze of circular conduits can also develop turbulence at very high speeds, high enough to confine the viscous diffusion to a very narrow layer next to the solid boundaries of the porous medium. In this case, the analogue to turbulence in a pipe, the turbulence intensity should be high when the conduits have large cross section area (high porosity). Narrowing the flow space by reducing the conduit cross section area, for the same fluid viscosity and speed, should lead to a reduction in turbulence intensity.

Microscopic turbulence should then be dependent on the geometric structure of the solid matrix of the porous medium, besides being affected by the porosity. The porosity affects the turbulence intensity in a way similar to the way in which the equivalent diameter of the solid particle or of the pore affects the turbulence intensity of the flow around a bluff body or along a channel, respectively.

Mickley et al. [59] reported velocity profiles and turbulence parameters obtained experimentally with a hot-wire anemometer within the voids of a packed (rhombohedral or body-centered cubic) bed of spheres (diameter d = 0.0381 m, porosity ϕ = 0.32). This porous medium falls into the bluff body class of porous media discussed above. The geometry of the pores in a bed of spheres presents strong variation of the cross section area perpendicular to the main flow direction along the flow path, typical of flow around bluff bodies. Packing the spheres, however, narrows the distance between the bodies and so restricts the form (separation) effect. In fact, in an earlier study using a similar bed, Prausnitz and Wilhelm [60] observed no occurrence of regular eddy shedding in this matrix.

The flow (air) velocity measurements made by Mickley et al. [59] were done at the seventh column of spheres from the entrance (four columns from the exit), with the probe located at one sphere from the upper wall and at the center of the column. The anemometer was positioned at two locations in the smallest cuspidal cross section area perpendicular to the main flow direction between four spheres. They observed, for $Re = U_0 d/\nu$ (U_0 is the Darcy velocity) equal to 4780 and 7010, turbulence intensity $0.23 \leq u^*/U \leq 0.53$, where u^* is the root-mean-square of the fluctuating velocity component and U is the time averaged local velocity. These Re values are translated to permeability-based Re values, $Re_K = U_0 K^{1/2}/\nu$, by estimating the permeability of the medium via the Kozeny equation:

$$K = \frac{d^2 \phi^3}{180\left(1-\phi\right)^2} \tag{15}$$

The result is $K = 5.72 \times 10^{-7}$ m^2, leading to $Re_K = 95$ and 140, respectively. We can notice that the flow regime is clearly beyond the transition criterion for the Darcy flow ($Re_K = 0.1$, for spheres). Assuming isotropic turbulence in the pore space, the ratio of turbulence kinetic energy and mean flow kinetic energy can be written as

$$\frac{E_t}{E} = \frac{\frac{3}{2}\rho u^{*2}}{\frac{1}{2}\rho U^2} \tag{16}$$

therefore, $0.15 < E_t/E < 0.85$. Clearly, turbulence energy is not negligible. Observe also that the location chosen for measuring turbulence intensity is the narrowest flow section along the bed, and therefore the location most prone to pipe-like turbulence – but not necessarily the most prone to bluff-body like turbulence. Separation along spheres, between entrance and exit of a packed bed, was shown to occur at approximately 130 degrees [61], measured from the stagnation point of a single sphere.

Relative turbulence intensity increases by approximately 25 percent when the Reynolds number increases from 4780 to 7010, and the relative turbulence intensity is greater at the center of the cusp section. Unfortunately, their results do not indicate the turbulence level near the transition from linear to quadratic flow regime. Nevertheless, some important conclusions were drawn from the results: (1) turbulence intensity does not seem to affect either radial or axial dispersion; (2) there was no indication of vortex shedding in this particular medium – therefore the viscous diffusion effect is the predominant phenomenon affecting turbulence; (3) turbulence intensity one sphere diameter downstream of the bed does not correlate with the turbulence inside the bed.

In a very interesting study, Jolls and Hanratty [62] investigated the transition to turbulent flow within a randomly packed bed of spheres (sphere diameter $d = 0.0254$ m, porosity equal to 0.41, bed diameter $D = 0.3$ m and length $L = 1.5$ m). Their results indicated a sharp transition from steady to unsteady flow at Reynolds number ($Re = U_0 d/\nu$) around 110-150. Transition was observed visually using dye (Sudan III) in a methyl-benzoate solution (to match the refraction index of the glass spheres) and confirmed by local and instantaneous surface measurements of mass flux, via an electrochemical technique using a weak concentration electrolyte aqueous solution. The measuring sphere was placed at the center of the packed column 0.3 m from the exit section. Results of electrode readings at two surface positions showed that transition to time dependent process at the microscopic scale is not followed by a significant change in the time-averaged mass transfer coefficient. Unfortunately, it is not possible to determine the geometrical surroundings condition (i.e. the location of the neighboring spheres, location of point-of-contact) of the measuring sphere because of the random packing of the spheres.

Again, using the Kozeny expression for permeability, equation (15), the equivalent permeability is $K = 7.1 \times 10^{-7}$ m^2 and the permeability based Reynolds number of transition to time dependent flow is equal to 3.6-5.0, which is still higher than the general transition from linear to quadratic flow regime.

van der Merwe and Gauvin [63] obtained turbulence intensity results for Reynolds number 2.5×10^3 to 2.7×10^4. Their measurements were at a plane transverse to the main flow direction in the void located behind each bank (a bank is a plane of spheres perpendicular to the main flow direction) and along the centerline of a bed with ten banks. The spheres were arranged in a regular (simple cubic) fashion with axis linking contact points of consecutive spheres longitudinal and transverse to the main flow direction, and in a skewed arrangement with the three axis linking contact points of consecutive spheres forming the same angle with the main flow direction.

In the regular sphere arrangement case, entrance effects are revealed by an increase in the turbulence intensity along the main flow direction. These effects are present up to four banks downstream of the bed, beyond which the turbulence intensity flattens out, being relatively uniform until the last bank. Mean velocity measurements, along the void plane, indicate a bluff-body-type effect with the velocity profile being flatter for $Re = 2.5 \times 10^3$ than for $Re = 2.7 \times 10^4$.

Moreover, a strong (20 to 30 percent of the mean velocity) recirculation zone near the sphere contact point is observed for all Reynolds numbers. This recirculation zone is not detected in the skewed case. When positioned at the center of the void, the results from the anemometer indicate relative turbulence intensity of about 20 percent for $Re \geq 5 \times 10^3$, and about 12 percent for $Re = 2.5 \times 10^3$. These values are smaller than the values at the center of the void obtained by Mickley et al. [59], but are comparable to the values obtained at the location of the maximum speed (recall that the sphere configuration in Mickley et al. [59] is different, and their results indicated the maximum flow speed to be away from the center of the void). Spectral analysis does not indicate the existence of vortex shedding in any case.

Observe that the remark, about turbulence downstream of the bed exit, made by van der Merwe and Gauvin [63] contradicts conclusion number 3 of Mickley et al. [59]. It seems that the downstream turbulence is highly dependent on the structure of the medium, even for media with similar porosities. Hence, the suggestion made by Howell et al. [64], p. 128, that "... measurements made close to the downstream edge of a screen may provide a flow field similar to that in a highly porous ceramic foam.", should be considered very carefully. Moreover, extremely important entrance effects are not revealed by a test with a screen. Consequently, extending the downstream turbulence results of Hall and Hiatt [65] to infer even qualitatively the turbulence inside a ceramic foam seems unwarranted.

Modeling Macroscopic Turbulence in Permeable Media

One of the earliest theoretical considerations of turbulence in a permeable medium is that by Ahmed and Sunada [66]. They started their analysis with the Navier-Stokes equation modeling the (microscopic) flow in the pores, followed by the classical time-averaging procedure. The resulting equation was then integrated in a representative elementary volume. They proceeded by drawing an analogy between the viscous diffusion term and the linear viscous drag term of the macroscopic equation, and between the convective inertia term and the quadratic form drag term. In both cases the turbulence terms of the average equation were neglected on the assumption of negligible relative turbulence in a porous medium. The final result is the macroscopic Hazen-Dupuit-Darcy equation.

Lim and Mathews [67] showed that better agreement between numerical predictions and experimental results is obtained when accounting for the turbulence effects on combustion. We note in passing that their turbulence κ-ε model does not account for the additional drag (viscous and form) effects imposed by the porous solid matrix to the flow.

Turbulence within the mushy (porous) zone of a solidifying binary alloy can play a significant role in the quality of the final solid metal. Prescott and Incropera [68] proposed a κ-ε model for simulating the momentum transport within the mushy zone by adding a sink term in the κ-equation of turbulent clear flow. This sink term artificially damps turbulence in the mushy zone. Although showing good agreement with their experimental results, their turbulence model is of limited use, as indicated by Lage [69], because it does not take into consideration buoyancy and other drag effects imposed by the porous region. Moreover, the mushy zone is not uniform but, rather, it presents increasing porosity when approaching the unobstructed fluid region. In situations in which this porosity variation is smooth, one would expect the turbulence energy of the unobstructed flow to penetrate deeper into the mushy zone, thus affecting the phase-change process.

Rudraiah studied turbulent natural convection in porous media using the Hazen-Darcy equation which is extended by the inclusion of the convective acceleration term (Rudraiah [70] and Rudraiah et al. [71]). Although the inclusion of the full convective acceleration term is still a controversial topic (Nield [72, 73], and Vafai and Kim [74]), the absence of the quadratic drag term when modeling high speed flow in porous media is difficult to justify.

Gratton et al. [75] presented a randomly varying morphology-based turbulence model obtained by the hierarchical modeling methodology and an advanced averaging technique developed by the same authors. Subsequently, Travkin and Catton [76] introduced a two-temperature heat and momentum transport model in which the morphology of the medium plays a fundamental role. Their momentum transport model, considering the elementary volume morphology as that of straight channels, does not include the convective inertia term. Moreover, for irregular media morphology, several complications arise with the inclusion of morphology-specific differential and integral terms in the average transport equations making the model unpractical.

The model proposed by Lee and Howell [77] was derived by including a turbulence eddy diffusivity in the viscous diffusion term of the general momentum equation. They did not time-average the general momentum equation, and therefore their model neglects an additional contribution of the Reynolds stresses. Furthermore, the absolute value of the velocity vector appearing in the quadratic form drag term is treated as turbulence independent, an approximation that simplifies the derivation of the κ and ε equations tremendously.

Wang and Takle [78] presented a turbulence model for the boundary-layer flow near porous obstacles. They considered the important phenomenon of turbulent air flow (wind) through forests and crops. This kind of flow is generally modeled as flow through porous media. The geometric dimensions and fluid speed are such that turbulence can be achieved.

Their methodology consisted of splitting the transport quantities into two terms: one which is averaged within small wavelengths (space averaged quantities) that are capable of capturing large wavelength variations only, and another that represents the small wavelength variations. They proceeded by integrating the microscopic transport equations within a volume that scales with the characteristic dimension of the solid obstacles, i.e. the averaging procedure smoothed-out only the effects caused by the obstacles. The closure of their integro-differential space averaged momentum equations was obtained by analogy between drag and form effects. The final step towards the derivation of their turbulence model is the classical time averaging Reynolds procedure.

Unfortunately a conceptual error is made during the time averaging step, that of considering the absolute value of the vector velocity in the form drag term as turbulence independent (similarly to Lee and Howell [77]). Moreover, because their closure to the space averaged problem does not account for the linear viscous drag effect, their model is restricted to high velocity and high porosity (disperse) matrix cases. They showed that the form drag effect is to increase the dissipation of turbulence kinetic energy.

Antohe and Lage [79] derived a macroscopic turbulence κ-ε model for incompressible flow in porous media. The turbulence model was derived by time-averaging the general equation for porous media with all terms: time acceleration, convective inertia, pressure gradient, fluid viscous shear, viscous drag, and from drag. It is possible to conclude from their model that the viscous drag effect of the porous structure is to cause an additional decay in the turbulence kinetic energy. The effect of the form drag is undetermined as expected because this drag effect is of *form* origin, consequently the shape of the medium can enhance or smooth out the turbulence.

Antohe and Lage [79] considered also a simplified turbulence model for steady unidirectional fully developed flow which indicates that *macroscopic turbulence* can not persist in a porous medium once this flow configuration is achieved. Their explanation for this result is that even when having turbulence at the pore level (known to exist), the space averaging of the velocity fluctuations measured at several points within the pore space might lead to a negligible turbulence intensity. This observation highlights the importance of distinguishing between *macroscopic* and *microscopic* turbulence.

The only known attempt to obtain macroscopic turbulence quantities from microscopic quantities is that by Kuwahara et al. [80]. Using a low-Reynolds two-equation turbulence model they simulated numerically the microscopic flow field within a two-dimensional periodic permeable medium. Their results show, for unidirectional flow, the microscopic turbulence level within the pores. The macroscopic pressure gradient obtained by volume averaging the microscopic pressure follows closely the one obtained via the Hazen-Dupuit-Darcy equation. It is worth noting that the theoretical prediction of Antohe and Lage's model of zero macroscopic turbulence kinetic energy gradient is exactly what Kuwahara et al. [80] imposed in their simulations: observe that they used periodic boundary conditions for microscopic κ, something which leads to zero gradient for macroscopic κ.

It is important to point out that time averaging does not commute with volume averaging in general. Wang and Takle [78] stated this aspect very clearly by writing: "Time averaging followed by spatial averaging implies that obstacle elements interact only with time-averaged flow ... turbulence energy-cascade process is precluded under these assumptions.". In reality, the final form of any macroscopic turbulence model depends on the order in which time and spatial averaging are taken. This unavoidable problem is caused by the closure required after each space and time averaging process. Even if one does not close the problem after each averaging, the final model obtained by the sequence time- and space-averaging will present different terms when compared with the terms obtained by space- and time-averaging the equations, so the closure problem will be different in each case.

Therefore, one can not use the results of microscopic κ and ε presented by Kuwahara et al. [80] to compute the macroscopic turbulence kinetic energy because the microscopic κ represents the local time-averaged turbulence intensity. To get the real macroscopic κ one has to obtain the local values of the fluctuation velocity within a representative elementary volume (rev), take the volume average of it, and then time average the result. Notice that the fluctuation velocity can take positive or negative values within a rev, in contrast with the always positive microscopic κ.

Antohe and Lage's [79] model suffers from similar difficulties. When time averaging the balance equations after closing the space averaging problem, the model inevitably loses the information of how local (in space) quantities contribute individually to the time-averaged quantities. This is in line with the concept of space averaging, and the result does not preclude turbulence induced by the porous matrix in a space average sense. It does, however, preclude the accurate characterization of turbulence induced by the porous matrix in a microscopic sense. This is the main deficiency of any general turbulence model obtained from the space-averaged equations.

ACKNOWLEDGMENTS

Professor Ingham suggested that a review of Darcy's law would be an ideal first chapter for this book. I am pleased for his suggestion as indicative of his belief on my scholarship. Some intricacies of the 19th-century French language were elucidated by my father, Mr. R. Lage, when discussing Darcy's manuscript over a (couple of) glass(es) of French wine. My doctorate student, Mr. V. Koulich, helped translating some of the papers written in German (without wine!). I am grateful also for Professor Nield's review of a draft of this manuscript. Support for this effort was provided by Southern Methodist University via the *J. L. Embrey Professorship in Mechanical Engineering*.

REFERENCES

1. H. P. G. Darcy, *Les Fontaines Publiques de la Ville de Dijon*. Victor Dalmont, Paris (1856).

2. H. Rouse and S. Ince, *History of Hydraulics*, 169-177. Dover, New York (1957).
3. C. L. M. H. Navier, Mémoire sur les lois du mouvement des fluids, *Mémoires de l'Académie Royale des Sciences de l'Institut de France* **6**, Paris (1822).
4. J. L. M. Poiseuille, *Le mouvement des Liquides dans de Tubes de Petits Diametres*. Paris (1844).
5. A. Hazen, Some physical properties of sand and gravels with special reference to their use in filtration, *Massachusetts State Board of Health, Twenty-fourth Annual Report*, p. 541 (1893).
6. A. E. Scheidegger, *The Physics of Flow through Porous Media*, p. 87. University of Toronto Press (1960).
7. E. Krüger, Die Grundwasserbewegung, *Internationale Mitteilungen fur Bodenkunde* **8**, p. 105 (1918).
8. F. Zunker, Das allgemeine Grundwasserfliessgesetz, *Journal fur Gasbeleuchtung und Wasserversorgung* (1920).
9. J. Kozeny, Ueber Grundwasserbewegung, *Wasserkraft und Wasserwirtschaft* **22**, p. 67 (1927).
10. P. G. Nutting, Physical analysis of soil sands, *Bulletin of the American Association of Petroleum Geologists* **14**, 1337-1349 (1930).
11. M. Muskat, *The Flow of Homogeneous Fluids through Porous Media*. McGraw Hill, New York (1937).
12. F. T. Mavis and E. F. Wilsey, A study of the permeability of sand, *University of Iowa Studies in Engineering*, Bulletin 7 (1936).
13. G. H. L. Hagen, Ueber [Über] die Bewegung des Wassers in engen c[z]ylindrischen Röhren, *Poggendorfs Annalen der Physik und Chemie* **16** (1839).
14. J. L. M. Poiseuille, Récherches experimentales sur le mouvement des liquides dans les tubes de très petits diamètres, *Comptes Rendus* **11**, 961-967 (1840).
15. M. Kaviany, *Principles of Heat Transfer in Porous Media*. Springer, New York (1991).
16. D. A. Nield and A. Bejan, *Convection in Porous Media*. Springer, New York (1992).
17. J. Szekely and R. G. Carr, On nonisothermal flow of gases through packed beds, *Transactions of the Metallurgical Society of AIME* **242**, 918-921 (1968).
18. J. X. Ling and A. Dybbs, The effect of variable viscosity on forced convection over a flat plate submersed in a porous medium, *ASME Journal of Heat Transfer* **114**, 1063-1065 (1992).
19. D. C. Porneala and J. L. Lage, Experimental verification of the Hazen-Darcy flow model for forced convection of a liquid with temperature dependent viscosity, *International Journal of Heat and Mass Transfer*, submitted for publication (1997).
20. D. A. Nield and J. L. Lage, Convection effect on the porous medium flow of a fluid with temperature dependent viscosity, *ASME Journal of Heat Transfer*, submitted for publication (1997).
21. J. le R. d'Alembert, *Nouvelles Expériences sur la Resistance des Fluids*. Jombert, Paris (1777).
22. A. J. E. J. Dupuit, *Études Théoriques et Pratiques sur le Mouvement des aux dans les Canaux Découverts et a Travers les Terrains Perméables*. Victor Dalmont, Paris (1863).
23. Prony, R. de, *Recherches Physico-mathématiques sur la Théorie des Eaux*

Courants, Paris (1804).

24. P. Forchheimer, Wasserbewegung durch Boden, *Zeitschrift des Vereines Deutscher Ingenieure* **45**, 1736-1741 & 1781-1788 (1901).
25. J. C. Ward, Turbulent flow in porous media, *ASCE Journal of the Hydraulics Division*, **90**, HY 5, 1-12 (1964).
26. G. S. Beavers, E. M. Sparrow and D. E. Rodenz, Influence of bed size on the flow characteristics and porosity of randomly packed beds of spheres, *Journal of Applied Mechanics* **40**, 655-660 (1973).
27. B. V. Antohe, J. L. Lage, D. C. Price and R. M. Weber, Experimental determination of permeability and inertia coefficients of mechanically compressed aluminum porous matrices, *ASME Journal of Fluids Engineering* **119**, 404-412 (1997).
28. S. Ergun, Fluid flow though packed columns, *Chemical Engineering Progress* **48**, 89-94 (1952).
29. G. S. Beavers and E. M. Sparrow, Compressible gas flow through a porous material, *International Journal of Heat and Mass Transfer* **14**, 1855-1859 (1971).
30. R. D. Wyckoff, H. G. Botset, M. Muskat and D. W. Reed, Measurement of permeability of porous media, *Bulletin of the American Association of Petroleum Geologists* **18**, 161-190 (1934).
31. R. M. Fand, B. Y. K. Kim, A. C. C. Lam and R. T. Phan, Resistance to the flow of fluids through simple and complex porous media whose matrices are composed of randomly packed spheres, *ASME Journal of Fluids Engineering* **109**, 268-274 (1987).
32. I. F. Macdonald, M. S. El-Sayed, K. Mow and F. A. L. Dullien, Flow through porous media – Ergun equation revisited, *Industrial Engineering Chemistry Fundamentals* **18**, 199-208 (1979).
33. L. W. Zhang, S. Balachandar, D. K. Tafti and F. M. Najjar, Heat transfer enhancement mechanisms in inline and staggered parallel-plate fin heat exchangers, *International Journal of Heat and Mass Transfer* **40**, 2307-2325 (1997).
34. J. Bear and Y. Bachmat, *Introduction to Modelling of Transport Phenomena in Porous Media*. Kluwer Academic, Dordrecht (1990).
35. V. Stanek and J. Szekely, Three-dimensional flow of fluids through nonuniform packed beds, *AIChE Journal* **20**, 974-980 (1974).
36. P. Knupp and J. L. Lage, Generalization of the Forchheimer-extended Darcy flow model to the tensor permeability case via a variational principle, *Journal of Fluid Mechanics* **299**, 97-104 (1995).
37. B. M. Liu, R. C. Xin and W. Q. Tao, Experimental and numerical investigations on flow resistance in porous medium composed of randomly packed spheres, *Transport Phenomena in Thermal Engineering* **1** (Eds. J. S. Lee, S. H. Chung, and K. Y. Kim), 231-236, Begell House, New York (1993).
38. P. A. Davis, N. E. Olague and M. T. Goodrich, Application of a validation strategy to Darcy's experiment, *Advances in Water Resources* **15**, 175-180 (1992).
39. J. L. Lage, B. V. Antohe and D. A. Nield, Two types of drag variation in transition to turbulent incompressible flow through a porous medium, *ASME Journal of Fluids Engineering* **119**, in press (1997).

40. H. C. Brinkman, A calculation of viscous force exerted by a flowing fluid on a dense swarm of particles, *Applied Science Research* **A1**, 27-34 (1947).
41. K. Vafai and C. L. Tien, Boundary and inertia effects on flow and heat transfer in porous media, *International Journal of Heat and Mass Transfer* **24**, 195-203 (1981).
42. C. T. Hsu and P. Cheng, Thermal dispersion in a porous medium, *International Journal of Heat and Mass Transfer* **33**, 1587-1597 (1990).
43. R. C. Givler and S. A. Altobelli, A determination of the effective viscosity for the Brinkman-Forchheimer flow model, *Journal of Fluid Mechanics* **258**, 355-370 (1994).
44. S. M. Hassanizadeh and W. G. Gray, High velocity flow in porous media, *Transport in Porous Media* **2**, 521-531 (1987).
45. A. Z. Barak, Comments on 'High velocity flow in porous media', *Transport in Porous Media* **2**, 533-535 (1987).
46. S. M. Hassanizadeh and W. G. Gray, Reply to comments by Barak on High velocity flow in porous media, *Transport in Porous Media* **3**, 319-321 (1988).
47. J. Geertsma, Estimating the coefficient of inertial resistance in fluid flow through porous media, *SPE Journal*, 445-450 (1974).
48. C. R. Dudgeon, An experimental study of the flow of water through coarse granular media, *La Houille Blanche* **7**, 785-801 (1966).
49. J. Bear, *Dynamics of Fluids in Porous Media.* Elsevier, New York (1972). [corrected reprint, Dover, New York, NY, 1988]
50. T. Masuoka and Y. Takatsu, Turbulence model for flow through porous media, *International Journal of Heat and Mass Transfer* **39**, 2803-2809 (1996).
51. D. A. Nield, Comments on 'Turbulence model for flow through porous media', *International Journal of Heat and Mass Transfer* **40**, p. 2499 (1997).
52. M. J. Hall and J. P. Hiatt, Exit flows from highly porous media, *Physics of Fluids* **6**, 469-479 (1994).
53. Y. Tsuji, Y. Morikawa and H. H. Shiomi, LDV measurements of an air-solid two-phase flow in a vertical pipe, *Journal of Fluid Mechanics* **139**, 417-442 (1984).
54. S. Schreck and S. J. Kleis, Modification of grid-generated turbulence by solid particles, *Journal of Fluid Mechanics* **249**, 665-688 (1993).
55. B. A. Goeckner, D. R. Helmich, T. A. McCarthy, J. M. Arinez, T. E. Peard, J. E. Peters, M. Q. Brewster and R. O. Buckius, Radiative heat transfer augmentation of natural gas flames in radiant tube burners with porous ceramic inserts, *Experimental Thermal Fluid Sciences* **5**, 848-860 (1992).
56. S. K. Sinha and T. Sundararajan, A variable property analysis of alloy solidification using the anisotropic porous medium approach, *International Journal of Heat and Mass Transfer* **35**, 2865-2877 (1992).
57. R. W. MacDonald, R. F. Griffiths and S. C. Cheah, Field experiments of dispersion through regular arrays of cubic structures, *Atmospheric Environment* **31**, 783-795 (1997).
58. M. K. Wang, Mobil E&P Technical Center, Reservoir Engineering, Dallas-Texas, private communication (1996).
59. H. S. Mickley, K. A. Smith and E. I. Korchak, Fluid flow in packed beds, *Chemical Engineering Science* **20**, 237-246 (1965).
60. J. M. Prausnitz and R. H. Wilhelm, Turbulent concentration fluctuations in a packed bed, *Industrial and Engineering Chemistry* **49**, 978-984 (1957).

61. D. F. van der Merwe and W. H. Gauvin, Pressure drag measurements for turbulent air flow through packed bed, *AIChE Journal* **17**, 402-409 (1971).
62. K. R. Jolls and T. J. Hanratty, Transition to turbulence for flow through a dumped bed of spheres, *Chemical Engineering Science* **21**, 1185-1190 (1966).
63. D. F. van der Merwe and W. H. Gauvin, Velocity and turbulence measurements of air flow through a packed bed, *AIChE Journal* **17**, 519-528 (1971).
64. J. R. Howell, M. J. Hall and J. L. Elizey, Combustion of hydrocarbon fuels within porous inert media, *Progress in Energy Combustion Science* **22**, 121-145 (1996).
65. M. J. Hall and J. P. Hiatt, Exit flows from highly porous media, *Physics of Fluids* **6**, 469-479 (1994).
66. N. Ahmed and D. K. Sunada, Nonlinear flow in porous media, *ASCE Journal of the Hydraulics Division* **95**, HY 6, 1847-1857 (1969).
67. I. G. Lim and R. D. Matthews, Development of a model for turbulent combustion within porous inert media, *Transport Phenomena in Thermal Engineering* **1** (Eds. J. S. Lee, S. H. Chung and K. Y. Kim), 631-636, Begell House, New York (1993).
68. P. J. Prescott and F. P. Incropera, The effect of turbulence on solidification of a binary metal alloy with electromagnetic stirring, *ASME J. Heat* Transfer **117**, 716-724 (1995).
69. J. L. Lage, Comments on 'The effect of turbulence on solidification of a binary metal alloy with electromagnetic stirring', *ASME J. Heat Transfer* **118**, 996-997 (1996).
70. N. Rudraiah, Turbulent convection in porous media with non-Darcy effects, *ASME Proceedings HTD 96* **1**, 747-754 (1988).
71. N. Rudraiah, G. Siddheshwar and T. Masuoka, Turbulent convection in a high porosity porous medium with convective acceleration, *Journal of Mathematical and Physical Sciences* **19**, 93-117 (1985).
72. D. A. Nield, The limitations of the Brinkman-Forchheimer equation in modeling flow in a saturated porous medium and at an interface, *International Journal of Heat and Fluid Flow* **12**, 269-272 (1991).
73. D. A. Nield, Modelling high speed flow of a compressible fluid in a saturated porous medium, *Transport Porous Media* **14**, 85-88 (1994).
74. K. Vafai and S. J. Kim, On the limitations of the Brinkman-Forchheimer-extended Darcy equation, *International Journal of Heat and Fluid Flow* **16**, 11-15 (1995).
75. L. Gratton, V. S. Travkin and I. Catton, Numerical solution of turbulent heat and mass transfer in a stratified geostatistical porous layer for high permeability media, *ASME Proceedings HTD 41*, 1-14 (1994).
76. V. Travkin and I. Catton, A two-temperature model for turbulent flow and heat transfer in a porous layer, *Journal of Fluids Engineering* **117**, 181-188 (1995).
77. K. Lee and J. R. Howell, Forced convective and radiative transfer within a highly porous layer exposed to a turbulent external flow field, *Proceedings 1987 ASME-JSME Thermal Engineering Joint Conference* **2**, 377-386 (1987).
78. H. Wang and E. S. Takle, Boundary-layer flow and turbulence near porous obstacles, *Boundary Layer Meteorology* **74**, 73-88 (1995).
79. B. V. Antohe and J. L. Lage, A general two equation turbulence model for incompressible flow in porous media, *International Journal of Heat and Mass Transfer* **40**, 3013-3024 (1997).

80. F. Kuwahara, Y. Kameyama, S. Yamashita and A. Nakayama, Numerical modeling of turbulent flow in porous media using a spatially periodic array, *Journal of Porous Media*, to appear (1997).

TRANSPORT PHENOMENA IN ENCLOSED POROUS CAVITIES

C. K. CHEN[1] and S. W. HSIAO[2]

[1]Department of Mechanical Engineering, National Cheng Kung University, Tainan, Taiwan 70101, R.O.C.
[2]Department of Industrial Design, National Cheng Kung University, Tainan, Taiwan 70101, R.O.C.

INTRODUCTION

Natural convection within enclosures filled with porous media has been studied extensively during the past two decades due to its widespread engineering applications, including geothermal systems, underground spread of pollutants, storage of nuclear waste materials, solidification of casting, thermal insulation, electronic cooling, petroleum reservoir modeling, burying of drums containing heat generating chemicals in the earth, design of chemical catalytic reactors, powder metallurgy, ceramic engineering, food and medical industries and so on. Most of the early theoretical studies were based on Darcy's law with the assumption of a uniform porosity medium, see Cheng [1, 2]. Based on this model, it has been found that the Nusselt number for natural convection extended to heated surfaces in an infinite medium is a function of the media Rayleigh number only. For natural convection in enclosed porous medium, numerical results show that the Nusselt number is a function of the media Rayleigh number and the aspect ratio of the enclosure. Experimental results [3-5] on natural convection in water/glass-spheres systems having a small dimensionless particle diameter (i.e. the ratio of the particle diameter to characteristic length) tend to confirm the results of the early analyses of Cheng [1]. However, other experiments for natural convection using different porous media and fluids at high Rayleigh numbers show that the Nusselt number depends on additional parameters, such as the Prandtl number of the fluid, the dimensionless particle diameter, and the thermal conductivity ratio of the solid to fluid phase [6-9]. The discovery of the dependency of additional parameters in the measured Nusselt number for natural convection in porous media have prompted researchers to study a number of effects which have been neglected in early analyses. These include the effects of boundary friction, inertia, acceleration, variable porosity and thermal dispersion. The boundary friction, inertia, and acceleration effects (collectively called the non-Darcian effects) tend to decrease the heat transfer rate, see Prasad and Kladias [9] and Hsiao [10].

On the other hand, Hsiao et al. [11, 12] proved that the variable porosity and thermal dispersion effects tend to enhance the heat transfer rate .

As mentioned previously, natural convection in porous cavities are widely applied in engineering and therefore how to get accurate results for a system is very important to the design of such a system. Recent work on the investigation of the flow and heat transfer characteristics for natural convection in porous enclosures will be the subject of discussion in this chapter, in which the non-Darcian effects, the effects of variable porosity and thermal dispersion, and different geometric types of heat sources are considered.

GOVERNING EQUATIONS

The macroscopic conservation equations of mass, momentum, and energy for natural convective heat transfer in a variable porosity medium with the Boussinesq approximation are

$$\nabla \cdot \mathbf{v}^* = 0 \tag{1}$$

$$\rho_f \left[\frac{\partial \mathbf{v}^*}{\partial t^*} + \nabla \cdot \left(\frac{\mathbf{v}^* \mathbf{v}^*}{\varepsilon} \right) \right] = -\nabla p^* + \mu_f \nabla^2 \mathbf{v}^* - \frac{\mu_f \varepsilon \mathbf{v}^*}{K} - \rho_f \frac{F|\mathbf{v}^*|\mathbf{v}^* \varepsilon}{\sqrt{K}} + \rho_f \beta_f (T^* - T_\infty^*)\varepsilon g \tag{2}$$

and

$$\frac{\partial}{\partial t^*}\left[(\rho_f C_{pf})\varepsilon + (\rho_s C_{ps})(1-\varepsilon) \right] T^* + (\rho_f C_{pf}) \nabla \cdot (\mathbf{v}^* T^*) = \nabla \cdot \left[k_e \nabla T^* \right] \tag{3}$$

where $\mathbf{v}^*$ and p^* are the volume-averaged velocity vector and pressure; ρ_f, μ_f, and β_f are the density, viscosity, and thermal expansion coefficient of the fluid; ρ_s is the density of the solid phase, C_{pf} and C_{ps} are the specific heats of the fluid and solid phases at constant pressure; F is the inertial coefficient; ε and K are the porosity and permeability of the porous medium, and k_e is the effective thermal conductivity of the saturated porous medium, which is a superposition of the stagnant thermal conductivity (k_m) and the dispersive conductivity (k_d), i.e. $k_e = k_m + k_d$. For a porous medium composed of packed spheres, the porosity ε varies from 0.36 far from the wall to nearly unity at the wall. As shown by Vortmeyer and Schuster [13], the variation of the porosity can be approximated by an exponential function of the form

$$\varepsilon = \varepsilon_\infty + (\varepsilon_o - \varepsilon_\infty)\exp(-N_1 x^*/d_p) \tag{4}$$

where d_p is the particle diameter, x^* is the distance from the wall; N_1 is an empirical constant and ε_∞ and ε_o are the porosities at locations far away and on the wall, respectively. The value of $N_1 = 7$ was used by Hsiao et al. [11] while Hsu and Cheng [14, 15] have suggested that a value of $N_1 = 5–7$ should be used instead. For a packed-sphere bed, the permeability and the Forchhemier coefficient are related to the porosity by

$$K = \frac{d_p^{\,2}\varepsilon^3}{A(1-\varepsilon)^2} \quad \text{and} \quad F = \frac{B}{\sqrt{A}\varepsilon^{3/2}} \tag{5}$$

where A and B are empirical constants. Ergun [16] found that A = 150 and B = 1.75, while Macdonald et al. [17] found A= 182 and B= 1.92. More recently, Fand et al. [18] performed some experiments on the pressure drop in tubes packed with different sizes of glass spheres, and determined that A = 182 and B = 1.92, based on the Ergun equation with a constant porosity equal to 0.36. In a recent paper, Cheng et al. [19] reanalyzed the data of Fand et al. [18] based on a hydrodynamically fully developed flow in a packed tube with a variable porosity model; they found that the calculated pressure drop matches with the data of Fand et al. [18] if the values of A = 215 and B = 1.92 are used in equation (5).

The value of the stagnant thermal conductivity of the saturated porous medium can be computed according to the following expression, given by Zehner and Schluender [20], as

$$\frac{k_m}{k_f} = \left(1-\sqrt{(1-\varepsilon)}\right) + \frac{2\sqrt{(1-\varepsilon)}}{1-\lambda B_o} \times \left[\frac{(1-\lambda)B_o}{(1-\lambda B_o)^2}\ln\left(\frac{1}{\lambda B_o}\right) - \frac{B_o+1}{2} - \frac{B_o-1}{1-\lambda B_o}\right] \tag{6}$$

where $B_o = 1.25\left[(1-\varepsilon)/\varepsilon\right]^{10/9}$ and $\lambda = k_f / k_s$ with k_f and k_s denoting the thermal conductivity of the fluid and the solid phase, respectively. The value of k_s for glass beads is calculated based on the investigation of Fand and Phan [21]:

$$k_s = 1.00416 + 1.6736 \times 10^{-3} T^* - 4.184 \times 10^{-6} T^{*2} \tag{7}$$

where T* is the temperature in °C and k_s is expressed in W m^{-1} °C^{-1}.

The thermal dispersion conductivity for flow through a porous medium is given by Hsu and Cheng [14, 15], namely,

$$k_d = C(\rho C_p)_f |w^*| \iota d_p \tag{8}$$

where ι is the dispersive length, which is given by

$$\iota = \frac{1-\varepsilon}{\varepsilon} \tag{9}$$

while $w^* = (u^{*2} + v^{*2})^{1/2}$ and the value of the dispersivity C= 0.02~0.04 was determined by Hsu and Cheng [14, 15] by a comparison of theory and experiments for forced convection in packed columns. Equations (1)-(9) are the governing equations for natural convection in a non-uniform porosity medium for considering the thermal dispersion effect and non-Darcian effects. This formulation has been used by Hsiao et al. [10, 11, 22] for numerical studies of natural convection in porous media.

NATURAL CONVECTION IN A HORIZONTALLY ECCENTRIC POROUS ANNULUS

The problem of natural convection in a porous annulus between horizontal cylinders has important applications to thermal insulation engineering and energy storage system. Most of the early theoretical studies on this problem assume that the horizontal cylinders are concentric and that Darcy's law is applicable. Recently, Bau et al. [23, 24] studied the problems of steady natural convection in eccentric porous annuli at low and high Rayleigh numbers based on Darcy's law and using perturbation methods. They found that eccentric insulation may be more efficient than concentric insulation. Recent work [25-29] has shown that Darcy's law is inaccurate for the prediction of heat transfer characteristics in a porous medium under certain conditions. In particular, Hong et al. [25] have shown that the non-Darcian effects have significant influence in a high-porosity medium, and reduce the heat transfer rate as compared with that predicted based on Darcy's law. Hsu and Cheng [26], as well as Tong and Subramanian [27], have shown that Darcy's law does not accurately predict natural convection characteristics in porous media if the thermal boundary-layer is thinner than the viscous sublayers. Beckermann et al. [28], as well as Prasad and Tuntomo [29], have shown that at high Darcy numbers, the inertia and the boundary friction terms must be included simultaneously to obtain realistic predictions of the Nusselt numbers. More recently, Kaviany [30] has studied the influences of inertia, boundary friction, and velocity square terms on steady natural convection in porous concentric annuli. He found that all of these effects reduce the heat transfer rate, with the boundary friction terms being the most significant.

In practical applications, two objectives are required for thermal insulation engineering. One is to prevent heat loss from a pipe which is transporting heated fluid. The other is to prevent heat input to a pipe which is transporting cooled fluid. The present section discusses the problem of transient natural convection in a porous annulus between concentric and eccentric horizontal cylinders at uniform, but different, temperatures, including the two cases mentioned above.

Problem Description

Consider two horizontal cylinders of radii r_o and r_i centered at o and o′, respectively, as shown in Figure 1. The eccentricity of the inner cylinder is measured by the distance E and the angle ϕ measured clockwise from the upward vertical. The sign of E is positive for $0 \le \phi \le \pi/2$ and negative for $\pi/2 \le \phi \le \pi$. The space between the inner and outer cylinders is filled with a constant-porosity medium. Initially, the porous annulus is at a uniform temperature, and a quiescent state is assumed. The following four cases are considered:

Case 1: Initially, the porous annulus is at a uniform temperature T_o^*. At $t^* > 0$ the temperature of the inner cylinder is suddenly changed to a higher temperature T_i^*. Thus $T_i^* > T_o^*$.

Case 2: Initially, the porous annulus is at a uniform temperature T_i^*. At $t^* > 0$ the temperature of the outer cylinder is suddenly changed to a lower temperature T_o^*. Thus $T_i^* > T_o^*$.

Case 3: Initially, the porous annulus is at a uniform temperature T_o^*. At $t^* > 0$ the temperature of the inner cylinder is suddenly changed to a lower temperature T_i^*. Thus $T_i^* < T_o^*$.

Case 4: Initially, the porous annulus is at a uniform temperature T_i^*. At $t^* > 0$ the temperature of the outer cylinder is suddenly changed to a higher temperature T_o^*. Thus $T_i^* < T_o^*$.

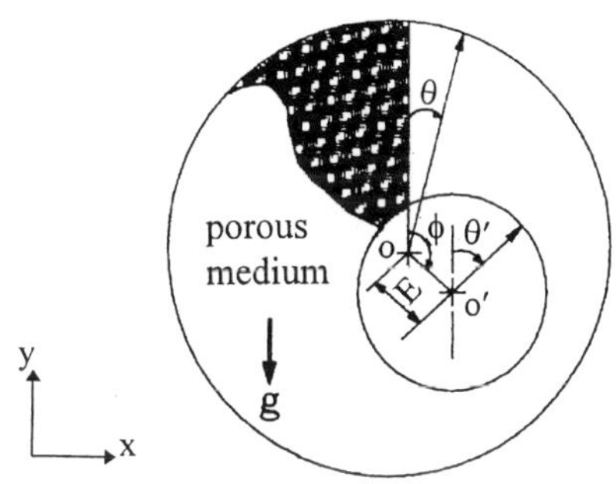

Figure 1 Physical model and coordinate system

The subsequent convective motion and the associated heat transfer characteristics in the eccentric porous annulus of the four cases will be the subjects of this section.

Problem Solving

Taking the following dimensionless variables

$$x = \frac{x^*}{D_i}, \quad y = \frac{y^*}{D_i}, \quad u = \frac{u^* D_i}{\alpha_e}, \quad v = \frac{v^* D_i}{\alpha_e}$$

$$w = \frac{w^* D_i}{\alpha_e}, \qquad \Theta = \frac{T^* - T_o^*}{T_i^* - T_o^*}, \qquad t = \frac{\alpha_e t^*}{D_i^2} \tag{10}$$

with D_i denoting the inner diameter of the annulus, the governing equations in terms of the dimensionless stream function, vorticity, and temperature in terms of the body-fitted coordinates (ξ, η) proposed by Thomas and Middlecoff [31] were derived by Chen and Hsiao [32] as

$$\tilde{\nabla}^2 \psi = -\Omega \tag{11}$$

$$\Omega_t + \frac{1}{\varepsilon J}(\psi_\eta \Omega_\xi - \psi_\xi \Omega_\eta) = -\frac{F\varepsilon}{J(Da)^{1/2}}[y_\eta(\tau_1)_\xi - y_\xi(\tau_1)_\eta - x_\xi(\tau_2)_\eta + x_\eta(\tau_2)_\xi]$$
$$-\frac{Pr\,\varepsilon}{Da}\Omega + Pr\,\tilde{\nabla}^2\Omega \pm \frac{Ra\,Pr\,\varepsilon}{J}(y_\eta \Theta_\xi - y_\xi \Theta_\eta) \tag{12}$$

$$\Theta_t + \frac{1}{\sigma J}(\psi_\eta \Theta_\xi - \psi_\xi \Theta_\eta) = \frac{1}{\sigma}\tilde{\nabla}^2\Theta \tag{13}$$

where $\tilde{\nabla}^2$ is the Laplace operator in the transformed coordinates, $Pr = \nu/\alpha_e$ is the Prandtl number of a saturated porous medium, $Da = K / D_i^2$ is the Darcy number, $Ra = g\beta_f |T_i^* - T_o^*| D_i^3 / \nu_f \alpha_e$ is the Rayleigh number, and the positive and negative signs in front of the buoyancy force term (i.e. the last term) refer to the case of $T_i^* > T_o^*$ (cases 1 and 2) and $T_i^* < T_o^*$ (cases 3 and 4), respectively. The constant F = 0.55 is the Forchheimer coefficient; α_e is the equivalent thermal diffusivity of the saturated porous medium defined as $\alpha_e = k_e / \rho_f C_{pf}$, with k_e denoting the stagnant thermal conductivity of the saturated porous medium; σ is the specific-heat ratio defined as $\sigma = [\varepsilon \rho_f C_{pf} + (1-\varepsilon)\rho_s C_{ps}] / (\rho_f C_{pf})$ and

$$\tau_1 = |w| v = [(x_\xi \psi_\eta - x_\eta \psi_\xi)^2 + (y_\xi \psi_\eta - y_\eta \psi_\xi)^2]^{1/2} \frac{y_\xi \psi_\eta - y_\eta \psi_\xi}{J^2} \tag{14a}$$

$$\tau_2 = |w| u = [(x_\xi \psi_\eta - x_\eta \psi_\xi)^2 + (y_\xi \psi_\eta - y_\eta \psi_\xi)^2]^{1/2} \frac{x_\xi \psi_\eta - x_\eta \psi_\xi}{J^2} \tag{14b}$$

The transformed initial conditions are

$$\psi(\xi, \eta, 0) = \Omega(\xi, \eta, 0) = \Theta(\xi, \eta, 0) = 0, \quad \text{for cases 1 and 3} \tag{15a}$$

$$\psi(\xi, \eta, 0) = \Omega(\xi, \eta, 0) = 0, \quad \Theta(\xi, \eta, 0) = 1, \quad \text{for cases 2 and 4} \tag{15b}$$

As a result of the transformation, the annulus is transformed into a rectangular domain as shown in Figure 2. The transformed boundary conditions for the four cases are

$$\psi(\xi,\eta_{min},t) = 0, \quad \Theta(\xi,\eta_{min},t) = 1, \quad \Omega(\xi,\eta_{min},t) = -\frac{\delta\psi_{\eta\eta}}{J^2}(\xi,\eta_{min},t) \tag{16a, b, c}$$

$$\psi(\xi,\eta_{max},t) = \Theta(\xi,\eta_{max},t) = 0, \quad \Omega(\xi,\eta_{max},t) = -\frac{\delta\psi_{\eta\eta}}{J^2}(\xi,\eta_{max},t) \tag{17a, b, c}$$

where δ is the transformation coefficient and J is the Jacobian of the transformation given by

$$\delta = x_\xi^{\ 2} + y_\eta^{\ 2}, \qquad J = x_\xi y_\eta - x_\eta y_\xi \tag{18a, b}$$

and η_{max} and η_{min} are the outer and inner boundaries in the transformed plane , Figure 2, respectively.

Equations (11)-(14) are discretized by the finite-difference method with the convective terms approximated by upwind differences. The finite-difference equations for the stream function were solved by the SOR method, while the finite-difference equations for the vorticity and energy equations were solved by the alternative direction implicit (ADI) method. The local and mean Nusselt numbers at the inner and outer radii are defined as follows:

$$Nu_i = -\frac{\ln R}{2}\left[r\frac{\partial\Theta}{\partial r}\right]_{r=1} = -\frac{\ln R}{2}\left[\frac{\delta\Theta_\eta - \beta\Theta_\xi}{\sqrt{\delta}}\right]_{\eta=\eta_{min}} \tag{19}$$

$$Nu_o = -\frac{\ln R}{2}\left[r\frac{\partial\Theta}{\partial r}\right]_{r=R} = -\frac{\ln R}{2}\left[\frac{\delta\Theta_\eta - \beta\Theta_\xi}{\sqrt{\delta}}\right]_{\eta=\eta_{max}} \tag{20}$$

$$\overline{Nu_i} = \frac{1}{2\pi}\int_0^{2\pi} Nu_i d\theta = \int_{\xi_{min}}^{\xi_{max}} Nu_i\sqrt{\delta}d\xi \Big/ \int_{\xi_{min}}^{\xi_{max}} \sqrt{\delta}d\xi \tag{21}$$

$$\overline{Nu_o} = \frac{1}{2\pi}\int_0^{2\pi} Nu_o d\theta = \int_{\xi_{min}}^{\xi_{max}} Nu_o\sqrt{\delta}d\xi \Big/ \int_{\xi_{min}}^{\xi_{max}} \sqrt{\delta}d\xi \tag{22}$$

where $r = r^*/r_i^*$ and $R = r_o^*/r_i^*$ with r_i^* and r_o^* denoting the radii of the inner and outer cylinders respectively; β ($= x_\xi x_\eta + y_\xi y_\eta$) is the transformation coefficient, 2/lnR is the heat conduction in a concentric annulus and θ is the angle measured clockwise from the upward vertical axis located at o and o' for the outer and inner cylinders, respectively. The values of ξ_{min} and ξ_{max} correspond to θ or $\theta' = 0$ and θ or $\theta' = 2\pi$ on the surfaces of the cylinder. The integrals in equations (21) and (22) were computed with the trapezoidal rule.

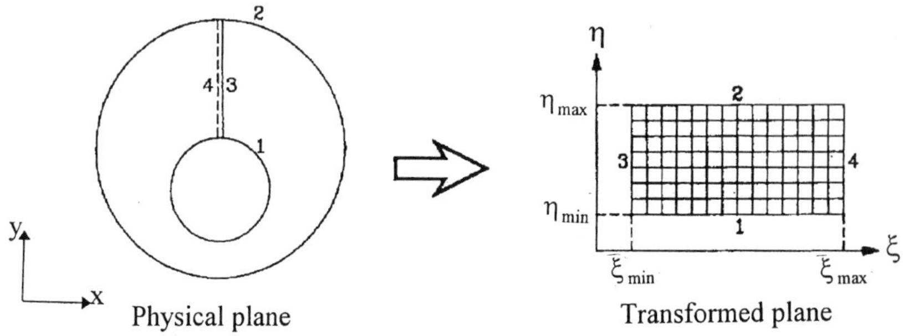

Figure 2 Coordinate transformation

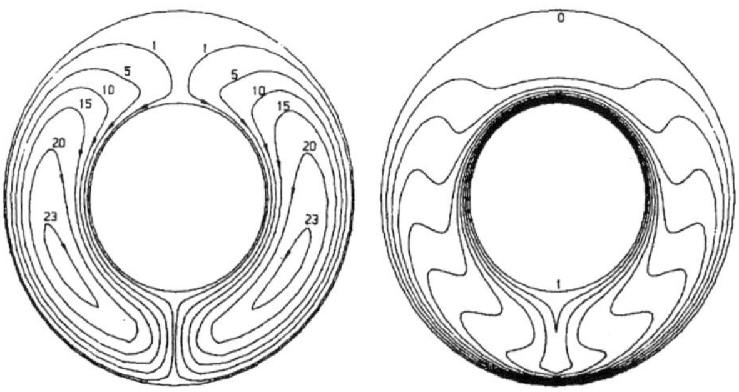

Figure 3 Steady-state streamlines (left) and isotherms (right) in a concentric porous annulus with R = 2, ε = 0.9, Pr = 0.5, Da = 0.1 and Ra = 10^6 for $T_i^* < T_o^*$

Heat Transfer Characteristics in a Concentric Porous Annulus

Since the governing equations and boundary conditions are identical for cases 1 and 2, the steady-state heat transfer characteristics for these cases are also identical. Similar conclusions hold for cases 3 and 4. Moreover, since cases 1 and 3 at steady state differ only on the sign of the buoyancy force term in the momentum equation, the convective pattern and the isotherms are the mirror image of each other. Figure 3 shows the steady streamlines and isotherms in a concentric porous annulus with R = 2, ε = 0.9, Pr = 0.5, Da = 0.1 and Ra = 10^6 for $T_i^* < T_o^*$ (cases 3 and 4). For this case, where heating is provided by the outer cylinder, the fluid will rise along the the outer cylinder because of buoyancy force and move downward along the colder inner cylinder. Temperature inversion is pronounced at a Rayleigh number of 10^6, as evident from Figure 3. For the case with $T_i^* > T_o^*$, where heating is provided by the inner cylinder, the direction of streamlines and the shape of isotherms are the mirror images of those in Figure 3. The steady mean Nusselt number of a concentric annulus is identical for the four cases under consideration, as shown in Figure 4. Figure 4 shows the effects of the radii ratio on the mean steady Nusselt number of a porous concentric annulus at Da = 0.1 for three different values of the Rayleigh number. It is shown that the mean Nusselt number increases with the value of the Rayleigh number as well as the radii ratio. The steady mean Nusselt number apporaches the values of 9.02, 5.05 and 2.29 for the three Rayleigh numbers under consideration. These asymptotic values are for the case of a heated horizontal cylinder embedded in an infinite porous medium.

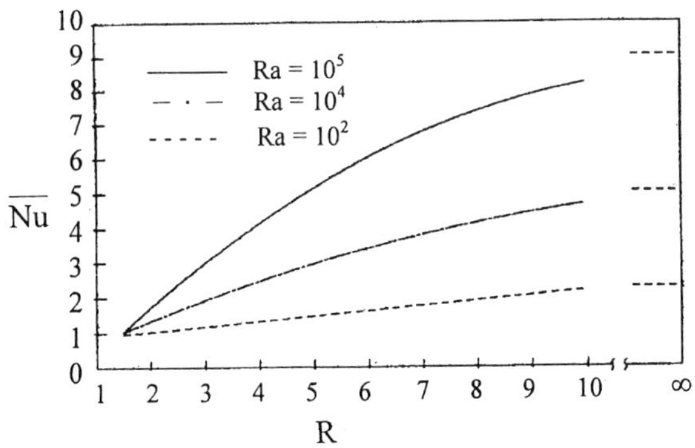

Figure 4 Steady mean Nusselt number versus radii ratio for a concentric porous annulus with ε = 0.9, Pr = 0.5 and Da = 0.1

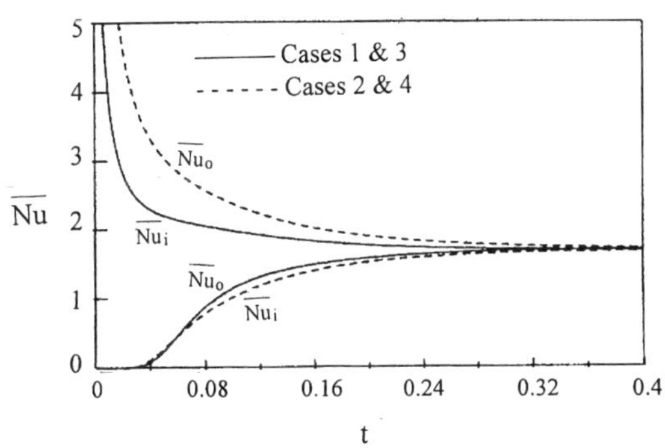

Figure 5 Transient mean Nusselt number of a concentric porous annulus with Ra = 10^5, R = 2, ε = 0.9, Pr = 0.5 and Da = 0.1

Figure 5 shows the transient mean Nusselt numbers of a concentric porous annulus with R = 2, ε = 0.9, Pr = 0.5, Da = 0.1, Ra = 10^5 and σ = 3 for the four cases under consideration. For cases 1 and 3, where the porous annulus is initially at a uniform temperature T_o* and the temperature of the inner cylinder is suddenly changed to T_i*, the mean Nusselt number of the inner cylinder ($\overline{Nu}_i$) thus decreases drastically with time, while that of the outer cylinder ($\overline{Nu}_o$) increases slowly from zero with time. For cases 2 and 4, where the porous annulus is initially at a uniform temperature T_i* and the temperature of the outer cylinder is suddenly changed to T_o*, the value of Nu_o thus decreases drastically with time, while that of Nu_i increases slowly from zero with time. The values of Nu_i and Nu_o approach each other at steady state, which occurs approximately at t = 0.25 for Ra = 10^5.

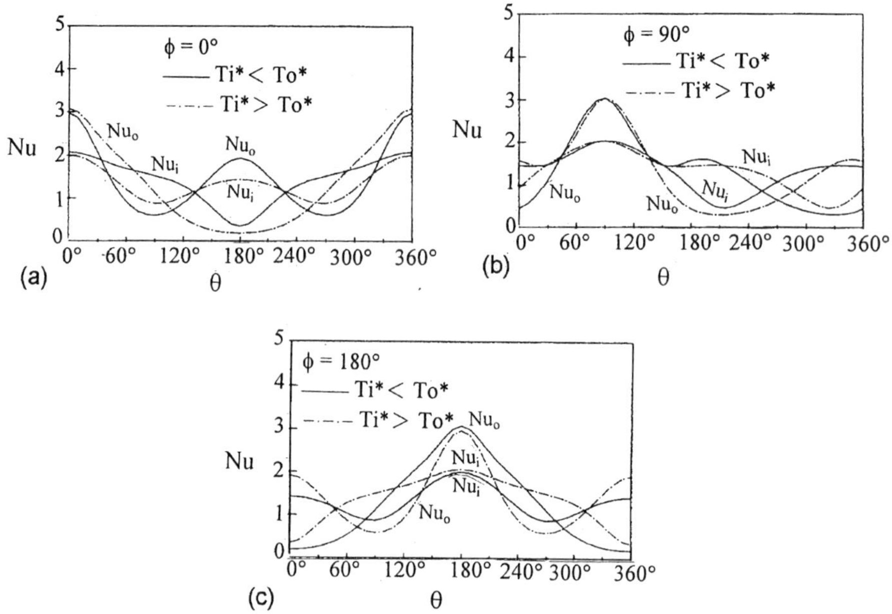

Figure 6 The variations of steady local Nusselt number in an eccentric porous annulus with R = 2, ε = 0.9, Pr = 0.5, Ra = 10^4, Da = 0.1 and E_{cc} = 0.6 for $T_i^* > T_o^*$ and $T_i^* < T_o^*$

Heat Transfer Characteristics in an Eccentric Porous Annulus

An eccentric annulus is characterized by the eccentric angle of the inner cylinder, ϕ. To study the effect of eccentric angle on the natural convection characteristics in an eccentric annulus, computations were carried out for R = 2, ε = 0.9, Pr = 0.5, Ra = 10^4, Da = 0.1 and E_{cc} = 0.6 at various ϕ for cases 1 and 2 ($T_i^* > T_o^*$). The eccentricity, E_{cc}, is definied as $E_{cc} = E/D_i$.The results of the computations for t = 0.3 at steady state are presented in Figures 6 and 7. The value of the eccentric angle has an important effect on the Nusselt number, see Figure 6, which also shows that, the local Nusselt numbers are not symmetrical with respect to the vertical axis except for the cases of ϕ = 0° or 180°.

The extreme values of the local Nusselt numbers of an eccentric porous annulus are no longer fixed at $\theta = 0°$ or $180°$ and it can be concluded, from Figure 6, that the maximum values of Nu_i and Nu_o always occur near the eccentric angle. Figure 7 shows the effect of eccentricity on the mean Nusselt number at steady state of a porous annulus with $R = 2$, $\varepsilon = 0.9$, $Pr = 0.5$ and $Da = 0.1$, at $\phi = 0°$ (with $E_{cc} > 0$) and $\phi = 180°$ (with $E_{cc} < 0$) for two values of Ra. For the case of $Ra = 10^4$, where heat conduction is dominant, the minimum value of the mean Nusselt number occurs at $E_{cc} = 0.4$ for $T_i^* > T_o^*$ and at $E_{cc} = -0.4$ for $T_i^* < T_o^*$.

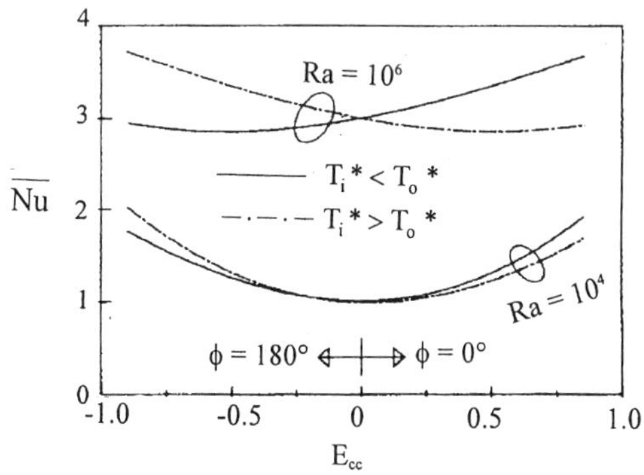

Figure 7 Steady mean Nusselt number versus eccentricities for cases $T_i^* > T_o^*$ and $T_i^* < T_o^*$ with $R = 2$, $\varepsilon = 0.9$ and $Pr = 0.5$

NATURAL CONVECTION ABOUT A CORRUGATED PLATE EMBEDDED IN A POROUS CAVITY

The problem of heat losses from buried pipes has been studied by Thiyagarajan et al. [33], Bau and Sadhal [34] and Difelice and Bau [35] who assumed that heat losses may be calculated using a conduction model. Thermal convection in the porous medium around buried pipes has been taken into account in experimental and theoretical works by Fernandez and Schrock [36], Farouk and Shayer [37], Bau [38], Himasekhar and Bau [39], Fand et al. [7] and Hsiao et al. [11, 32]. Another numerical and experimental investigation of the problem of natural convection associated with a cylindrical heat source embedded in a box has been performed by Himasekhar and Bau [40]. However the same problem, with a more complicated configuration of heat source in a porous cavity, which is often encountered in casting engineering is rarely considered. Here, several investigation results of Hsiao [41, 42] on the problem of transient natural convection about a heated corrugated plate embedded in an enclosed porous medium, which can be applied in the casting engineering, are presented.

Mathematical Formulation

Consider a corrugated plate of length H_2, thickness C_2, and corrugated angle θ embedded in an enclosed constant-porosity medium as shown in Figure 8. The corrugated plate is kept at a uniform temperature T_h*, while the wall temperature of the cavity is kept at room temperature T_c* where $T_h^* > T_c^*$.

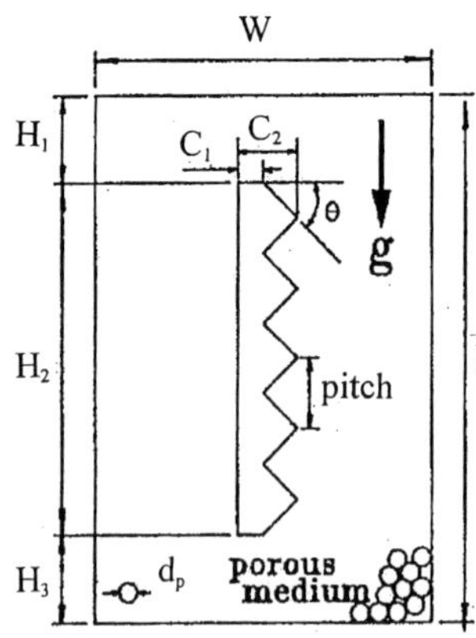

Figure 8 Physical model and notation

Taking the length of the plate, H_2, as the characteristic length for the dimensionless variables, the resulting nondimensional equations for the stream function, vorticity and temperature in terms of the body-fitted coordinates are also in the forms of equations (11)-(13), except the minus sign "$-$" of the buoyancy-force term in equation (12) is neglected, and the Rayleigh number is defined as $Ra = g\beta_f(T_h{}^* - T_c{}^*)H_2{}^3 / \nu_f\alpha_e$, and the Darcy number as $Da = K / H_2{}^2$. The local and mean Nusselt numbers at the plate are defined as

$$Nu = -\frac{\partial T}{\partial n} = \left[\frac{1}{J\sqrt{\delta}}(-\delta\Theta_\eta + \beta\Theta_\xi)\right]_{\eta=\eta_{min}} \tag{23}$$

$$\overline{Nu} = \int_{\xi_{min}}^{\xi_{max}} Nu\, d\xi \Big/ \int_{\xi_{min}}^{\xi_{max}} d\xi \tag{24}$$

Steady State Flow and Heat Transfer Characteristics

Figure 9 shows the steady isotherms (left) and streamlines (right) for a corrugated plate embedded in a porous cavity with $W/H_2 = H/H_2 = 5$, $C_2/H_2 = 1/8$, $H_1/H_2 = H_3/H_2 = 2$, $pitch/H_2 = 1/8$, $\theta = 45°$, $Pr = 0.7$, $\varepsilon = 0.9$ and $Da = 10^{-4}$. As shown in Figure 9(a) for $Ra_m = 30$, the heat transfer mechanism inside the cavity is conduction dominated, as evident from the weak circulation around the plate. The centers of the two counter-rotating vortices are located slightly above the heated plate. The flow and temperature fields inside the cavity are almost symmetrical with the vertical axis of the cavity. The Darcy-Rayleigh number, Ra_m, is defined as $Ra_m = g\beta_f(T_h^* - T_c^*)KH_2 / \nu_f\alpha_e$.

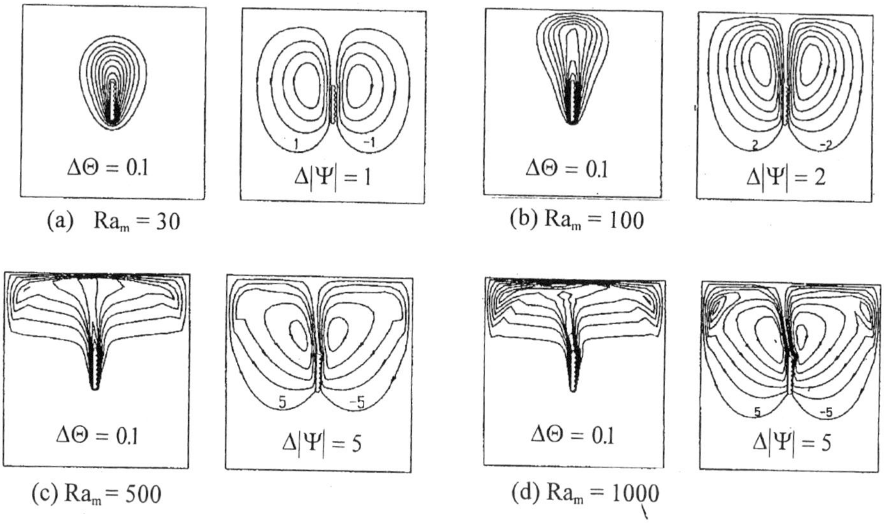

Figure 9 Steady state isotherms (left) and streamlines (right) for a corrugated plate with $W/H_2 = H/H_2 = 5$, $C_2/H_2 = 1/8$, $H_1/H_2 = H_3/H_2$, $pitch/H_2 = 1/8$, $\theta = 45°$, $Pr = 0.7$, $\varepsilon = 0.9$ and $Da = 10^{-4}$

As the Darcy-Rayleigh number is increased to 100, Figure 9(b), the plume clearly rises above the plate, and the centers of the vortices move upward with increasing strength. Due to the influence of the corrugated configuration, the flows on the right hand side of the plate are slightly slower than those on the left hand side. In this case, the flow pressure on the right portion of the cavity is larger than that on the left. As a result of the pressure difference, an extra horizontal driving force is created, which then acts on the fluid from right to left. As a result, the flow field is no longer symmetrical with the vertical axis of the cavity. With further increases in Ra_m to 500, Figure 9(c), the strength of the plume rising above the plate is increased and consequently increases in the velocity of the rising fluid. Due to the higher velocity, the thermal boundary-layer developed around the corrugated plate becomes thinner and the pressure difference between the right and left portion of the cavity is also increased. Thus, the distortion of the streamlines and isotherms is more pronounced. When the Darcy-Rayleigh number is increased further to 1000, Figure 9(d), secondary vortices begin to appear downstream of the heated plate in the upper corners of the porous cavity. This is due to the viscous effect associated with the high velocities of the upward flowing fluids. These secondary vortices cause the instability in the porous cavity. Figure 10 shows the effect of plate position on the steady mean Nusselt number for a corrugated plate and the results are computed by varying the value of H_3/H_2 and keeping the other parameters with $W/H_2 = H/H_2 = 5$. The results show that $\overline{Nu}$ decreases if the plate moves downward along the vertical axis of the cavity.

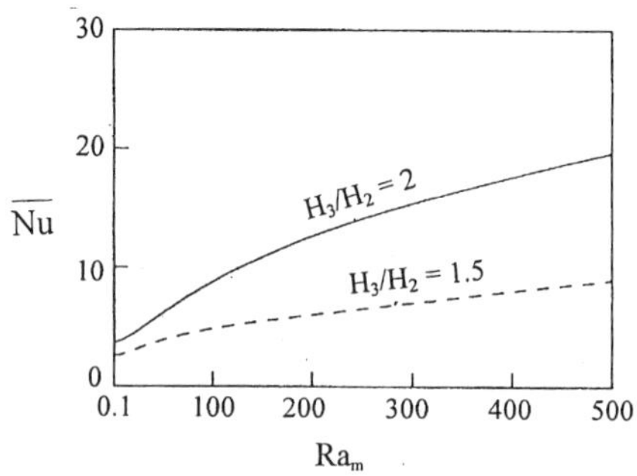

Figure 10 Effect of plate position on steady mean Nusselt number for a corrugated plate with $W/H_2 = H/H_2 = 5$, $C_2/H_2 = 1/8$, $H_1/H_2 = H_3/H_2 = 2$, $Pitch/H_2 = 1/8$, $\theta = 45°$, $Pr = 0.7$, $\varepsilon = 0.9$ and $Da = 10^{-4}$

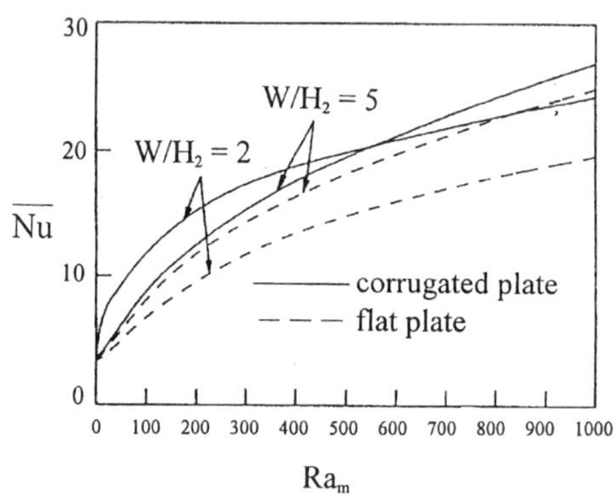

Figure 11 Effect of cavity aspect ratio on steady mean Nusselt number for $H/H_2 = 5$, $C_2/H_2 = 1/8$, $H_1/H_2 = H_3/H_2 = 2$, $Pitch/H_2 = 1/8$, $\theta = 45°$, $Pr = 0.7$, $\varepsilon = 0.9$ and $Da = 10^{-4}$

Figure 11 presents the effect of the aspect ratio of the cavity on the steady mean Nusselt number for a corrugated plate and a flat plate embedded in a porous cavity. For a corrugated plate (solid lines), the steady mean Nusselt number increases as the aspect ratio is reduced when the Darcy-Rayleigh number is less than 550, while there is an opposite effect as the Darcy-Rayleigh number exceeds 550.

Transient Characteristics

The transient behavior of the mean Nusselt numbers for a corrugated plate with $W/H_2 = H/H_2 = 5$, $C_2/H_2 = 1/8$, $H_1/H_2 = H_3/H_2 = 2$, $Pitch/H_2 = 1/8$, $\theta = 45°$, $Pr = 0.7$, $\varepsilon = 0.9$ and $Da = 10^{-4}$ at different Darcy-Rayleigh number are shown in Figure 12(a). It is shown that the transient mean Nusselt number increases with increase in the Darcy-Rayleigh number. Initially, the sharp temperature gradients between the heated plate and porous medium are responsible for a rapid rate of heat transfer and their corresponding high Nusselt numbers. As shown in Figure 12(a), for $Ra_m \leqq 500$, the Nusselt number asymptotically approaches a steady state value. As Ra_m increases to 1000, the Nusselt number oscillates with a very small amplitude. The oscillations continue to $t = 0.09$ and then keep at a stable value until $t = 0.18$. As time increases to $t \geqq 0.18$, oscillations again occur and the Nusselt number finally approaches a stable value. With further increases in Ra_m, the frequency of these oscillations increase and no stable solution can be obtained.

Comparison of a Corrugated Plate with a Flat Plate

In order to see the corrugated configuration effect on the flow and heat transfer characteristics, computations were performed with $Pr = 0.7$, $\varepsilon - 0.9$ and plate aspect ratio $(C_2/H_2) = 1/8$ at various Darcy-Rayleigh numbers and cavity aspect ratios. Figure 11 shows the effect of the cavity aspect ratio on the steady mean Nusselt number for both a corrugated plate and a flat plate embedded in a porous cavity. As mentioned earlier, for $Ra_m < 550$, the steady mean Nusselt number for a corrugated plate is increased as the aspect ratio of the cavity (W/H) is reduced from 1 to 0.4 while keeping the value of H/H_2 at 5, and an opposite effect is obtained as $Ra_m > 550$. For a flat plate, however, the steady mean Nusselt number always decreases with decreasing the value of W/H. This is due to the fact that the flow field is influenced by the corrugated configuration.

Figures 12(a) and 12(b) present typical transient Nusselt numbers of the corrugated plate and flat plate, respectively. The results are obtained with $W/H_2 = H/H_2 = 5$, $C_2/H_2 = 1/8$, $H_1/H_2 = H_3/H_2 = 2$, $\sigma = 3$, $Pr = 0.7$, $\varepsilon = 0.9$ and $Da = 10^{-4}$. It can be seen that, for both plates, the transient behavior of the Nusselt number is very stable for $Ra_m < 1000$ but as Ra_m increases to 1000, the transient Nusselt number of a corrugated plate begins to oscillate with a small amplitude as previously discussed, while that of a flat plate remains stable.

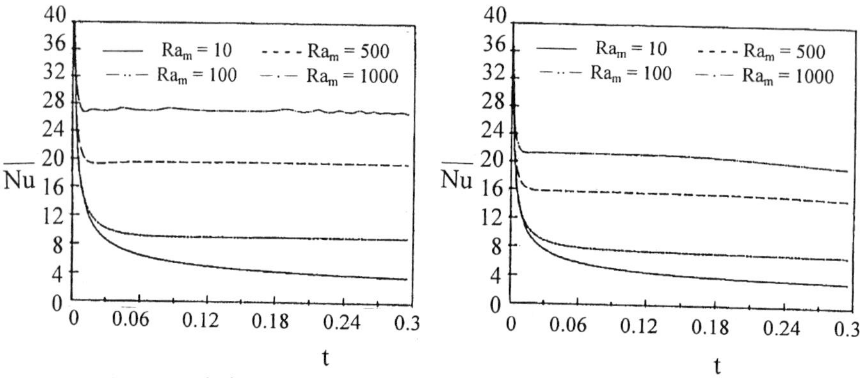

(a) Corrugated plate (b) Flat plate

Figure 12 Time history of the Nusselt number for $W/H_2 = H/H_2 = 5$, $C_2/H_2 = 1/8$, $H_1/H_2 = H_3/H_2 = 2$, $\sigma = 3$, $Pr = 0.7$, $\varepsilon = 0.9$ and $Da = 10^{-4}$ at various Darcy-Rayleigh numbers, (a) corrugated plate ($\theta = 45°$, pitch/H_2 = 1/8), (b) flat plate

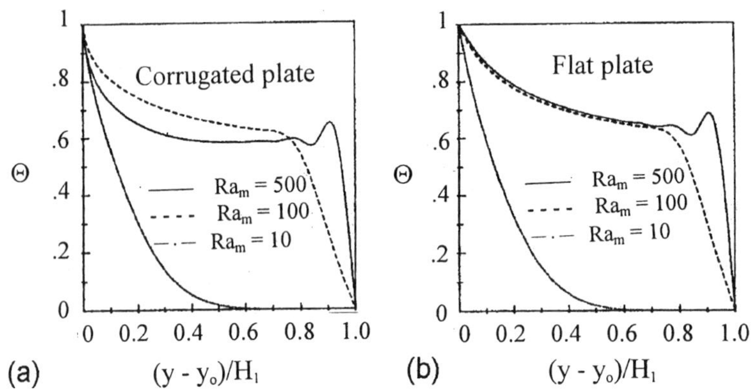

Figure 13 Temperature distribution in the center of the thermal plume with $W/H_2 = H/H_2 = 5$, $C_2/H_2 = 1/8$, $H_1/H_2 = H_3/H_2 = 2$, $Pr = 0.7$, $\varepsilon = 0.9$ and $Da = 10^{-4}$ for various Ra_m

The steady state temperature distribution along the plume axis, depicted as a function of the vertical location, for various values of Ra_m for $W/H_2 = H/H_2 = 5$ and $Pr = 0.7$ are shown in Figure 13, where y_o represents the y coordinate at the top of the plate and H_1 stands for the distance between the top surface of the cavity and the plate. Since the heat transfer mechanism inside the enclosed cavity is conduction dominated (Ra_m = 10 and 100), the temperature distributions induced by a corrugated plate are very close to those induced by a flat plate. However, when the heat transfer is dominated by the convection mode (Ra_m = 500), the deviation of the temperature profiles induced by these two plates becomes more pronounced. For a corrugated plate, Figure 13(a), the temperatures above the plate are less at $Ra_m = 500$ than at $Ra_m = 100$ if the dimensionless vertical distance is less than 0.75. The opposite tendency is found for distances above 0.75. However, for a flat plate, Figure 13(b), the vertical temperature above the plate is always greater at high Ra_m than at a low Ra_m. This is explained by the fact that the flow field is distributed by an extra horizontal driving force induced by the corrugated configurations.

NATURAL CONVECTION ABOUT A HEATED HORIZONTAL CYLINDER IN AN ENCLOSED POROUS MEDIUM

In recent years the problem of natural convection about a heated horizontal cylinder embedded in a porous medium has attracted a great deal of attention because of its wide range of applications in engineering and geophysics. Based on Darcy's law and the boundary-layer approximation. Merkin [43] obtained a similarity solution for natural convection about axisymmetric and two-dimensional bodies of arbitrary shape in a constant porosity medium of infinite extent at high Rayleigh numbers. Using the same approach, Cheng [2] has obtained a similarity solution for natural convection about a heated horizontal cylinder at uniform temperature, with the following expression for the average Nusselt number:

$$\overline{Nu}_m = 0.565 Ra_m^{1/2} \tag{25}$$

In the above equation, Ra_m and $\overline{Nu}_m$ are the media Rayleigh number and the average media Nusselt number, which are defined as $Ra_m = K_\infty g\beta_f\rho_f D\Delta T^* / \mu_f\alpha_m$, and $\overline{Nu}_m = \bar{h}D / k_{m\infty}$, respectively, where D is the diameter of the cylinder; $\bar{h}$ is the average heat transfer coefficient; ΔT^* is the difference between the wall temperature of the cylinder and the ambient temperature; $k_{m\infty}$, α_m and K_∞ are the stagnant thermal conductivity, the effective thermal diffusivity, and the permeability of the porous medium, which are assumed to be constant in deriving equation (25). The first experimental investigation on natural convection about a horizontal cylinder (at uniform temperature) embedded in a porous medium was carried out by Fernandez and Schrock [36], who also performed a numerical solution of the problem based on the Darcy-Brinkman model in a constant porosity medium. Subsequently, Fand et al. [7] conducted a similar experimental investigation with different sizes of glass spheres saturated with water or silicone oil. A comparison of equation (25) with the experimental data of Fand et al. [7] shows that the measured Nusselt numbers are higher than those predicted by the theory of Cheng [2]. Numerical and experimental investigations of natural convection about a heated horizontal cylinder with constant heat flux have recently been performed by Himasekhar and Bau [40]. They found that at high Rayleigh numbers, the measured average Nusselt numbers are higher than those obtained from their numerical solutions, which are based on the assumptions that Darcy's law is applicable, and the effects of variable porosity and thermal dispersion are negligible. The discrepancy between theory and experiments at high Rayleigh numbers is attributed to the transition to a three-dimensional, time-dependent flow that they observed in their experiments. Similar discrepancies between theory and experiments have also been observed for natural convection heat transfer from other bodies embedded in a porous medium .

Mathematical Formulation

Consider a horizontal circular cylinder of diameter D that is embedded in an enclosed porous medium at T_c^*, see Figure 14. The wall temperature of the cylinder is suddenly raised to T_h^* at $t > 0$ and the convective heat transfer characteristics at $t > 0$ are interested. Basing on the governing equations (1)-(9) and the method used in previous sections, the problem can be solved. The relationship between the fluid Rayleigh number, Ra_f, and the medium Rayleigh number, Ra_m, is defined as

$$Ra_f = Ra_m Da_\infty \left(\alpha_e / \alpha_f\right) \tag{26}$$

where Ra_f is the fluid Rayleigh number and Da_∞ is the bulk Darcy number, which are defined as

$$Ra_f = \frac{g\beta_f \left(T_h * - T_c *\right) D^3}{\alpha_f \nu_f} \tag{27}$$

and

$$Da_\infty = K_\infty / D^2 = \frac{\gamma^2 \varepsilon_\infty^{\ 3}}{A\left(1 - \varepsilon_\infty\right)^2} \tag{28}$$

where $\gamma = d_p / D$ is the dimensionless particle diameter. The local Darcy number, Da, is related to the bulk Darcy number, Da_∞, by

$$Da = Da_\infty \frac{K}{K_\infty} = Da_\infty \left(\frac{\varepsilon}{\varepsilon_\infty}\right)^3 \left(\frac{1 - \varepsilon_\infty}{1 - \varepsilon}\right)^2 \tag{29}$$

The local and the mean fluid Nusselt numbers, Nu_f and $\overline{Nu_f}$, are defined as

$$Nu_f = \frac{hD}{k_f} = -\left(\frac{\partial \Theta}{\partial r}\right)_{r=1}, \qquad \overline{Nu_f} = \frac{1}{\pi}\int_0^\pi Nu_f d\theta \tag{30a, b}$$

where θ is measured clockwise from the upward vertical axis. The local and average media Nusselt numbers are defined as

$$Nu_m = \frac{hD}{k_{m\infty}}, \qquad \overline{Nu_m} = \frac{1}{\pi}\int_0^\pi Nu_m d\theta \tag{31a, b}$$

It follows that

$$Nu_m = \frac{k_f}{k_{m\infty}} Nu_f \quad \text{and} \quad \overline{Nu_m} = \frac{k_f}{k_{m\infty}} \overline{Nu_f} \tag{32a, b}$$

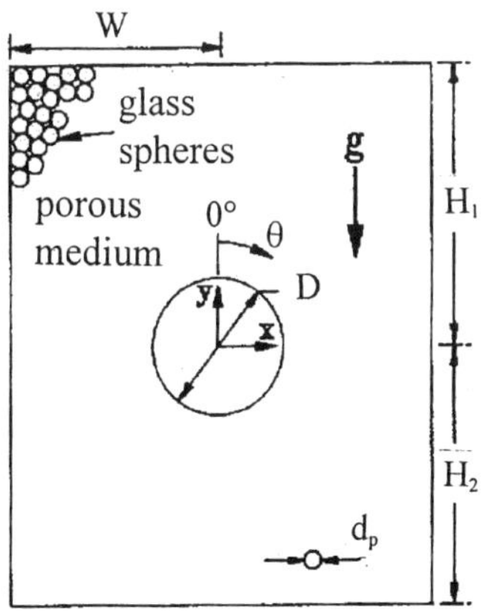

Figure 14 Physical model and coordinate system

Comparison with the Data of Fand et al. [7]

The heat transfer characteristics will be influenced by the parameters H_1/D, H_2/D, W/D, $Ra_f(Ra_m)$, Pr_f, λ, σ and Da_∞ (or γ). All of the numerical results presented were calculated with $\varepsilon_\infty = 0.36, \varepsilon_0 = 0.9, N_1 = 7$, $A = 215$ and $B = 1.92$. A parametric study with different values of parameters was performed and the results for surface heat flux will be presented either in terms of fluid Nusselt number, $\overline{Nu_f}$, or media Nusselt number, $\overline{Nu_m}$, as a function of the fluid Rayleigh number, Ra_f, or media Rayleigh number, Ra_m, whichever is appropriate. Computations were also carried out for the cases corresponding to the experimental conditions of Fand et al. [7].

Table 1 Values of Ra_f and Ra_m used in the numerical computations

d_p (mm)	γ	$Ra_m = 15$	$Ra_m = 273$	$Ra_f = 10^6$
2	0.175	$Ra_f = 9.3 \times 10^5$	$Ra_f = 1.69 \times 10^7$	$Ra_m = 16.1$
3	0.262	$Ra_f = 3.9 \times 10^5$	$Ra_f = 7.18 \times 10^6$	$Ra_m = 38.0$
4	0.349	$Ra_f = 2.3 \times 10^5$	$Ra_f = 4.23 \times 10^6$	$Ra_m = 64.5$
6	0.524	$Ra_f = 1.03 \times 10^5$	$Ra_f = 1.88 \times 10^6$	$Ra_m = 145.0$

Table 1 lists the values of Ra_f corresponding to Ra_m= 15 and 273 for D = 11.45 mm and different sizes of glass spheres saturated with water, which were used in the experiments of Fand et al. [7]. The following symbols are used in the presentation of results: VPWD is the variable porosity with thermal dispersion effect, VPND the variable porosity without thermal dispersion effect; CPWD the constant porosity with thermal dispersion effect, and CPND the constant porosity without thermal dispersion effect.

Natural Convection in a Constant-porosity Medium

From experimental observations, Fand et al. [7] have concluded that the enclosed medium shown in Figure 14 can be considered as infinite if $H_1/D = H_2/D = 11.1$ and $W/D = 8.8$. This conclusion was also proved by Hsiao et al. [11]. Thus, the numerical computations were carried out with this size to simulate the case of a heated horizontal cylinder embedded in an infinite medium.

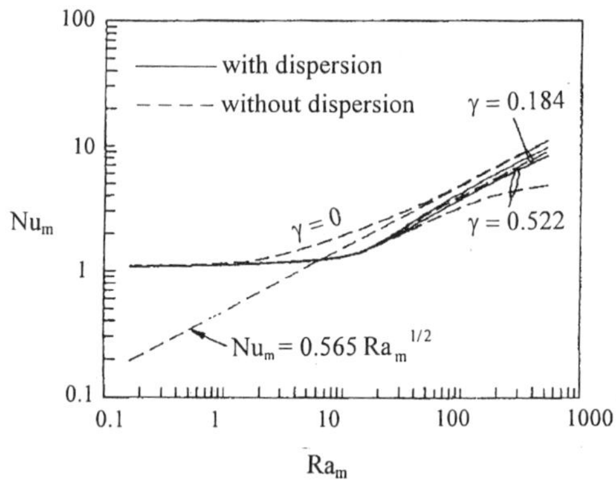

Figure 15 Effects of thermal dispersion on steady average Nusselt number based on the CPWD and CPND models

The effects of dimensionless particle diameter, thermal dispersion and no-slip boundary condition on the steady average media Nusselt number as a function of the media Rayleigh number in a constant porosity medium are shown in Figure 15. Results based on Darcy's law without the boundary-layer approximation and thermal dispersion are represented by dashed lines with $\gamma = 0$. Results for $\gamma = 0.184$ and 0.522 based on the Darcy-Brinkman model with and without thermal dispersion are also presented for comparison purposes. It is relevant to note that the assumption of a continuum may not be valid for the case $\gamma = 0.522$. From this figure it can be concluded that (i) the boundary-layer approximation is valid if $Ra_m > 30$, (ii) the thermal dispersion effect is important only at high Rayleigh numbers with a high dimensionless particle diameter, and (iii) the onset of free convection begins at $Ra_m = 1.9$ in a constant porosity medium if Darcy's law is used as the momentum equation.

Natural Convection in a Non-uniform Porosity Medium

To investigate the effects of reservoir width on the heat transfer rate from the horizontal heated cylinder embedded in a variable porosity medium, computations were also carried out for smaller values of W/D while keeping $H_1/D = H_2/D = 11.1$. The results of the computation for the steady average fluid Nusselt number, $\overline{Nu_f}$, are presented in Figure 16, which shows that the value of $\overline{Nu_f}$ decreases as the value of W/D decreases from 11.1. However, the effects of thermal dispersion are small in constant and variable porosity models. Figure 17(a) shows a comparison of Fand et al.'s measured Nusselt numbers for d_p = 4 mm and 2 mm with calculated Nusselt numbers based on the VPWD model. The results for dp = 3 mm are also plotted for comparison purposes. The results in Figure 17(a) are replotted in terms of the media Nusselt number as a function of the media Rayleigh number in Figure 17(b). It is shown that (i) the predicted average Nusselt numbers with variable porosity and thermal dispersion effects taken into consideration agree with experimental data, (ii) the Nusselt number in the convection regime increases as the particle diameter increases, and (iii) the Nusselt number in the conduction regime is relatively independent of the particle diameter.

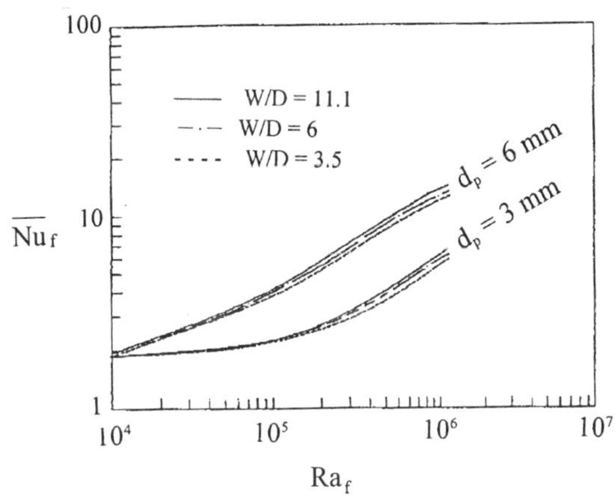

Figure 16 Effects of reservoir width on Nusselt number based on the VPWD model

The effects of variable porosity and particle diameter on the average media Nusselt number, $\overline{Nu_m}$, as a function of time are depicted in Figure 18(a) for $Ra_m = 15$ and in Figure 18(b) for $Ra_m = 273$. It can be seen that the variable porosity effect is negligible at small times and its influence increases with particle diameter and Rayleigh number. The effects of variable porosity and thermal dispersion on the tangential velocity and temperature profiles are shown in Figure 19 for $Ra_m = 247$, $Pr_f = 3.2$ and $d_p = 3$ mm at θ

= 45°. It is shown that both the variable porosity and thermal dispersion effects increase the tangential velocity and the temperature gradient, leading to an enhancement of surface heat flux. The variable porosity effect tends to reduce the thermal boundary-layer thickness while the thermal dispersion has the opposite effect. Thus, near the wall the calculated temperature with thermal dispersion is lower than those without thermal dispersion and this behavior reverses away from the wall.

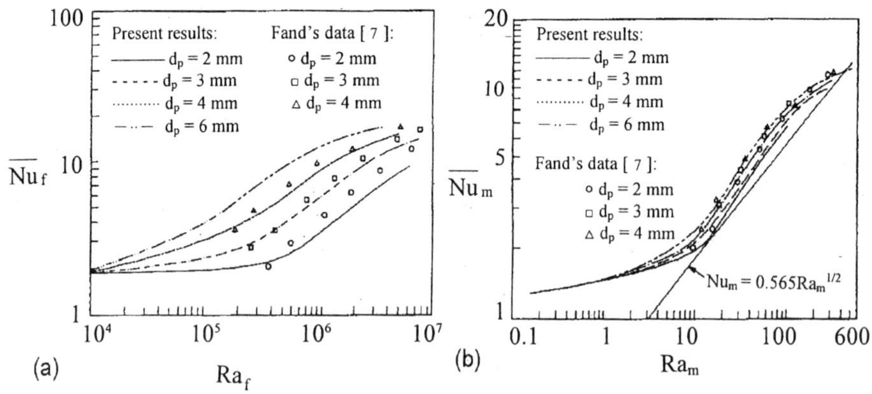

Figure 17 A comparison of measured and calculated steady Nusselt numbers of a horizontal cylinder embedded in an infinite porous medium: (a) $\overline{Nu}_f$ as a function of Ra_f; (b) $\overline{Nu}_m$ as a function of Ra_m

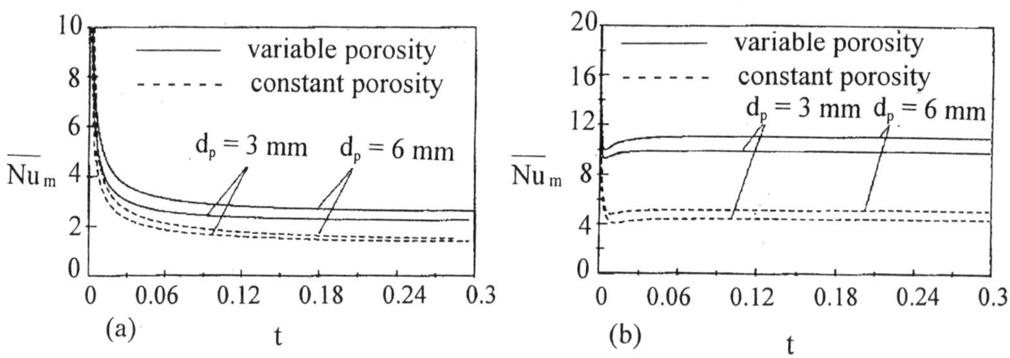

Figure 18 Effects of variable porosity and particle diameter on transient average Nusselt number based on the CPWD and VPWD models: (a) $Ra_m = 15$; (b) $Ra_m = 273$

The effects of non-uniform porosity and thermal dispersion on streamlines (left) and isotherms (right) at $Ra_f = 10^6$ and $\gamma = 0.268$ (dp = 3 mm) are illustrated in Figure 20.

With both non-uniform porosity and thermal dispersion effects taken into consideration, the strength of the vortices is the largest and the thermal energy is dispersed further away from the heated cylinder.

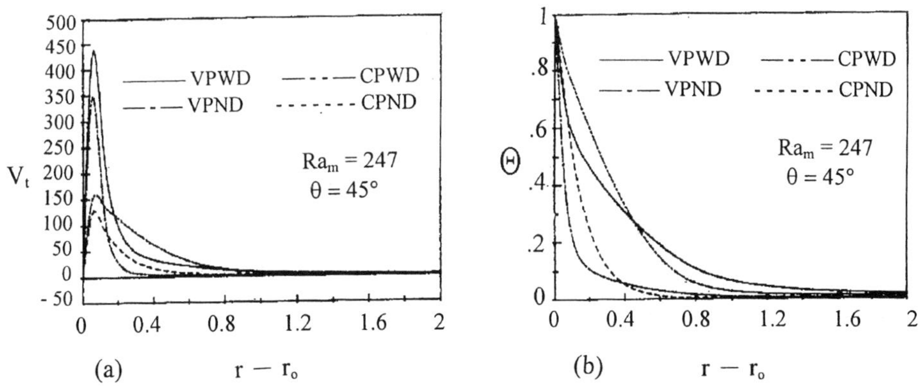

Figure 19 Effects of variable porosity and thermal dispersion on (a) tangential fluid velocity; (b) temperature distribution

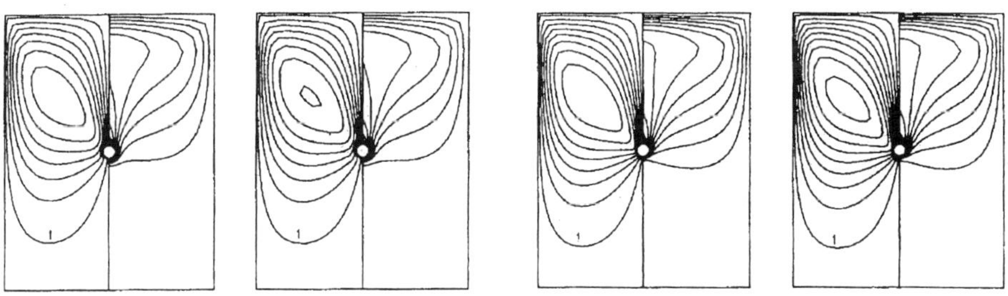

(a) CPND (b) CPWD (c) VPND (d) VPWD

Figure 20 Isotherms (right) and streamlines (left) for steady natural convection about a horizontal cylinder embedded in a porous cavity at $Ra_f = 10^6$ and $\gamma = 0.268$

REFERENCES

1. P. Cheng, *Heat Transfer in Geothermal Systems*, Adv. Heat Transfer **14**, 1-105 (1978).
2. P. Cheng, Natural Convection in Porous Media: External Flow, in *Natural Convection: Fundamentals and Applications*, 475-513, edited by S. Kakac, W. Aung and R. Viskanta, Martinus Nijhoff Publisher, The Haque, The Netherlands

(1985).
3. T. Kaneko, M. F. Mohtadi and K. Aziz, An Experimental Study of Natural Convection in Inclined Porous Media, *Int. J. Heat Mass Transfer* **17**, 485-496 (1974).
4. P. Cheng and C. L. Ali, An Experimental Investigation of Free Convection about a Heated Inclined Surface in a Porous Medium, Paper 81-HT-85, 20th National Heat Transfer Conference, (August 1981).
5. P. Cheng, C. L. Ali and A. K. Verma, An Experimental Study of Non-Darcian Effects in Free Convection in a Saturated Porous Medium, *Letters in Heat and Mass Transfer* **8**, 261-266 (1981).
6. N. Seki, S. Fukusako and H. Inaba, Heat Transfer in a Confined Rectangular Cavity Packed with Porous Media, *Int. J. Heat Mass Transfer* **21**, 985-989 (1978).
7. R. M. Fand, T. Steiberger and P. Cheng, Natural Convection Heat Transfer from a Horizontal Cylinder Embedded in a Porous Medium, *Int. J. Heat Mass Transfer* **29**, 119-133 (1986).
8. V. Prasad, F. A. Kulacki and A. V. Kulkarni, Free Convection in a Vertical Porous Annulus with Constant Heat Flux on the Inner Wall-experimental Results, *Int. J. Heat Mass Transfer* **29**, 713-723 (1986).
9. V. Prasad and N. Kladias, Non-Darcy Natural Convection in Saturated Porous Media, in *Convective Heat and Mass Transfer in Porous Media*, 173-224, edited by S. Kakac, B. Kilkis, F. A. Kulacki and F. Arinc, Kluwer Academic Publishers, Dordrecht (1991).
10. S. W. Hsiao, Natural Convection in an Inclined Porous Cavity with Variable Porosity and Thermal Dispersion Effects, *Int. J. Num. Meth. Heat Fluid Flow*, in press.
11. S. W. Hsiao, P. Cheng and C. K. Chen, Non-uniform Porosity and Thermal Dispersion Effects on Natural Convection about a Heated Horizontal Cylinder in an Enclosed Porous Medium, *Int. J. Heat Mass Transfer* **35**, 3407-3418 (1992).
12. P. Cheng, S. W. Hsiao and C. K. Chen, Natural Convection in Porous Media with Variable Porosity and Thermal Dispersion Effects, in *Convective Heat and Mass Transfer in Porous Media*, 173-224, edited by S. Kakac, B. Kilkis, F. A. Kulacki and F. Arinc, Kluwer Academic Publishers, London (1991).
13. D. Vortmeyer and J. Schuster, Evaluation of Steady Flow Profiles in Rectangular and Circular Packed Beds by a Variational Method, *Chem. Engng Sci.* **38**, 1691-1699 (1983).
14. C. T. Hsu and P. Cheng, Closure Schemes of the Macroscopic Energy Equation for Convective Heat Transfer in Porous Media, *Int. Commun. Heat Mass Transfer* **15**, 689 (1988).
15. C. T. Hsu and P. Cheng, Thermal Dispersion in Porous Media, *Int. J. Heat Mass Transfer* **33**, 1587-1597 (1990).
16. S. Ergun, Fluid Flow through Packed Columns, *Chem. Engng Prog.* **48**, 89-94 (1952).
17. I. F. Macdonald, M. S. El-Sayed, K. Mow and F. A. Dullien, Flow through Porous

Media-The Ergun Equation Revisited, *Ind. Engng Chem. Fundam.* **18**, 199-207 (1979).

18. R. M. Fand, B.Y. Kim, A.C.C. Lam and R.T.Phan, Resistance to the Flow of Fluids through Simple and Complex Porous Media Whose Matrices are Composed of Randomly Packed Spheres, *J. Fluid Engng* **109**, 268-274 (1987).
19. P. Cheng, A. Chowdbury and C. T. Hsu, Forced Convection in Packed Tubes and Channels with Variable Porosity and Thermal Dispersion Effects, *Proc. NATO Advanced Study Institute on Convective Heat and Mass Transfer in Porous Media*, M. 361-404 (1990).
20. P. Zehner and E. U. Schluender, Waermeleitfahigkeit von Schuettungen bei Massigen Temperaturen, *Chemie-Ingr. Tech.* **42**, 933-941 (1970).
21. R. M. Fand and R. T. Phan, Combined Forced and Natural Convection Heat Transfer from a Horizontal Cylinder Embedded in a Porous Medium, *Int. J. Heat Mass Transfer* **30**, 1351-1358 (1987).
22. S. W. Hsiao, C. K. Chen and P. Cheng, A Numerical Solution for Natural Convection in an Inclined Porous Cavity with a Discrete Heat Source on One Wall, *Int. J. Heat Mass Transfer* **37**, 2193-2201 (1994).
23. H. Bau, Low Rayleigh Number Thermal Convection in Saturated Porous Medium Bounded by Two Horizontal Eccentric Cylinders, *J. Heat Transfer* **106**, 166-175, (1984).
24. K. Himasekhar and H. Bau, Large Rayleigh Number Convection in a Horizontal Eccentric Annulus Containing Saturated Porous Media, *Int. J. Heat Mass Transfer* **29**, 703-712 (1986).
25. J. T. Hong, C. L. Tien and M. Kaviany, Non-Darcian Effects on Vertical-Plate Natural Convection in Porous Media with High Porosities, *Int. J. Heat Mass Transfer* **28**, 2149-2157 (1985).
26. C. T. Hsu and P. Cheng, Brinkmann Model for Natural Convection about a Vertical Plate in a Porous Medium, *Int. J. Heat Mass Transfer* **28**, 683-697 (1985).
27. T. W. Tong and E. Subramanian, A Boundary Layer Analysis for Natural Convection in Vertical Porous Enclosures-Use of the Brinkmann-Extended Darcy Model, *Int. J. Heat Mass Transfer* **28**, 563-571 (1985).
28. C. Beckermann, R. Viskanta and S. Ramadhyani, A Numerical Study of Non-Darican Natural Convection in a Vertical Enclosure Filled with a Porous Medium, *Numer. Heat Transfer* **10**, 557-570 (1986).
29. V. Prasad and A. Tuntomo, Inertia Effects on Natural Convection in a Vertical Porous Cavity, *Numer. Heat Transfer* **11**, 295-320, (1987).
30. M. Kaviany, Non-Darcian Effects on Natural Convection in Porous Media Confined between Horizontal Cylinders, *Int. J. Heat Mass Transfer* **29**, 1513-1519, (1986).
31. J. F. Middlecoff and P.D. Thomas, Direct Control of the Grid Point Distribution in Meshes Generated by Elliptic Equations, *AIAA J.* **18**, 652-656 (1980).
32. C. K. Chen, S. W. Hsiao and P. Cheng, Transient Natural Convection in an Eccentric Porous Annulus between Horizontal Cylinders, *Numerical Heat Transfer*,

Part A **17**, 431-448 (1990).

33. R. Thiyagarajan and M. M. Yovanovich, Thermal Resistance of Buried Cylinder with Constant Flux Boundary Condition, *J. Heat Transfer* **96**, 249-250 (1974).
34. H. H. Bau and S. S. Sadhal, Heat Losses from a Fluid Flowing in a Buried Pipe, *Int. J. Heat Mass Transfer* **25**, 1621-1629 (1982).
35. R. F. Difelice and H. H. Bau, Conductive Heat Transfer between Eccentric Cylinders with Boundary Conditions of the Third Kind, *J. Heat Transfer* **105**, 678-680 (1983).
36. R. T. Fernandez and V. E. Schrock, Natural Convection around a Heated Cylinder Buried in a Saturated Porous Medium, *Proc. Int. Heat Transfer Conf.* Munich **2**, 335-340 (1982).
37. B. Farouk and H. Shayer, Natural Convection around a Heated Cylinder Buried in a Saturated Porous Medium, *J. Heat Transfer* **110,** 642-648 (1988).
38. H. H. Bau, Convective Heat Losses from a Pipe Buried in a Semi-infinite Porous Medium, *Int. J. Heat Mass Transfer* **27**, 2047-2056 (1984).
39. K. Himasekhar and H. H. Bau, Thermal Convective Associated with Hot/Cold Pipes Buried in a Semi-infinite Saturated Porous Medium, *Int. J. Heat Mass Transfer* **30**, 263-273 (1987).
40. K. Himasekhar and H. H. Bau, Thermal Convection around a Heat Source Embedded in a Box Containing a Saturated Porous Medium, *J. Heat Transfer* **110**, 649-654 (1988).
41. S. W. Hsiao, A Numerical Study of Transient Natural Convection about a Corrugated Plate Embedded in an Enclosed Porous Medium, *Int. J. of Numerical Methods for Heat and Fluid Flow* **5**, 629-645 (1995).
42. S. W. Hsiao and C. K. Chen, Natural Convection Heat Transfer form a Corrugated Plate Embedded in an Enclosed Porous Medium, *Numerical Heat Transfer*, Part A **25**, 331-345 (1994).
43. J. H. Merkin, Free Convection Boundary Layers on Axisymmetric and Two-dimensional Bodies of Arbitrary Shape in Saturated Porous Medium, *Int. J. Heat Mass Transfer* **22**, 1461-1462 (1979).

HEAT CONDUCTION

PING CHENG and CHIN-TSAU HSU

Department of Mechanical Engineering, The Hong Kong University of Science and Technology, Clear Water Bay, Kowloon, Hong Kong

INTRODUCTION

In this chapter, we discuss conduction heat transfer in a porous medium. For steady heat conduction under local thermal equilibrium conditions it will be shown that the governing equation for heat conduction in a porous medium can be rewritten in a form similar to that of the classical heat conduction equation with an effective stagnant thermal conductivity consisting of two components: the first component represents the volumetric averaging of the thermal conductivities of the solid and the fluid phases while the second component represents the tortuosity effect due to the undulating thermal path across the fluid-solid interface. Exact numerical evaluations of the thermal conductivities of some idealized two-dimensional, spatially periodic media will be obtained based on closure modeling, or solving the two-dimensional conjugate heat conduction problem in the fluid and solid phase of a unit cell. Various analytical models for the calculation of the stagnant thermal conductivity will be reviewed.

GOVERNING EQUATIONS

The microscopic energy equations for the fluid and the solid phases under the no flow condition are

$$\rho_f C_{pf} \frac{\partial T_f}{\partial t} = \nabla \cdot (k_f \nabla T_f) \tag{1a}$$

$$\rho_s C_{ps} \frac{\partial T_s}{\partial t} = \nabla \cdot (k_s \nabla T_s) \tag{1b}$$

with the boundary conditions at the fluid-solid interface being given by

$$T_f = T_s \qquad \text{on } A_{fs} \tag{2}$$

$$\mathbf{n}_{fs} \cdot k_f \nabla T_f = \mathbf{n}_{fs} \cdot k_s \nabla T_s \qquad \text{on } A_{fs} \tag{3}$$

If the volumetric averaging process is applied to equation (1), the phase averaged equations are, see Nozad et al. [1]

$$\rho_f C_{pf} \frac{\partial(\phi \overline{T}_f)}{\partial t} = \overline{\nabla} \cdot \left[k_f \overline{\nabla}(\phi \overline{T}_f)\right] + \overline{\nabla} \cdot \left(\frac{k_f}{V} \iint_{A_{fs}} T_f \, \mathbf{ds} \right) + Q_{fs} \tag{4}$$

$$\rho_s C_{ps} \frac{\partial[(1-\phi)\overline{T}_s]}{\partial t} = \overline{\nabla} \cdot \left\{k_s \overline{\nabla}[(1-\phi)\overline{T}_s]\right\} - \overline{\nabla} \cdot \left(\frac{k_s}{V} \iint_{A_{fs}} T_s \mathbf{ds} \right) - Q_{fs} \tag{5}$$

where $\overline{T}_f$ and $\overline{T}_s$ are the volumetric temperatures, ϕ is the porosity, V is the representative elementary volume, $\mathbf{ds}$ is the outward normal vector of the surface element from the fluid into the solid, i.e. $\mathbf{ds} = \mathbf{n}_{fs} ds$ where s is the interphase surface area, and Q_{fs} is the interphase heat transfer per unit volume which is a source term in equation (4) but becomes a sink term in equation (5).

LOCAL THERMAL EQUILIBRIUM

We now assume that the local thermal equilibrium condition prevails such that

$$\overline{T}_f = \overline{T}_s = \overline{T} \tag{6}$$

With this assumption, and assuming that the porosity is constant, equations (4) and (5) can be added to give

$$(\rho C_p)_m \frac{\partial \overline{T}}{\partial t} = \overline{\nabla} \cdot (k_m \overline{\nabla} \overline{T}) + \overline{\nabla} \cdot \left[\frac{(k_f - k_s)}{V} \iint_{A_{fs}} T \, \mathbf{ds} \right] \tag{7}$$

where

$$(\rho C_p)_m = \phi \rho_f C_{pf} + (1-\phi) \rho_s C_{ps} \tag{8}$$

and

$$k_m = \phi k_f + (1-\phi) k_s \tag{9}$$

are the volumetric averaged thermal capacity and conductivity, respectively. At steady state condition, equation (7) reduces to

$$\overline{\nabla}\cdot\left(k_m\overline{\nabla T}\right)+\overline{\nabla}\cdot\left[\frac{(k_f-k_s)}{V}\iint_{A_{fs}}T\,ds\right]=0 \tag{10}$$

The area integrals in the above equations represent the tortuosity effect due to the undulating thermal path across the fluid-solid interface.

CLOSURE MODELING FOR STEADY HEAT CONDUCTION

In this section, we shall discuss the closures for equation (10). The microscopic energy equations for steady heat conduction in the fluid and solid phases are

$$\nabla\cdot\left(k_f\nabla T_f\right)=0 \qquad \text{in the fluid} \tag{11}$$

and

$$\nabla\cdot\left(k_s\nabla T_s\right)=0 \qquad \text{in the solid} \tag{12}$$

with the boundary conditions also given by equations (2) and (3).

We now decompose T_f and T_s into

$$T_f=\overline{T}_f+T_f' \tag{13a}$$

$$T_s=\overline{T}_s+T_s' \tag{13b}$$

where T_f' and T_s' are the temperature deviations from the volumetric temperatures of the fluid and the solid phases. Substituting equation (13) into equations (11)-(12) and boundary conditions (2)-(3) yields

$$\nabla\cdot\left(k_f\nabla T_f'\right)=\frac{1}{\phi}\overline{\nabla}\cdot\left(\frac{k_f}{V}\iint_{A_{fs}}T_f'ds\right)+\frac{Q_{fs}}{\phi} \tag{14a}$$

$$\nabla\cdot\left(k_s\nabla T_s'\right)=-\frac{1}{1-\phi}\overline{\nabla}\cdot\left(\frac{k_s}{V}\iint_{A_{fs}}T_s'ds\right)-\frac{Q_{fs}}{1-\phi} \tag{14b}$$

subject to the boundary conditions

$$T_f'=T_s' \qquad \text{on } A_{fs} \tag{15}$$

$$\mathbf{n}_{fs}\cdot k_f\nabla T_f'=\mathbf{n}_{fs}\cdot k_s\nabla T_s'+\mathbf{n}_{fs}\cdot\left(k_s-k_f\right)\overline{\nabla T} \qquad \text{on } A_{fs} \tag{16}$$

In the above derivation, we have invoked the local thermal equilibrium condition and the constant porosity assumptions. The boundary condition (16) suggests that the deviation of the temperature from the phase averaged value is caused by the macroscopic thermal gradient incorporated with the difference in the thermal conductivities of the solid and the fluid phases. Therefore, it is plausible to assume that

$$T_f' = \mathbf{f} \cdot \overline{\nabla}\overline{T} \tag{17a}$$

$$T_s' = \mathbf{g} \cdot \overline{\nabla}\overline{T} \tag{17b}$$

where $\mathbf{f}$ and $\mathbf{g}$ are vector functions in the microscopic coordinates. It can be shown that the right hand side of equation (14) is smaller than the left hand side. Thus, equation (14) reduces to

$$\nabla \cdot \left(k_f \nabla T_f'\right) = 0 \tag{18a}$$

$$\nabla \cdot \left(k_s \nabla T_s'\right) = 0 \tag{18b}$$

Substituting equation (17) into equation (18) and equations (15)-(16) leads to

$$\nabla^2 \mathbf{f} = 0 \tag{19a}$$

$$\nabla^2 \mathbf{g} = 0 \tag{19b}$$

subject to the boundary conditions

$$\mathbf{f} = \mathbf{g} \qquad \text{on } A_{fs} \tag{20}$$

$$\mathbf{n}_{fs} \cdot \nabla \mathbf{f} = \Lambda\, \mathbf{n}_{fs} \cdot \nabla \mathbf{g} + (\Lambda - 1)\, \mathbf{n}_{fs} \cdot \underline{\underline{\mathbf{I}}} \qquad \text{on } A_{fs} \tag{21}$$

where $\Lambda = k_s / k_f$ and $\underline{\underline{\mathbf{I}}}$ is a unit tensor. With the aid of equation (17a), the tortuosity integral can be evaluated as

$$\frac{1}{V} \iint_{A_{fs}} T' ds = \underline{\underline{\mathbf{G}}} \cdot \overline{\nabla}T \tag{22a}$$

where

$$\underline{\underline{\mathbf{G}}} = \frac{1}{V} \iint_{A_{fs}} \mathbf{f}\, \mathbf{n}_{fs}\, ds \tag{22b}$$

Thus, equation (10) can be written as

$$\overline{\nabla} \cdot \left(\underline{\underline{\mathbf{K}}}_e \cdot \overline{\nabla} \mathrm{T} \right) = 0 \tag{23}$$

where the effective stagnant thermal conductivity $\underline{\underline{\mathbf{K}}}_e$ is a tensor given by

$$\frac{\underline{\underline{\mathbf{K}}}_e}{k_f} = \left[\phi + (1-\phi)\Lambda \right] \underline{\underline{\mathbf{I}}} + (1-\Lambda) \underline{\underline{\mathbf{G}}} \tag{24}$$

which is also the expression obtained by Nozad et al.[1]. For isotropic media, $\underline{\underline{\mathbf{G}}} = \mathrm{G}\, \underline{\underline{\mathbf{I}}}$ such that

$$\underline{\underline{\mathbf{K}}}_e = k_e \underline{\underline{\mathbf{I}}} \tag{25a}$$

where

$$k_e = \left[\phi + (1-\phi)\Lambda \right] + (1-\Lambda)\, \mathrm{G} \tag{25b}$$

is the scalar effective stagnant thermal conductivity.

THE EFFECTIVE STAGNANT THERMAL CONDUCTIVITY

It should be noted that equation (23) with the effective conductivity $\underline{\underline{\mathbf{K}}}_e$ given by equation (25) is the standard steady heat conduction equation for isotropic media, provided that the stagnant thermal conductivity k_e can be measured or evaluated. It will be shown that the magnitude of the effective stagnant thermal conductivity of a porous medium depends not only the porosity, the thermal conductivities of the solid and the fluid phases, but also on whether the particles are in good contact, point contact or no contact.

Experimental Investigations

A large number of experiments have been carried out to measure stagnant thermal conductivities of porous media saturated with fluids. A review of the literature on the methods used for the determination of the stagnant thermal conductivities is given by Tsotsas and Martin [2]. A compilation of early experimental data by a number of investigators [3-11] was given by Kunii and Smith [12], Krupiczka [13], Zehner and Schlunder [14], as well as by Crane and Vachon [15]. Most of these measurements were carried out for materials with solid/fluid thermal conductivity ratio in the range of $1 < k_s / k_f < 10^3$ under atmospheric conditions. Thermal conductivities of porous materials with higher values of k_s / k_f were obtained by Swift [16] and Nozad et al. [1] while those with lower values of k_s / k_f were obtained by Prasad et al. [17].

Numerical Solutions

Nozad et al. [1] considered the stagnant thermal conductivity of arrays of touching square cylinders shown in Figure 1, where c/a is the touching parameter and the non-touching square cylinders is obtained by setting this parameter to zero. After numerical solutions of equations (19), subject to the boundary conditions (20) and (21), for the touching and non-touching square cylinders were obtained, Nozad et al. [1] determined the thermal conductivities of these geometries from equations (24) and (22b). Nozad et al. [1] found that their model with a touching parameter of 0.01 matched with existing experimental data for packed-sphere beds.

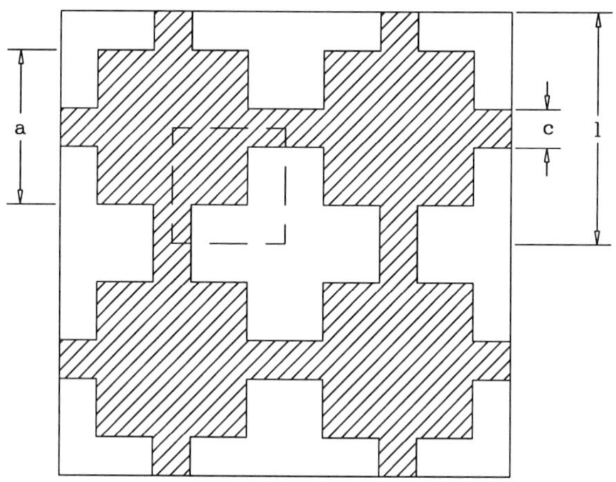

Figure 1 Arrays of touching square cylinders

Another approach to obtain the stagnant thermal conductivity is to solve numerically the multi-dimensional conjugate heat conduction problem in the fluid phase and in the solid phase in a unit cell. After the temperature distribution and the total heat flux have been computed, the effective thermal conductivity is determined based on the total heat flux through the unit cell. Deissler and Boegli [11] obtained a finite-difference solution for a cubic packing of spheres (ϕ = 0.476). Similar numerical solutions were obtained for an orthorhombic structure (ϕ = 0.259) by Wakao and Kato [18], as well as by Wakao and Vortmeyer [19], assuming that both point contacts and finite area contacts have a contact radius r_s. Wakao and Vortmeyer [19] found that the dimensionless contact area, defined as $D = \left(r_s/d_p\right)^2$ where d_p denotes the diameter of the solid particles, has an important effect on the stagnant thermal conductivity when the solid/fluid conductivity is high.

Analytical Models

We shall now discuss the analytical models of the stagnant thermal conductivities of porous media consisting of particles with no contact, point contact and good contact.

Maxwell's Model and Its Extensions Assuming a sufficiently dilute dispersion of spheres, Maxwell [20] obtained the following expression for the stagnant thermal conductivity of a packed-sphere bed :

$$\frac{k_e}{k_f} = \frac{2\phi + \Lambda(3-2\phi)}{3-\phi+\Lambda\phi} \tag{26}$$

Equation (26), valid in the limit of the porosity approaches unity ($\phi \to 1$), provides a lower bound for the stagnant thermal conductivity of a packed-sphere bed. Subsequently, numerous attempts have been made to extend Maxwell's solution to higher order in (1-ϕ). A review of these early non-touching models has been given by Churchill [21].

Composite-Layer Models The simplest model for the calculation of the effective stagnant thermal conductivity is to assume that the porous medium consists of a fluid layer and a solid layer, either in parallel or in series with respect to the temperature gradient. As pointed out by Deisser and Boegli [11], these two arrangements represent the maximum and the minimum values of the effective stagnant thermal conductivity of a porous medium.

If the fluid and the solid layer are in parallel with respect to the thermal gradient, it can be shown that the effective stagnant thermal conductivity is

$$k_e = k_f\phi + k_s(1-\phi) \tag{27a}$$

or

$$k_e/k_f = \phi + (1-\phi)k_s/k_f \tag{27b}$$

It should be noted that the effective stagnant thermal conductivity, given by equation (27), is identical to the first term of equation (24). Thus, under the parallel-layer model, the tortuosity effect is presumably neglected. The fluid and the solid layers are perpendicular to the temperature gradient, the effective stagnant thermal conductivity is

$$\frac{1}{k_e} = \frac{\phi}{k_f} + \frac{(1-\phi)}{k_s} \tag{28a}$$

or

$$k_e/k_f = \frac{k_s/k_f}{(1-\phi)+\phi k_s/k_f} \tag{28b}$$

Equations (27) and (28) are presented in Figure 2 as a function of the thermal conductivity ratio. It is seen that the values computed from equations (27) and (28) differ slightly if k_s is almost equal to k_f and they differ substantially if k_s is considerably different from k_f.

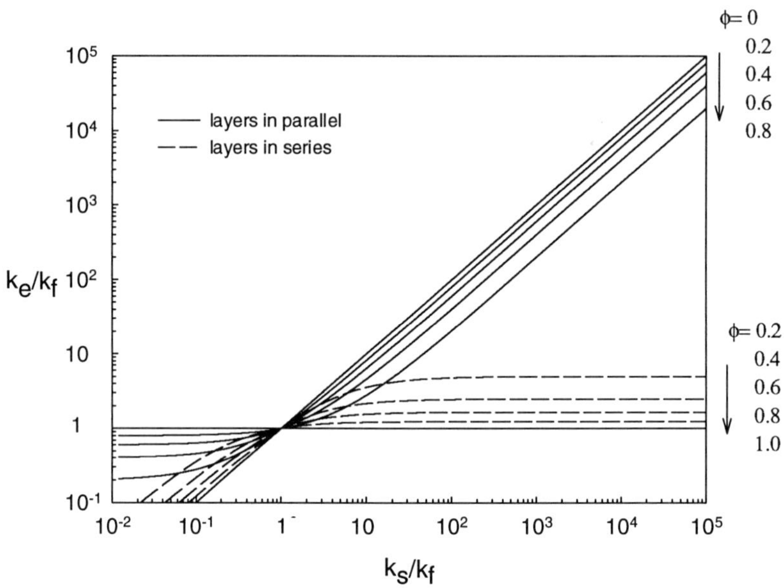

Figure 2 Maximum and minimum values of the effective stagnant thermal conductivity

Kunii and Smith's Model Kunii and Smith [12] obtained the following approximate expression for the stagnant thermal conductivity of a packed-sphere bed:

$$\frac{k_e}{k_f} = \phi + \frac{(1-\phi)}{\left(\varepsilon + \frac{2}{3}\lambda\right)} \tag{29}$$

where $\lambda = k_f/k_s$ is the fluid/solid conductivity ratio and $\varepsilon = \varepsilon_2 + (\phi - 0.259)(\varepsilon_1 - \varepsilon_2)/0.217$ is a linear interpolation between ε_1 and ε_2, which are the values of ε corresponding to a cubic packing of uniform spheres (ϕ = 0.476) and to a tetrahedral packing of uniform spheres (ϕ = 0.259), respectively. The values of ε_1 and, ε_2 as a function of the solid/fluid thermal conductivity ratio is presented graphically by Kunii and Smith [12].

Zehner and Schlunder's Model An analytical expression for the effective stagnant thermal conductivity of a porous medium has been obtained by Zehner and Schlunder [14] who considered 1/8 of the cylinder, see Figure 3, instead of a cube as a unit cell. The unit cell consists of an outer cylinder with radius R and an inner cylinder with unit radius. Fluid is filled between the inner and outer cylinder while the inner cylinder consisting of both the solid and fluid phase with its interface A_{fs} described by

$$r^2 + \frac{z^2}{\left[B-(B-1)z\right]^2} = 1 \tag{30}$$

where B is the shape factor to parametrize the geometrical effect of the solid particles. As shown in Figure 3, equation (30) assumes that the particles are in point contact.

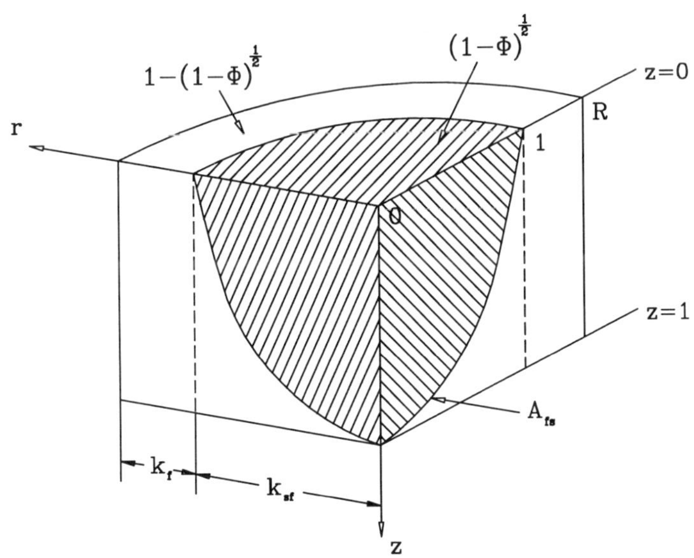

Figure 3 Unit cell used by Zehner and Schlunder [14]

Based on this model, Zehner and Schlunder [14] obtained the following expression for the effective thermal conductivity of a packed bed:

$$\frac{k_e}{k_f} = 1 - \sqrt{1-\phi} + \frac{2\sqrt{1-\phi}}{1-\lambda B}\left[\frac{(1-\lambda)B}{(1-\lambda B)^2}\ln\frac{1}{\lambda B} - \frac{B+1}{2} - \frac{B-1}{1-\lambda B}\right] \tag{31}$$

where the value of B is approximated by

$$B = C\left(\frac{1-\phi}{\phi}\right)^m \tag{32}$$

with m = 10/9. The value of C depends on the shape of the particles and Zehner and Schlunder [14] suggested that C = 1.25, 1.4 and 2.5 for spheres, broken particles and cylinders, respectively. The stagnant thermal conductivity given by equations (31) and (32), as a function of k_s/k_f, for different values of porosity are presented in Figure 4. For comparison, the experimental data obtained by Nozad et al. [1] are also plotted in the same graph. It is noted from this graph that the Zehner-Schlunder model [14] with $\phi = 0.4$ is in good agreement with the experimental data for $k_s/k_f < 10^3$. For higher values of the thermal conductivity ratio ($k_s/k_f > 10^3$), the Zehner-Schlunder model underpredicts the stagnant thermal conductivity, see Kaviany [22].

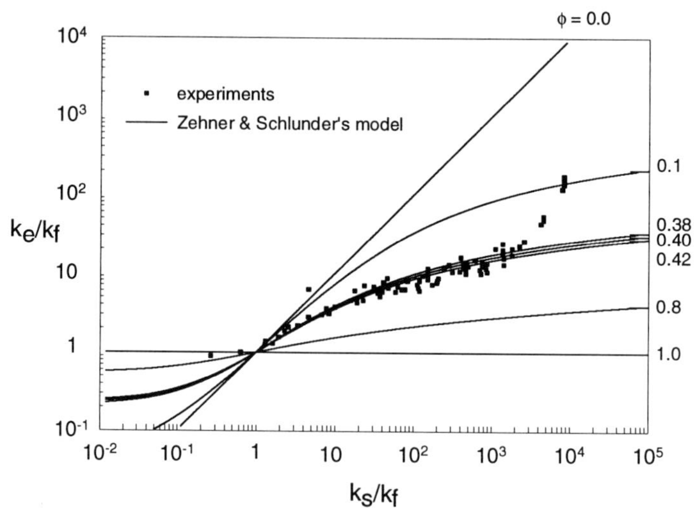

Figure 4 Comparison of Zehner and Schlunder's model with experimental data of Nozad et al. [1]

Area-Contact Model Since finite area contacts between spheres are likely to occur in a packed-sphere bed, Hsu et al. [23] postulated that the reason for the Zehner-Schlunder model underpredicting the stagnant thermal conductivity of a packed-sphere bed at high solid/fluid thermal conductivity ratios is due to the fact that their model assumes point contacts between spheres. To take into consideration the finite area contacts between spheres, Hsu et al. [23] modified the solid/fluid interface A_{fs} by the following equation

$$r^2 + \frac{z^2}{\left[(1+\alpha)B-(B-1)z\right]^2} = 1 \tag{33}$$

where α is the deformation factor. The truncated sphere is indicated as the shaded area in Figure 5 where r_s is the radius of the contact area which can be expressed in terms of α and B by imposing the condition: $r = r_s$, $z = 1$. It follows from equation (33) that

$$r_s^2 = 1 - \frac{1}{(1+\alpha B)^2} \tag{34}$$

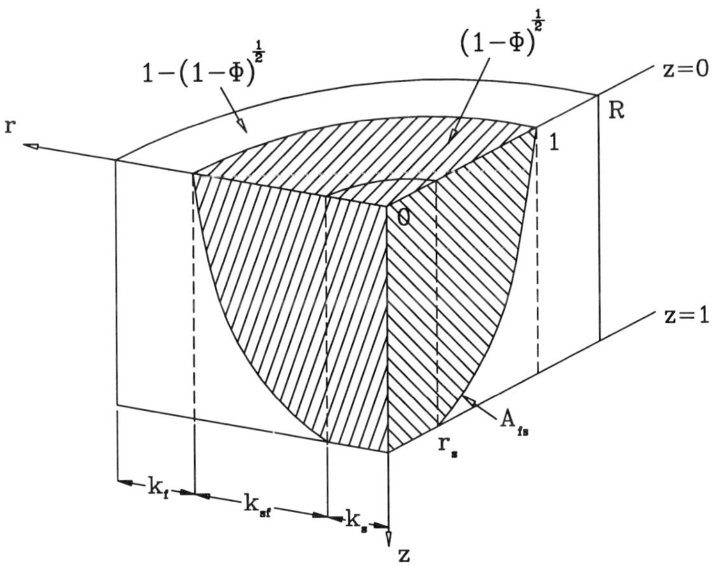

Figure 5 Unit cell for the area-contact model

Based on the modification of Zehner and Schlunder's approach, Hsu et al. [23] obtained the following expression for the effective stagnant thermal conductivity of a packed sphere bed with finite contact areas:

$$\frac{k_e}{k_f} = \left(1-\sqrt{1-\phi}\right) + \frac{\sqrt{1-\phi}}{\lambda}\left[1 - \frac{1}{(1+\alpha B)^2}\right]$$
$$+ \frac{2\sqrt{1-\phi}}{\left[1-\lambda B+(1-\lambda)\alpha B\right]}\left\{\frac{(1-\lambda)(1+\alpha)B}{\left[1-\lambda B+(1-\lambda)\alpha B\right]^2}\ln\frac{1+\alpha B}{(1+\alpha)B\lambda}\right.$$
$$\left.-\frac{B+1+2\alpha B}{2(1+\alpha B)^2} - \frac{(B-1)}{\left[1-\lambda B+(1-\lambda)\alpha B\right](1+\alpha B)}\right\} \qquad (35)$$

It should be noted that equation (35), with α = 0, reduces to equation (31) given by Zehner and Schlunder [14]. The value of B in equation (35) is also approximated by equation (32) with C and m being a function of α as shown in Figure 6.

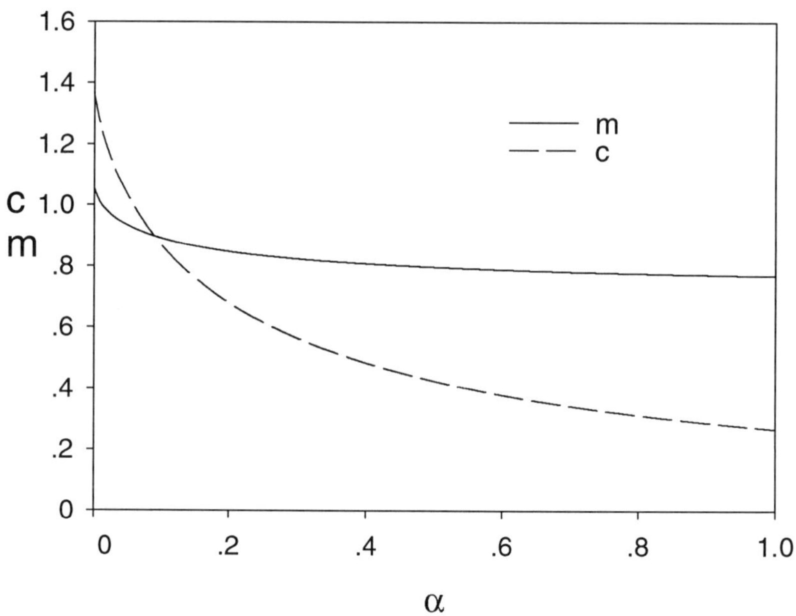

Figure 6 Variations of C and m with the deformation factor α

Computations were carried out for equation (35) with various values of α and ϕ and it was found that equation (35), with α = 0.002 and ϕ = 0.42, matched well with the experimental data of Nozad et al. [1]. Results based on equation (35), with α = 0.002 and ϕ = 0.42, are compared with the experimental data of Nozad et al. [1] in Figure 7, and the computed results based on the two-dimensional model of Nozad et al. [1] are also presented in the same plot for comparison. It is shown that results based on equation (35) are slightly lower than Nozad et al.'s model for $k_s / k_f > 10^3$.

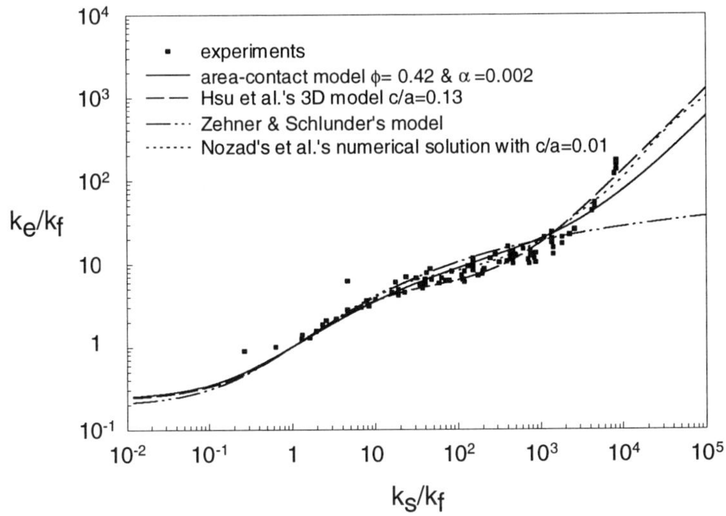

Figure 7 Comparison of the area-contact model with other theoretical models and experimental data, see Hsu et al. [24]

Lumped Parameter Models Hsu et al. [24] proposed a lumped parameter model to obtain simple expressions for the effective stagnant thermal conductivities spatially periodic media. The essence of the method is to select a unit cell and assume one-dimensional heat conduction in the cell. Next divide the unit cell into layers in parallel to the heat flow direction and these layers may consist of fluid phase, solid phase or composite phase. The equivalent thermal resistance in the composite layer is obtained by assuming that the thermal resistances in the solid and the fluid phases are in series. A number of geometries, which had been considered previously, were obtained based on the lumped parameter method, and it is found that the results based on the lumped parameter models are in excellent agreement with existing numerical solutions. A review of the lumped parameter model has been given recently by Cheng and Hsu [25].

Square Cylinders Consider the stagnant thermal conductivity of an array of touching square cylinders as shown in Figure 1. Based on the lumped parameter method,

Hsu et al. [24] obtained the following expression for the effective stagnant thermal conductivity in the direction perpendicular to the cylinders:

$$\frac{k_e}{k_f}=\frac{\gamma_a\gamma_c}{\lambda}+\frac{\gamma_a(1-\gamma_c)}{1+(\lambda-1)\gamma_a}+\frac{(1-\gamma_a)}{1+(\lambda-1)\gamma_a\gamma_c} \tag{36}$$

where $\gamma_a = a/\ell$ and $\gamma_c = c/a$. Equation (36) is presented in Figure 8 for ϕ= 0.36 and for different values of the touching parameter c/a. The results show that the effect of the touching parameter c/a is important when the solid/fluid thermal conductivity ratio is high. For comparison, the experimental data of Nozad et al. [1], as well as Prasad et al. [17] and the numerical results of Nozad et al. [1] for c/a = 0.01 are also presented in Figure 8. It is shown that results computed based on the lumped parameter model for c/a = 0.01 are in excellent agreement with the experimental data.

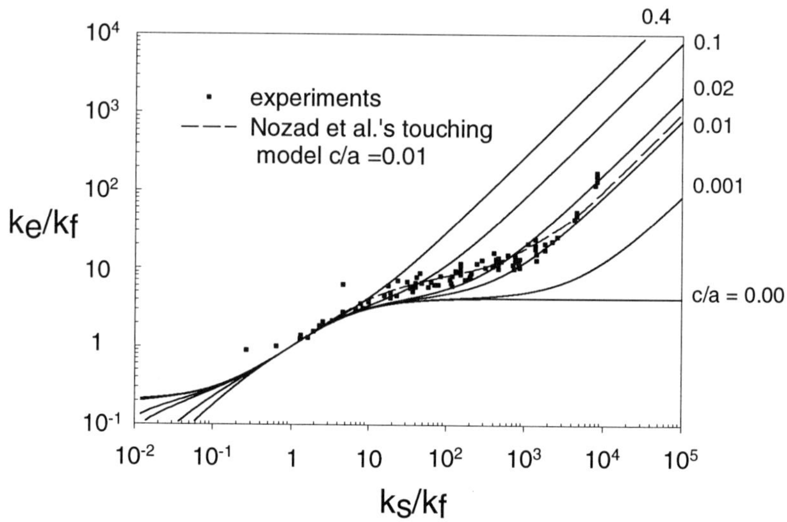

Figure 8 Comparison of stagnant thermal conductivity of touching square cylinders based on the lumped-parameter model with experimental data, Hsu et al. [24]

Since non-touching square cylinders is a special case of touching squares, the thermal conductivity of the former case can be obtained by letting c = 0 and consequently, $\gamma_c = 0$ and γ_a = 1 -ϕ. Under this condition, equation (36) becomes

$$\frac{k_e}{k_f} = \left(1-\sqrt{1-\phi}\right) + \frac{\sqrt{1-\phi}}{1+(\lambda-1)\sqrt{1-\phi}} \tag{37}$$

Equation (37) for ϕ = 0.2, 0.4, and 0.7 are presented in Figure 9. For comparison, the numerical results of Nozad et al. [1] for ϕ = 0.1, 0.36, 0.437 and 0.64 are also presented in the same graph. It is shown that they are in excellent agreement.

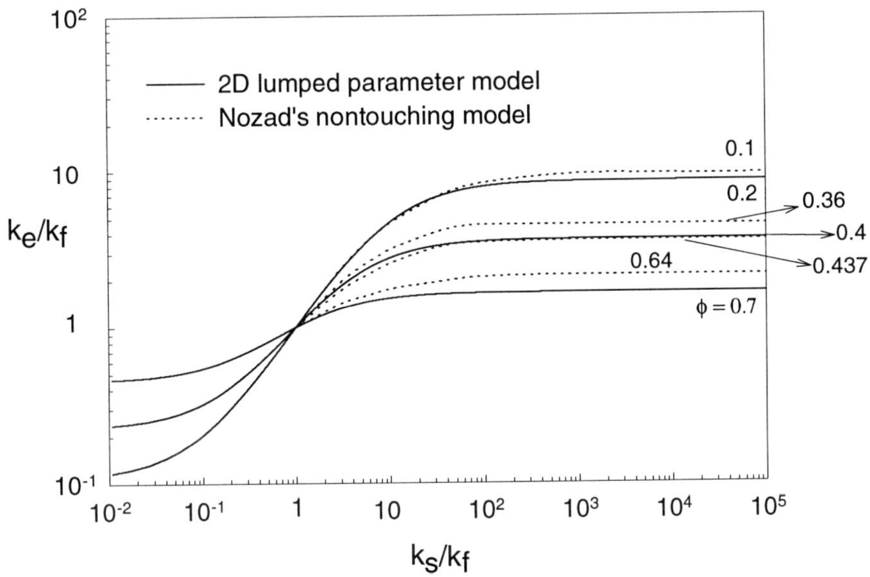

Figure 9 Comparison of stagnant thermal conductivity of non-touching square cylinders based on the lumped-parameter model with numerical results of Nozad et al. [1]

Circular Cylinders Based on the lumped parameter method, Hsu et al. [24] obtained simple expressions for the stagnant thermal conductivity of in-line circular cylinders of diameter a with connecting width c, see Figure 10. Figure 11 is a comparison of the numerical results of Nozad et al. [1] for touching square cylinders with c/a = 0.01 and those based on the lumped parameter method for touching square and circular cylinders as well as the experimental data of a packed-sphere bed. It is shown that the stagnant thermal conductivity of the circular cylinders are slightly higher than that of square cylinders for $k_s/k_f > 1$ and the behavior reverses itself when $k_s/k_f < 1$. As shown in Figure 12, similar characteristics are observed for non-touching circular and non-touching square cylinders with ϕ = 0.4, 0.5, and 0.6. From Figure 12, we also see that the results from the present model agree excellently with the numerical results of Sabraoui and Kaviany[26].

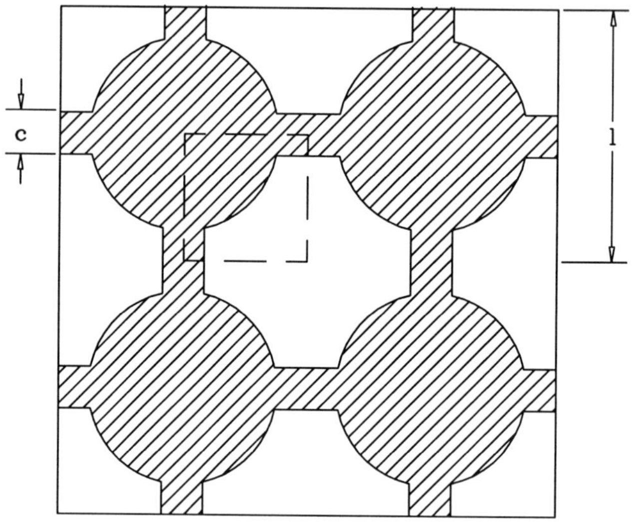

Figure 10 Arrays of touching circular cylinders

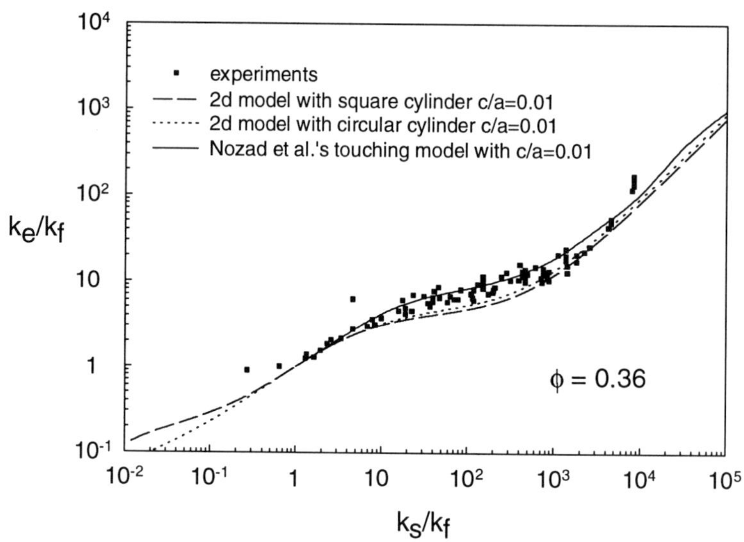

Figure 11 Comparison of stagnant thermal conductivity of touching square and touching circular cylinders, Hsu et al. [24]

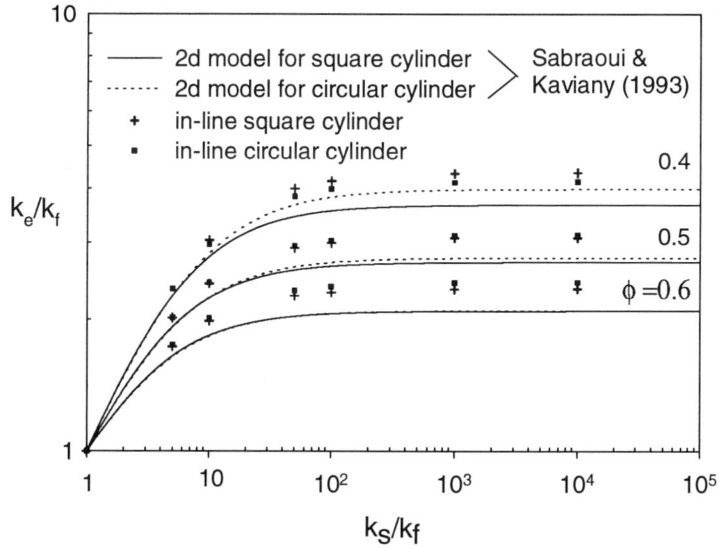

Figure 12 Stagnant thermal conductivity of non-touching square and circular cylinders based on the lumped-parameter model, see Hsu et al. [24]

<u>Cubes</u> The stagnant thermal conductivity of an array of in-line cubes, with connecting bars shown in Figure 13, is also given by Hsu et al. [24] based on the lumped parameter method. They obtained the following expression for the stagnant thermal conductivity:

$$\frac{k_e}{k_f} = 1-\gamma_a^2 - 2\gamma_a\gamma_c + 2\gamma_c\gamma_a{}^2 + \frac{\gamma_c^2\gamma_a^2}{\lambda} + \frac{\gamma_a^2(1-\gamma_c^2)}{1+\gamma_a(\lambda-1)} + \frac{2\gamma_a\gamma_c(1-\gamma_a)}{1+\gamma_c\gamma_a(\lambda-1)} \tag{38}$$

For non-touching cubes, where $\gamma_c \to 0$ and $\gamma_a \to (1-\phi)^{1/3}$, the above equation reduces to

$$\frac{k_e}{k_f} = \left[1-(1-\phi)^{2/3}\right] + \frac{(1-\phi)^{2/3}}{1+(\lambda-1)(1-\phi)^{1/3}} \tag{39}$$

Figure 14 is a comparison of experimental data of the thermal conductivity of packed-sphere beds with models based on touching cubes (with c/a = 0.13) and touching square cylinders with two values of the touching parameters (c/a = 0.01 and 0.02). It is shown that the results based on the touching cubes model agree better with experimental data than that of touching square cylinders model. For comparison, the results for non-touching square cylinders and non-touching cubes are also plotted in Figure 14. It is shown that the stagnant thermal conductivity of the three-dimensional non-touching cubes model is higher than that of the two-dimensional model of non-touching square cylinders.

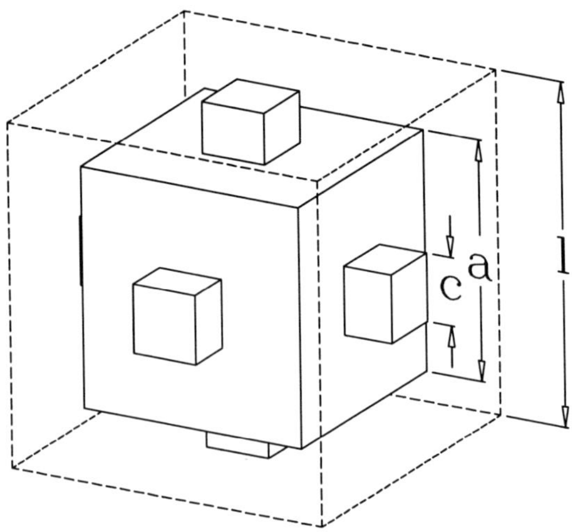

Figure 13 In-line touching cubes

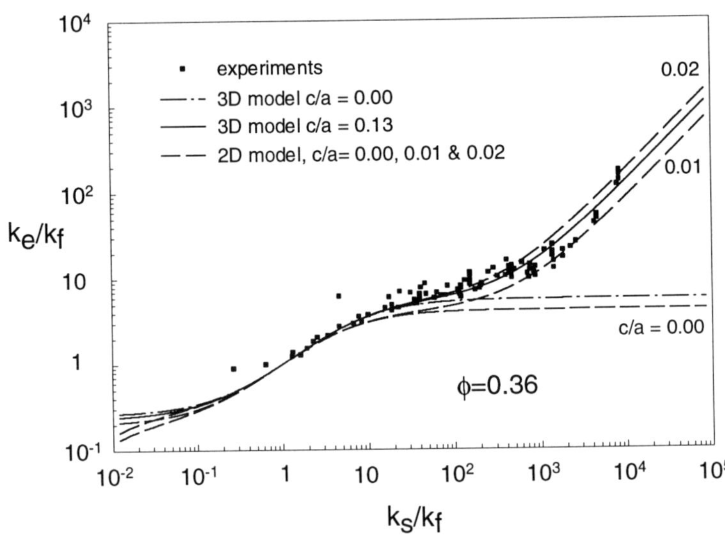

Figure 14 Comparison of the theoretical models with the experimental data

CONCLUDING REMARKS

Conduction heat transfer in porous media has been reviewed in this chapter. Our attention has been confined to steady heat conduction under local thermal equilibrium conditions. Particular attention has been given to the effective stagnant thermal conductivity of a porous medium which is saturated with a single component fluid. Further work is required for transient heat conduction in porous media under non-local thermal equilibrium conditions, and for the thermal conductivity of a porous medium which is saturated with a multi-component fluid.

References

1. S. Nozad., R. G. Carbonell and S. Whitaker, Heat conduction in multi-phase systems, I: Theory and experiments for two-phase systems, *Chem. Engg. Sci.*, **40**, 843-855 (1985).
2. E. Tsotsas, and H. Martin, Thermal conductivity of packed beds: A review, *Chem. Eng. Process*, **22**, 19-37 (1987).
3. T. E. W. Schuman, and V. Voss, Heat flow through granulated materials, *Fue*l, **13**, 249-256 (1934).
4. G. Kling, Das Warmeleitvermogen eines Kugelhaufwerks in ruhenden Gasen, *Forsch. Geb. Ingenierw*, **9**, 28-34 (1938).
5. R. H. Wilheim, W. C. Johnson, R. Wynkoop, and D. W. Collier, Reaction rate, heat transfer and temperature distribution in fixed bed catalytic converters, *Chem. Eng.Progr*. **44**, 105-116 (1948).
6. A. L. Waddems, The flow of heat through granular materials, *J. Soc. Chem. Ind.* **63**, 337-340 (1944).
7. S. Hatta, S. and S. Maeda, Heat transfer in beds of granular catalyst I, *Chem. Engg.* (Japan), **12**, p. 56 (1948).
8. H. Vershoor and G. C. A. Schuit, Heat transfer to fluid flowing through a bed of granular solids, *Appl. Science Res.*, **42**, p.97 (1950) .
9. D. A. DeVries, *Physics of Plant Environment*, 2nd edition, p.210, North Holland, Amsterdam (1966).
10. F. W. Preston, Mechanism of heat transfer in unconsolidated porous media at low flow rates, Ph.D. Dissertation, Penn. State University (1957).
11. R. G. Deisser and J. S. Boegli, An investigation of effective thermal conductivites of powders in various gases, *ASME Transactions,* **80**, 1417-1425 (1958).
12. D. Kunii and J. M. Smith, Heat transfer characteristics of porous rocks, *A.I.Ch.E. J.* **6**, 71-78 (1960).
13. R. Krupiczka, Analysis of thermal conductivity in granular materials, *Int. Chem. Engg.*, **7**, 122-144 (1967).
14. P. Zehner and E. U. Schlunder, Thermal conductivity of granular materials at moderate temperatures (in German), *Chemie. Ingr.-Tech.*, **42**, 933-941 (1970).

15. R. A. Crane and R. I. Vachon, A prediction of the bounds on the effective thermal conductivity of granular materials, *International Heat & Mass Transfer,* **20**, 711-723 (1977).
16. D. L. Swift, The thermal conductivity of spherical metal powders including the effect of an oxide coating, *Int. J. Heat Mass Transfer*, **9**, 1061-1073 (1966).
17. V. Prasad, N. Kladas, A. Bandyopadhay and Q. Tian, Evaluation of correlations for stagnant thermal conductivity of liquid-saturated porous beds of spheres, *Int. J. Heat Mass Transfer*, **32**, 1793-1796 (1989).
18. N. Wakao and K. Kato, Effective thermal conductivity of packed beds, *J. Chem. Engg.*, Japan, **2**, 24-320 (1969).
19. N. Wakao and D. Vortmeyer, Pressure dependency of effective thermal conductivity of packed beds, *Chem. Engg. Sci.*, **26**, 1753-1765 (1971).
20. J. C. Maxwell, *A Treatise on Electricity and Magnetism*, **41**, 365, Clarenden Press, Oxford (1873).
21. S. W. Churchill, The thermal conductivity of dispersions and packed beds -- an illustration of the unexploited potential of limiting solutions for correlation, *Advances in Transport Processes*, **4**, 394-418, edited by A. S. Mujumdar and R. A. Mashelkar, Wiley Eastern, New Delhi (1986).
22. M. Kaviany, *Principles of Heat Transfer in Porous Media*, Springer Verlag, New York Inc. (1991).
23. C. T. Hsu, P. Cheng and K. W. Wong, Modified Zehner-Schlunder model for stagnant thermal conductivity of porous media, *Int. J. Heat Mass Transfer,* **37**, 2751-2759 (1994).
24. C. T. Hsu, P. Cheng and K. W. Wong, A lumped parameter model for stagnant thermal conductivity of spatially periodic porous media, *J. Heat Transfer*, **117**, 264-269 (1995).
25. P. Cheng and C. T. Hsu, The effective stagnant thermal conductivity of porous media with periodic structures, Proceedings of the *International Conference on Porous Media and Their Applications in Science, Engineering and Industry,* edited by K. Vafai and P. N. Shivakumar, 288-319 (1996). To appear in *J. of Porous Media* (1997).
26. M. Sabraoui and M. Kaviany, Slip and non-slip temperature boundary conditions at interface of porous plain media, *Int. J. Heat Mass Transfer*, **36**, 1019-1033 (1993).

ONSET OF OSCILLATORY CONVECTION IN A POROUS MEDIUM

S. KIMURA

Department of Mechanical Systems Engineering, Faculty of Engineering
Kanazawa University
2-40-20, Kodatsuno, Kanazawa 920, JAPAN

INTRODUCTION

Onset of oscillatory convection in a fluid saturated porous layer heated from below has been investigated extensively in the last two decades. This is partly due to mathematical interest of nonlinear mechanics, where a mathematical system modeling thermal convection in porous media represents one of the simplest nonlinear systems. The system exhibits a series of bifurcations from a steady conductive solution to a complex oscillatory state as the Rayleigh number is increased. Therefore, several critical Rayleigh numbers are encountered in a course of bifurcating processes of a convective system. The critical value of the Rayleigh number depends on the specific system that one looks at. This paper reviews steady-to-oscillatory bifurcating processes that take place in a fluid saturated porous medium heated from below, as well as in a system with different thermal and geometrical boundary conditions, and their subsequent super critical evolution. Aside from the above fundamental and academic interest, thermal convection in porous media and its heat and mass transfer characteristics have a wide range of applications from various thermal engineering designs to geophysical problems. Nonetheless, in this review we are primarily concerned with the fundamental aspect of thermal convection in a porous medium, i.e. a transition from steady convective flow to unsteady oscillating states and their subsequent evolution.

Historically speaking, studies of thermal convection in a fluid saturated porous layer heated from below were first put forward by Horton and Rogers [1] and Lapwood [2]. Since it is Lapwood who carried out a stability analysis of the conductive state and suggested the critical Rayleigh number $4\pi^2$ above which convection occurs. This convective system, which is analogous to Bénard convection in a clear fluid system, is often referred to as the Lapwood convection. This critical value of Rayleigh number has been tested and proved experimentally by numerous researchers: among others Schneider [3], Katto and Masuoka [4], Combarnous [5], Bories [6] and Combarnous and Bories [7] should be mentioned in this regard. These authors extended their experiments beyond the critical Rayleigh number, and presented average heat transfer characteristics as a Nusselt number - Rayleigh number correlation. An occurrence of oscillatory convection was first observed in experiments conducted by Combarnous and Le Fur [8], and further by Caltagirone, Cloupeau and Combarnous [9], when convection is sufficiently vigorous and the Rayleigh number is about 7 times greater than the critical value for the onset of convection (Ra_c = 190-390). This somewhat large deviation observed for the critical Rayleigh number was attributed to different structures of the porous matrix used for experiment. The transition from steady convection to unsteady oscillating convection was first recognized in a layer of great horizontal extent. However, subsequently it was demonstrated that this transition also takes place in a cellular cavity, and that these oscillatory flows are two-dimensional (Caltagirone [10]).

Inspired by the existence of the second transition, several theoretical studies have been advanced. Two-dimensional numerical calculations were carried out by Horne and O'Sullivan [11], and independently by Caltagirone [10]. Horne and O'Sullivan considered a square cross section with uniform top wall temperature and either uniform or non uniform bottom wall temperature; the side walls are adiabatic. A finite-difference technique has been employed to solve the momentum and the energy equations. When the bottom wall is uniformly heated to a constant temperature, their results show that the onset of permanently oscillating convection sets in at a Rayleigh number somewhere between 250 and 375. With heating at a constant temperature in a fractional portion on the bottom wall, however, the critical Rayleigh number is greater and lies between 375 and 500. The oscillation frequencies for both cases generally increase with the Rayleigh number. They also presented experimental results obtained for a Hele-Shaw cell. The Hele-Shaw cell experiment conducted for a half-heated bottom boundary condition at a rather high Rayleigh number (Ra =1600) indicates that the convecting flow is permanently fluctuating, and that the oscillation is regular, periodically generating a'tongue-like' instability along the heated bottom boundary. On the other hand, Caltagirone performed a finite-difference calculation and a linear stability analysis based on the Galerkin method in order to determine a critical value for the transition. It seems that the author was more interested in an effect of the aspect ratio of the cross section on the criticality. In general, the tall cross sectional area increases the critical Rayleigh number, and the horizontally elongated cross section lowers the critical value. His linear stability analysis, based on the Galerkin method, determined the critical value Ra_c = 385 for the aspect ratio of unity. Schubert and Straus [12] employed a Galerkin technique in order to solve the governing equations. They determined the critical Rayleigh number to lie somewhere between 300 and 320. Later, recalculating with higher truncation numbers (more Fourier expansion terms), Schubert and Straus [13] claimed $380<Ra_c<400$. Further pursuing this line, but with a much more efficient numerical technique, Kimura *et al.* [14] established the critical value Ra_c = 389.7 and the nondimensional oscillation frequency f_c = 82.5 at the onset of oscillations. Independently, Aidun and Steen [15], using an eigenfunction expansion in conjunction with a numerical branch-tracing technique, obtained Ra_c = 390.7 with the onset frequency f_c = 82.8. These critical conditions have been also confirmed by Riley and Winters [16], who used a Galerkin-based finite element method in order to solve the nonlinear governing system, and applied a generalized eigenvalue formulation to determine the critical Rayleigh number.

In a two-dimensional square cross section, a further increase of the Rayleigh number beyond the criticality transforms a simply periodic state to more complex oscillatory convection. The first attempt to delineate this process was made by Schubert and Straus [13], who found a series of transitions from a simply periodic state to a quasi-periodic state, and back to a simply periodic state again. Kimura *et al.* [17] confirmed the earlier results by Schubert and Straus [13] with much higher resolution, and were able to compute a chaotic state at Rayleigh numbers greater than 1000. In a series of papers Riley and Winters [16,18,19] have focused on a modal exchange mechanism and they shed light on the complex nature of multiplicity of solutions for a given value of the Rayleigh number. More recently, Graham and Steen [20] have reconsidered a super-critical regime, and made a detailed analysis with an attempt to gain a better understanding of bifurcating processes. They presented a somewhat different evolution scenario from the result by Kimura *et al.* [17] on the super critical regime. Recently Kimura *et al.* [21] replaced the top and bottom thermal boundary conditions by a constant flux condition in a horizontally-elongated two-dimensional space. The steady convecting flow, which can be closely approximated as a fully developed counter flow, exhibits a transition from steady to oscillatory convection. Assuming the fully-developed temperature and velocity profiles, a linear stability analysis determines the onset of oscillatory convection at $Ra_c = 506$.

So far our discussion has been limited to strictly two-dimensional problems. However, if one thinks convecting processes in a real physical situation, one needs to concern about its realizability of strictly two-dimensional convection in a three-dimensional space. Based on a linear stability analysis, Straus [22] has shown that steady single-cell, two-dimensional convection is unstable to infinitesimal disturbances and can not be realized in infinitely long square cylinders for the Rayleigh number larger than about 200; no two-dimensional convection is possible when the Rayleigh number is larger than 380. The work, however, was done for the infinite horizontal extent, and bears only indirectly on finite domain cases. In order to clarify this point, Straus and Schubert [23] analyzed the stability of two-dimensional convection in square cylinders with finite length. They found that the two-dimensional convection can exist only if the length of the cylinder is sufficiently short relative to the dimension of the square cross section.

Onset of convection in a three-dimensional space was first considered by Beck [24]. Three-dimensional convection, a superposition of two orthogonally oriented two-dimensional cells in a cube, was studied by Zebib and Kassoy [25] in a slightly supercritical regime, using the method of weakly nonlinear analysis (Palm *et al.* [26]). Relatively high Rayleigh number cases have been investigated numerically by Holst and Aziz [27], Horne [28], Straus and Schubert [29] and Schubert and Straus [12]. Among them Horne [28] and Schubert and Straus [12] carried out their calculations up to sufficiently high Rayleigh numbers to result in oscillatory solutions. Schubert and Straus [12] have found that a 'strictly three-dimensional' convection pattern, which was referred to as a (1,1,1) mode by Straus and Schubert [29], becomes unstable and moves in an ever-oscillating state at a Rayleigh number whose value is about the same as that in two-dimensional convection; the critical Rayleigh number estimated is about 300. Employing a much more efficient numerical algorithm Kimura *et al.* [30] determined that this 'strictly three-dimensional convection becomes oscillatory at a much higher value of the Rayleigh number, namely in a range $550<Ra_c<575$. Due to a much heavier computing load than that of the two-dimensional counterpart, investigations on three-dimensional convection have been very few. Kimura *et al.*[30] proceeded their three-dimensional computations beyond the criticality, and demonstrated that the simply-periodic oscillation eventually gives way to a quasi periodic state at a higher value of the Rayleigh number.

There have been several other problems of convection in porous media in which permanently oscillatory convection is observed. Therefore, it would be worthy to make a brief note on such problems. One such configuration is thermal convection in a horizontal, concentric annulus with a heated inner cylinder at constant temperature. It was Caltagirone [31] who first did a systematic analytical and experimental study on this problem. He showed that two-dimensional steady convection at a relatively low Rayleigh number becomes three-dimensional and fluctuating. However, if the convecting flow is in a strictly two-dimensional space, it remains steady. He determined experimentally that the bifurcation takes place at about $Ra_C = 65$ for a outer-to-inner radius ratio 2. He also performed a linear stability analysis of steady two-dimensional flow to perturbations in the third dimension (the cylinder's axial direction), and determined $Ra_C = 66.96$ for the same outer-to-inner radius ratio. Himasekhar and Bau [32] showed a possibility that the bicellular mode in two-dimensional space first becomes multicellular, and then bifurcates to an oscillatory mode when the Rayleigh number increases for a fixed outer-to-inner radius ratio. Another example demonstrated by Kimura [33] is the situation in which a vertical wall of a two-dimensional square cavity is heated along the lower half portion and cooled along the upper half portion. This thermal boundary condition generates a collision of ascending hot fluid and descending cold fluid at the midheight. It is important in this problem that the cold boundary temperature must be lower than that of the fluid in the cavity in order to drive a cold current along the upper half portion of the vertical wall. When the Rayleigh number is small, steady horizontally layered two convecting cells are formed However, as the Rayleigh number increases, the steady convection experiences a Hopf bifurcation and gives way to an oscillatory solution. The critical Rayleigh number is about 160.

In the following, we describe a mathematical formulation to express the conservation laws of mass, momentum and energy in a fluid-saturated porous medium. Then the onset of oscillatory convection in a two-dimensional cavity heated from below is reviewed in some detail. A discussion on Hopf bifurcation of three-dimensional convection in a cube then follows. Finally, Hopf bifurcations of convection in a horizontal concentric cylinder which is heated with an inner cylinder, in a horizontal layer heated from below by a constant flux and in a two-dimensional cavity with an unstably-heated vertical wall is discussed.

MATHEMATICAL FORMULATION

Governing Equations to Describing the Conservation Laws

Fluid flows and heat transport through a porous matrix is a microscopic process in nature, whereas what is of interest in many engineering and geophysical applications is a transport process of fluid mass and heat in a macroscopic sense. Since direct formulation from a microscopic view is practically unrealistic, due to the complexity of fluid paths formed in a porous matrix, it requires some kind of averaging technique to derive a set of equations which govern the macroscopic transport processes. Detailed discussions of this matter may be found in a review paper by Cheng [34] and a monograph by Nield and Bejan [35]. Depending upon the degree of approximations performed in the averaging process, various forms of governing equations can be deduced. In this chapter, however, we restrict ourselves to one of the simplest formulations in heat and fluid flows in a fluid-saturated porous medium (for example, see Bejan [36]).

The basic assumptions underlying this formulation may be reviewed here briefly. First the solid matrix is assumed to be homogeneous, isotropic, nondeformable and stationary with respect to the appropriate coordinate systems. The flow through a porous matrix is described by Darcy's law and the Darcy number is small enough to neglect the inertia terms in the momentum balance. The thermal diffusion time across matrix elements of pore scale is smaller than a characteristic time scale for the fluid to flow through the pores, so that the solid matrix and the fluid are always in thermal equilibrium. The temperature differences imposed across the boundaries are small, and consequently the Boussinesq approximation is valid.

We assume that the porous medium filling the space of interest has a porosity ϕ, permeability K and thermal capacity $(\rho C)_s$. The saturating fluid has its kinematic viscosity ν, thermal expansion coefficient β and thermal capacity $(\rho C)_f$. It is also assumed that the thermal properties of the fluid-saturated porous medium is characterized by an effective thermal conductivity $k_e = \phi k_f + (1-\phi)k_s$ and by an effective thermal capacity $(\rho C)_e = \phi(\rho C)_f + (1-\phi)(\rho C)_s$. The thermal diffusivity of porous medium is defined by $\alpha = k_e /(\rho C)_f$. Although we will later discuss a problem where it is more appropriate to apply a cylindrical coordinate system, all the other problems in this chapter can be described by the Cartesian coordinate system. Therefore, we write the governing equations in terms of the Cartesian coordinate system and a schematic diagram of the representative problem, namely a cube heated from below and the coordinate system is shown in Figure 1. Upon invoking the Boussinesq approximation and the linear Darcy's law, the non-dimensional equations for mass, momentum and energy conservation laws are

$$\nabla \cdot \mathbf{V} = 0, \tag{1}$$

$$\nabla p + \mathbf{V} - Ra T \mathbf{k} = 0, \tag{2}$$

$$\frac{\partial T}{\partial t} + \mathbf{V} \cdot \nabla T = \nabla^2 T, \tag{3}$$

where $\mathbf{V}$ is the Darcy velocity vector, $\mathbf{k}$ is the unit vector in the vertical direction, T is the temperature and p is the pressure. The non-dimensionalization has been performed in accordance with

$$(x,y,z) = \frac{(x^*,y^*,z^*)}{H}, \quad \mathbf{V} = \frac{\mathbf{V}^*}{\alpha / H}, \quad T = \frac{T^* - T_C}{T_H - T_C}, \quad p = \frac{p^*}{\mu\alpha / K}, \quad t = \frac{t^*}{\sigma H^2 / \alpha}, \tag{4}$$

where H is a characteristic length scale, μ is the fluid viscosity, K is the permeability of the porous medium and σ is the thermal capacity ratio of the porous medium defined by

$$\sigma = (\phi(\rho C)_f + (1-\phi)(\rho C)_s)/(\rho C)_f. \tag{5}$$

and the Rayleigh number is defined by

$$Ra = \frac{g\beta K (T_H - T_C) H}{\alpha \nu}, \tag{6}$$

with g is the magnitude of the gravitational acceleration. The non-dimensional boundary conditions corresponding to Figure 1 are written as

$$u = 0, \quad \partial T / \partial x = 0 \quad \text{on } x = 0,\ 1 \tag{7}$$
$$v = 0, \quad \partial T / \partial y = 0 \quad \text{on } y = 0,\ 1 \tag{8}$$
$$w = 0, \quad T = 1 \quad \text{on } z = 0 \tag{9}$$
$$w = 0, \quad T = 0 \quad \text{on } z = 1 \tag{10}$$

where (u, v, w) are the (x, y, z) components of the Darcy velocity $\mathbf{V}$.

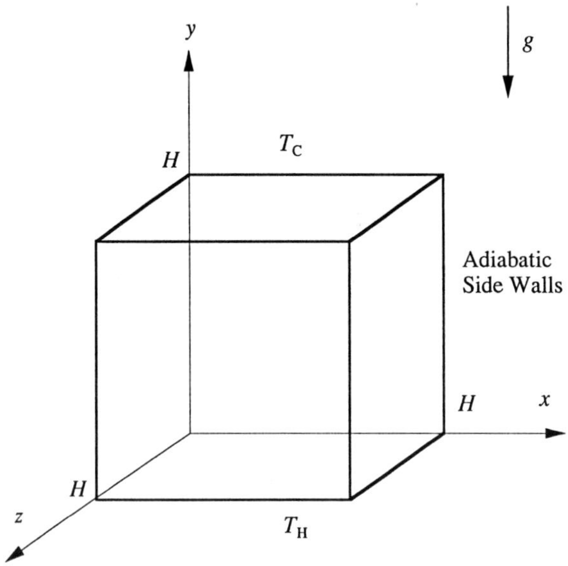

Figure 1: A cube filled with a saturated porous medium heated from below and its coordinates system

Convenient Forms for Numerical Calculation

For numerical calculations it is sometimes convenient to modify the original governing equations (equations (1) and (2)). In two-dimensional calculations, *i.e.* the *x*-*y* cross sectional area, it is a common practice to introduce the non-dimensional stream function to satisfy the continuity equation and to remove the pressure terms in the momentum equation. The stream function in the *x*-*y* plane may be defined as

$$u = \frac{\partial \psi}{\partial y}, \quad v = -\frac{\partial \psi}{\partial x}. \tag{11}$$

Eliminating the pressure terms by cross-differentiation between the two component equations of the momentum equation (2), and substituting equation (11) into the resulting single equation yields the following Poisson equation for ψ,

$$\nabla^2 \psi = -Ra \frac{\partial T}{\partial x} . \tag{12}$$

Equation (12) and the energy equation (3) forms a set of governing equations. The boundary condition for ψ in the cavity flow may be simply written as

$$\psi = 0 \text{ on the all boundaries.} \tag{13}$$

The governing equations can be discretized by an appropriate finite-difference method; the resulting nonlinear algebraic equations may be solved by marching in time.

For three-dimensional case the following *pressure formulation* is one possibility. The formulation can be obtained by eliminating the velocity between equations (1) and (2), after taking the divergence of equation (2) and use of equation (1), gives way to

$$\nabla^2 p = Ra\frac{\partial T}{\partial y} . \tag{14}$$

The velocity vector may also be eliminated in the energy equation to yield

$$\frac{\partial T}{\partial t} = \nabla^2 T + \nabla p \cdot \nabla T - RaT\frac{\partial T}{\partial y} . \tag{15}$$

The boundary conditions for the pressure in the present cavity problem may be stated as

$$\frac{\partial p}{\partial n} = 0 \quad \text{on all the boundary surfaces.} \tag{16}$$

Alternatively, Holst and Aziz [27], and later Horne [28], used a different formulation for their numerical solution. Due to the solenoidal nature of $\boldsymbol{V}$, as given in equation (1), it is possible to introduce a *vector potential* Φ defined by

$$\frac{1}{Ra}\mathbf{V} = \nabla\times\Phi . \tag{17}$$

Taking the curl of the momentum equation and then substituting the definition of Φ, we have

$$\nabla^2\Phi = \left(\frac{\partial T}{\partial z},\ 0,\ \frac{\partial T}{\partial x} \right). \tag{18}$$

In the above process we make use of the fact that the potential is arbitrary to the gradient of a scalar, and hence its divergence can be put to zero. The boundary conditions on each component of Φ for the present cavity problem may be written as

$$\begin{aligned} \partial\phi_1 / \partial x &= \phi_2 = \phi_3 = 0 , \\ \partial\phi_2 / \partial y &= \phi_1 = \phi_3 = 0 , \\ \partial\phi_3 / \partial z &= \phi_1 = \phi_2 = 0 . \end{aligned} \tag{19}$$

Due to a consequence of the above boundary conditions and equation (18), the second component of the potential is identically zero everywhere in the cavity. Thus the energy equation may be expressed in terms of only two components of the vector potential:

$$\frac{\partial T}{\partial t} = \nabla^2 T - Ra\left[\frac{\partial(T,\phi_3)}{\partial(x,y)} + \frac{\partial(T,\phi_1)}{\partial(y,z)}\right], \tag{20}$$

where we have used the notation for the Jacobian operator. Horne [28] claimed that the vector formulation is superior to the first pressure formulation in both speed and accuracy. A finite-difference technique can be equally applied for the three-dimensional governing equations.

Straus and Schubert [29] showed yet another formulation, which has an interesting property, because a single governing equation is resulted of the introduction of the velocity potential. In order to describe this formulation, we first define the non-dimensional temperature as a perturbation from a conductive state,

$$\theta = \frac{T - T_C - \Delta T + \Delta T y / H}{\Delta T}. \tag{21}$$

Taking the curl of the momentum equation, it can be seen that the vertical component of the vorticity is zero, i.e.

$$\frac{\partial w}{\partial x} - \frac{\partial u}{\partial z} = 0. \tag{22}$$

This is identically satisfied if one introduces a velocity potential defined by the following manner,

$$u = \Phi_{xy}, \quad w = \Phi_{yz}, \quad v = -\Phi_{xx} - \Phi_{zz}. \tag{23}$$

It can be shown that the velocity potential Φ also satisfies the continuity equation. Elimin:ting the temperature θ between the momentum equation expressed by Φ and the energy equation, then a single equation for Φ can be derived,

$$\frac{\partial}{\partial t}\nabla^2\Phi + \Phi_{xy}\nabla^2\Phi_x - (\Phi_{xx} + \Phi_{zz})\nabla^2\Phi_y + \Phi_{yz}\nabla^2\Phi_z = \nabla^4\Phi + Ra(\Phi_{xx} + \Phi_{zz}). \tag{24}$$

The boundary conditions on Φ that satisfy both the velocity and temperature conditions are

$$\Phi_{xx} + \Phi_{zz} = \Phi_{yy} = 0 \qquad \text{on } y = 0, 1$$

$$\Phi_{xy} = \nabla^2\Phi_x = 0 \qquad \text{on } x = 0, 1 \tag{25}$$

Furthermore, the solution that satisfy the above boundary conditions can be expressed as a product of Fourier series, namely,

$$\Phi = \sum_{i=0}^{\infty} \sum_{j=1}^{\infty} \sum_{k=0}^{\infty} \phi_{ijk}(t)\sin(j\pi y)\cos(i\pi x)\cos(k\pi z). \tag{26}$$

Upon substituting the Fourier series solution into equation (23), and making use of the orthogonality relations, a set of coupled nonlinear ordinary differential equations in time can be obtained. The solution may be obtained numerically for a truncated system; $N \geq i+j+k$, where N is the truncation number. One deficiency of this method is the rapid increase of computation time as the truncation number becomes large. This is due to nonlinear terms, where the calculation requires to evaluate a product of the two series. In order to avoid the direct evaluation of these nonlinear terms, Kimura *et al.* [17, 30] used a pseudo-spectral method. The method is much more efficient, particularly at large truncation numbers.

OSCILLATORY CONVECTION IN TWO-DIMENSION

Onset of Oscillatory Convection in a Two-Dimensional Square Cross Section

As we have seen in the introduction, oscillatory thermal convection in a porous medium heated from below was first found experimentally by Combarnous and Le Fur [8] and by Caltagirone *et al.* [9]. The value of the critical Rayleigh number for a two-dimensional square cavity, Ra = 390, was determined by Kimura *et al.* [14], and independently by Aidum and Steen [15]. Kimura *et al.* [14] performed a successive linear stability analysis to the steady state sub-critical solutions. Based on the last formulation described in the preceding section, a single governing equation for the velocity potential in two dimensions may be derived by ignoring the third dimension in equation (24). Similarly, the solution can be expressed by forming a Fourier series expansion;

$$\Phi(t,x,y) = \sum_{j=1}^{\infty} \sum_{i=0}^{\infty} \phi_{ij}(t)\cos(i\pi x)\sin(j\pi y). \tag{27}$$

Substituting the above expression into the two-dimensional governing equation, and setting the time derivative of the left hand side to be identically zero, the resulting nonlinear algebraic equation for the Fourier coefficients can be solved by a Newton-Raphson scheme. The solution for this system provides a base state on which small perturbations are superimposed. The process can be described as

$$\phi_{ij} = \phi_{ij}^{0} + \gamma_{ij} e^{(\sigma_r + i\sigma_i)t}, \tag{28}$$

where ϕ_{ij}^{0} represents a steady base state and γ_{ij} is a small perturbation amplitude. Substitution of this expression into the series solution (25) and then into the two-dimensional governing equation and linearizing yields a matrix whose eigenvalues are the σ 's. The linear system may be expressed formally as

$$\dot{\varepsilon}_{ij} = \mathbf{J}_{ijkl}\gamma_{kl}, \tag{29}$$

where $\varepsilon_{ij} = \gamma_{ij}\exp(\sigma_{ij}t)$ is a time-evolution perturbation and $\mathbf{J}_{ijkl}$ is the Jacobian matrix evaluated from the base state. Therefore, the real parts of the eigenvalues of the Jacobian matrix determine the growth or decay of the perturbation with time, and the imaginary parts give frequencies of possible oscillatory solutions. The transition from steady to oscillatory convection is indicated by a change in sign of σ_r from negative to positive, and the onset of oscillation frequency is related to its imaginary part by $f=\sigma_i/2\pi$ (a Hopf bifurcation). Figure 2 shows how two of the eigenvalues of small real part in absolute value shift in the σ_r-σ_i plane as the Rayleigh number increases from 370 to 390. These results were obtained with the truncation number $N = 26$. It is observed that a single pair of eigenvalues crosses the imaginary axis at Ra = 389.7. The imaginary part of the corresponding eigenvalue is about 520, which gives the frequency, at the onset of oscillation, $f=\sigma_i/2\pi$ = 82.5. These results are in excellent agreement with those computed independently by Aidun and Steen [15], who determined $Ra_C = 390.7$ and $f = 82.8$.

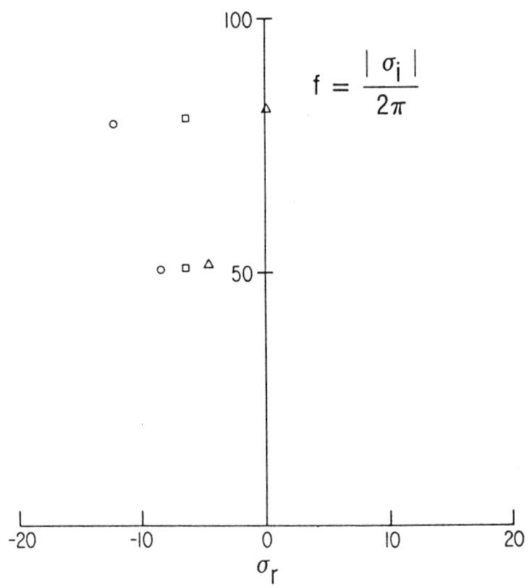

Figure 2: Eigenanalysis for a Hopf bifurcation in a square cross section: ○ Ra = 370, □ Ra = 380, △ Ra = 390 (From Kimura *et al.*[14])

Steen and Aidun [37] further extended their work to a single cell in a non-square rectangular cross section whose width-to-height aspect ratio lies between 0.5 and 1.5, and analyzed the thermal-disturbance structure at the onset of oscillatory motion and the corresponding critical Rayleigh number. They found a decrease of the critical Rayleigh number as the aspect ratio increases. On the other hand, Riley and Winters [16] investigated single cell solutions in an even larger aspect ratio, and found a rather complex behavior of a single cell solution when the aspect ratio is greater than about 2.4. They also computed the eigenvector at the critical condition for a square cross section, and it is shown in Figure 3, in order to see what is the structure from a disturbance whose eigenvalue has a positive real part at the transition. These plots clearly indicate that instabilities are arising along the top and bottom boundary layers. Applying a method of generalized eigenvalue system and a pseudo-arclength continuation technique, they followed the paths of the points of the onset and destabilization of the steady unicellular flow. It was shown that the path of the Hopf bifurcation develops a curly line for the aspect ratio greater than about 2.1. This behavior is plotted in Figure 4(a) for the critical Rayleigh number and Figure 4(b) for the onset frequency. These results also confirm the earlier computations made by Steen and Aidun [37]. This curious behavior was argued in detail in Riley and Winters [16]. Nonetheless, it can be generally said that both the critical Rayleigh number and the onset frequency decrease as the aspect ratio increases. However, the unicellular flow terminates its existence when the aspect ratio is greater than about 2.7.

Postcritical Transitions of a Unicellular Flow in a Square Cross Section

Again restricting our interest to a two-dimensional square cavity, we next describe how a unicellular oscillatory flow which sets in at *Ra*=390 evolves with the Rayleigh number. Further transitions in a super critical regime were computed by Schubert and Straus [13] up to *Ra*=650. They observed a transition from a simply periodic flow to a doubly periodic one, which is characterized by the introduction of a subfrequency in addition to a primary frequency. Further increase of the Rayleigh number turns the system back to a simply periodic state. These results were confirmed by Kimura *et al.* [17] with a much higher

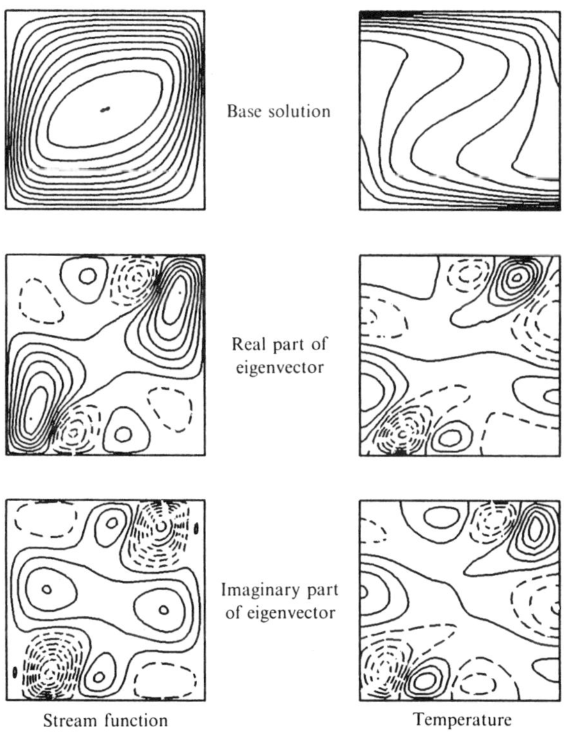

Figure 3: Streamlines and isotherms for the steady solution and the real and imaginary parts of the critical eigenvector at the Hopf bifurcation point (From Riley and Winters [16], courtesy of Cambridge University Press)

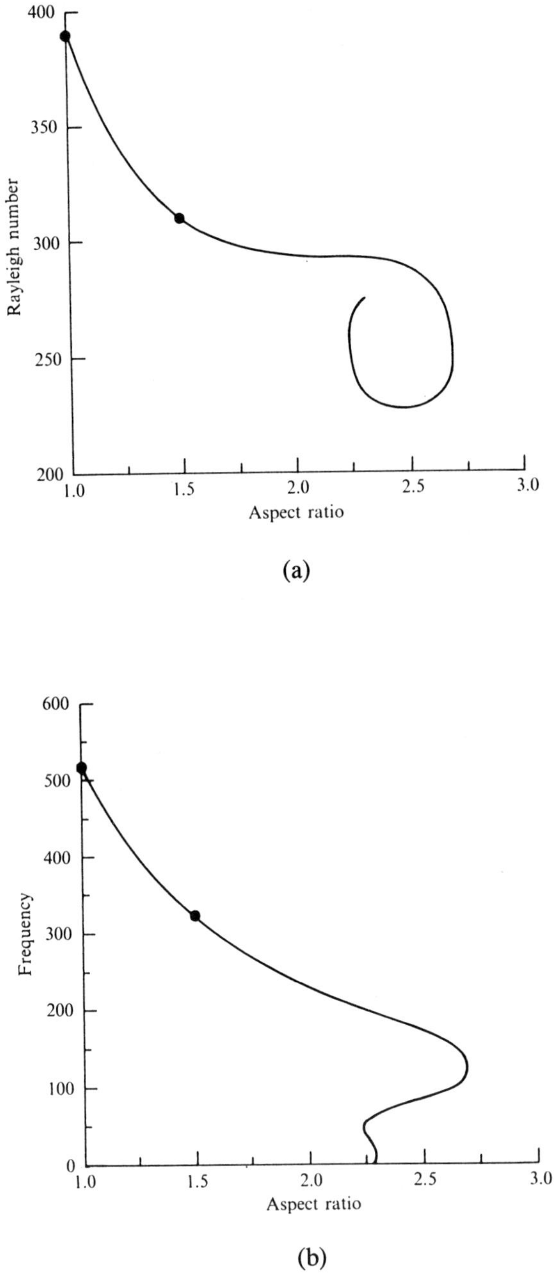

Figure 4: Path of Hopf bifurcations (a) and the corresponding onset angular frequency (b). The full circles are taken from Steen and Aidun [37] (From Riley and Winters [16], courtesy of Cambridge University Press)

resolution. Thereby, the corresponding critical Rayleigh numbers determined are 510 and 565 for the, respective, transitions. The second simply periodic regime persists, at least up to $Ra = 850$. At $Ra = 1000$ notable changes in the oscillating flows are observed. One of them is a centro-symmetry breaking; a centro-symmetry flow, which has been always the case, ceases to exist. Another point to note is a broadband noise in the frequency spectrum diagram. Although still a few peaks are identified in the diagram, the background noise level becomes significantly larger in comparison with the quasiperiodic regime. Thus, an evolution scenario determined for a unicellular flow is

$$\mathbf{S} \longrightarrow \mathbf{P} \longrightarrow \mathbf{QP} \longrightarrow \mathbf{P} \longrightarrow \mathbf{NP}\,. \tag{30}$$

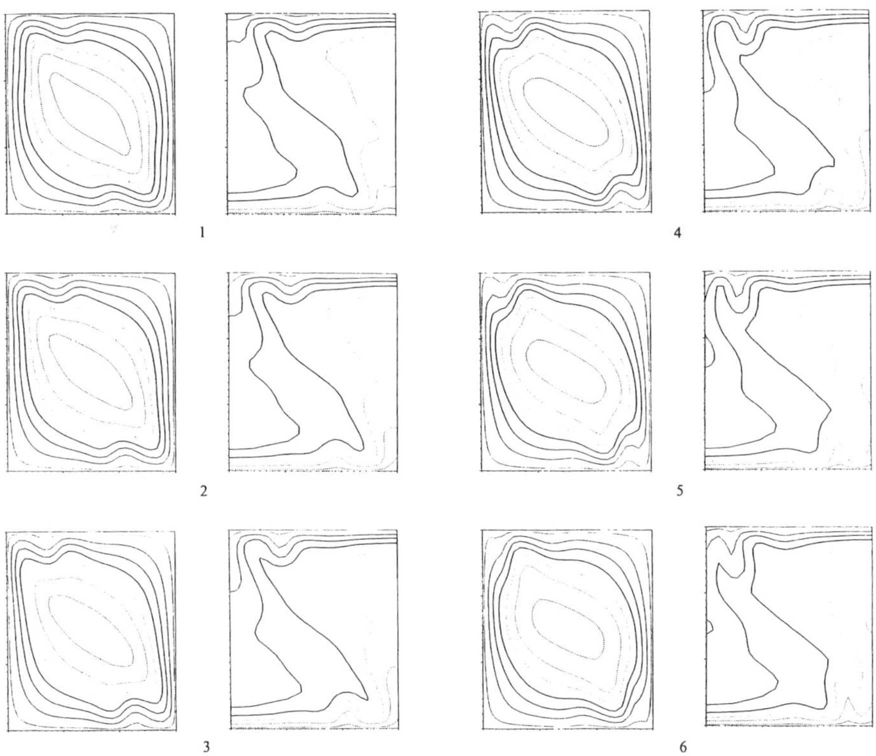

Figure 5: Streamline and isotherm patterns at equal intervals of time for $Ra = 650$. The non-dimensional time interval between successive figures is 0.00056 and the total elapsed time is 0.0044 ($1/f_1 = 0.0050$). The isotherm contour interval is 0.15. The streamfunction Φ_z contour interval is 3.3 (from Kimura *et al.* [17])

where, **S**, **P**, **QP**, **NP** represent steady, periodic, quasiperiodic and nonperiodic states, respectively. The representative streamlines and isotherms of the second simply periodic regime and the non-periodic regime are shown in Figures 5 and 6, In Figure 5 the thermal boundary-layer along the top and bottom boundaries are visible. As the fluid travels along the bottom boundary, it experiences an upward instability due to the buoyancy effect. When the tongue-like thermal blob in the boundary-layer arrives at the vertical adiabatic wall, it is released periodically along the vertical wall. Figure 6 shows isotherms and streamlines for $Ra = 1000$. The thermal boundary-layer becomes even thinner in comparison with $Ra = 850$, and it is seen that more than one thermal blobs are present in the lower boundary-layer, whilst there is only one blob in the upper boundary-layer at an instantant of time. The generation and the release of the thermal blobs along the vertical walls, therefore,

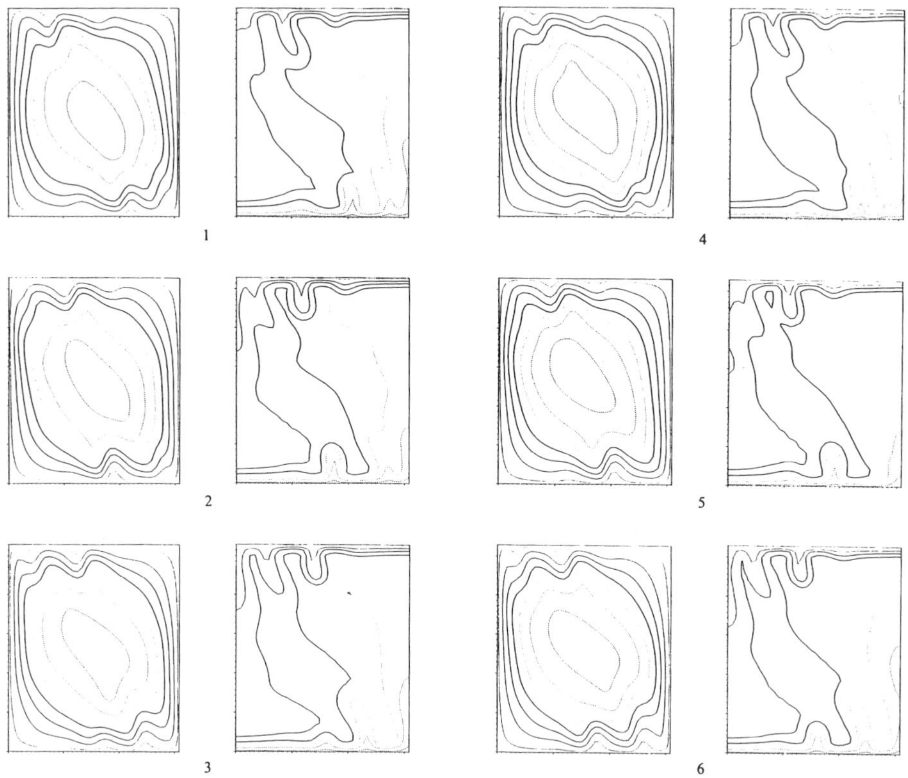

Figure 6: Streamlines and isotherms at equal time intervals for Ra = 1000. The non-dimensional time interval between successive figures is 0.0016 and the total elapsed time is 0.0094 ($1/f_1 = 0.0022$, $1/f_2 = 0.0053$, $1/f_3 = 0.053$). The isotherm contour interval is 0.15 and the streamfunction contour interval is 5 (from Kimura *et al.* [17])

are not periodic any more. Since the flow and temperature structure in the core regime is similar to that in the second simply periodic state, it is believed that this random behavior in the boundary-layer is largely responsible for a high level of background noise in the spectram diagram. Also the thermals are generated at a much faster rate in the NP regime than they are at lower values of *Ra*. The convecting velocity also becomes about 1.5 times greater than that for *Ra* = 850.

The average Nusselt numbers in the second simply periodic regime and the non-periodic regime grow consistently with a law of

$$\overline{Nu} \propto Ra \tag{31}$$

This proportionality between the average Nusselt number and the Rayleigh number, found in equation (29), is in fact in accord with a scaling analysis by Bejan [36]. In Figures (7) and (8) we show the time series of average Nusselt number for *Ra* = 540 and 1000 and the corresponding power spectra. At *Ra* = 540 it is found that there are two incommensurate peaks in the power spectrum, and the other peaks can be identified as sums or differences of the two. However, at *Ra* = 1000, there are still a few visible peaks, the rise of the background noise is also obvious. In Figure 9 we assemble all the incommensurate peaks found in the power spectrum diagrams of Nusselt number oscillation. It is interesting to see that the primary frequency in the first simply periodic state fits to the power law

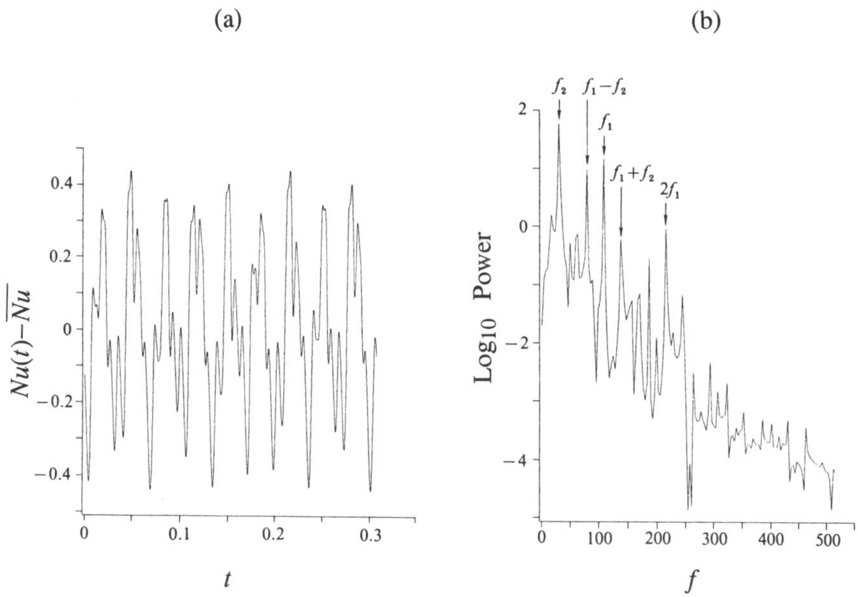

Figure 7: (a) The time series $Nu(t)-\overline{Nu}$, and (b) its power spectrum at Ra = 540 (from Kimura, *et al.* [17])

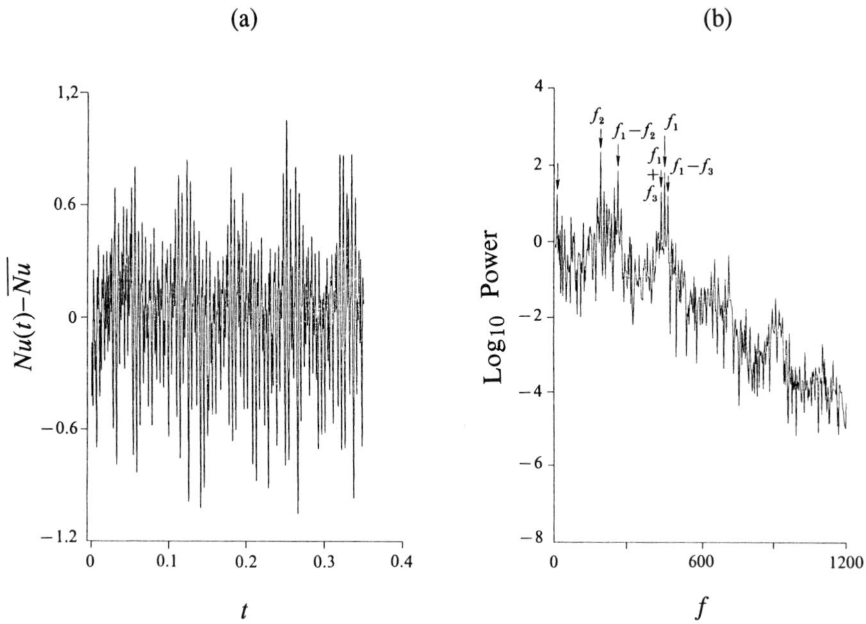

Figure 8: (a) The time series $Nu(t)-\overline{Nu}$ and (b) its power spectrum at Ra=1000 (from Kimura *et al.* [17])

$Ra^{7/8}$, whilst the primary frequency in the second single-frequency regime follows the trend $Ra^{3/2}$, or probably a little higher power if one respects a trend in only the high Rayleigh number regime ($Ra \geq 600$). Graham and Steen [20] argue that these transitions in the unicellular oscillatory convecting flows are due to different wave lengths of the traveling wave arising from the boundary-layer instability. They showed that the system in the vicinity $Ra = 1000$, for instance, can admit both centrosymmetric and non-centrosymmetric solutions, depending upon a particular integer wavenumber of the traveling wave which is present in the cavity. In this regime the modal exchange and resonance processes of the two solutions are responsible for generating a mixed mode of the two; a quasi-periodic state and a weakly chaotic state.

OSCILLATORY CONVECTION IN THREE-DIMENSION

A Strictly Three-Dimensional (1,1,1) Mode in a Cube

Due to increasing spatial freedom, convection in a three-dimensional cavity is far more complex than that in a two-dimensional cross section. Therefore, as studied by Kimura *et al.* [30], the consideration here is restricted to a particular convection pattern referred to as the (1,1,1) mode; it involves coupled modes in all three orthogonal directions. The solution of this mode is best characterized by the dominance of a particular Fourier coefficient $\Phi_{111}(t)$ over the other coefficients in equation (26). This convection mode is known to set

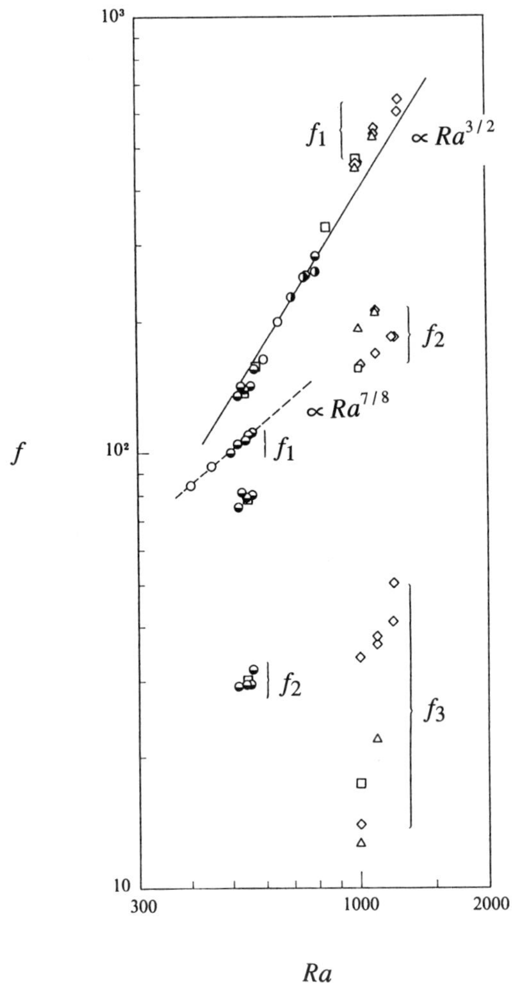

Figure 9: Dimensionless frequencies, f ,of the incommensurate peaks in the power spectra of the Nusselt number oscillation. Various symbols indicate different truncation numbers in numerical calculations (from Kimura *et al.* [17])

in at $Ra_C = 4.5\pi^2$ in a cube, but later Steen [40] showed that the (1,1,1) mode is unstable in the range $4.5\pi^2 < Ra < 4.87\pi^2$. Nonetheless, several investigations, namely Schubert and Straus [12], Horne [28] and Straus and Schubert [29] on this flow pattern proved that the steady (1,1,1) mode stably exists up to at least $Ra = 300$.

In Figure 10 we show a typical vertical velocity distribution of the steady state (1,1,1) mode on a horizontal cross section at six different vertical positions. Near the bottom region a strong rising flow, which has a higher temperature and extends diagonally, is dominant (see Figures 10a and 10b), whilst a strong descending flow of a lower temperature, formed along the other diagonal, is observed in the upper space (see Figures 10e and 10f).

These two rising and falling currents with hot and cold fluids, respectively, become broader in width as they proceed, and eventually collide in the midheight. This collision divides the region into four of equal areas, and each of them is occupied by hot rising or cold falling fluids alternatively (see Figures 10c and 10d). According to the distribution of Φ_y , in horizontal cuts (not shown here) it is seen that horizontal component of the fluid motion is towards the diagonals in the top and bottom boundary layers. On the other hand the horizontal components at the midheight are outward from the center $(x, y, z) = (0.5, 0.5, 0.5)$. The symmetries about the two diagonals in each horizontal plane and the antisymmetries with respect to the reflection plane at $y = 0.5$ are characteristics of the convection patterns in the regime.

Oscillatory Convection in a (1,1,1) mode in a Cube

Schubert and Straus [12] found that the steady (1,1,1) mode becomes oscillatory in the vicinity of $Ra = 300$ when the Rayleigh number is increased gradually. With much larger truncation numbers (higher resolution) Kimura *et al.* [30] showed that the transition from a steady to an oscillatory solution takes place in the range $550<Ra_C<575$. Preceding to this transition, the solution exhibits a partial symmetry breaking at $Ra = 550$; a loss of symmetry with respect to the diagonals in each horizontal plane. In these steady state flows, the rising and sinking motion near the lower and upper boundaries shifts slightly to off-diagonal positions as shown in Figure 11. At $Ra = 575$ the solution is oscillatory, and it is simply periodic. We show the vertical velocity distribution in this regime ($Ra = 625$) in the horizontal cross sections in Figure 12. The contours of the vertical velocity in the planes at b, c, d and e show almost identical features to those in Figure 11. Within the top and bottom boundary layers, however, ripple-like disturbances are evident. This flow structure indicates that the boundary-layer instability may be the cause of the oscillation in the Nusselt number. It should be mentioned that the temperature and the horizontal velocity fields in the boundary layers are surprisingly smooth and do not show any irregularity corresponding to these ripples.

When the Rayleigh number is increased further, the solution becomes quasi-periodic with two or three incommensurate peaks at Ra = 650. It is much more difficult to characterize this regime by spectral power diagrams obtained from the Nusselt number time series than it's two-dimensional counter part . The solutions in this regime often exhibit intermittence of two or more different distinctive states, each of which lasts for 0.2 - 0.3 in the non-dimensional time and is characterized by different frequency peaks and different average Nusselt numbers. In the narrow range around $Ra = 700$ it is observed that the quasi-periodic states revert, with increasing *Ra,* to another simply periodic state with a drastic reduction in the variance of the Nusselt number. The increase of the Rayleigh number to $Ra = 740$ transforms the solution again to a quasi-periodic state. The vertical velocity distributions on the horizontal planes in this regime are shown in Figure 13 and several new features are visible in these figures. Disturbances originating from the boundary layers penetrate more deeply into the core flow than was the case at $Ra = 625$. They are identified by hot rising, or cold descending, spots in the horizontal mid planes (plates b-e in Figure 13). In comparison with $Ra = 625$, disturbances within the boundary layers are of smaller scales, probably due to the thinner boundary-layers. Despite this complex vertical flow structure, the antisymmetry about $y = 0.5$ still holds. The evolution process of oscillatory convection up to $Ra = 740$ may be given as

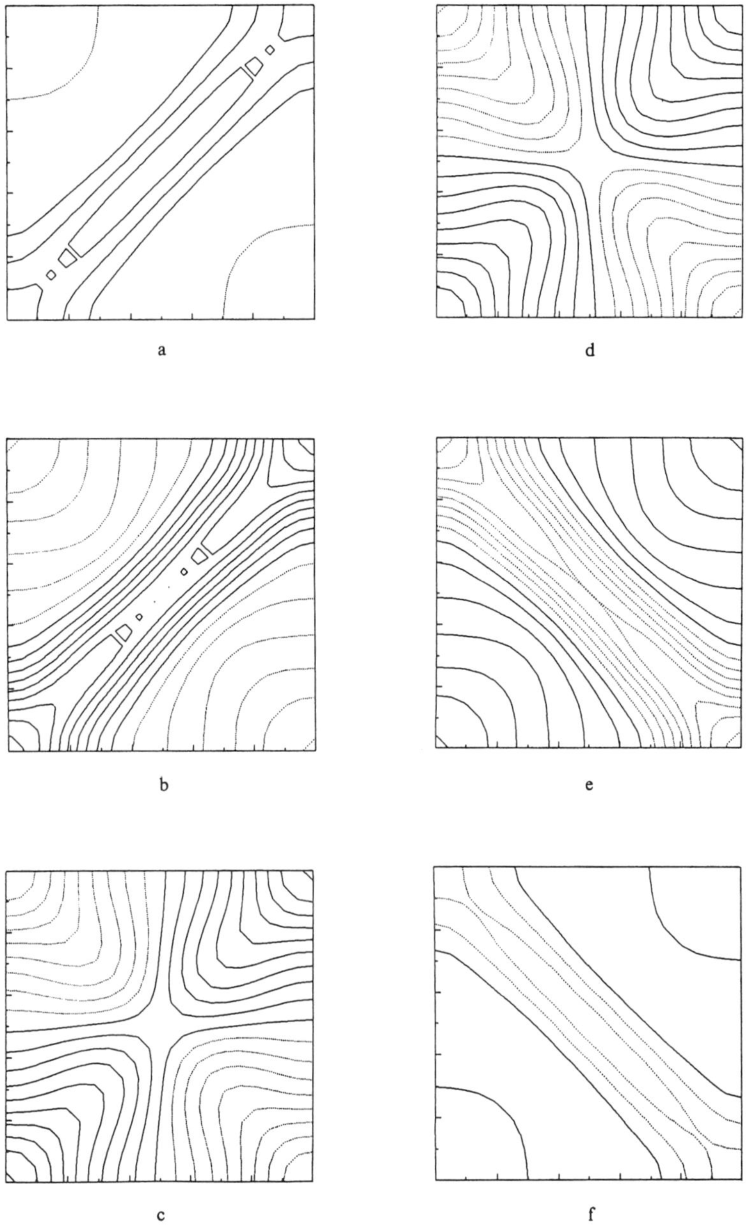

Figure 10: Contours of the verical velocity at each horizontal cut for Ra=250. The contour interval is 8. Upward flows are shown solid and downward ones are shown dotted. Plate (a) is located near the bottom of the cube and plate (f) is near the top. (from Kimura *et al.*[30])

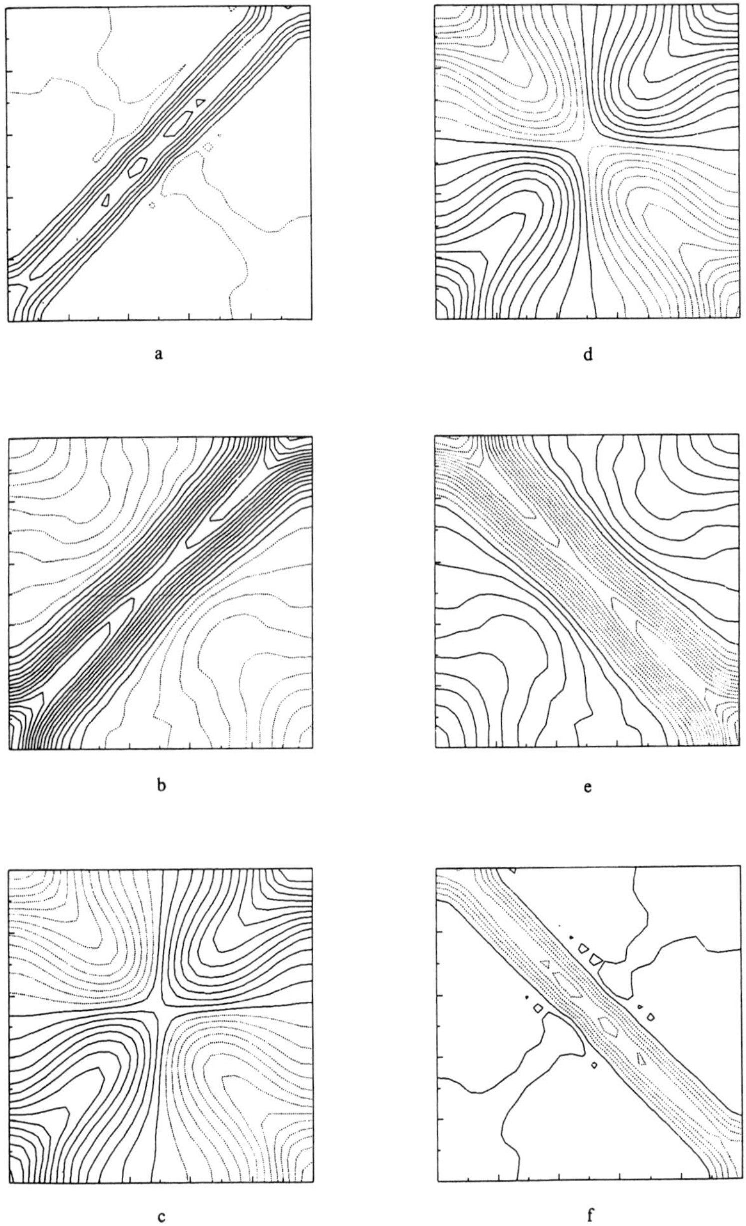

Figure 11: Contours of the vertical velocity on horizontal planes at various heights for *Ra*=550 and the contour intervals are 10 (from Kimura *et al.* [30])

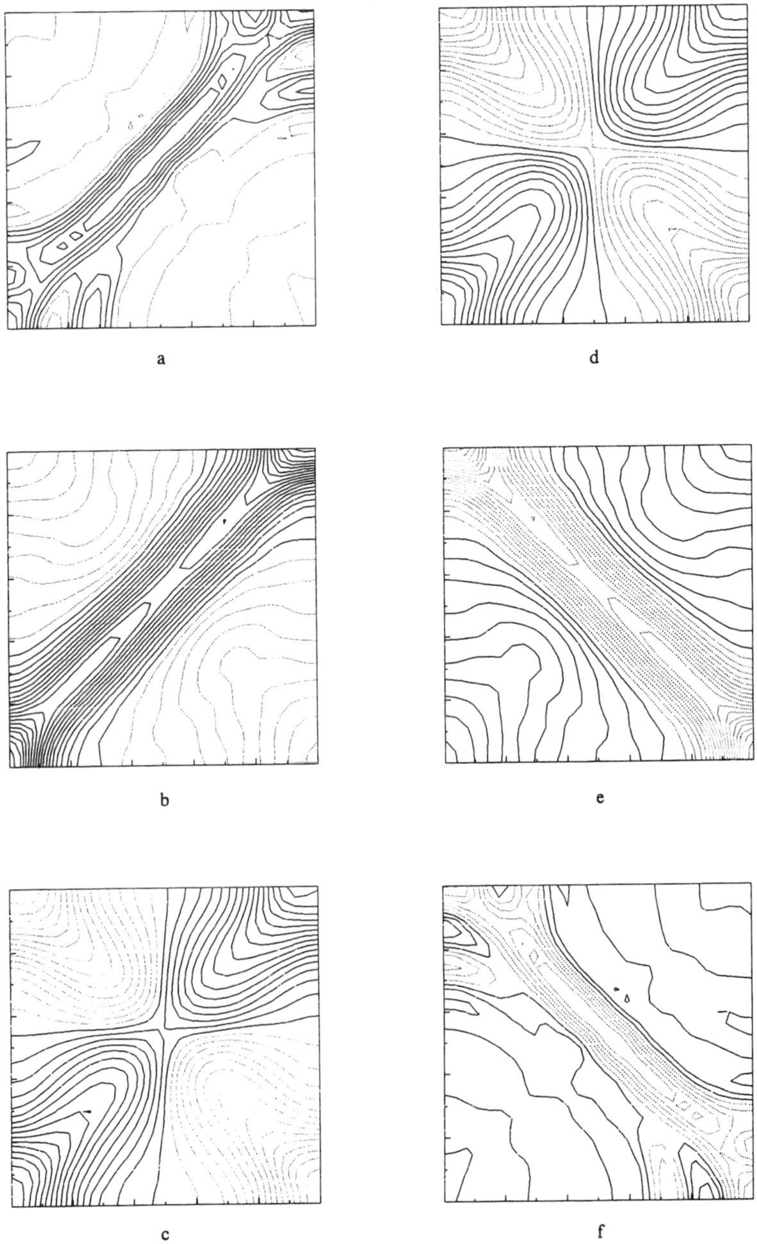

Figure 12: Contours of the verical velocity on horizontal planes at various heights for Ra=625, a simply periodic state. The contour intervals are 10 (from Kimura *et al.* [30])

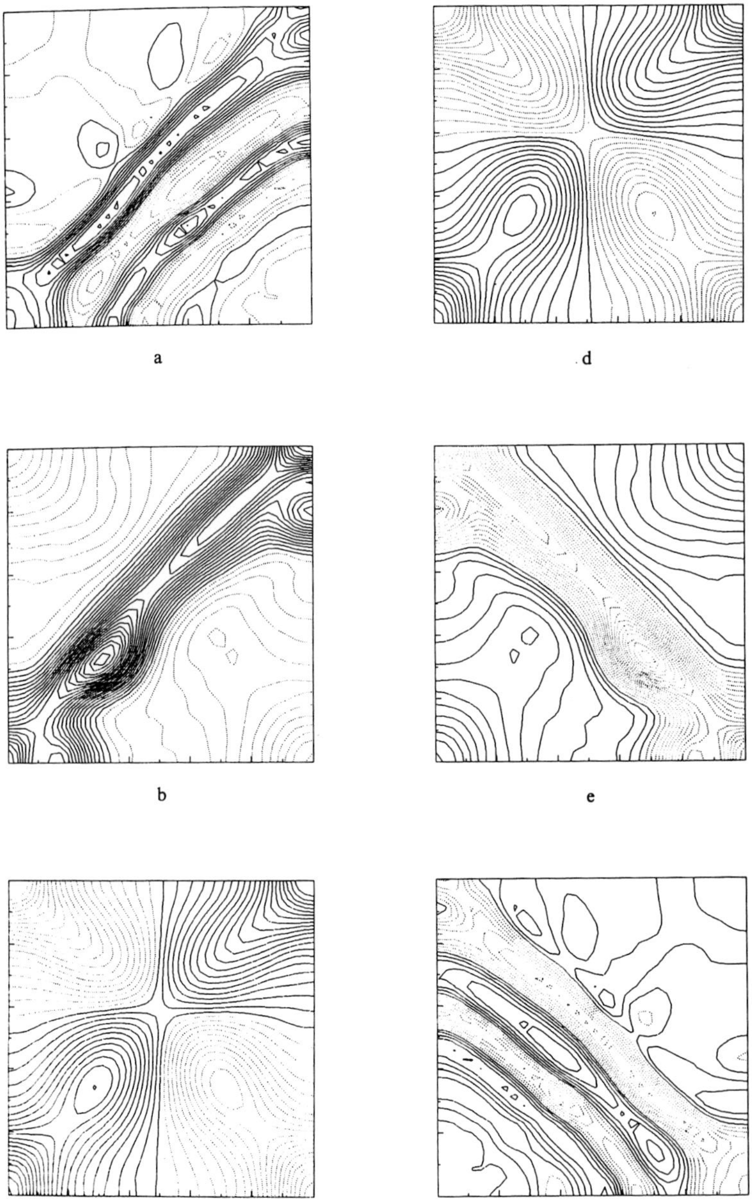

a d b e c f

Figure 13: Contours of the vertical velocity at each horizontal plane for Ra=740. Convection is a quasi-periodic state. The contour intervals are 10 (from Kimura *et al.* [30])

$$\mathbf{S} \longrightarrow \mathbf{P} \longrightarrow \mathbf{QP} \longrightarrow \mathbf{(P)} \longrightarrow \mathbf{QP}\,. \qquad (32)$$

In the simply periodic state, the frequency increases with Ra according to $f \propto Ra^2$. Despite the further transitions in the higher Rayleigh number regime, it appears that the increase of the highest frequency peak is consistent with the above law and extends into the quasi-periodic states.

OSCILLATORY CONVECTION IN OTHER PROBLEMS

Convection in a Horizontal Concentric Annulus

Caltagirone [31] studied, both theoretically and experimentally, natural convection in a porous layer bounded by two horizontal concentric cylinders. Based on his word, he claimed that the evolution process from steady to periodic state is the two-dimensional bicellular solution being replaced by a three-dimensional fluctuating solution at criticality. The experiments defined the critical value for $Ra_C = 65$ for an outer-to-inner radius ratio $R = 2$ and his stability analysis is in close agreement with a value $Ra_C = 69$ for the instability of the two-dimensional solution. In the two-dimensional calculation of the relatively low Rayleigh numbers, he observed that the solutions are all steady. Therefore, he concluded that the three-dimensionality is a necessary condition for the onset of oscillations. More recently, Himasekhar and Bau [32] have investigated a similar problem over a greater range of values of R and the Rayleigh number. They showed that oscillatory convection can occur in a strict two-dimensional space, depending upon R and the convecting patterns. According to them the two-dimensional bicellular convection first bifurcates to multicellular convection with increasing the Rayleigh number. The further increase of the Rayleigh number leads to oscillatory convection.

Horizontal Layer Heated from Below by Uniform Heat Flux

Kimura *et al.* [21] investigated a case where a horizontal porous layer is heated from below and cooled at the top with constant fluxes. In a strictly two-dimensional space, a horizontal layer with a large height-to-horizontal length aspect ratio admits a fully-developed counter flow, in which the velocity and temperature field can be expressed in analytical form. A stability analysis on the fully-developed counter flow in an infinitely long two-dimensional horizontal layer was conducted and it was predicted that the steady state solution bifurcates to oscillatory convection at the critical value of the Rayleigh number $Ra_C = 506$ with a critical oscillation frequency $f_C = 22.1$ in the diffusion time. On the other hand, the corresponding two-dimensional numerical solution indicates that the critical Rayleigh number is highly dependent of the layer aspect ratio. For instance, the critical Rayleigh number is between 730 and 750 for a cavity of aspect ratio 4, whilst it is between 630 and 650 for an aspect ratio 8.

Unstably Heated-and-Cooled Vertical Plate

Kimura [33] studied the natural convection along an unstably heated-and-cooled vertical wall in a two-dimensional square cavity; the upper-half portion is cooled and the lower-half is heated along the same vertical wall. The cold current descends along the upper-half cold plate, whilst the warm current rises along the lower-half hot plate. The two currents collide at the midheight, and change their flow directions. Therefore, two convecting cells are formed in the square cavity. Numerical simulations reveal that the horizontally-layered two cell solution is steady, at least up to Ra = 150, but at Ra = 200 it is found that the solution is periodically oscillating. The Nusselt number is simply periodic at the onset of oscillations. The temperature distribution in the midheight region, near to the vertical wall, is more or less conductive at small Rayleigh numbers and the isotherms radiate radially from the temperature discontinuity point. At large Rayleigh number, however, the isotherms are horizontal and the region forms a gravitationally-unstable layer. It is believed that the formation of the unstable layer is responsible for the onset of oscillatory convection.

CONCLUDING REMARKS

In this chapter, Hopf bifurcations in convection in a fluid-saturated porous medium have been reviewed. For a single cell the convection in a two-dimensional square cross section heated from below it has been well established that the critical Rayleigh number is 390 with the critical oscillation frequency being about 82. However, it is a somewhat difficult task to draw a single scenario regarding the post-critical evolution processes of the oscillating convection. This is primarily due to the fact that the system admits multiple solutions for a given value of Ra, and they are easily interchangeable in a process of raising or decreasing the Rayleigh number. This makes it extremely difficult to stay on a single branch of the solution which is currently under investigation. Therefore, it is often not clear whether a series of numerically generated solutions represent a trace of a single branch curve of the evolution process. Nonetheless, it can be said that a single cell solution in a square cavity repeats a simply-periodic to quasi-periodic transition and the reverse transition before it becomes a chaotic state at Rayleigh numbers greater than 1000.

Three-dimensional convection in a cube also exhibits a Hopf bifurcation and the critical Rayleigh number possibly depends on a mode of the convecting pattern. Since a very limited amount of effort has been devoted to three-dimensional convection, the entire picture of the bifurcating process is less clear. Nonetheless it was determined that a (1,1,1) mode convection, which sets in at $Ra_C=4.5\pi^2$ from the conductive state, becomes oscillatory somewhere between Ra = 550 and 575 when the Rayleigh number is increased gradually. The criticality is much greater than the previously computed value Ra_C = 300-330 which has much less resolution. The oscillation at criticality is simply periodic with a frequency of about f_C = 177. A further increase in the Rayleigh number transforms the system to a quasi-periodic state, often with intermittence. Therefore, it is difficult to characterize a quasi-periodic state with frequency peaks in a spectral power diagram.

A few other problems in which transitions from steady to time-dependent fluctuating convection take place have been also mentioned.

REFERENCES

1. C.W. Horton and F.T. Rogers, Convection currents in a porous medium, *J. Appl.Phys.* **16**, 367- 370 (1945).
2. E.R. Lapwood, Convection of a fluid in a porous medium, *Proc. Cambridge Phil.Soc.* **44**, 508-521 (1948).

3. K.J. Schneider, Investigation of the influence of free convection on heat transfer through granular material, *Proc. Int. Congr. Refrig., 11th, Munich.* Pap. 11-4(1963).
4. Y. Katto and T. Masuoka, Criterion for onset of convective flow in a fluid in a porous medium, *Int. J. Heat Mass Transfer* **10**, 297-309 (1967).
5. M. Combarnous, Convection naturelle et convection mixte dans une poreuse horizontale, *Rev. Gén. Therm.* **9**, 1335-1355 (1970).
6. S. Bories, Comparis on des prévisions d'une théorie non linéaire et des résultats expérimentaux en convection naturelle dans une couche poreuse saturée horizontale, *C.R. Acad. Sci., Ser. B* **271**, 269-272 (1970).
7. M.A. Combarnous and S.A. Bories, Hydrothermal thermal convection in saturated porous media, *Adv. in Hydroscie.* **10**, 231-307 (1975).
8. M. Combarnous and B. Le Fur, Transfert de chaleur par convection naturelle dans une couche poreuse horizontale, *C.R. Acad. Sci. Ser.B* **269**, 1009-1012(1969).
9. J.P. Caltagirone, M. Cloupeau and M. Combarnous, Convection naturelle fluctuante dans une couche poreuse horizontale, *C.R. Acad. Sci., Ser. B* **273**, 833-836 (1971).
10. J.P. Caltagirone, Thermoconvective instabilities in a horizontal porous layer, *J.Fluid Mech.* **72**, 269-287 (1975).
11. R.N. Horne and M.J. O'Sullivan, Oscillatory convection in a porous medium heated from below, *J. Fluid Mech.* **66,** 339-352 (1974).
12. G. Schubert and J.M. Straus, Three-dimensional and multicellular steady and unconvection in fluid-saturated porous media at high Rayleigh number, *J. Fluid Mech.* **66**, 339-352 (1979).
13. G. Schubert and J.M. Straus, Transitions in time-dependent thermal convection in fluid-saturated porous media, *J. Fluid Mech.* **121**, 301-313 (1982).
14. S. Kimura, G. Schubert and J.M. Straus, Instabilities of steady, periodic and quasiperiodic modes of convection in porous media, *J. Heat Transfer* **109**, 350-355 (1987).
15. C.K. Aidun and P.H. Steen, Transition to oscillatory convection in a fluid-saturated porous medium, *J. Thermophs. Heat Transfer* **1**, 268-273 (1987).
16. D.S. Riley and K.H. Winters, Time-periodic convection in porous media: the evolution of Hopf bifurcations with aspect ratio, *J. Fluid Mech.* **223**, 457-474(1991).
17. S. Kimura, G. Schubert and J.M. Straus, Route to chaos in porous-medium thermal convection, *J. Fluid Mech.* **166**, 305-324 (1986).
18. D.S. Riley and K.H. Winters, Modal exchange mechanisms in Lapwood convection, *J. Fluid Mech.* **204**, 325-358 (1989).
19. D.S. Riley and K.H. Winters, A numerical bifurcation study of natural convection in a tilted two-dimensional porous cavity, *J. Fluid Mech.* **215**, 309-329 (1990).
20. M.D. Graham and P.H.Steen, Plume formation and resonant bifurcations in porous-media convection, *J. Fluid Mech.* **272**, 67-89 (1994).
21. S. Kimura, M. Vynnycky and F. Alavyoon, Unicellular natural circulation a shallow horizontal porous layer heated from below by a constant flux, *J. Fluid Mech.* **294**, 231-257 (1995).
22. J.M. Straus, Large amplitude convection in porous media, *J. Fluid Mech.* **64**, 51-63 (1974).
23. J.M. Straus and G. Schubert, On the existence of three-dimensional convection in a rectangular box of fluid-saturated porous material, *J. Fluid Mech.* **87**, 385-394 (1978).
24. L.J. Beck, Convection in a box of porous material saturated with fluid, *Phys. Fluids*

15, 1377-1383 (1972).
25. A. Zebib and D.R. Kassoy, Three-dimensional natural convection motion in a confined porous medium, *Phys. Fluids* **21**, 1-3 (1978).
26. E. Palm, J.E. Weber and O. Kvernvold, On steady convection in a porous medium, *J. Fluid Mech.* **54**, 153-161 (1972).
27. P.H. Holst and K. Aziz, Transient three-dimensional natural convection motion in a confined porous medium, *Int. J. Heat Mass Transfer.* **15**, 73-90 (1972).
28. R.N. Horne, Three-dimensional natural convection motion in a confined porous medium heated from below, *J. Fluid Mech.* **92**, 751-766 (1979).
29. J.M. Straus and G. Schubert, Three-dimensional convection in a cubic box of fluid-saturated porous material, *J. Fluid Mech.* **91**, 155-165 (1979).
30. S. Kimura, G. Schubert and J.M. Straus, Time-dependent convection in a fluid-saturated porous cube heated from below, *J. Fluid Mech.* **207**, 153-189 (1989).
31. J.P. Caltagirone, Thermoconvective instabilities in a porous medium bounded by two concentric horizontal cylinders, *J. Fluid Mech.* **76**, 337-362 (1976).
32. K. Himasekhar and H.H. Bau, Two-dimensional bifurcation phenomena in thermal convection in horizontal, concentric annuli containing saturated porous media, *J. Fluid Mech.* **187**, 267-300 (1988).
33. S. Kimura, Time-dependent phenomena in porous media convection, *Heat and Mass Transfer in Porous Media* (Ed. M. Quintard and M. Todorovic), pp. 277-292. Elsevier, Amsterdam (1992).
34. P. Cheng, Heat transfer in geothermal systems, *Adv. in Heat Transfer* **14**, 1-105 (1978).
35. D.A. Nield and A. Bejan, *Convection in Porous Media*, Springer-Verlag, New York (1992).
36. A. Bejan, *Convection Heat Transfer*, Wiley, New York (1984).
37. P.H. Steen and C.K. Aidun, Time-periodic convection in porous media: transition mechanism, *J. Fluid Mech.* **196**, 263-290 (1988).

THERMAL NONEQUILIBRIUM FORCED CONVECTION IN POROUS MEDIA

A.V. KUZNETSOV

Christian Doppler Laboratory for Continuous Solidification Processes
Institute of Fluid Mechanics and Heat Transfer, Technical University Vienna
Wiedner Hauptstrasse 7/322, A-1040 Vienna, Austria

1. INTRODUCTION

Packed beds are often used for the storage of heat energy and in the chemical industry. For many industrial applications, for example for applications involving high-speed flows, it is important to take account of thermal nonequilibrium effects. This explains the considerable number of publications devoted to this subject. To account for the thermal nonequilibrium effects it is necessary to use a model which includes two energy equations, one for the fluid and one for the solid phase. Using the two energy equation model requires the knowledge of an extra parameter to be determined experimentally, namely the heat transfer coefficient between the fluid and solid phases.

Most of the analytical studies of thermal nonequilibrium were concentrated on the Schumann model of a packed bed suggested in Schumann [1]. In the Schumann model a flow of incompressible fluid through a packed bed is considered and longitudinal conduction terms in both the fluid and solid phase energy equations are neglected. Analytical and numerical solutions for this model are obtained in refs. [2-5]. A comparison between the analytical solutions for the two energy equation model and for the one energy equation model (this model corresponds to a local thermal equilibrium assumption) is presented in Riaz [6]. In White and Korpela [7] and in Spiga and Spiga [8] exact solutions for the case when the inlet fluid temperature is a function of time and the initial temperature of the packed bed is a function of the space variable are obtained. Analytical investigations of the heat and mass transfer in cylindrical packed beds with internal heat generation are presented in Amundson [9-10].

It is worth noting that inclusion of longitudinal thermal conductivity essentially complicates the problem. This can be viewed as follows. Physically, for the Schumann model, information is carried downstream from the inlet plane by the flow. However if longitudinal thermal conductivity is included, information can also travel upstream. In other words, if we consider a steady problem, the Schumann model would result in a parabolic set of equations, while inclusion of longitudinal conductivity results in an elliptic problem. The problem with longitudinal thermal conductivity included is analytically analyzed in Kuznetsov [11] and it is found that for such a formulation the temperature difference between the fluid and solid phases exhibits wave properties.

All the above studies are limited to the case when the fluid flow is governed by Darcy's law. In the numerical investigations reported in [12-16], non-Darcian and thermal nonequilibrium effects, as well as longitudinal thermal dispersion effects, are accounted for. In these papers a set of volume-averaged governing equations for non-thermal equilibrium, condensing, forced fluid flow through sensible and latent heat storage beds is presented and comprehensive numerical analyses of the phenomena are carried out. These papers provide important data for a better understanding of these complicated phenomena. For example, for all the cases computed in refs. [12-16] the temperature difference between the fluid and solid phases is found to be small compared to the difference between the inlet temperature of the fluid phase and the initial temperature of the packed bed. This suggests that equations governing thermal nonequilibrium forced flow through a packed bed contain a small parameter.

2. MATHEMATICAL FORMULATION

First a problem on thermal nonequilibrium Darcian flow of an incompressible fluid through a packed bed with a uniform porosity is considered. It is assumed that the fluid flow is one-dimensional and occurs with a constant velocity in the longitudinal direction only. The mathematical formulation of the problem includes two volume averaged energy equations for the fluid and solid phases which are given by, respectively,

$$\varepsilon\langle\rho_f\rangle^f(c_p)_f\frac{\partial\langle T_f\rangle^f}{\partial t}+\langle\rho_f\rangle^f(c_p)_f\langle v_f\rangle\cdot\nabla\langle T_f\rangle^f=\lambda_{feff}\nabla^2\langle T_f\rangle^f+h_{sf}a_{sf}\left[\langle T_s\rangle^s-\langle T_f\rangle^f\right] \quad (1)$$

$$(1-\varepsilon)\langle\rho_s\rangle^s(c_p)_s\frac{\partial\langle T_s\rangle^s}{\partial t}=\lambda_{seff}\nabla^2\langle T_s\rangle^s-h_{sf}a_{sf}\left[\langle T_s\rangle^s-\langle T_f\rangle^f\right] \quad (2)$$

where a_{sf} is the specific surface area common to the solid and fluid phases, c_p is the specific heat at constant pressure, h_{sf} is the fluid-to-particle heat transfer coefficient between the solid and fluid phases, t is the time, T is the temperature, $\langle v_f\rangle$ is the superficial average velocity of the fluid phase, ε is the porosity, λ_{feff} and λ_{seff} are the effective thermal conductivities of the fluid and solid phases, respectively, and ρ is the density. Subscripts f and s denote quantities associated with the fluid and solid phase, respectively, operators $\langle\ldots\rangle^f$ and $\langle\ldots\rangle^s$ denote intrinsic phase average of a quantity, and operator $\langle\ldots\rangle$ denotes local volume average of a quantity.

The coefficients λ_{feff} and λ_{seff} can be estimated using the correlations given in Sözen and Vafai [14], namely,

$$\lambda_{feff}=\varepsilon\lambda_f \quad (3a)$$

$$\lambda_{seff}=(1-\varepsilon)\lambda_s \quad (3b)$$

where λ_f and λ_s are thermal conductivities of the fluid and solid phases, respectively.

To find a small parameter in these equations we consider correlations for the heat transfer coefficient between the fluid and solid phases and for the specific surface area of the packed bed. The fluid-to-solid heat transfer coefficient h_{sf} in equations (1) and (2) can be calculated based on correlations established in Dixon and Cresswell [17]:

$$\frac{1}{h_{sf}} = \frac{d}{Nu_{sf}\lambda_f} + \frac{d}{\beta\ \lambda_s} \tag{4}$$

where d is the average particle diameter in the packed bed and $\beta = 10$ if the porous bed particles are of a spherical nature. The fluid-to-solid Nusselt number for $Re_p > 100$ is well correlated by the expression presented in Handley and Heggs [18]:

$$Nu_{sf} = \frac{0.255}{\varepsilon} Pr^{1/3} Re_p^{2/3} \tag{5}$$

where $Nu_{sf} = h_{sf} d / \lambda_f$, $Pr = \mu_f (c_p)_f / \lambda_f$, and $Re_p = \langle \rho_f \rangle^f \langle v_f \rangle d / \mu_f$.

Estimates of Nu_{sf} for low values of Re_p vary between 0.1 and 12.4 based on the data supplied in Miyauchi et al. [19] and in Wakao et al. [20].

It follows from Dullien [21] that the specific surface area of the packed bed for the fluid phase may be expressed in the following form:

$$a_{sf} = \frac{6(1-\varepsilon)}{d} \tag{6}$$

For many practical cases, for example when the average particle diameter d is small, the product $h_{sf}a_{sf}$ takes on large values and therefore it is convenient to introduce a dimensionless small parameter as follows:

$$\delta = \frac{1}{h_{sf}a_{sf}} \frac{\varepsilon\left[\langle \rho_f \rangle^f (c_p)_f\right]^3 \langle v_f \rangle^2}{\left[\varepsilon \langle \rho_f \rangle^f (c_p)_f + (1-\varepsilon)\langle \rho_s \rangle^s (c_p)_s\right](\lambda_{feff} + \lambda_{seff})} \tag{7}$$

To apply the perturbation technique to the set of equations (1)-(2), the following dimensionless variables are introduced:

temperature

$$\Theta = \frac{T_{in} - T}{T_{in} - T_0} \tag{8}$$

where T_{in} is the inlet temperature of the fluid phase and T_0 is the initial temperature of the packed bed,

coordinates

$$\xi_j = \frac{\langle\rho_f\rangle^f (c_p)_f \langle v_f\rangle}{\lambda_{feff} + \lambda_{seff}} x_j \tag{9}$$

time

$$\tau = \frac{\left[\langle\rho_f\rangle^f (c_p)_f \langle v_f\rangle\right]^2}{\left[\varepsilon\langle\rho_f\rangle^f (c_p)_f + (1-\varepsilon)\langle\rho_s\rangle^s (c_p)_s\right](\lambda_{feff} + \lambda_{seff})} t \tag{10}$$

where x_j are Cartesian coordinates (j=1,2,3).

If the product $h_{sf}a_{sf}$ takes on large values, the heat exchange between the fluid and solid phases is very intensive, and the temperature difference between the fluid and solid phases is small. Then the dimensionless solid phase temperature can be represented as

$$\Theta_s = \Theta_f + \delta\Delta\Theta \tag{11}$$

where $\Delta\Theta$ is dimensionless difference between temperatures of the fluid and solid phases.

Utilizing equation (11), the set of equations (1)-(2) can be put into the following form

$$\frac{\partial\Theta_f}{\partial\tau} + \frac{\partial\Theta_f}{\partial\xi_1} = \nabla^2\Theta_f \tag{12}$$

$$\Delta\Theta = (1-\Lambda_2)\frac{\partial\Theta_f}{\partial\tau} + (\Lambda_1 - \Lambda_2)\frac{\partial\Theta_f}{\partial\xi_1} \tag{13}$$

where ξ_1 is a longitudinal coordinate,

$$\Lambda_1 = \frac{\varepsilon\langle\rho_f\rangle^f (c_p)_f + (1-\varepsilon)\langle\rho_s\rangle^s (c_p)_s}{\varepsilon\langle\rho_f\rangle^f (c_p)_f}$$

and

$$\Lambda_2 = \frac{\lambda_f\left[\varepsilon\langle\rho_f\rangle^f(c_p)_f + (1-\varepsilon)\langle\rho_s\rangle^s(c_p)_s\right]}{\varepsilon\langle\rho_f\rangle^f(c_p)_f\left[\lambda_{feff} + \lambda_{seff}\right]}$$

Equation (12) is obtained by combining equations (1) and (2) and equation (13) is equation (1) in the dimensionless form with regard to equation (11). Terms which are $O(\delta)$ are neglected in equations (12) and (13).

Equations (12) and (13) reduce the problem of thermal nonequilibrium forced convection to the solution of the same problem assuming local thermal equilibrium. Equation (12) shows that the fluid phase temperature can be determined from the same energy equation as though the problem was solved under the local thermal equilibrium assumption. Once this problem is solved, the solution for the temperature difference between the fluid and solid phases follows from equation (13).

3. SOLUTIONS FOR DIFFERENT GEOMETRY OF THE PACKED BED: THE PRODUCT SOLUTIONS

Consider the problem of heating a packed bed filled with an incompressible fluid and which is, initially, at a uniform temperature, T_0. In the dimensionless form this initial condition can be expressed as

$$\Theta_f\Big|_{\tau=0} = 1 \tag{14}$$

At the instant τ=0, fluid at a higher temperature, T_{in}, is suddenly allowed to flow through the packed bed and for all the problems considered in this Section, the same inlet and outlet boundary conditions are assumed. At the inlet, a constant temperature of the fluid phase, T_{in}, is assumed. Using dimensionless variables, this condition may be expressed as

$$\Theta_f\Big|_{\xi_1=0} = 0 \tag{15}$$

At the outlet boundary, it is assumed that the temperature of the fluid does not change after the fluid leaves the packed bed and there is no temperature jump at the outlet boundary. This results in the zero temperature gradient at the outlet, $\xi_1 = R_1$, namely,

$$\left.\frac{\partial\Theta_f}{\partial\xi_1}\right|_{\xi_1=R_1} = 0 \tag{16}$$

Solution of this problem for a different geometry of the packed bed and for different boundary conditions at the peripheral walls can be obtained, after some algebraic manipulations, using the separation of variables technique. In this section we consider problems with such boundary conditions at the peripheral walls, when the solution of equation (12) can be represented as the product of the solutions of the pertinent transient one-dimensional problems, see Carslaw and Jaeger [22]. A case when the product solution is impossible is considered in Section 4.

3.1. Three-dimensional packed bed with convection at the peripheral walls

Consider a rectangular parallelepiped ($0 \le \xi_1 \le R_1$, $-R_2 \le \xi_2 \le R_2$, $-R_3 \le \xi_3 \le R_3$) filled with porous medium. The peripheral walls of the parallelepiped ($\xi_2 = \pm R_2$ and $\xi_3 = \pm R_3$) are heated by a fluid which is at the inlet temperature, T_{in}. The boundary conditions at the peripheral walls can be then expressed as

$$\frac{\partial \Theta_f}{\partial \xi_2}(\xi_1, \pm R_2, \xi_3, \tau) = \mp \alpha \Theta_f(\xi_1, \pm R_2, \xi_3, \tau) \tag{17a}$$

$$\frac{\partial \Theta_f}{\partial \xi_3}(\xi_1, \xi_2, \pm R_3, \tau) = \mp \alpha \Theta_f(\xi_1, \xi_2, \pm R_3, \tau) \tag{17b}$$

where $\alpha = \dfrac{h_W}{\langle \rho_f \rangle^f (c_p)_f \langle v_f \rangle}$ is the dimensionless heat transfer coefficient at the peripheral walls, h_W is the heat transfer coefficient at the peripheral walls, $R_j = \dfrac{\langle \rho_f \rangle^f (c_p)_f \langle v_f \rangle}{\lambda_{feff} + \lambda_{seff}} L_j$ are dimensionless dimensions of the packed bed (j=1,2,3), L_1 is the length of the packed bed in the longitudinal (streamwise) direction, and L_2 and L_3 are one half of the width and one half of the height of the packed bed, respectively.

The limiting case of $\alpha \to \infty$ corresponds to the fluid inlet temperature being kept at the peripheral wall temperature and the limiting case of $\alpha \to 0$ corresponds to insulated peripheral walls.

Solutions for the dimensionless temperature of the fluid phase and for the dimensionless temperature difference between the fluid and solid phases are then given by the following equations, see Kuznetsov [23]:

$$\Theta_f = \sum_{n,m,k=1}^{\infty} 4\alpha^2 C_n \exp\left[\frac{\xi_1}{2} - \left(\frac{1}{4} + b_n^2 + \gamma_m^2 + \gamma_k^2\right)\tau\right]$$

$$\times \frac{\sin(b_n \xi_1)\cos(\gamma_m \xi_2)\cos(\gamma_k \xi_3)}{\left[(\alpha^2 + \gamma_m^2)R_2 + \alpha\right]\left[(\alpha^2 + \gamma_k^2)R_3 + \alpha\right]\cos(\gamma_m R_2)\cos(\gamma_k R_3)} \tag{18}$$

and

$$\Delta\Theta = \sum_{n,m,k=1}^{\infty} \frac{4\alpha^2 C_n \exp\left[\frac{\xi_1}{2} - \left(\frac{1}{4} + b_n^2 + \gamma_m^2 + \gamma_k^2\right)\tau\right]\cos(\gamma_m \xi_2)\cos(\gamma_k \xi_3)}{\left[(\alpha^2 + \gamma_m^2)R_2 + \alpha\right]\left[(\alpha^2 + \gamma_k^2)R_3 + \alpha\right]\cos(\gamma_m R_2)\cos(\gamma_k R_3)}$$

$$\times\left\{\left[-\frac{1}{4} + \frac{\Lambda_1}{2} - \frac{\Lambda_2}{4} + (\Lambda_2 - 1)(b_n^2 + \gamma_m^2 + \gamma_k^2)\right]\sin(b_n \xi_1) + (\Lambda_1 - \Lambda_2) b_n \cos(b_n \xi_1)\right\} \quad (19)$$

where the coefficients b_n, C_n, γ_m and γ_k are determined by the following equations

$$C_n = \frac{\dfrac{b_n}{1/4 + b_n^2}}{\dfrac{R_1}{2} - \dfrac{\sin(2b_n R_1)}{4b_n}} \quad (20)$$

b_n are different positive solutions of the following transcendental equation

$$\tan b_n R_1 = -2b_n \quad (21)$$

γ_m are different positive solutions of the following transcendental equation

$$\gamma_m \tan \gamma_m R_2 = \alpha \quad (22a)$$

and γ_k are different positive solutions of the following transcendental equation

$$\gamma_k \tan \gamma_k R_3 = \alpha \quad (22b)$$

It is worth noting that although α^2 is present in the numerators of both equations (18) and (19), for $\alpha \to 0$ the solution given by equations (18) and (19) does not vanish because for this case the denominator in these equations also tends to zero. In the numerical procedure, if the case with insulated peripheral walls has to be studied, then instead of using the zero value of α then a sufficiently small, but non-zero value, has to be utilized in equations (18) and (19).

Figure 1 depicts the space distributions of the dimensionless temperature difference calculated utilizing equation (19) for $R_1 = 4$, $R_2 = 2.5$, $R_3 = 5$, $\alpha = 1$, $\Lambda_1 = 13$ and $\Lambda_2 = 2.6$. These values of Λ_1 and Λ_2 correspond to the following ratios of the thermophysical properties: $(c_p)_f \langle\rho_f\rangle^f = 0.25 (c_p)_s \langle\rho_s\rangle^s$, $\lambda_{feff} = 0.25\lambda_{seff}$ and $\varepsilon = 0.25$. This figure clearly shows the wave-like behaviour of the temperature difference and the temperature difference has a local maximum which moves downstream. The amplitude of the temperature difference decreases as its maximum moves downstream.

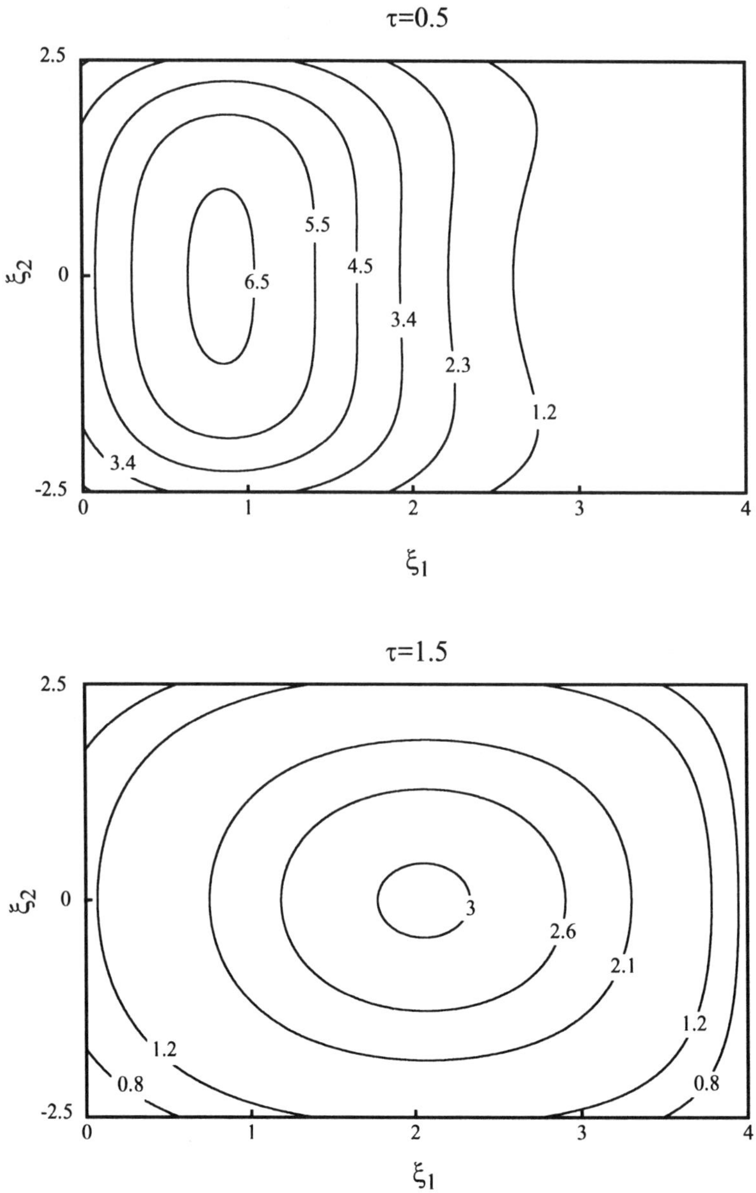

Figure 1 Isopleths of the dimensionless temperature difference between the fluid and solid phases, $\Delta\Theta$, in the three-dimensional packed bed for $\xi_3=0$

3.2. Cylindrical packed bed with convection at the cylindrical surface

Consider a finite circular cylinder ($0 \le \xi_1 \le R_1$, $0 \le \xi_r \le R_o$) filled with a porous medium whose cylindrical surface is heated by a fluid with the inlet temperature, T_{in}. The boundary conditions at the cylindrical surface can be then expressed as

$$\frac{\partial \Theta_f}{\partial \xi_r}(\xi_1, R_o, \tau) = -\alpha \Theta_f(\xi_1, R_o, \tau) \tag{23}$$

where $R_1 = \frac{\langle \rho_f \rangle^f (c_p)_f \langle v_f \rangle}{\lambda_{feff} + \lambda_{seff}} L_1$ is the dimensionless length of the cylinder and $R_o = \frac{\langle \rho_f \rangle^f (c_p)_f \langle v_f \rangle}{\lambda_{feff} + \lambda_{seff}} r_o$ is its dimensionless outer radius. The dimensionless longitudinal coordinate, ξ_1, for this problem is determined according to equation (9) and the dimensionless radial coordinate, ξ_r, is determined as

$$\xi_r = \frac{\langle \rho_f \rangle^f (c_p)_f \langle v_f \rangle}{\lambda_{feff} + \lambda_{seff}} r$$

where r is the radial coordinate.

Solutions for the dimensionless temperature of the fluid phase and for the dimensionless temperature difference between the fluid and solid phases are then given by the following equations, see Kuznetsov [24]:

$$\Theta_f = \sum_{n,m=1}^{\infty} C_n \exp\left[\frac{\xi_1}{2} - \left(\frac{1}{4} + b_n^2 + \frac{\gamma_m^2}{R_o^2}\right)\tau\right] \frac{2\alpha R_o \sin(b_n \xi_1)\, J_0\left(\frac{\xi_r \gamma_m}{R_o}\right)}{(\gamma_m^2 + \alpha^2 R_o^2) J_0(\gamma_m)} \tag{24}$$

and

$$\Delta\Theta = \sum_{n,m=1}^{\infty} \frac{2\alpha R_o\, C_n \exp\left[\frac{\xi_1}{2} - \left(\frac{1}{4} + b_n^2 + \frac{\gamma_m^2}{R_o^2}\right)\tau\right] J_0\left(\frac{\xi_r \gamma_m}{R_o}\right)}{(\gamma_m^2 + \alpha^2 R_o^2) J_0(\gamma_m)} \Bigg\{ \sin(b_n \xi_1)$$

$$\times \left[-\frac{1}{4} - b_n^2 - \frac{\gamma_m^2}{R_o^2} + \frac{\Lambda_1}{2} - \frac{\Lambda_2}{4} + \Lambda_2 b_n^2 + \Lambda_2 \left(\frac{\gamma_m}{R_o}\right)^2\right] + \cos(b_n \xi_1) b_n [\Lambda_1 - \Lambda_2] \Bigg\} \tag{25}$$

where J_ν is Bessel function of the first kind of the order ν, the coefficients b_n and C_n are determined by equations (20) and (21) and the coefficients γ_m are different positive solutions of the following transcendental equation

$$\gamma_m J_1(\gamma_m) = \alpha R_o J_0(\gamma_m) \tag{26}$$

3.3. Hollow cylindrical packed bed with the outer and inner cylindrical surfaces kept at the inlet temperature

Consider a concentric tube annulus ($0 \le \xi_1 \le R_1$, $R_i \le \xi_r \le R_o$) filled with a porous medium whose outer and inner cylindrical surfaces are kept at the fluid inlet temperature, T_{in}. The boundary conditions at the cylindrical surfaces can be then expressed as

$$\Theta_f(\xi_1, R_i, \tau) = 0, \quad \Theta_f(\xi_1, R_o, \tau) = 0 \tag{27}$$

where $R_1 = \dfrac{\langle\rho_f\rangle^f (c_p)_f \langle v_f\rangle}{\lambda_{feff} + \lambda_{seff}} L_1$ is dimensionless length of the annulus, $R_i = \dfrac{\langle\rho_f\rangle^f (c_p)_f \langle v_f\rangle}{\lambda_{feff} + \lambda_{seff}} r_i$ is its dimensionless inner radius, and $R_o = \dfrac{\langle\rho_f\rangle^f (c_p)_f \langle v_f\rangle}{\lambda_{feff} + \lambda_{seff}} r_o$ is its dimensionless outer radius.

Solutions for the dimensionless temperature of the fluid phase and for the dimensionless temperature difference between the fluid and solid phases are then given by the following equations, see Kuznetsov [25]:

$$\Theta_f = \sum_{n,m=1}^{\infty} C_n \pi \exp\left[\frac{\xi_1}{2} - \left(\frac{1}{4} + b_n^2 + \gamma_m^2\right)\tau\right] \frac{J_0(R_i\gamma_m)\sin(b_n\xi_x)U_0(\xi_r\gamma_m)}{J_0(R_i\gamma_m) + J_0(R_o\gamma_m)} \tag{28}$$

and

$$\Delta\Theta = \sum_{n,m=1}^{\infty} \frac{C_n \pi \exp\left[\frac{\xi_1}{2} - \left(\frac{1}{4} + b_n^2 + \gamma_m^2\right)\tau\right] J_0(R_i\gamma_m) U_0(\xi_r\gamma_m)}{J_0(R_i\gamma_m) + J_0(R_o\gamma_m)}$$

$$\times\left\{\sin(b_n\xi_1)\left[-\frac{1}{4} - b_n^2 - \gamma_m^2 + \frac{\Lambda_1}{2} - \frac{\Lambda_2}{4} + \Lambda_2 b_n^2 + \Lambda_2\gamma_m^2\right] + \cos(b_n\xi_1) b_n[\Lambda_1 - \Lambda_2]\right\} \tag{29}$$

where

$$U_0(\xi_r\gamma_m) = J_0(\xi_r\gamma_m)Y_0(\gamma_m R_o) - J_0(\gamma_m R_o)Y_0(\xi_r\gamma_m)$$

and Y_ν is the Bessel function of the second kind of the order ν, the coefficients b_n and C_n are determined by equations (20) and (21) and the coefficients γ_m are different positive solutions of the following transcendental equation

$$J_0(\gamma_m R_i)Y_0(\gamma_m R_o)-J_0(\gamma_m R_o)Y_0(\gamma_m R_i)=0 \tag{30}$$

4. RECTANGULAR PACKED BED WITH THE PERIPHERAL WALLS KEPT AT AN ARBITRARY CONSTANT TEMPERATURE

Consider a two-dimensional rectangular packed bed ($0\le\xi_1\le R_1$, $0\le\xi_2\le R_2$) whose peripheral walls are kept at an arbitrary constant temperature, T_W. For this problem the dimensionless fluid temperature previously given by equation (8) is redefined as

$$\Theta=\frac{T-T_W}{T_{in}-T_W} \tag{31}$$

Equations (7), (9)-(13) remain unchanged. The initial condition for this problem is given by

$$\Theta_f(\xi_1,\xi_2,0)=\Theta_0 \tag{32}$$

where $\Theta_0=\dfrac{T_0-T_W}{T_{in}-T_W}$ and can be either positive or negative or zero.

Boundary conditions at the inlet and at the outlet are given by the following equations:

$$\Theta_f(0,\xi_2,\tau)=1 \tag{33}$$

$$\frac{\partial\Theta_f}{\partial\xi_1}(R_1,\xi_2,\tau)=0 \tag{34}$$

The boundary conditions at the peripheral walls for this problem can be then expressed as

$$\Theta_f(\xi_1,0,\tau)=0 \tag{35}$$

$$\Theta_f(\xi_1,R_2,\tau)=0 \tag{36}$$

where $R_1=\dfrac{\rho_f(c_p)_f\langle v_f\rangle}{\lambda_{feff}+\lambda_{seff}}L_1$ and $R_2=\dfrac{\rho_f(c_p)_f\langle v_f\rangle}{\lambda_{feff}+\lambda_{seff}}L_2$ are dimensionless length and width of the packed bed, respectively.

The solution of equation (12), with the initial condition as given in equation (32) and the boundary conditions of equations (33)-(36), cannot be obtained simply as a product of the solutions of the pertinent one-dimensional problems. As recommended in Carslaw and Jaeger [22], this problem was reduced in Kuznetsov [26] to two simpler problems, namely a problem with a steady temperature and a problem with a zero surface temperature/heat flux. After considerable algebraic manipulations, solutions for the dimensionless temperature of the fluid phase and for the dimensionless temperature difference between the fluid and solid phases were obtained as

$$\Theta_f(\xi_1,\xi_2,\tau)=u(\xi_1,\xi_2)+w(\xi_1,\xi_2,\tau) \tag{37}$$

and

$$\Delta\Theta=(\Lambda_1-\Lambda_2)\left[\frac{u}{2}+\sum_{n=1}^{\infty}b_n C_n \exp\left(\frac{\xi_1}{2}\right)\sin\left(\frac{\pi n}{R_2}\xi_2\right)\right.$$

$$\left.\times\left\{\left(\frac{1}{2}+b_n\right)\exp\left[b_n(R_1-\xi_1)\right]+\left(\frac{1}{2}-b_n\right)\exp\left[-b_n(R_1-\xi_1)\right]\right\}\right]-(1-\Lambda_2)$$

$$\times\sum_{n,m=1}^{\infty}\left[\frac{1}{4}+a_m^2+\left(\frac{\pi n}{R_2}\right)^2\right]D_{n,m}\exp\left\{\frac{\xi_1}{2}-\left[\frac{1}{4}+a_m^2+\left(\frac{\pi n}{R_2}\right)^2\right]\tau\right\}\sin(a_m\xi_1)\sin\left(\frac{\pi n}{R_2}\xi_2\right)$$

$$+(\Lambda_1-\Lambda_2)\left[\frac{w}{2}+\sum_{n,m=1}^{\infty}a_m D_{n,m}\exp\left\{\frac{\xi_1}{2}-\left[\frac{1}{4}+a_m^2+\left(\frac{\pi n}{R_2}\right)^2\right]\tau\right\}\cos(a_m\xi_1)\sin\left(\frac{\pi n}{R_2}\xi_2\right)\right] \tag{38}$$

where

$$u=\sum_{n=1}^{\infty}C_n\exp\left(\frac{\xi_1}{2}\right)\sin\left(\frac{\pi n}{R_2}\xi_2\right)$$

$$\times\left[-\left(\frac{1}{2}+b_n\right)\exp\left[b_n(R_1-\xi_1)\right]+\left(\frac{1}{2}-b_n\right)\exp\left[-b_n(R_1-\xi_1)\right]\right] \tag{39}$$

$$w=\sum_{n,m=1}^{\infty}D_{n,m}\exp\left\{\frac{\xi_1}{2}-\left[\frac{1}{4}+a_m^2+\left(\frac{\pi n}{R_2}\right)^2\right]\tau\right\}\sin(a_m\xi_1)\sin\left(\frac{\pi n}{R_2}\xi_2\right) \tag{40}$$

$$b_n=\left[\frac{1}{4}+\left(\frac{\pi n}{R_2}\right)^2\right]^{1/2} \tag{41}$$

$$C_n = \frac{2\left[1-(-1)^n\right]}{\pi n\left[-\left(\frac{1}{2}+b_n\right)\exp(b_n R_1)+\left(\frac{1}{2}-b_n\right)\exp(-b_n R_1)\right]} \tag{42}$$

$$D_{n,m} = \frac{\left[1-(-1)^n\right]\left\{\Theta_0 \frac{R_2}{\pi n}\frac{4a_m}{1+4a_m^2} - \frac{a_m L}{\pi n\left(a_m^2+b_n^2\right)}\right\}}{\frac{R_2}{2}\left[\frac{R_1}{2}-\frac{\sin(2a_m R_1)}{4a_m}\right]} \tag{43}$$

and a_m are different positive solutions of the following transcendental equation

$$\tan(a_m R_1) = -2a_m \tag{44}$$

Figure 2 depicts the space distribution of the dimensionless temperature difference calculated according to equation (38) for $\Theta_0 = -3$, $R_1 = 4$, $R_2 = 5$, $\Lambda_1 = 13$ and $\Lambda_2 = 2.6$. Since for this problem the inlet temperature differs from the temperature kept at the peripheral walls of the packed bed, there is a sharp increase in the temperature difference at the peripheral walls near the inlet boundary. The solution given by equations (37)-(44) is obtained utilizing the perturbation technique and is valid everywhere except in a small neighborhood of the inlet boundary, where the assumption that the temperature difference takes on small values fails. It can be seen that the temperature difference consists of the steady and transient components. The steady component comes from the steady solution for the function u and describes the temperature difference at $\tau = \infty$. In contrast to the results presented in Figure 1, the temperature difference for this problem does not approach zero as the time approaches infinity. This is because in this case the fluid inlet temperature differs from the temperature at the walls of the packed bed. The steady component is localized near the inlet boundary while the transient component comes from the transient solution for the function w and describes a wave propagating in the ξ_1-direction from the fluid inlet boundary. As in Figure 1, the amplitude of this wave quickly decreases as the wave propagates downstream.

5. THERMAL NONEQUILIBRIUM, NON-DARCIAN FORCED CONVECTION UNDER STEADY CONDITIONS

For many industrial applications of porous media it is important to account for non-Darcian effects which can be divided into inertial and boundary effects. For this reason, in recent years it has become very popular to use the Brinkman-Forchheimer-extended Darcy equation as the momentum equation for a flow in a porous medium. Because the Brinkman-Forchheimer-extended Darcy equation is a non-linear equation, there are only few analytical solutions available for this flow model. An important contribution to modeling forced convection in situations when Darcy's law is inapplicable was made in Vafai and Kim [27] and in Vafai and Thiyagaraja [28]. These references present analytical solutions for a fully developed steady flow at the interface region of a porous medium. The problem of forced convection in a

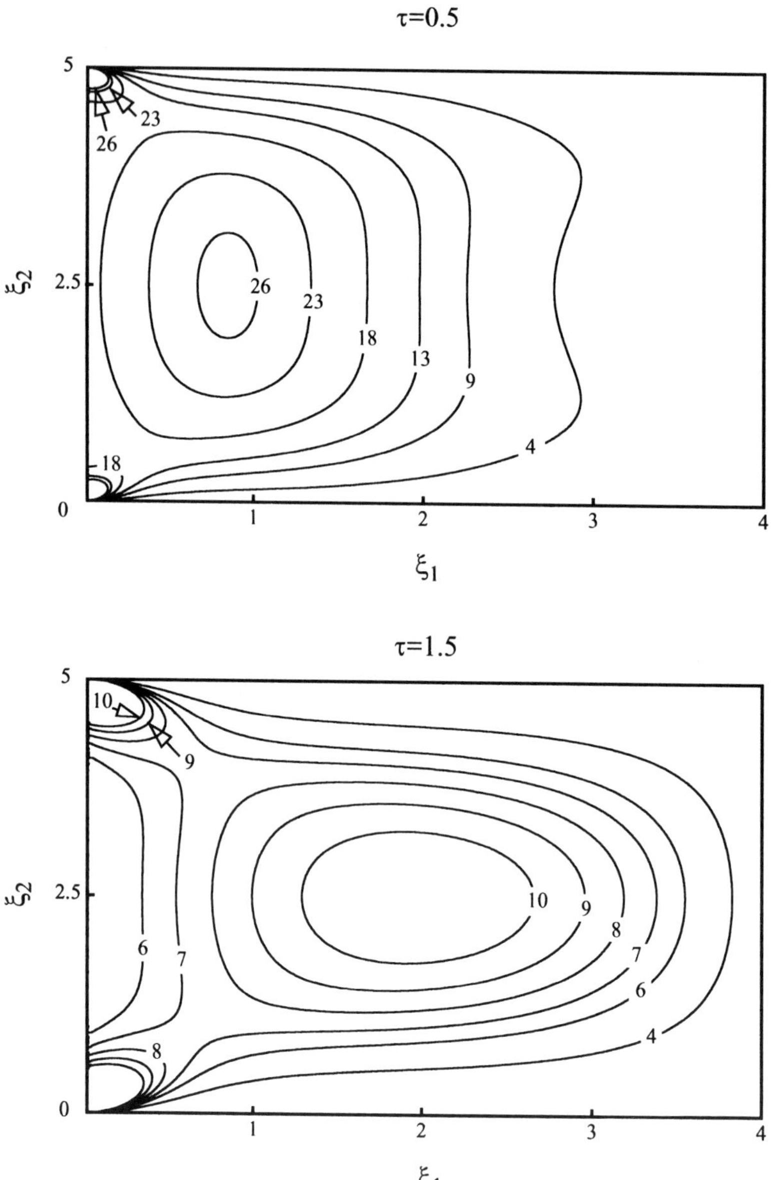

Figure 2 Isopleths of the dimensionless temperature difference between the fluid and solid phases, $\Delta\Theta$, in the two-dimensional packed bed

channel filled with a fluid saturated porous medium for the situation when the fluid flow in a channel is described by the Brinkman-Forchheimer-extended Darcy equation was first considered in Cheng et al. [29]. Another step forward in this problem was made in Vafai and Kim [30]. Considering a fully developed region, and utilizing the boundary layer approximation, Vafai and Kim [30] obtained an elegant solution for both the fluid velocity and temperature distributions in a channel. Comparison of this solution with a full numerical solution was presented in Vafai and Kim [31], where some typos made in Vafai and Kim [30] were also corrected. Very recently, Nield et al. [32] have presented a solution to the same problem but without using the boundary layer approximation.

In refs. [27-32] the assumption of local thermal equilibrium between the fluid and solid phases is utilized. In addition to accounting for the inertial and boundary effects by utilizing the Brinkman-Forchheimer-extended Darcy model, Kuznetsov [33] also takes into account the local thermal nonequilibrium effects. In this reference a channel which is bounded by two parallel plates is considered. The width of the channel is 2H ($-H \le x_2 \le H$), the channel is filled with a fluid saturated porous medium of uniform porosity and there is a constant heat flux, q_W, at its boundaries.

A steady forced convection problem in the fully developed region of this channel is analyzed. Since non-Darcian effects are taken into account, and in contrast to the problems considered in Sections 3 and 4, the seepage velocity, $\langle v_f \rangle$, is not a constant but is a function of the transverse coordinate, x_2. The governing equations for this problem can be presented as

$$-\frac{d\langle p_f \rangle^f}{dx_1} + \mu_{eff} \frac{d^2 \langle v_f \rangle}{dx_2^2} - \frac{\mu_f}{K} \langle v_f \rangle - \frac{\langle \rho_f \rangle^f \varepsilon F}{K^{1/2}} \langle v_f \rangle^2 = 0 \tag{45}$$

$$\langle \rho_f \rangle^f (c_p)_f \langle v_f \rangle \frac{\partial \langle T_f \rangle^f}{\partial x_1} = \lambda_{feff} \frac{\partial^2 \langle T_f \rangle^f}{\partial x_2^2} + h_{sf} a_{sf} \left[\langle T_s \rangle^s - \langle T_f \rangle^f \right] \tag{46}$$

$$0 = \lambda_{seff} \frac{\partial^2 \langle T_s \rangle^s}{\partial x_2^2} - h_{sf} a_{sf} \left[\langle T_s \rangle^s - \langle T_f \rangle^f \right] \tag{47}$$

where F is the Forchheimer coefficient, K is the permeability of the porous medium, $\langle p_f \rangle^f$ is the intrinsic average pressure, x_1 is the streamwise coordinate, x_2 is the transverse coordinate, μ_f is the fluid dynamic viscosity, and μ_{eff} is the effective viscosity in the Brinkman term.

Equation (45) is a momentum equation (the Brinkman-Forchheimer-extended Darcy equation) and equations (46) and (47) are the energy equations for the fluid and solid phases, respectively. In equations (46) and (47) it is assumed that the longitudinal heat conduction term is negligible and at the boundaries ($x_2 = \pm H$) the no-slip condition and a constant heat flux are utilized. Following Vafai and Kim [30], the effective viscosity in the Brinkman term, μ_{eff}, is estimated as

$$\mu_{eff} = \mu_f / \varepsilon \tag{48}$$

As in Sections 2-4, it is assumed that the solid phase temperature differs from the fluid phase temperature by a small perturbation. Then applying a perturbation approach similar to that considered above it was shown in Kuznetsov [33] that the fluid phase temperature can be determined from the same energy equation as if the problem were solved under the local thermal equilibrium assumption

$$\langle\rho_f\rangle^f (c_p)_f \langle v_f\rangle \frac{\partial\langle T_f\rangle^f}{\partial x_1} = (\lambda_{feff} + \lambda_{seff}) \frac{\partial^2\langle T_f\rangle^f}{\partial x_2^2} \tag{49}$$

The temperature difference between the fluid and solid phases is then determined by the following equation

$$\langle T_s\rangle^s - \langle T_f\rangle^f = \frac{q_w}{h_{sf}a_{sf}} \frac{\lambda_{seff}}{\lambda_{feff} + \lambda_{seff}} \frac{1}{H} \frac{\langle v_f\rangle}{v_m} \tag{50}$$

where

$$v_m = \frac{1}{H}\int_0^H \langle v_f\rangle dx_2 \tag{51}$$

is the mean velocity in the channel.

By applying a perturbation technique, the problem of thermal nonequilibrium forced convection through a channel with isoflux boundaries is reduced to the solution of the same problem assuming that there is local thermal equilibrium given by equations (45) and (49). Once this problem is solved, the solution for the temperature difference between the fluid and the solid phases follows from equation (50). It worth noting that equation (50) has a very clear physical meaning: it shows that the temperature difference is proportional to the ratio of the fluid flow velocity to the mean velocity.

The solutions obtained assuming that there is local thermal equilibrium are given in Vafai and Kim [30-31] and in Nield et al. [32]. In these papers the equations for both the fluid velocity and the temperature distributions are obtained. According to Vafai and Kim [30-31], the velocity distribution is given by the following equation

$$\frac{\langle v_f\rangle}{v_\infty} = 1 - \frac{A+B}{A}\operatorname{sech}^2\left[D\left(\frac{x_2}{H} + C_1\right)\right] \tag{52}$$

where v_∞ is the fluid velocity outside the momentum boundary layer and

$$A = \frac{2}{3}\Lambda_3 \varepsilon^{1/2} Da^{-1/2}, \qquad B = \varepsilon Da^{-1} + \frac{4}{3}\Lambda_3 \varepsilon^{1/2} Da^{-1/2}, \qquad D = \frac{\sqrt{A+B}}{2},$$
$$\Lambda_3 = \varepsilon^{3/2} F \frac{v_\infty H}{v_f}, \qquad C_1 = -\frac{1}{D}\operatorname{arcsech}\sqrt{\frac{A}{A+B}} - 1 \tag{53}$$

and the Darcy number is defined as $Da = \frac{K}{H^2}$.

Outside the momentum boundary layer the Brinkman (viscous) term in equation (45) is negligible and equation (45) reduces to the following equation

$$-\frac{d\langle p_f\rangle^f}{dx_1} = \frac{\mu_f}{K} v_\infty + \frac{\langle \rho_f\rangle^f \varepsilon F}{K^{1/2}} v_\infty^2 \tag{54}$$

which is a quadratic equation for the fluid velocity outside the momentum boundary layer, v_∞.

From equations (50) and (52) it follows that

$$\langle T_s\rangle^s - \langle T_f\rangle^f = \frac{q_w}{h_{sf} a_{sf}} \frac{\lambda_{seff}}{\lambda_{feff} + \lambda_{seff}} \frac{1}{H} \frac{1 - \frac{A+B}{A} \operatorname{sech}^2\left[D\left(\frac{x_2}{H} + C_1\right)\right]}{1 - \frac{A+B}{A} \frac{\tanh[D(1+C_1)] - \tanh[DC_1]}{D}} \tag{55}$$

The solution presented in Nield et al. [32] does not utilize the boundary layer approximation and is of a much more complicated form than the solution obtained by Vafai and Kim [30-31] and it requires numerical integration and the numerical solution of a transcendental equation. According to Nield et al. [32], the $\tilde{x}_2$-dependence of the dimensionless velocity, $\tilde{v}$, can be obtained from the following equation

$$\left(\frac{2\tilde{F}\mu_f}{3\mu_{eff}}\right)^{1/2} \tilde{x}_2 = \int_{\tilde{v}}^{b_2} \left[(t-b_1)(t-b_2)(t-b_3)\right]^{-1/2} d\vartheta \tag{56}$$

where $\tilde{F} = -F \frac{d\langle p_f\rangle^f}{dx_1} \frac{\langle \rho_f\rangle^f H^4}{K^{1/2}\mu_f^2}$ is a scaled Forchheimer coefficient, $\tilde{v} = -\frac{\mu_f \langle v_f\rangle}{H^2 d\langle p_f\rangle^f / dx_1}$ is the dimensionless velocity, $\tilde{x}_2 = \frac{x_2}{H}$ is the dimensionless transverse coordinate, and the values of the constants b_1 and b_3 are given in terms of b_2 by the following equations:

$$b_1 = \frac{1}{4\tilde{F}}\left\{-\frac{3}{Da} - 2\tilde{F}b_2 - \left[\frac{9}{Da^2} + 48\tilde{F} - \frac{12\tilde{F}b_2}{Da} - 12\tilde{F}^2 b_2^2\right]^{1/2}\right\} \tag{57}$$

$$b_3 = \frac{1}{4\tilde{F}}\left\{-\frac{3}{Da} - 2\tilde{F}b_2 + \left[\frac{9}{Da^2} + 48\tilde{F} - \frac{12\tilde{F}b_2}{Da} - 12\tilde{F}^2 b_2^2\right]^{1/2}\right\} \tag{58}$$

and the value of the constant b_2 can be found from the following equation

$$\left(\frac{2\tilde{F}\mu_f}{3\mu_{eff}}\right)^{1/2} = \int_0^{b_2}\left[(t-b_1)(t-b_2)(t-b_3)\right]^{-1/2} d\vartheta \tag{59}$$

The temperature difference, $\langle T_s\rangle^s - \langle T_f\rangle^f$, can now be determined by substituting the velocity function found from equation (56) into equation (50).

A comparison between the solutions for the temperature difference, based on the results obtained in Vafai and Kim [30-31] and in Nield et al. [32], was carried out in Kuznetsov [33]. This comparison has shown that for small values of the Darcy number, which is the case of most practical porous media, the boundary layer approximation utilized in Vafai and Kim [30-31] is valid. If this is the situation, the temperature difference between the fluid and the solid phases is given by equation (55). However, for materials with very high permeabilities, the value of the Darcy number can be close to, or even larger than, unity and examples of such materials are given in Nield et al. [32]. For those materials it is necessary to use the velocity distribution in equation (50) which follows from the solution obtained in Nield et al. [32] and given by equations (56)-(59).

6. OPTIMIZATION PROBLEMS FOR THERMAL NONEQUILIBRIUM FORCED CONVECTION: AN APPLICATION OF THE MINIMUM PRINCIPLE OF PONTRYAGIN

Optimization problems in heat and mass transfer have recently received considerable attention, see refs. [34-37]. This is because these problems are interesting from the fundamental point of view and also they are relevant to many practical applications.

In this section we consider the application of the minimum principle of Pontryagin to the solution of two optimal control problems for a one-dimensional porous packed bed heated/cooled by a flow of an incompressible fluid. For these two problems, the Schumann model of the packed bed which neglects the longitudinal conduction terms in both the fluid and the solid phase energy equations is adopted. Under these assumptions, it follows from Carslaw and Jaeger [22] that the equations governing the solid and fluid temperature distributions can be expressed in the following nondimensional form:

$$\frac{\partial\theta}{\partial t} = \phi - \theta \tag{60}$$

$$\Lambda\frac{\partial\phi}{\partial t} + \frac{\partial\phi}{\partial z} = \theta - \phi \tag{61}$$

where

$$\Lambda = \frac{\varepsilon\langle\rho_f\rangle^f (c_p)_f}{(1-\varepsilon)\langle\rho_s\rangle^s (c_p)_s} \tag{62}$$

Here the dimensionless temperature of the solid phase is defined as

$$\theta(z,t)=\frac{\langle T_s\rangle^s-T_1}{T_2-T_1} \tag{63}$$

and the dimensionless temperature of the fluid phase is defined as

$$\phi(z,t)=\frac{\langle T_f\rangle^f-T_1}{T_2-T_1} \tag{64}$$

where T_1 and T_2 are reference temperatures chosen to suitably normalize the initial and boundary conditions.

The dimensionless time and coordinate in equations (60) and (61) are defined as follows:

$$t=\frac{h_{sf}\,a_{sf}\,t'}{(1-\varepsilon)\langle\rho_s\rangle^s(c_p)_s} \tag{65}$$

and

$$z=\frac{h_{sf}\,a_{sf}\,z'}{\varepsilon\langle\rho_f\rangle^f(c_f)_f\langle v_f\rangle} \tag{66}$$

where t' and z' are the time and the streamwise Cartesian coordinate, respectively.

An exact solution of equations (60)-(61) for the case when the entrance fluid temperature is a function of time and the initial temperature of the packed bed is a function of the space variable was obtained in White and Korpela [7] and in Spiga and Spiga [8]. Here we proceed from the solution obtained in White and Korpela [7] where the following initial and boundary conditions are utilized

$$\theta(z,0)=\theta_0(0) \tag{67}$$

$$\phi(0,t)=\phi_{in}(t) \tag{68}$$

Upon a simple rearrangement, the solution obtained in White and Korpela [7] can be put into the following form:

$$\theta(z,t)=\exp(-z)\int_0^{t-\Lambda z}\phi_{in}(t-\Lambda z-\tau)\exp(-\tau)I_0\left[(4\tau z)^{1/2}\right]d\tau$$

$$+\exp(\Lambda z-t)\left[\theta_0(z)+\int_0^{z}\theta_0(z-\xi)\exp(-\xi)\left(\frac{t-\Lambda z}{\xi}\right)^{1/2}I_1\left[\{4\xi(t-\Lambda z)\}^{1/2}\right]d\xi\right] \tag{69}$$

$$\phi(z,t)=\exp(\Lambda z-t)\int_0^z \theta_0(z-\xi)\exp(-\xi)I_0\left[\{4\xi(t-\Lambda z)\}^{1/2}\right]d\xi$$

$$+\exp(-z)\left[\phi_{in}(t-\Lambda z)+\int_0^{t-\Lambda z}\phi_{in}(t-\Lambda z-\tau)\exp(-\tau)\left(\frac{z}{\tau}\right)^{1/2}I_1\left[(4\tau z)^{1/2}\right]d\tau\right] \quad (70)$$

Equations (69)-(70) determine the temperature of the solid and the fluid phases at a particular point in the porous bed with the position z', or the corresponding dimensionless coordinate z, after this point is reached by the temperature front moving from the fluid inlet boundary with a velocity v, i.e. when $t' \geq z'/\langle v_f \rangle$. In dimensionless coordinates this condition is $t \geq \Lambda z$. Because the thermal conductivity within both the solid and the fluid phases is neglected, for $t < \Lambda z$ the temperature of the solid phase at this point equals the initial temperature determined by the function $\theta_0(z)$ in equation (67).

6.1. Optimization of the inlet temperature

In the first problem we consider that case when the inlet fluid flow temperature is taken to be the control and the amount of heat energy stored in the slab is utilized as the optimization criterion. It is necessary to maximize this quantity under the following constraints: (a) a given amount of heat can be supplied by the incoming fluid flow, (b) there is a given duration of the process, t_f, and (c) the inlet fluid flow temperature is subject to a variation between a given minimum value, u_{min}, and a given maximum value, u_{max}. The minimum value corresponds to the fluid temperature in the "cold tap" and the maximum value corresponds to the fluid temperature in the "hot tap".

A one-dimensional porous slab of the length L' is considered and then the dimensionless length of the slab is defined as

$$L=\frac{h_{sf}\,a_{sf}\,L'}{\varepsilon\langle\rho_f\rangle^f(c_p)_f\langle v_f\rangle} \quad (71)$$

It is assumed that the initial temperature of the slab is uniform and constant. In this case it is always possible to select the reference temperatures, T_1 and T_2, in equations (63) and (64) so that the dimensionless initial temperature of the solid phase is zero. This essentially simplifies equations (69) and (70), because in this case the second term on the right-hand side of equation (69), and the first term on the right-hand side of equation (70), are zero.

It is assumed that the dimensionless fluid temperature at the entrance to the porous slab is given by some function of time, $\phi_{in}(t)$. Since there is a temperature difference between the fluid and the solid phases, the outlet fluid temperature is higher than the temperature of the solid phase at the fluid outlet. In other words, a part of the heat that could be stored in the porous slab leaves the slab with the fluid flow. To increase the efficiency of the heat storage in the packed bed it is important to find a way to minimize this loss of heat energy.

The mathematical formulation of this problem is as follows. It is necessary to determine the optimal inlet temperature $\hat{\phi}_{in}(t)$ that maximizes the following performance functional

$$\Phi(\phi_{in}) = \int_0^L \theta(z,t_f)dz \rightarrow \max \tag{72}$$

where the function $\theta(z,t_f)$ is determined by equation (69), under the following constraints:

$$\int_0^{t_f} \phi_{in}(\tau)d\tau = E = \text{constant} \tag{73}$$

and

$$u_{min} \le \phi_{in}(t) \le u_{max} \tag{74}$$

To put the problem, given by equations (72)-(74), in the form of an optimal-control problem it is necessary to rearrange the functional (72). To accomplish this, equation (69) for the function $\theta(z,t)$ is first rearranged by the following change of the integration variable

$$\tau^* = t_f - \tau - \Lambda z \tag{75}$$

Then, accounting for the assumption that the initial temperature of the slab is uniform, equation (69) at the time $t = t_f$ can be written as

$$\theta(z,t_f) = \exp(-z)\int_0^{t_f-\Lambda z} \phi_{in}(\tau^*)\exp(-t_f + \tau^* + \Lambda z)I_0\left[\left\{4(t_f - \tau^* - \Lambda z)z\right\}^{1/2}\right]d\tau^* \tag{76}$$

For $t_f < \Lambda L$, the temperature front has not yet reached the outlet boundary and no heat has yet been lost to the fluid leaving the slab. Therefore we consider only the case $t_f \ge \Lambda L$.

Consider the function

$$\Psi(z,\tau) = \begin{cases} \exp(-z - t_f + \tau + \Lambda z)I_0\left[\left\{4(t_f - \tau - \Lambda z)z\right\}^{1/2}\right] & \text{if } 0 \le \tau \le t_f - \Lambda z \\ 0 & \text{if } \tau > t_f - \Lambda z \end{cases} \tag{77}$$

Then, utilizing equations (76) and (77), and changing the order of the integration, equation (72) can be recast as

$$\Phi(\phi_{in}) = \int_0^L \theta(z,t_f)dz = \int_0^{t_f} \phi_{in}(\tau)\Xi(\tau)d\tau \rightarrow \max \tag{78}$$

where

$$\Xi(\tau) = \int_0^L \Psi(z,\tau)dz \tag{79}$$

The problem given by equations (73), (74) and (78) is an optimal-control problem and it has been solved in Kuznetsov [38] utilizing the minimum principle of Pontryagin [39-40]. Application of this principle leads to the following requirement

$$\hat{\phi}_{in}(t)[\lambda_1 - \Xi(t)] \rightarrow \min \tag{80}$$

where λ_1 is the Lagrange multiplier.

Equation (80), when applied to the constraint (74), makes it possible to determine the optimal temperature, $\hat{\phi}_{in}(t)$, as

$$\begin{aligned} \hat{\phi}_{in}(t) &= u_{min} \ \text{ if } \ \lambda_1 - \Xi(t) > 0 \\ \hat{\phi}_{in}(t) &= u_{max} \ \text{ if } \ \lambda_1 - \Xi(t) < 0 \end{aligned} \tag{81}$$

To make use of equations (81) it is necessary to calculate the value of the Lagrange multiplier, λ_1. To do this, the transcendental equation (73) needs to be solved taking into account equation (81). To solve this problem, we first selected a segment that unequivocally contains the desired value of λ_1 and then an algorithm for finding a root of a transcendental equation on a given segment was applied to equation (73).

6.2. Optimization of the initial temperature distribution - rate of "forgetting" the initial conditions

The second problem we have considered is that of a porous packed bed which is cooled by the flow of an incompressible fluid. The initial temperature of the packed bed is considered as the control and the amount of heat transferred to the fluid phase is utilized as the optimization criterion. It is necessary to maximize this quantity under the following constraints: (a) a given amount of heat is initially stored in the packed bed, (b) there is a given duration of the process, t_f, and (c) the initial temperature is subject to a variation between a given minimum value, u_{min}, and a given maximum value, u_{max}.

Again, a one-dimensional porous slab of the length L' is considered and the dimensionless length of the slab is then defined by equation (71) and it is assumed that the inlet fluid phase temperature is constant during the process. In this case it is always possible to select the reference temperatures, T_1 and T_2, in equations (63) and (64), so that the dimensionless inlet temperature is zero. This essentially simplifies equations (69) and (70), because in this case the first term on the right-hand side of equation (69) and the second term on the right-hand side of equation (70) are zero.

The mathematical formulation of this problem is as follows. It is necessary to determine the optimal initial temperature distribution $\hat{\theta}_0(z)$ that maximizes the following functional

$$\Phi(\theta_0) = \int_{\Lambda L}^{t_f} \phi(L,t)\,dt \;\rightarrow\; \max \tag{82}$$

where the function $\phi(L,t)$ is determined by equation (70), under the following constraints:

$$\int_0^L \theta_0(\xi)\,d\xi = E = \text{constant} \tag{83}$$

and

$$u_{min} \leq \theta_0(z) \leq u_{max} \tag{84}$$

The lower integration limit in equation (82) corresponds to the instant of time when the fluid front reaches the outlet boundary, and the upper limit corresponds to the duration of cooling. We consider the case when $t_f > \Lambda L$.

The problem given by equations (82)-(84) was rearranged in Kuznetsov [41] in a similar manner to that presented in Section 6.1 and it was transformed to an optimal control problem. It was then solved in Kuznetsov [41] by the minimum principle of Pontryagin which leads to the following requirement

$$\hat{\theta}_0(z)[\lambda_1 - \Psi(z)] \;\rightarrow\; \min \tag{85}$$

where

$$\Psi(z) = \exp(\Lambda L + z - L)\int_{\Lambda L}^{t_f} \exp(-t)\, I_0\left[\{4(L-z)(t-\Lambda L)\}^{1/2}\right]dt \tag{86}$$

and λ_1 is the Lagrange multiplier.

Equation (85), when applied to the constraint (84), makes it possible to determine the optimal control, $\hat{\theta}_0(z)$, as

$$\begin{aligned} \hat{\theta}_0(z) &= u_{min} \quad \text{if} \quad \lambda_1 - \Psi(z) > 0 \\ \hat{\theta}_0(z) &= u_{max} \quad \text{if} \quad \lambda_1 - \Psi(z) < 0 \end{aligned} \tag{87}$$

which requires the calculation of the value of the Lagrange multiplier, λ_1. To do this, the transcendental equation (83) needs to be solved taking into account equation (87) and to solve this problem we first took a segment that unequivocally contains the desired value of λ_1. Then an algorithm for finding a root of a transcendental equation on a given segment was applied to equation (83).

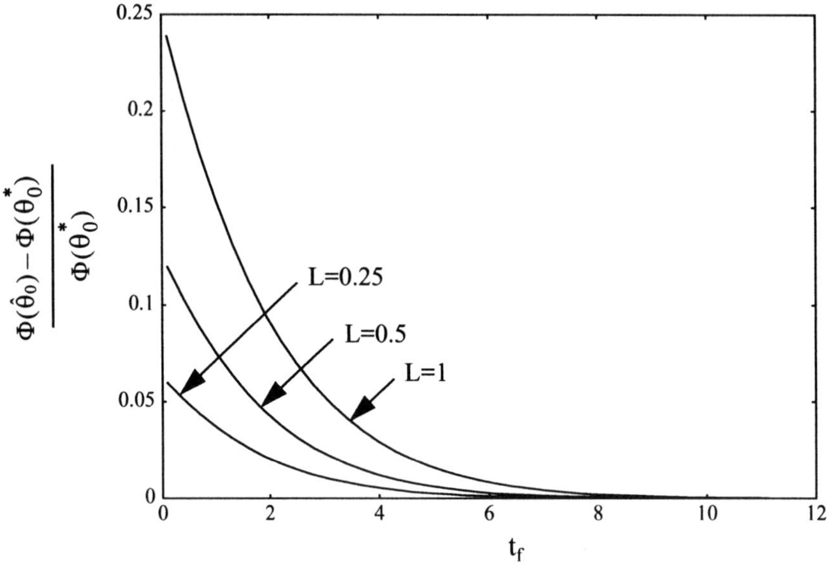

Figure 3 The time dependence on the gain in the amount of heat energy transfered to the fluid phase when the optimal initial distribution is utilized instead of the constant initial temperature of the duration of the process

It is interesting to compare the value of the performance functional $\Phi(\theta_0)$ at the optimal functions determined by equations (87) and (83) and at the functions $\theta_0^*(z) \equiv E/L$ =constant. These functions, $\theta_0^*(z)$, correspond to a constant uniform initial temperature distribution in the slab and clearly satisfy constraint (83). The ratio $\left[\Phi(\hat{\theta}_0)-\Phi(\theta_0^*)\right]/\Phi(\theta_0^*)$ characterizes the gain in the amount of the heat energy transferred to the fluid phase when the optimal initial temperature distribution is utilized, instead of the constant uniform initial temperature. The dependence of this ratio on the duration of the process for $u_{min}=0$, $u_{max}=1$, $E=L/2$, $\Lambda=0.05$ is depicted in Figure 3. It can be seen that the optimal initial temperature distribution makes it possible to considerably improve the discharging characteristics of the packed bed (up to approximately 25% for L=1) for short durations of the process. The influence of the initial temperature on the discharging efficiency quickly decreases with an increase of the duration and for $t_f > 10$ it becomes so small that it is hardly visible in Figure 3. This conclusion agrees with the well known property of heat transfer processes to forget initial conditions, see Prigogine [42]. According to Figure 3, the longer is the duration of the process, the smaller the amount of heat transferred to the fluid phase by $t = t_f$ is influenced by the choice of the initial temperature distribution in the packed bed. For $t_f > 10$ this quantity, which can be interpreted as a characteristic of a final state of the process, is practically independent on whether the optimal or the uniform temperature distribution is utilized as the initial condition. The rate at which the curves depicted in Figure 3 tends to zero characterizes the rate at which the process forgets its initial condition.

ACKNOWLEDGMENT

The support provided by the Christian Doppler Laboratory For Continuous Solidification Processes and by the AvHumboldt Foundation is gratefully acknowledged and appreciated. The author is also grateful to Prof. W.Schneider whose support and advice made this work possible. Constructive comments by Prof. D.A.Nield are also appreciated.

REFERENCES

1. T.E.W.Schumann, Heat transfer: liquid flowing through a porous prism, *Journal of the Franklin Institute* **208,** 405-416 (1929).
2. V.S.Arpaci and J.A.Clark, Dynamic response of fluid and wall temperatures during pressurized discharge for simultaneous time-dependent inlet gas temperature, ambient temperature, and/or ambient heat flux, *Advances in Cryogenic Engineering* **7,** 419-432 (1962).
3. F.T.Hung and R.G.Nevins, Unsteady-state heat transfer with a flowing fluid through porous solids, ASME paper No.65-HT-10 (1965).
4. W.J.Jang and C.P.Lee, Dynamic response of solar heat storage systems, ASME Paper No. 74-WA/HT-22 (1974).
5. D.M.Burch, R.W.Allen and B.A.Peavy, Transient temperature distributions within porous slabs subjected to sudden transpiration heating, *ASME Journal of Heat Transfer* **98,** 221-225 (1976).
6. M.Riaz, Analytical solutions for single- and two-phase models of packed-bed thermal storage systems, *ASME Journal of Heat Transfer* **99,** 489-492 (1977).
7. H.C.White and S.A.Korpela, On the calculation of the temperature distribution in a packed bed for solar energy applications, *Solar Energy* **23,** 141-144 (1979).
8. G.Spiga and M.Spiga, A rigorous solution to a heat transfer two phase model in porous media and packed beds, *Int. J. Heat Mass Transfer* **24,** 355-364 (1981).
9. N.R.Amundson, Solid-fluid interactions in fixed and moving beds. Fixed beds with large particles, *Industrial and Engineering Chemistry* **48,** 26-35 (1956).
10. N.R.Amundson, Solid-fluid interactions in fixed and moving beds. Fixed beds with small particles, *Industrial and Engineering Chemistry* **48,** 35-43 (1956).
11. A.V.Kuznetsov, An investigation of a wave of temperature difference between solid and fluid phases in a porous packed bed, *Int. J. Heat Mass Transfer* **37,** 3030-3033 (1994).
12. K.Vafai and M.Sözen, Analysis of energy and momentum transport for fluid flow through a porous bed, *ASME Journal of Heat Transfer* **112,** 690-699 (1990).
13. K.Vafai and M.Sözen, An investigation of a latent heat storage porous bed and condensing flow through it, *ASME Journal of Heat Transfer* **112,** 1014-1022 (1990).
14. M.Sözen and K.Vafai, Analysis of the non-thermal equilibrium condensing flow of a gas through a packed bed, *Int. J. Heat Mass Transfer* **33,** 1247-1261 (1990).
15. M.Sözen, K.Vafai and L.A.Kennedy, Thermal charging and discharging of sensible and latent heat storage packed beds, *J. Thermophysics* **5,** 623-625 (1991).
16. A.Amiri and K.Vafai, Analysis of dispersion effects and non-thermal equilibrium, non-Darcian, variable porosity incompressible flow through porous media, *Int. J. Heat Mass Transfer* **37,** 939-954 (1994).

17. A.G.Dixon and D.L.Cresswell, Theoretical predictions of effective heat transfer parameters in packed beds, *AIChE Journal* **25,** 663-676 (1979).
18. D.Handley and P.J.Heggs, Momentum and heat transfer mechanisms in regular shaped packings, *Trans. Inst. Chem. Engrs.* **46,** T251-T264 (1968).
19. H.Miyauchi, H.Kataoka and T.Kikuchi, Gas film coefficients of mass transfer in low Peclet number region for sphere packed beds, *Chem. Engn. Sci.* **31,** 9-13 (1976).
20. N.Wakao, K.Tanaka and H.Nagai, Measurement of particle-to-gas mass transfer coefficients from chromatographic adsorption experiments, *Chem. Engn. Sci.* **31,** 1109-1113 (1976).
21. F.A.L.Dullien, Porous Media Fluid Transport and Pore Structure, Chap.3, Acad. Press, New York (1979).
22. H.S.Carslaw and J.C.Jaeger, Conduction of heat in solids, Oxford University Press, Oxford (1959).
23. A.V.Kuznetsov, A perturbation solution for a nonthermal equilibrium fluid flow through a three-dimensional sensible heat storage packed bed, *ASME Journal of Heat Transfer* **118,** 508-510 (1996).
24. A.V.Kuznetsov, Investigation of a non-thermal equilibrium flow of an incompressible fluid in a cylindrical tube filled with porous media, *ZAMM* **76,** 411-418 (1996).
25. A.V.Kuznetsov, Analysis of a non-thermal equilibrium fluid flow in a concentric tube annulus filled with a porous medium, *Int. Comm. Heat Mass Transfer* **23,** 929-938 (1996).
26. A.V.Kuznetsov, A perturbation solution for heating a rectangular sensible heat storage packed bed with a constant temperature at the walls, *Int. J. Heat Mass Transfer* **40,** 1001-1006 (1997).
27. K.Vafai and S.J.Kim, Fluid mechanics of the interface region between a porous medium and a fluid layer - an exact solution, *Int. J. Heat and Fluid Flow* **11,** 254-256 (1990).
28. K.Vafai and R.Thiyagaraja, Analysis of flow and heat transfer at the interface region of a porous medium, *Int. J. Heat Mass Transfer* **30,** 1391-1405 (1987).
29. P.Cheng, C.T.Hsu and A.Chowdhury, Forced convection in the entrance region of a packed channel with asymmetric heating, *ASME Journal of Heat Transfer* **110,** 946-954 (1988).
30. K.Vafai and S.J.Kim, Forced convection in a channel filled with a porous medium: an exact solution, *ASME Journal of Heat Transfer* **111,** 1103-1106 (1989).
31. K.Vafai and S.J.Kim, Discussion of the paper by A.Hadim "Forced convection in a porous channel with localized heat sources," *ASME Journal of Heat Transfer* **117,** 1097-1098 (1995).
32. D.A.Nield, S.L.M.Junqueira and J.L.Lage, Forced convection in a fluid saturated porous medium channel with isothermal or isoflux boundaries, *Journal of Fluid Mechanics* **322,** 201-214 (1996).
33. A.V.Kuznetsov, Thermal nonequilibrium, non-Darcian forced convection in a channel filled with a fluid saturated porous medium - A perturbation solution, *Applied Scientific Research* , in press
34. A.Bejan, The optimal spacing for cylinders in crossflow forced convection, *ASME Journal of Heat Transfer* **117,** 767-770 (1995).
35. A.Bejan and Al.M.Morega, The optimal spacing of a stack of plates cooled by turbulent forced convection, *Int. J. Heat Mass Transfer* **37,** 1045-1048 (1994).

36. A.Bejan, How to distribute a finite amount of insulation on a wall with nonuniform temperature, *Int. J. Heat Mass Transfer* **36,** 49-56 (1993).
37. S.Mereu, E.Sciubba and A.Bejan, The optimal cooling of a stack of heat generating boards with fixed pressure drop, flowrate or pumping power, *Int. J. Heat Mass Transfer* **36,** 3677-3686 (1993).
38. A.V.Kuznetsov, Optimal control of the heat storage in a porous slab, *Int. J. Heat Mass Transfer* **40**, 1720-1723 (1997)
39. L.S.Pontryagin, V.Boltyanskii, R.Gamkrelidze and E.Mishchenko, The Mathematical Theory of Optimal Processes, Interscience Publishers, New York (1962).
40. M.Athans and P.L.Falb, Optimal Control : An Introduction to the Theory and Its Applications, McGraw-Hill, New York (1966).
41. A.V.Kuznetsov, Determination of the optimal initial temperature distribution in a porous bed, *Acta Mechanica* **120,** 61-69 (1997).
42. I.Prigogine, From being to becoming: time and complexity in the physical sciences, W.H.Freeman, San Francisco (1980).

MATHEMATICAL MODELS FOR HEAT AND MASS TRANSPORT IN GEOTHERMAL SYSTEMS

R. McKIBBIN

Department of Mathematics, Massey University
Private Bag 11 222, Palmerston North, New Zealand

INTRODUCTION

Geothermal systems are an important energy source in many countries, both for direct use of hot groundwater for cooking, bathing, heating and chemical processes as well as usage of higher-enthalpy fluids for electricity generation, see for example Freeston [1] and Huttrer [2]. While the former uses have been made for centuries through tapping of surface outflows, the latter had to wait for the suitable drilling, piping, machinery and materials technology of this century.

The mathematical modelling of heat and mass flows within underground geothermal systems was motivated by desires to discover the effects of exploitation of such resources on a large scale and also to predict how systems would react locally to proposed usage. The predictive capabilities of quantitative models would lead to their being able to be used in the design process, and indeed they are playing an essential role in planning new energy developments and in improving current ones.

As the relatively near-surface regions are better understood, so attention is now being focussed on the deeper zones which underlie geothermal reservoirs, and which provide a link between their bases and the magmatic heat sources further below. In this regard, a direct connection between the work and knowledge of volcanologists and those of geothermalists is becoming closer to reality.

PHYSICAL PROCESSES

Models of flow processes in geothermal systems have to take into account the strong coupling between heat and mass transport. The usual conceptual models on which such quantitative mathematical models are based involve motion of a single-phase fluid (liquid or gas) or a flow of two fluid phases which are in thermodynamic equilibrium, within a stationary porous rock matrix. The dominant fluid component is water, with solutes and gases in relatively small concentrations.

The rock structure is thermally conductive, and the heat contained in it contributes significantly to the energy reserve. In contrast to the flow of hydrocarbons within an oil reservoir, the fluid in a geothermal structure serves more significantly as a medium to transport the heat energy contained within the rock matrix, rather than as the only container of energy in the system.

The earliest models of geothermal systems were based on pure water as the saturating fluid; consequently, consideration of the thermodynamics of pure water substance was paramount. The determination of suitable correlations for its properties, in forms easily converted for use in computer subroutines, was essential to the quantification of mass and heat flows by numerical simulation.

Development of further sophistication through consideration of chemicals dissolved in the liquid phase, and of various gas components contributing to the total gas pressure, has led recently to more complicated formulations, and also to interest in modelling the transport of minerals and the deposition of solid phases of solutes within the rock matrix. The porous matrix has received attention, too, with more detail being placed on the fractured nature of geothermal rocks, in contrast to the early models which were based on typical groundwater aquifers, considered to be homogeneous, but not necessarily isotropic, sedimentary structures.

CONSERVATION EQUATIONS

The description of a geothermal system is largely based on the continuum hypothesis, using a Representative Elementary Volume (REV) formulation. It is assumed that the scale of description is large enough for volume-averaged quantities to be statistically valid, while being small compared with the macroscopic dimensions of the large geophysical structure, see for example Bear [3], Bear and Bachmat [4] and Nield and Bejan [5]. System descriptions are based on averaged quantities.

The distribution of phases is calculated using volume fractions, while the phase properties are based on mass units. The volume fraction of the system occupied by solid rock is $1 - \phi$, where ϕ, the effective porosity, is a local measure of the pore space available to the geothermal fluid. The volume fractions of the pore space occupied by the liquid and gas phases are denoted S_ℓ and S_g ($= 1 - S_\ell$), respectively.

Conservation of mass

Equations derived from the principle of mass conservation for each of the system fluid components i (i = water, solute, non-condensible gas, etc.), which may be distributed within both liquid and gas phases, are usually written in the following form, see for example Bear and Bachmat [4]:

$$\frac{\partial A_m^{(i)}}{\partial t} + \nabla \cdot \mathbf{Q}_m^{(i)} = q_m^{(i)} \tag{1}$$

Here the mass per unit formation volume for component i is given by

$$A_m^{(i)} = \phi \, [\, X_\ell^{(i)} S_\ell \rho_\ell + X_g^{(i)} S_g \rho_g \,] \tag{2}$$

where ρ_ℓ and ρ_g are the densities of the liquid and gas phases respectively, while $X_\ell^{(i)}$ and $X_g^{(i)}$ are the mass fractions of component i present in each of the separate fluid phases. The specific mass flux, or mass flux of component i per unit cross-sectional area of the formation, is $\mathbf{Q}_m^{(i)}$, defined below, and $q_m^{(i)}$ is a source term for component i in units of mass flux per unit formation volume. This last term can be used to model extraction or injection of fluid via boreholes, and to model precipitation and/or dissolution of chemicals contained in the fluid onto/from the pore surfaces of the rock matrix.

Summation of the mass conservation equations (1), (2) over all fluid components gives:

$$\frac{\partial A_m}{\partial t} + \nabla \cdot \mathbf{Q}_m = q_m \tag{3}$$

where the total fluid mass per unit formation volume is given by

$$A_m = \phi \, [S_\ell \rho_\ell + S_g \rho_g] \tag{4}$$

while $\mathbf{Q}_m$ is the total specific mass flux and q_m is the total mass source term.

Conservation of linear momentum (Darcy's law)

Known as Darcy's law, the simplest model for fluid flow in a porous medium is derived by considering, because fluid velocities are small, terms representing inertial forces in the momentum conservation equation are negligible compared with those for body and viscous forces (i.e. Re << 1), see Nield and Bejan [5]. This approximation seems reasonable for geothermal flows, except perhaps near boreholes; the latter, however, form only very small regions of the total formation volumes considered.

When only one fluid phase is present in the pores, one momentum equation is required, but for a general two-phase formulation, there are separate equations for the mass flux per unit cross-sectional formation area of the liquid and gas phases $\mathbf{Q}_{m\ell}$ and $\mathbf{Q}_{mg}$. These are given, respectively, by:

$$\mathbf{Q}_{m\ell} = \rho_\ell \mathbf{v}_\ell = \rho_\ell \frac{k_{r\ell}}{\mu_\ell} \mathbf{k} \otimes [- \nabla p_\ell + \rho_\ell \mathbf{g}] \tag{5a}$$

$$\mathbf{Q}_{mg} = \rho_g \mathbf{v}_g = \rho_g \frac{k_{rg}}{\mu_g} \mathbf{k} \otimes [- \nabla p_g + \rho_g \mathbf{g}] \tag{5b}$$

where $\otimes$ is the tensor product operator, and $\mathbf{v}_\ell$ and $\mathbf{v}_g$ are the specific volume fluxes (volume flux per unit cross-sectional formation area) for the liquid and gas phases, respectively. These are also known as the Darcy velocities of the separate phases. The mass fluxes for component i in the liquid and gas phases are given by

$$\mathbf{Q}_{m\ell}^{(i)} = X_\ell^{(i)} \mathbf{Q}_{m\ell} \tag{6a}$$

$$\mathbf{Q}_{mg}^{(i)} = X_g^{(i)} \mathbf{Q}_{mg} + D_w^{(i)} \rho_g [- \nabla X_g^{(i)}] \tag{6b}$$

The total mass flux for component i [as used in Equation (1)] is then given by:

$$\mathbf{Q}_m^{(i)} = \mathbf{Q}_{m\ell}^{(i)} + \mathbf{Q}_{mg}^{(i)} \tag{7}$$

Equation (6b) reflects the transport of mass in the gas phase by diffusion as well as advection. Parameters not already defined include the following: $\mathbf{k}$ is the formation intrinsic permeability tensor, $k_{r\ell}$ and k_{rg} are permeability reduction factors (or relative permeabilities) which depend on the liquid saturation S_ℓ, μ_ℓ and μ_g are the dynamic viscosity of each phase, p_ℓ and p_g are the liquid and gas phase pressures, while $D_w^{(i)}$ is the mass diffusion coefficient for component i in water vapour in the gas phase. Further definitions and comments about these parameters will be given below.

Note that when only one fluid phase is present, a single equation can be retrieved by setting to zero either all the $X_\ell^{(i)}$ (for gas phase only) or all the $X_g^{(i)}$ (for liquid phase only).

Conservation of energy

As already mentioned, geothermal fluid flow speeds are small. It is then usual to assume that all components and phases are in local thermodynamic equilibrium, and in particular are at the same temperature. The equation which reflects conservation of energy may be written as follows, see for example Bear and Bachmat [4]:

$$\frac{\partial A_e}{\partial t} + \nabla \cdot \mathbf{Q}_e = q_e \tag{8}$$

where the energy per unit formation volume is

$$A_e = (1 - \phi)\, \rho_r\, u_r + \phi\, [\, S_\ell\, \rho_\ell\, u_\ell + S_g\, \rho_g\, u_g \,] \tag{9}$$

and the specific energy flux (flux per unit cross-sectional formation area) is

$$\mathbf{Q}_e = \sum_i [\, h_\ell^{(i)}\, \mathbf{Q}_{m\ell}^{(i)} + h_g^{(i)}\, \mathbf{Q}_{mg}^{(i)} \,] + \mathbf{K} \otimes (-\nabla T) \tag{10}$$

Equation (10) reflects the transport of heat by both advection and thermal conduction. In these equations, q_e is a source term in units of energy flux per unit formation volume, while ρ_r and u_r are the density and specific internal energy of the rock particles; u_ℓ and u_g are the specific internal energies and $h_\ell^{(i)}$ and $h_g^{(i)}$ are the component specific enthalpies of the liquid and gas fluid phases respectively. The local temperature of all components and phases is T, while $\mathbf{K}$ is the effective thermal conductivity tensor of the rock-fluid mixture.

Formation parameters

Early models of geothermal reservoirs regarded the formations as essentially consisting of (maybe several) isotropic sedimentary-type layers, with anisotropy arising from the layering itself. More recent modelling recognises that thermal fracturing of the volcanic rocks which form such reservoirs have an inherent anisotropy. The rock material between the fractures may also have a secondary permeability, with flow into and out of the surrounding fractures.

The parameters used in the conservation equations above are discussed in more detail here. The discussion relates only to geothermal modelling application of the more general theory of flow in porous media, and represents approximations which are necessarily made to model large-scale systems about which little detailed data is available.

Effective porosity ϕ. This is the local fraction of the formation which consists of connected pore space which is available to fluid. Isolated pore space, assumed relatively small, is not taken into account. While the pointwise porosity may vary from zero in a rock particle to 100% in a pore, ϕ is the spatially-averaged value over a REV; it may vary with position in a reservoir, but is more commonly assumed to be of uniform value within specified sub-regions of the formation being considered. The quantity ϕ is dimensionless.

Effective thermal conductivity tensor $\mathbf{K}$. The effective or overall thermal conductivity of the rock-fluid mixture may be expressed approximately in terms of the separate rock and fluid values by

$$\mathbf{K} = (1 - \phi)\,\mathbf{K}_r + \phi\,\mathbf{K}_f \tag{11}$$

Since porosity ϕ is usually small, the rock value dominates in this expression. Thermal conductivity of such material is isotropic and $\mathbf{K}$ is usually approximated by an isotropic constant tensor with $\mathbf{K}_{ii} = K$ (i = 1, 2, 3), with the scalar value K in the range 2.0 – 2.5 W m^{-1} K^{-1}, typical values for solid rock material.

Rock density ρ_r. The density of the solid rock particles is usually taken to lie in the range 2000 – 3000 kg m^{-3}, depending on the rock type.

Rock specific internal energy u_r. The specific heat c_r of rock varies very little with either temperature or pressure, so the specific internal energy of the rock particles may be written:

$$u_r = c_r\,T \tag{12}$$

where c_r has a constant value of approximately 1000 kJ kg^{-1} K^{-1} and T is measured in °C. (Note that the zero datum for energy values is at 0 °C.)

Intrinsic permeability tensor $\mathbf{k}$. The permeability is a measure of how easily a single phase fluid moves in a porous medium under the influence of a dynamic pressure gradient (absolute pressure gradient adjusted for gravitational effects). It is a property of the porous matrix only; fluid property effects are incorporated in Darcy's law through the fluid density and dynamic viscosity [see Equations (5a, b)]. The principal axes of $\mathbf{k}$ are influenced mostly by the bedding of the porous matrix, with little variation in the horizontal plane, and in most cases a significantly smaller value in the vertical direction. With respect to a cartesian coordinate system with position vector components (x, y, z) where the gravitational acceleration vector is represented by $\mathbf{g} = (0, 0, -g)$, the principal components of $\mathbf{k}$ can be written (k_h, k_h, k_v), where the vertical permeability k_v is smaller, by up to an order of magnitude, than the horizontal value k_h. Typical values of the components of $\mathbf{k}$ for geothermal systems lie in the range $10^{-15} - 10^{-12}$ m^2 (i.e. 1 millidarcy – 1 darcy, where 1 darcy $\approx 10^{-12}$ m^2).

Relative permeabilities $k_{r\ell}$, k_{rg}. Known also as permeability reduction factors, these parameters allow for modelling the way that one fluid phase interferes with the motion of the other when two phases are present in the pores. Each reduction factor is a function of the liquid saturation S_ℓ. Most models reflect various experimental evidence that, if one of the phases is present in only relatively small amounts ($S_\ell < S_{\ell r}$ or $S_g < S_{gr}$, where $S_{\ell r}$ and S_{gr} are called the residual liquid and gas saturations, respectively), movement of that phase is almost completely inhibited by the other and the corresponding permeability reduction factor is effectively zero.

While there are a number of specific formulations of the relationship between the relative permeabilities and liquid or gas saturation, most follow a common structure. Typically, the functions are of the form $k_{r\ell}(S_\ell)$ and $k_{rg}(S_\ell)$, where both are continuous and monotonic, with,

$$\begin{array}{llll} \text{for } S_\ell \le S_{\ell r}, & k_{r\ell} = 0, & k_{rg} = 1 & \\ \text{for } S_{\ell r} < S_\ell < 1 - S_{gr}, & k_{r\ell} = f(S_\ell), & k_{rg} = g(S_\ell) & (13) \\ \text{for } S_\ell \ge 1 - S_{gr} \ (\text{or } S_g \le S_{gr}), & k_{r\ell} = 1, & k_{rg} = 0 & \end{array}$$

where f and g are monotonic functions of S_ℓ. An extra condition sometimes imposed is that $f(S_\ell) + g(S_\ell) = 1$. The simplest formulation uses straight-line functions:

$$f(S_\ell) = S_\ell^* = \frac{S_\ell - S_{\ell r}}{1 - S_{gr} - S_{\ell r}} \qquad g(S_\ell) = 1 - S_\ell^* \tag{14a}$$

while another commonly-used set of formulae is that first derived by Corey [6]:

$$f(S_\ell) = (S_\ell^*)^4 \qquad g(S_\ell) = (1 - S_\ell^*)^2 \left[1 - (S_\ell^*)^2\right] \tag{14b}$$

where S_ℓ^* is defined in Equation (14a). Note that Corey's formulae do not satisfy the condition that $f(S_\ell) + g(S_\ell) = 1$ for all S_ℓ.

Typical values of $S_{\ell r}$ and S_{gr} lie in the ranges 0 – 0.4 and 0 – 0.1, respectively. The two sets of relative permeability curves defined above are illustrated in Figure 1 for the case $S_{\ell r} = 0.3$, $S_{gr} = 0.05$.

Voidage ε. Deposition of solid precipitates from the fluid onto the walls of the rock pore spaces decreases the available pore volume available for the fluid. This may be measured by the voidage ε, the volume fraction of the rock pore space available for fluid. Equations (1) – (4), (8) and (9) above take $\varepsilon = 1$, but any consideration of the solid precipitate's contribution to mass and energy balances would need another parameter such as ε to account for it, while specific properties for the solid precipitate phase would also be required.

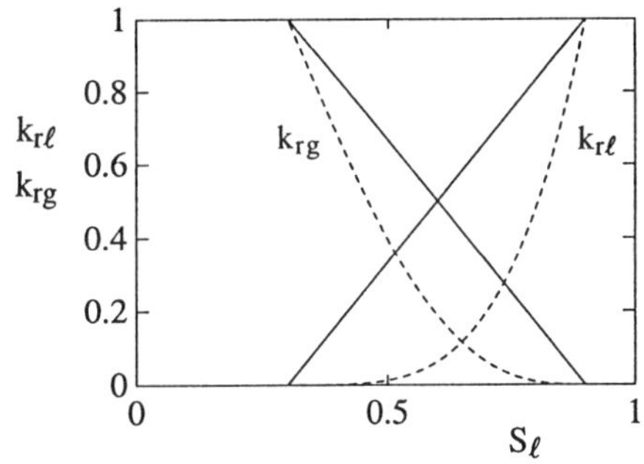

Figure 1: Relative permeability functions $k_{r\ell}(S_\ell)$ and $k_{rg}(S_\ell)$ for $S_{\ell r} = 0.3$, $S_{gr} = 0.05$: —— straight line functions [Equation (14a)], – – – Corey curves [Equation (14b)]

While large amounts of such deposition would inevitably alter the permeability by sealing pore connections, it is usually assumed that the permeability is not substantially altered by small amounts of precipitate. As will be described in the last main section of this work, recent developments in modelling deep high-pressure high-temperature brines, where solid salt precipitate may exist in equilibrium with solute-saturated fluids, explicitly include the solid precipitate fraction in mass and energy balances, see McKibbin and McNabb [7, 8].

Fluid parameters

Because water is the predominant fluid component in geothermal systems, accurate modelling of its thermodynamic properties is essential. Tables of the thermodynamic properties of water substance are readily available, usually given in terms of the independent state variable pair (p, T), with pressure p in bars absolute (1 bar = 10^5 Pa) and temperature T in °C (0 °C = 273.15 K).

For any given reservoir temperature T < 374.15 °C (the critical temperature), a boiling or saturation pressure exists, denoted by $p = p_{sat}(T)$, at which, under ideal conditions, both the liquid and gas (steam, or water vapour) phases of water can co-exist; conditions then are said to be saturated. The inverse function, which gives the saturation temperature in terms of pressure, is written $T = T_{sat}(p)$. For T > 374.15 °C, or p > 221.2 bars absolute, the conditions are said to be supercritical, and there is no pressure at which two phases become distinguishable.

For a fluid which is composed mainly of water, but with some disassociated salts and non-condensible gases dissolved in it, the boiling or saturation relationship will be slightly different from that for pure water. In general, the boiling pressure will be given by $p = p_{sat}(T, X_\ell^{(1)}, X_\ell^{(2)}, \ldots, X_\ell^{(N)})$, where N is the total number of components, and i = 1 corresponds to water.

Generally, two thermodynamic variables determine the state of a system for a pure substance; at saturated conditions, p and T are functionally related and so are no longer independent. The liquid saturation S_ℓ may be used as a replacement for either T or p; then either (p, S_ℓ) or (T, S_ℓ) is used as an independent state variable pair (together with concentrations of other components, if present in large enough mass fractions $X_\ell^{(i)}$ to cause significant effects). For $p < p_{sat}(T, X_\ell^{(i)})$ [or $T > T_{sat}(p, X_\ell^{(i)})$], superheated vapour forms the only phase (gas), while single-phase compressed liquid conditions prevail for $p > p_{sat}(T, X_\ell^{(i)})$ [or $T < T_{sat}(p, X_\ell^{(i)})$]. Generally, the presence of dissolved salts increases the boiling temperature, while non-condensible gases cause boiling-temperature depression, see O'Sullivan et al. [9].

Surface tension and adsorption effects may balance a small difference in liquid and gas phase pressures p_ℓ and p_g in two-phase fluids in porous systems; this is called vapour pressure lowering, and is discussed further below. Such effects are usually small under natural geothermal two-phase conditions, and the assumption that the pressures in the liquid and gas phases are the same produces little error. However, injection of colder water into a two-phase vapour-dominated system may produce significant effects, as shown through numerical simulation by Pruess [10].

Liquid saturation S_ℓ. For conditions corresponding to single-phase (compressed) liquid, where $p > p_{sat}(T)$, the liquid saturation $S_\ell = 1$, while the gas saturation $S_g = 0$. Similarly, for $T > T_{sat}(p)$, $S_\ell = 0$ while $S_g = 1$. Note that these are volume fractions, unlike the quantity known as the quality or dryness used in thermo-mechanical process calculations; that parameter, commonly denoted X, is the mass fraction of steam (water vapour) in a two-phase mixture. Problems associated with determining suitable values of the saturations when conditions are supercritical have been addressed by Kissling [11].

Fluid phase pressures p_ℓ, p_g. The phenomenon of vapour pressure lowering occurs when surface tension effects at the fluid phase interfaces balance a difference in the separate phase pressures. The thermodynamic properties of the fluid phases are then altered inside the porous medium by capillary forces and by adsorption of liquid on mineral phases. For pure water, the pressure difference may be expressed by

$$p_\ell - p_g = p_{suc}(S_\ell) \tag{15}$$

where p_{suc}, a function of the liquid saturation S_ℓ, is termed the suction pressure, see Pruess [10]. Vapour pressure above a liquid phase held by capillary or adsorptive forces is reduced in comparison to the saturation vapour pressure above the flat surface of a bulk liquid. The reduction is expressed in terms of a vapour pressure lowering factor, $f = p_g/p_{sat}(T)$, given by Kelvin's equation:

$$f = \exp\left[\frac{M_w\, p_{suc}}{\rho_\ell\, R\, (T + 273.15)}\right] \tag{16}$$

where M_w is the molecular weight of water, R is the universal gas constant; and all the other parameters have previously been defined.

Fluid phase densities ρ_ℓ, ρ_g. For pure water, these properties are well-known and are tabulated over a wide range of pressures and temperatures. Reliable correlation formulae are also available; these make computer calculations faster than if reliant on table lookup and interpolation.

When other components are present in the fluid, the pure water densities must be modified. Components other than water may be divided into two groups: chemical solutes and non-condensible gases. The former reside mainly in the liquid phase, while non-condensible gases have greatest effect in the gas phase. Over the normal range of conditions prevailing in a geothermal system, it is usual to assume that the liquid density is unaltered by the presence of non-condensible gases, but ρ_ℓ must be modified to take account of any solutes present.

In the gas phase, modifications are made for the non-condensible gas components, density $\rho_g^{(i)}$, which also each contribute a partial pressure $p_g^{(i)}$ to the total gas pressure p_g. The total gas phase density is expressed in terms of the component densities by Dalton's law. The mass fraction of each dissolved non-condensible gas component in the liquid phase may be related to an equivalent gas partial pressure through Henry's law. Both distribution laws are discussed further below. Density units are kg m^{-3}.

Fluid phase specific enthalpies h_ℓ, h_g. As for densities, the specific enthalpies for pure water may be calculated from correlation formulae. In a two-phase fluid mixture, the difference between the gas and liquid values is called the latent heat of vaporisation $h_{\ell g}$, given by

$$h_{\ell g} = h_g - h_\ell \tag{17}$$

Again, water specific enthalpy values are altered by the presence of other components. The overall gas specific enthalpy is calculated as a linear combination of the separate gas component values, weighted by their mass fractions:

$$h_g = \sum_i X_g^{(i)} h_g^{(i)} \tag{18}$$

The specific enthalpy of each of the dissolved non-condensible gases in the liquid phase is expressed in terms of its gas value plus its heat of solution:

$$h_\ell^{(i)} = h_g^{(i)} + h_{sol}^{(i)} \tag{19}$$

The overall liquid specific enthalpy is then given by a linear combination of the water value (i = 1), the contributions from the solutes and those from the non-condensible gas fractions [Equation (19)], as follows:

$$h_\ell = X_\ell^{(1)} h_\ell^{(1)} + \sum_{\text{solutes}} X_\ell^{(i)} h_\ell^{(i)} + \sum_{\text{gases}} X_\ell^{(i)} [h_g^{(i)} + h_{sol}^{(i)}] \tag{20}$$

Specific enthalpy values have units kJ kg^{-1}.

Fluid phase dynamic viscosities μ_ℓ, μ_g. Again, pure water values are readily available from correlation formulae over a wide range of (p, T) values. It is usually assumed that the liquid value μ_ℓ is altered negligibly from the pure water value due to dissolved non-condensible gases. However, large concentrations of chemical solutes affect the liquid viscosity; in particular, very saline liquids are significantly more viscous than water at the same (p, T) conditions. Correlations are available for such "pure" solutions, but are reliable over only limited ranges of (p, T, $X_\ell^{(i)}$) values.

The overall mixture viscosity for the gas phase is commonly estimated by a linear combination of the separate non-condensible gas component values:

$$\mu_g = \sum_i X_g^{(i)} \mu_g^{(i)} \tag{21}$$

Dynamic viscosity measurement units are kg m^{-1} s^{-1}.

Kinematic viscosities ν_ℓ, ν_g. These are defined for each phase in terms of the respective dynamic viscosities and densities by

$$\nu_\ell = \frac{\mu_\ell}{\rho_\ell} \quad \text{and} \quad \nu_g = \frac{\mu_g}{\rho_g} \tag{22}$$

with units m^2 s^{-1}. As will be shown in the section on one-dimensional flows below, for horizontal two-phase flows a total, or effective, kinematic viscosity can be defined, based on separate phase viscosities weighted by relative permeabilities.

Distribution laws

In two-phase conditions, the components are distributed within both phases. Usually salt concentrations within the gas phase are small, but non-condensible gases may dissolve in the liquid phase.

Molecular diffusion. Within any gas phase present, molecular diffusion of different gas molecules is taken into account through the last term in Equation (6b). The net transport of a particular component is proportional to the gradient of its concentration expressed as a mass fraction of the total gas mixture. Since water vapour is the dominant gas component, the transport of a minor component may be regarded as controlled predominantly by its binary diffusion rate in water vapour and is quantified by $D_w^{(i)}$, the mass diffusion coefficient of component i, i = 2, ..., N (i = 1 is water) in pure water vapour. The coefficient for minor component i may be expressed in the form:

$$D_w^{(i)}(p, T) = \tau \phi S_g \frac{D_w^{(i)}(p_0, T_0)}{p/p_0} \left[\frac{T + 273.15}{273.15}\right]^\theta \tag{23}$$

where τ is the tortuosity of the porous matrix, and $D_w^{(i)}(p_0, T_0)$ and θ are material parameters corresponding to standard conditions, e.g. (p_0, T_0) = (1 bar, 0 °C), see Pruess [12].

Diffusive fluxes of components in the liquid phase are regarded as very small compared with the advected transport, and are neglected

Dalton's law. Standard models of gas mixtures assume that each component is uniformly distributed throughout the mixtures, and that the total density is the sum of the densities of the parts:

$$\rho_g = \sum_i \rho_g^{(i)} \tag{24}$$

Each component contributes a partial pressure to the total gas pressure. Only non-condensible gas components make significant contributions. By modelling all components as ideal gases, the total pressure can be approximated by the sum of the partial pressures:

$$p_g = \sum_i p_g^{(i)} \tag{25}$$

This then allows a connection to be made between the concentrations of gas-phase components with their liquid-phase concentrations through Henry's law.

Henry's law. Non-condensible gas solubility in the liquid phase may be expressed in terms of Henry's law, which gives a relationship between the partial pressure of a component in the gas phase and its molar fraction in the liquid, see Perry et al. [13]. This can be expressed by:

$$p_g^{(i)} = K_H^{(i)} \frac{\frac{X_\ell^{(i)}}{M^{(i)}}}{\frac{X_\ell^{(1)}}{M_w} + \sum_{i\geq 2} \frac{X_\ell^{(i)}}{M^{(i)}}}$$

where component 1 is water and $M^{(i)}$ is the molecular weight of component i. For single components at small concentration, the relationship is almost linear:

$$p_g^{(i)} = K_H^{(i)} X_\ell^{(i)} \frac{M_w}{M^{(i)}} \tag{26}$$

Here, $K_H^{(i)}$ is Henry's constant for component i in pure water, and it is generally a function of temperature T. At high concentrations, the relationship is non-linear; however, as mentioned at the outset, geothermal systems are mainly water, with other components in relatively small quantities, and Equation (26) serves as a reasonable approximation even when there are several minor components.

Boundary conditions

To close the mathematical problem, boundary conditions are normally stated in terms of thermodynamic state variables, or in terms of mass and/or heat fluxes.

Prescription of temperatures or pressures implies knowledge of heat or fluid reservoirs at the boundary which are unaffected in properties by fluid outflow or recharge. Examples include surface water at atmospheric conditions, or recharge fluid available in groundwater aquifers adjacent to a geothermal system.

Impermeable boundaries may be modelled by specifying that normal mass flux is zero, while non-zero mass flows may be controlled by boundary system pressures relative to a specified local known exterior pressure.

STEADY ONE-DIMENSIONAL FLOWS

Horizontal flows: total viscosity and flowing enthalpy

Total, or effective viscosity. If diffusive effects are neglected, then for horizontal flows with liquid and gas pressures equal, use of Equations (5a, b) for the phase mass fluxes (all proportional to the horizontal pressure gradient $\nabla_h p$) gives the total horizontal specific mass flux $\mathbf{Q}_{mh}$ in the form

$$\mathbf{Q}_{mh} = \left[\frac{k_{r\ell}}{\nu_\ell} + \frac{k_{rg}}{\nu_g} \right] k_h \left(- \nabla_h p \right) \tag{27}$$

By comparison with the equation for horizontal flow of a single-phase fluid, an equivalent viscosity ν_t, called the total or effective kinematic viscosity, is defined by

$$\frac{1}{\nu_t} = \frac{k_{r\ell}}{\nu_\ell} + \frac{k_{rg}}{\nu_g} \tag{28}$$

Since the relative permeabilities are functions of liquid saturation S_ℓ, the effective viscosity itself depends on the relative proportions of liquid and gas present in the pores of the matrix.

The dependence of ν_t on S_ℓ is shown in Figure 2(a) for the case where T = 250 °C and for both sets of relative permeability functions described in Equations (14a) and (14b) with residual saturations $S_{\ell r} = 0.3$ and $S_{gr} = 0.05$. For $S_\ell \leq S_{\ell r} = 0.3$, ν_t takes the gas value ν_g, while for $S_g \leq S_{gr}$ ($S_\ell \geq 1 - S_{gr}$), $\nu_t = \nu_\ell$. For intermediate values, the dependence on the form of the relative permeability functions is clear.

Flowing enthalpy. The specific energy flux is expressed by Equation (10). If diffusive effects are neglected, then for steady horizontal flows with liquid and gas pressures equal, the total horizontal specific energy flux $\mathbf{Q}_{eh}$ is given by

$$\mathbf{Q}_{eh} = \left[\frac{k_{r\ell}\, h_\ell}{\nu_\ell} + \frac{k_{rg}\, h_g}{\nu_g} \right] k_h \left(- \nabla_h p \right) \tag{29}$$

A fluid-averaged enthalpy value h_f, termed the flowing enthalpy, is found by dividing the magnitude of the total specific energy flux $\mathbf{Q}_{eh}$ by that of the total convected specific mass flux $\mathbf{Q}_{mh}$ given in Equation (27). This averaged value is of the form:

$$h_f = \frac{h_\ell \frac{k_{r\ell}}{\nu_\ell} + h_g \frac{k_{rg}}{\nu_g}}{\frac{k_{r\ell}}{\nu_\ell} + \frac{k_{rg}}{\nu_g}} = \left[h_\ell \frac{k_{r\ell}}{\nu_\ell} + h_g \frac{k_{rg}}{\nu_g} \right] \nu_t \tag{30}$$

after using Equation (28). For single-phase liquid conditions, $S_\ell = 1$ and $k_{rg} = 0$, leading to $h_f = h_\ell$ as expected, while single-phase gas conditions lead to $h_f = h_g$. The dependence of h_f on S_ℓ is shown in Figure 2(b) for the same case as that in Figure 2(a) for ν_t. Comments similar to those for the behaviour of ν_t are pertinent (see above).

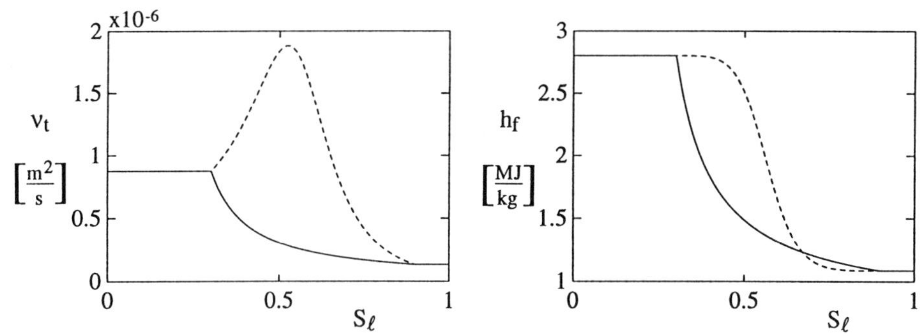

Figure 2: Variation of (a) effective viscosity ν_t and (b) flowing enthalpy h_f with liquid saturation S_ℓ for horizontal convection of two-phase water at T = 250 °C. The relative permeability functions have residual saturations $S_{\ell r} = 0.3$, $S_{gr} = 0.05$: ——— straight line functions [Equation (14a)], – – – Corey curves [Equation (14b)]

Steady vertical flows: heat pipes

Since heat flow within natural geothermal systems is predominantly in the upward direction and exceeds that which could ensue from conduction only, transfer of energy from deep regions to the surface by convective processes is important. Within two-phase regions, a phenomenon known as a heat pipe can take effect. Upward mass flux of the gas phase, with its relatively high specific enthalpy, is balanced by a downward counterflow of liquid at a similar mass flow rate, but with smaller specific enthalpy. The effect is a small net mass transfer with a large upward heat flux.

The process can be modelled by considering the steady-state, one-dimensional vertical flow equations, with no internal sources. To simplify the demonstration, it will be assumed that the fluids are pure water (note: no diffusive flux in the gas phase), although a model which includes a non-condensible gas has been investigated by McKibbin & Pruess [14]. In a rectangular (cartesian) coordinate system (x, y, z) where $\mathbf{g} = (0, 0, -g)$, the specific mass flux is of the form $\mathbf{Q}_m = (0, 0, Q_{mv})$. Then Equation (3) requires that $dQ_{mv}/dz = 0$, i.e. Q_{mv} is independent of vertical position.

Assuming that the liquid and vapour pressures are the same, Equations (5a, b) give the separate vertical liquid and vapour mass flows to be:

$$(Q_{m\ell})_v = k_v \frac{k_{r\ell}\,\rho_\ell}{\mu_\ell}\left[-\frac{dp}{dz} - \rho_\ell\, g\right] \tag{31a}$$

$$(Q_{mg})_v = k_v \frac{k_{rg}\,\rho_g}{\mu_g}\left[-\frac{dp}{dz} - \rho_g\, g\right] \tag{31b}$$

where k_v is the vertical permeability and the net vertical mass flux is then given by $Q_{mv} = (Q_{m\ell})_v + (Q_{mg})_v$. In general, the pressure decreases with height (increases with depth) in a geothermal system, so $dp/dz < 0$ [$- dp/dz = dp/d(-z) > 0$].

It is obvious that there are two special cases of interest. The case:

$$-\frac{dp}{dz} = \rho_\ell\, g \tag{32}$$

is called the hydrostatic, or liquid-static, gradient and occurs when there is no vertical movement of liquid; it is the vertical pressure gradient which occurs in a single-phase warm water reservoir. The second special case is

$$-\frac{dp}{dz} = \rho_g\, g \tag{33}$$

which is called vapour-static, or steam-static. It should be noted that $\rho_g < \rho_\ell$ and these two special cases separate three flow situations:

(i)
$$\rho_\ell\, g < -\frac{dp}{dz}$$

This means that the pressure gradients are steeper than liquid-static and, from Equations (31a, b), both phases move upwards. For smaller pressure gradients:

(ii)
$$\rho_g\, g < -\frac{dp}{dz} < \rho_\ell\, g$$

In this case the pressure gradient lies between liquid-static and vapour-static. Liquid moves downwards while vapour moves upwards; this is called *counterflow*. When the pressure change with depth is small enough,

(iii)
$$-\frac{dp}{dz} < \rho_g\, g$$

and both liquid and steam fall.

While it is possible to set $Q_{mv} \neq 0$, corresponding to a net vertical mass throughput (considered by McGuinness [15], for example), here the net mass flux is taken to be zero. Then the downward liquid mass flux is equal to the upward steam mass flux:

$$(Q_{m\ell})_v = -(Q_{mg})_v \tag{34}$$

The pressure gradient may then be determined explicitly from Equations (31a) and (31b):

$$\frac{dp}{dz} = - \frac{\frac{k_{r\ell}}{\nu_\ell}\rho_\ell + \frac{k_{rg}}{\nu_g}\rho_g}{\frac{k_{r\ell}}{\nu_\ell} + \frac{k_{rg}}{\nu_g}} g$$

which lies between the liquid-static ($S_\ell = 1$) and the vapour-static ($S_\ell = 0$) pressure gradients given in Equations (32) and (33) respectively [see Figure 3(a)]. Note that the pressure gradient so found is independent of the vertical permeability k_v. The vertical specific energy flux is then found from Equation (10):

$$Q_{ev} = (Q_{mg})_v (h_g - h_\ell) - K \frac{dT}{dz} \tag{35}$$

This shows that counterflow with zero net mass flux can transport considerable amounts of energy, even when no mass is moved, since the latent heat of vaporisation $h_{\ell g} = h_g - h_\ell$ is large (about 1800 kJ kg^{-1}) for water at typical reservoir temperatures.

The advected vertical specific energy flux is represented by the first term on the right-hand side of Equation (35): $(Q_{ev})_{adv} = (Q_{mg})_v (h_g - h_\ell)$. For a given temperature, this depends on the liquid saturation S_ℓ and is directly proportional to permeability k_v.

The dependence on S_ℓ of the vertical pressure gradient dp/dz and the ratio $(Q_{ev})_{adv}/k_v$ is shown in Figure 3 for the case where T = 250 °C and for both sets of relative permeability functions defined in Equations (14a) and (14b) with residual saturations $S_{\ell r} = 0.3$ and $S_{gr} = 0.05$ (conditions the same as in Figure 2). The choice of relative permeability functions has a marked effect on the calculated values,

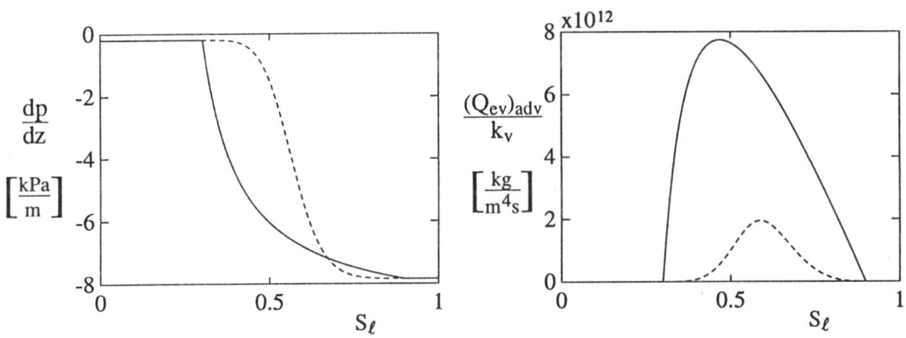

Figure 3: Variation of (a) the vertical pressure gradient dp/dz and (b) the ratio of the advected vertical specific energy flux to vertical permeability, $(Q_{ev})_{adv}/k_v$, with liquid saturation S_ℓ for vertical counterflow of two-phase water at T = 250 °C. The relative permeability functions have residual saturations $S_{\ell r} = 0.3$, $S_{gr} = 0.05$: ——— straight line functions [Equation (14a)], – – – Corey curves [Equation (14b)]

especially for the advected heat transfer. Note that for $S_\ell \leq S_{\ell r}$, the liquid relative permeability $k_{r\ell} = 0$. The pressure gradient is vapour-static and the gas phase does not move. Equation (35) indicates that the heat transfer is then conductive only. Similar remarks apply for the liquid-static case where $S_g \leq S_{gr}$ ($S_\ell \geq 1 - S_{gr}$).

HYDROLOGICAL MODELLING

Many warm water aquifers are exploited for direct use. Some such liquid hydrothermal systems can be modelled in the same way as hydrological aquifers. If the water in the system is at low temperatures (T < 100 °C, say) then it will remain as a single-phase liquid ($S_\ell = 1$) even when reduced to atmospheric pressure (about 1 bar absolute). In situations where the temperature gradients are small enough to inhibit buoyancy-driven convection, the flow in such aquifers may be regarded as isothermal flow of a liquid whose physical properties are nearly constant; the equation of conservation of energy is then not used. The appropriate form of the equation describing the mass flow of each component is, using Equations (1), (2), (5a), (6a), (7):

$$\frac{\partial}{\partial t}[\,\phi\, X^{(i)} \rho\,] + \nabla \cdot [\,X^{(i)} \frac{\rho}{\mu} \mathbf{k} \otimes (-\nabla p + \rho\, \mathbf{g})\,] = q_m{}^{(i)}$$

where the subscript ℓ has been dropped for convenience. Summation over all fluid components gives:

$$\frac{\partial}{\partial t}(\phi\, \rho) + \nabla \cdot \{\frac{\rho}{\mu} \mathbf{k} \otimes [-\nabla p + \rho\, \mathbf{g}\,]\} = q_m \tag{36}$$

For the typical case of the flow of water (ρ is nearly constant) in an aquifer with uniform properties, Equation (36) may be reduced to:

$$\phi\, \rho\, g\, C \frac{\partial h}{\partial t} + \nabla \cdot \{\, \mathbf{K}_{hyd} \otimes [-\nabla h\,]\} = q_v \tag{37}$$

where

$$\mathbf{K}_{hyd} = \frac{\rho\, g}{\mu} \mathbf{k} \qquad \text{and} \qquad C = \frac{1}{\phi\rho} \frac{d}{dp}(\phi\rho) \tag{38}$$

are, respectively, the hydraulic conductivity tensor (units: m s^{-1}) and the compressibility (units: bar^{-1}) of the saturated formation, and q_v (= q_m/ρ) is a water volume production (–ve: sink) or injection (+ve: source) rate per unit formation volume. The parameter h, defined in terms of the pressure and elevation z above some datum (mean sea level, for example) by

$$h = \frac{p}{\rho\, g} + z \tag{39}$$

is the total, or piezometric, head. If $\mathbf{K}_{hyd}$ is isotropic and uniform with principal value K_{hyd} and there is no volumetric source ($q_v = 0$), then Equation (37) reduces to a standard pressure diffusion equation:

$$\frac{\partial h}{\partial t} = \frac{K_{hyd}}{\phi \rho g C} \nabla^2 h \tag{40}$$

The method of treatment of hydrological problems is well-documented, see for example Bear and Verruijt [16]. Suffice it to say that hydrology is an important part of geothermal modelling insofar as no geothermal system is isolated from groundwater.

NUMERICAL SIMULATION

Several numerical simulation computer packages have been developed for solving the equations derived from conservation principles. In all techniques, the total formation region of interest is partitioned into a finite number N_{VE} of discrete, non-overlapping sub-regions, or volume elements. Algorithms are then based on discretised forms of the non-linear differential equations, using finite-difference, finite element, finite volume or integrated finite-difference techniques. The last is exemplified by the SHAFT-MULKOM-TOUGH sequence developed by Pruess and co-workers [12, 17, 18] at the Lawrence Berkeley Laboratory.

In the integrated finite-difference scheme, which is the technique that most closely models the REV formulation used to derive the equations described earlier in this work, thermodynamic conditions are assigned a uniform average value within each element. For any volume element V_n, n = 1, 2, ..., N_{VE}, the conservation of mass for component i is encapsulated in the integro-differential equation:

$$\frac{d}{dt} \int_{V_n} A_m^{(i)} \, dV = - \int_{S_n} \mathbf{Q}_m^{(i)} \cdot \mathbf{n} \, dS + \int_{V_n} q_m^{(i)} \, dV \tag{41}$$

where S_n is the boundary surface to the volume element and $\mathbf{n}$ is the outward-pointing normal to S_n, while $A_m^{(i)}$ is defined in Equation (2), $\mathbf{Q}_m^{(i)}$ in Equation (7) and other parameters have already been defined. [Note that this equation may be reduced to Equation (1) by standard calculus techniques.] Introduction of appropriate volume averages allows the mass accumulation for component i in volume element n, the first volume integral in Equation (41), to be written:

$$\int_{V_n} A_m^{(i)} \, dV = V_n M_n^{(i)} \tag{42}$$

where $M_n^{(i)}$, a function of time t, is the average value of $A_m^{(i)}$ over V_n.

The surface integrals are approximated by a sum of average fluxes between an element and its neighbours:

$$- \int_{S_n} \mathbf{Q}_m{}^{(i)} \cdot \mathbf{n} \, dS \;=\; \sum_j S_{nj} \, Q_m{}^{(i)}{}_{nj} \tag{43}$$

where S_{nj} is the surface area between element n and neighbouring element j. The areally-averaged mass flux $Q_m{}^{(i)}{}_{nj}$ across surface S_{nj} from element j into element n is obtained from the discretised form of Equations (5) – (7), and may be expressed as:

$$\begin{aligned} Q_m{}^{(i)}{}_{nj} \;=\; k_{nj} \Bigg\{ & X_\ell{}^{(i)}{}_{nj} \left(\frac{k_{r\ell}\,\rho_\ell}{\mu_\ell}\right)_{nj} \left[\frac{p_{\ell j} - p_{\ell n}}{d_{nj}} + \rho_{\ell nj}\, g_{nj}\right] + \\ & + X_g{}^{(i)}{}_{nj} \left(\frac{k_{rg}\,\rho_g}{\mu_g}\right)_{nj} \left[\frac{p_{gj} - p_{gn}}{d_{nj}} + \rho_{gnj}\, g_{nj}\right]\Bigg\} + \\ & + D_w{}^{(i)}{}_{nj}\, \rho_{gnj} \frac{X_g{}^{(i)}{}_j - X_g{}^{(i)}{}_n}{d_{nj}} \end{aligned} \tag{44}$$

The subscripts (nj) indicate that the quantity is to be evaluated at the interface between elements n and j, based on average values within V_n and V_j. Various weighting procedures are used to ensure stability. The distance d_{nj} between the nodal centres of elements n and j is used in calculation of gradients of pressure and mass fraction.

The conservation of energy within volume element n is given by:

$$\frac{d}{dt} \int_{V_n} A_e \, dV \;=\; - \int_{S_n} \mathbf{Q}_e \cdot \mathbf{n} \, dS \;+\; \int_{V_n} q_e \, dV \tag{45}$$

where A_e is defined in Equation (9), $\mathbf{Q}_e$ in Equation (10) and all the other parameters have already been defined. [This may be reduced to Equation (8) by standard calculus methods.] Analogously to the mass term above, the energy accumulation in volume element n is approximated by

$$\int_{V_n} A_e \, dV \;=\; V_n \, E_n \tag{46}$$

The discretised form of the areally-averaged energy flux from element j into element n may be deduced from Equation (10) and the forms of Equations (43) and (44).

For each of the N_{VE} volume elements, the mass accumulation terms $M_n{}^{(i)}$ for each of the N fluid components as well as the energy accumulation term E_n must be evaluated. These depend, through thermodynamic relationships, on a set of primary variables which may be chosen according to the problem and the phase composition. A common set is $\{T_n, p_n, X_n{}^{(1)}, X_n{}^{(2)}, \ldots, X_n{}^{(N)}\}$ when single-phase conditions exist in the element, or $\{T_n, S_{\ell n}, p_{gn}{}^{(1)}, p_{gn}{}^{(2)}, \ldots, p_{gn}{}^{(N)}\}$ when two phases are present. There is therefore a total of $(N + 1) \times N_{VE}$ primary variable quantities to be calculated from the same number of the discretised forms of the mass and energy balance equations.

An implicit time-stepping procedure is used to ensure stability, and the problem is reduced to a set of coupled algebraic equations in the set of derived unknown quantities $\{(E_n, M_n^{(1)}, M_n^{(2)}, \ldots, M_n^{(N)}), n = 1, 2, \ldots, N_{VE}\}$, all of which are functions of the primary variables. Fast algorithms for solving large sets of sparse linear equations are used, see Pruess [12]. Because the derived quantities are non-linear functions of the primary variables, Newton-Raphson iteration techniques are used to speed up the convergence at each time step.

Simulations

There is a large literature dealing with geothermal reservoir simulation, both from the theoretical viewpoint and as case studies. The latter are often contained in technical reports to companies or government agencies which need to know the possible effects of exploitation of a resource.

Theoretical studies are based on the sciences (physical chemistry, classical physics, geophysics, geology, thermodynamics, mathematics, statistics, computational methods) and establish commonality with other similar processes. An extensive list of references would take too much space here and therefore only a few published works are referred to here. The reader can work from the reference lists provided therein [9, 12, 17 – 24].

DEEP HIGH-PRESSURE HIGH-TEMPERATURE SYSTEMS

As mathematical modelling of geothermal systems at subcritical conditions has advanced, so more attention is now being focussed on modelling the deeper regions which supply some fluid and most of the heat to the base of geothermal reservoirs. The deep temperatures and pressures constitute conditions which may be regarded as supercritical for pure water, but since the deep fluids carry solutes and gases released from magma sources, such components need to be included in any models of mass and heat flows at depth. The thermodynamic state-space (phase-space) for such mixtures is not as simple as that for pure water.

No very deep wells have yet been drilled to sample fluids much below the bottom of geothermal reservoirs. However, geochemical evidence from nearer-surface fluid samples, as well as the chemical characteristics of surface discharge features of geothermal systems, indicate that the main dissolved salt is NaCl (perhaps 80% of total solutes) and the main non-condensible gas component is CO_2.

A first attempt at constructing models of deep flows is currently being made by McKibbin and McNabb [7, 8], with fluid properties based on a H_2O-NaCl-CO_2 system. Since system conditions involving a brine saturated with respect to chloride cannot be ruled out, contributions from the solid chloride precipitate phase are explicitly included in the mass and energy accumulation terms, and there are no internal mass or energy sources or sinks. If the volumes of liquid, gas, solid chloride and rock phases within a representative elementary formation volume V are denoted

V_ℓ, V_g, V_{sc}, and V_r respectively, then the porosity ϕ and the voidage ε (volume fraction of the pores not filled with solid precipitate and therefore available for the fluid phases) are defined by:

$$\phi = \frac{V_\ell + V_g + V_{sc}}{V} = \frac{V - V_r}{V} \qquad \text{and} \qquad \varepsilon = \frac{V_\ell + V_g}{V_\ell + V_g + V_{sc}} \tag{47}$$

Conceptual model. The model is built by regarding the non-condensible gas as effectively a component added to the brine, the properties of which are determined by the mass fractions of water and chloride relative to their own total mass, rather than to the total mass of water plus chloride plus non-condensible gas. In this regard, the basic fluid is a brine, rather than pure water as in the model already described in the sections above for geothermal systems. The brine is then treated as one "component" of the mixture, but is itself characterised by the overall mass fraction X of chloride in the brine. It is assumed that the water and any non-condensible gas reside only in the fluid phases.

For that reason, the mass fractions $X^{(i)}$ are defined slightly differently from those in the general development outlined in the first part of this work. The chloride mass fraction X is given by

$$X = \frac{A_m{}^c}{A_m{}^w + A_m{}^c} = \frac{A_m{}^c}{A_m{}^b} \tag{48}$$

where $A_m{}^b = A_m{}^w + A_m{}^c$ is the mass of brine, while the mass fraction Y of the non-condensible gas relative to the total non-matrix mass is defined by:

$$Y = \frac{A_m{}^n}{A_m{}^w + A_m{}^c + A_m{}^n} = \frac{A_m{}^n}{A_m{}^b + A_m{}^n} \tag{49}$$

It follows that the mass fraction of water in the brine, and that relative to the overall mixture, are given by

$$\frac{A_m{}^w}{A_m{}^b} = 1 - X \qquad \text{and} \qquad \frac{A_m{}^w}{A_m{}^b + A_m{}^n} = (1 - X)\,(1 - Y) \tag{50}$$

respectively, while the chloride mass fraction in the overall mixture is:

$$\frac{A_m{}^c}{A_m{}^b + A_m{}^n} = X\,(1 - Y) \tag{51}$$

Note that $A_m{}^n = 0$ gives $Y = 0$, and the analysis is then the same as that for a brine without a non-condensible gas.

The above ratios allow the masses of chloride and non-condensible gas to be expressed in terms of the mass of water present as follows:

$$A_m{}^c = \frac{X}{1-X} A_m{}^w \qquad \text{and} \qquad A_m{}^n = \frac{Y}{(1-X)(1-Y)} A_m{}^w \tag{52}$$

These relationships enable the flow of the various components in the system to be completely specified using the water flowrate $\mathbf{Q}_m{}^w$ together with mass fractions X and Y. The brine salinity X represents the NaCl equivalent concentration of all solutes, while Y is the CO_2 equivalent of all non-condensible gas components.

The ***mass conservation equations*** are written in the form:

$$\text{H}_2\text{O:} \qquad \frac{\partial A_m{}^w}{\partial t} + \nabla \cdot \mathbf{Q}_m{}^w = 0$$

$$\text{NaCl:} \qquad \frac{\partial A_m{}^c}{\partial t} + \nabla \cdot \mathbf{Q}_m{}^c = 0 \tag{53}$$

$$\text{CO}_2\text{:} \qquad \frac{\partial A_m{}^n}{\partial t} + \nabla \cdot \mathbf{Q}_m{}^n = 0$$

where

$$A_m{}^w = \phi\, \varepsilon\, [S_\ell(1-X_\ell)(1-Y_\ell)\rho_\ell + (1-S_\ell)(1-X_g)(1-Y_g)\rho_g]$$

$$A_m{}^c = \phi\, \{(1-\varepsilon)\rho_{sc} + \varepsilon\, [S_\ell X_\ell(1-Y_\ell)\rho_\ell + (1-S_\ell)X_g(1-Y_g)\rho_g]\}$$

$$A_m{}^n = \phi\, \varepsilon\, [S_\ell Y_\ell \rho_\ell + (1-S_\ell)Y_g\rho_g]\}$$

are the mass per unit volume of formation of the water, chloride and non-condensible gas respectively, given in terms of the chloride mass fractions X_ℓ and X_g and the CO_2 mass fractions Y_ℓ and Y_g within each fluid phase, the liquid, gas and solid chloride phase densities ρ_ℓ, ρ_g and ρ_{sc}, and in terms of other parameters already defined.

The total mass of material per unit volume of formation is given by the sum of the respective quantities for the rock matrix, water, chloride and non-condensible gas:

$$\begin{aligned} A_m &= A_m{}^r + A_m{}^w + A_m{}^c + A_m{}^n \\ &= (1-\phi)\rho_r + \phi\, \{(1-\varepsilon)\rho_{sc} + \varepsilon\, [S_\ell\rho_\ell + (1-S_\ell)\rho_g]\} \end{aligned} \tag{54}$$

where ρ_r is the density of the solid (rock) matrix. Note that the mass per unit formation volume of the chloride within the fluid phases only is given by

$$A_{mf}{}^c = \phi\, \varepsilon\, [S_\ell X_\ell(1-Y_\ell)\rho_\ell + (1-S_\ell)X_g(1-Y_g)\rho_g]\} \tag{55}$$

When no solid chloride is present, the voidage $\varepsilon = 1$, and $A_m{}^c = A_{mf}{}^c$.

The ***momentum conservation equations*** (Darcy's law) are as given in Equations (5a, b). Molecular diffusion is assumed small compared to the advective flux and is neglected. The specific mass fluxes are then given for each component by the sum of the mass fluxes associated with each fluid phase, as in Equations (6a), 6(b) and (7):

$$\mathbf{Q}_m{}^w = \mathbf{Q}_{m\ell}{}^w + \mathbf{Q}_{mg}{}^w = (1 - X_\ell)(1 - Y_\ell)\,\mathbf{Q}_{m\ell} + (1 - X_g)(1 - Y_g)\,\mathbf{Q}_{mg}$$

$$\mathbf{Q}_m{}^c = \mathbf{Q}_{m\ell}{}^c + \mathbf{Q}_{mg}{}^c = X_\ell\,(1 - Y_\ell)\,\mathbf{Q}_{m\ell} + X_g\,(1 - Y_g)\,\mathbf{Q}_{mg}$$

$$\mathbf{Q}_m{}^n = \mathbf{Q}_{m\ell}{}^n + \mathbf{Q}_{mg}{}^n = Y_\ell\,\mathbf{Q}_{m\ell} + Y_g\,\mathbf{Q}_{mg}$$

The equation of ***conservation of energy*** in the absence of sources or sinks, is:

$$\frac{\partial A_e}{\partial t} + \nabla \cdot \mathbf{Q}_e = 0 \tag{56}$$

where

$$A_e = (1 - \phi)\rho_r u_r + \phi\,\{(1 - \varepsilon)\rho_{sc} u_{sc} + \varepsilon\,[S_\ell \rho_\ell u_\ell + (1 - S_\ell)\rho_g u_g]\}$$

is the energy per unit volume of the fluid-filled formation with u_r, u_{sc}, u_ℓ and u_g being the specific internal energies for the rock matrix, solid chloride, liquid and gas phases respectively. The specific energy flux per unit area $\mathbf{Q}_e$ is given by:

$$\mathbf{Q}_e = \mathbf{Q}_{m\ell}\,h_\ell + \mathbf{Q}_{mg}\,h_g + \mathbf{K} \otimes (-\nabla T) \tag{57}$$

Rock properties are based on a stationary rock matrix. However, at near-magma depths, this assumption is probably not correct since the rock is not crystalised, and the model would then not apply. Otherwise, in regions where the solid matrix assumption does apply, it is assumed in the current model that any solute deposition does not alter the effective permeability. Improvements to the model would require some dependence of permeability on voidage to be included, and/or rock stress analysis to describe dilation of the formation as deposition continues.

Fluid properties are the focus of recent attention, see for example Kissling [11] and White and Kissling [25]. While there is some data on NaCl brine properties for the liquid phase at lower temperatures, there is little for the regions of the p-T-X state-space for a H_2O-NaCl brine mixture that would apply in deep systems. The p-T-X state-space itself is complicated, and mass flows in such a brine system trace state-paths through the space. Provided an approximate realisation of the H_2O-NaCl state-space region boundaries is made, and similarly approximate correlations are found for the density, energy and viscosity, then a first model for deep brine fluids has been found. Addition of a non-condensible gas then completes the essential ingredients of deep high-pressure high-temperature systems.

SUMMARY

An overview of currently-used mathematical models for geothermal heat and mass transfer has been given. The sets of partial differential equations which describe the principles of conservation of mass, momentum and energy of such multi-phase multi-component systems are further complicated by complex dependence of the various formation and fluid parameters on thermodynamic variables.

While little has been said here about the related disciplines of geology, geophysics, geochemistry and reservoir engineering, these use scientific methods to deduce the formation parameters for geothermal reservoirs, and also provide information about reservoir extent and likely boundary conditions for simulations. Without such a multidisciplinary approach, interaction and feedback about conceptual and mathematical models would not be possible.

There are few analytical solution methods which produce useful results, although some provide insights on a local scale. Numerical procedures are successful in producing solutions to the governing equations, but require sophisticated discretisation and matrix inversion methods.

Recent efforts are focussed more on deeper heat and mass flows in regions below geothermal reservoirs, in order to provide better understanding of the processes which transfer heat and chemicals from deep magmatic sources to the base of reservoirs and to surface discharge features. These investigations may eventually prove useful also in modelling ore formation and other deposition processes.

REFERENCES

1. D. H. Freeston, Direct uses of geothermal energy 1995 (Preliminary review). *Proc. World Geothermal Congress,* Florence, May 1995, pp. 15-25. International Geothermal Association, Auckland (1995).
2. G. W. Huttrer, The status of world geothermal power production 1990-1994. *Proc. World Geothermal Congress,* Florence, May 1995, pp. 3-14. International Geothermal Association, Auckland (1995).
3. J. Bear, *Dynamics of Fluids in Porous Media.* Dover, New York (1972).
4. J. Bear and Y. Bachmat, *Introduction to Modeling of Transport Phenomena in Porous Media.* Kluwer, Dordrecht (1991).
5. D. A. Nield and A. Bejan, *Convection in Porous Media.* Springer-Verlag, New York (1992).
6. A. T. Corey, The interrelation between gas and oil relative permeabilities. *Producers Monthly* **19**, 38-41 (1954).
7. R. McKibbin and A. McNabb, Modelling the phase boundaries and properties of the system H_2O-NaCl at high temperatures and pressures. *Proc. 15th New Zealand Geothermal Workshop 1993*, The University of Auckland, 267-273 (1993).
8. R. McKibbin and A. McNabb, From magma to groundwater: the brine connection. *Proc. World Geothermal Congress,* Florence, May 1995, pp. 1125-1130. International Geothermal Association, 1125-1130 (1995).

9. M. J. O'Sullivan, G. S. Bodvarsson, K. Pruess and M. R. Blakeley, Fluid and heat flow in gas-rich geothermal reservoirs. *Soc. Petroleum Engineers J.*, 215-226 (1985).
10. K. Pruess, Numerical simulation of water injection into vapor-dominated reservoirs. *Proc. World Geothermal Congress,* Florence, May 1995, pp. 1673-1679. International Geothermal Association, Auckland (1995).
11. W. M. Kissling, Extending MULKOM to supercritical temperatures and pressures. *Proc. World Geothermal Congress,* Florence, May 1995, pp. 1687-1690. International Geothermal Association, Auckland (1995).
12. K. Pruess, *Development of the General Purpose Simulator MULKOM.* Earth Sciences Division Annual Report 1982, Report LBL-15500, Lawrence Berkeley Laboratory, Berkeley, California (1983).
13. R. H. Perry, D. W. Green and J. O. Maloney, *Perry's Chemical Engineers' Handbook* (6th edn). McGraw-Hill, Singapore (1984).
14. R. McKibbin and K. Pruess, Some effects of non-condensible gas in geothermal reservoirs with steam-water counterflow. *Geothermics* **18**, 367-375 (1989).
15. M. McGuinness, Heat pipes and through-flows in geothermal reservoirs. *Proc. 18th New Zealand Geothermal Workshop 1996,* The University of Auckland, 285-290 (1996).
16. J. Bear and A. Verruijt, *Modeling Groundwater Flow and Pollution.* Reidel, Dordrecht (1987).
17. K. Pruess (Ed), *Proceedings of the TOUGH Workshop.* Report LBL-29710, Lawrence Berkeley Laboratory, Berkeley, California (1990).
18. K. Pruess (Ed), *Proceedings of the TOUGH Workshop '95.* Report LBL-37200, Lawrence Berkeley Laboratory, Berkeley, California (1995).
19. M. J. O'Sullivan and R. McKibbin, *Geothermal Reservoir Engineering: A Manual for Geothermal Reservoir Engineering courses* (2nd edn). Geothermal Institute, The University of Auckland, New Zealand (1988).
20. *Proceedings of the Workshop on Geothermal Reservoir Engineering.* Stanford Geothermal Program, Stanford University, Stanford, California (1976-).
21. *Proceedings of the New Zealand Geothermal Workshop.* Geothermal Institute, The University of Auckland (1989-).
22. *Geothermics.* Elsevier-Pergamon, Oxford (1972-).
23. M. A. Grant, I. G. Donaldson and P. F. Bixley, *Geothermal Reservoir Engineering.* Academic, New York (1982).
24. *Proceedings of the World Geothermal Congress,* Florence, May 1995. International Geothermal Association, Auckland (1995).
25. S. P. White and W. M. Kissling, Including chloride and CO_2 chemistry in large-scale reservoir models. *Proc. 18th New Zealand Geothermal Workshop 1996*, The University of Auckland, 295-300 (1996).

NATURAL CONVECTION IN A HORIZONTAL POROUS ANNULUS

M. C. CHARRIER-MOJTABI and A. MOJTABI

Institut de Mécanique des Fluides de Toulouse, UMR CNRS/INP/UPS n° 5502, Avenue du Professeur Camille Soula , 31400 Toulouse, France

INTRODUCTION

Natural convection in porous media is of interest in many applications and the cylindrical annular geometry has a wide variety of technological applications, such as thermal insulators, underground cable systems, storage of thermal energy, etc.

The problem of accurately determining the fluid flows and the heat transfer between the two cylinders, and the stability of the flows observed numerically and experimentally, has stimulated a large number of investigations over the past twenty years. Although a large number of papers have numerically dealt with two-dimensional natural convection in a horizontal porous annulus, very few deal with experimental or stability analysis of three-dimensional free convective flows. However, three-dimensional convective flows frequently occur within a porous annulus at low Rayleigh numbers, as demonstrated by Caltagirone [1]. In that investigation, Caltagirone [1] showed that in a cell of large longitudinal aspect ratio ($A = L/(r_o-r_i) = 20$ and radii ratio $R = r_o/r_i = 2$, where r_i and r_o denote the inner and outer radius of the cylinders, respectively, and L is the length of the cylindrical annular) three-dimensional perturbations appear at the top of the annulus even for low Rayleigh numbers. The presence of these three-dimensional perturbations increases the overall heat transfer compared to those obtained for two-dimensional flows.

Two-dimensional, unicellular flows have been studied by several numerical approaches, such as the finite-difference method, see for example Caltagirone [1] and Burns and Tien [2], the finite element method, see Mojtabi *et al.* [3]), or the Galerkin spectral method, see Charrier -Mojtabi and Caltagirone [4], Rao *et al.* [5] and Himasekhar and Bau [6]. For these flows, some asymptotic solutions have also been proposed by Caltagirone [1] and Himasekhar and Bau [6] but more recently by Mojtabi *et al.* [7] using a symbolic algebra code.

The Galerkin spectral method has also been used by several investigators [5, 6, 8, 9] to describe the two-dimensional multicellular flows which appear in the annulus, depending on the radii ratio of the two cylinders, for larger Rayleigh numbers. Charrier-Mojtabi *et al.* [8], have shown that the Fourier-Chebyshev method gives a better accuracy than does the full Fourier-Galerkin method for the description of these two-dimensional multicellular flows. Some authors, for example Himasekhar and Bau [10], have also studied the effects of the eccentricity on the overall heat transfer.

Using a finite-difference method, Fukuda *et al.* [11] analyzed the three-dimensional flows for the case of an inclined annuli, but their results could not be extended to the case of a horizontal annulus. Charrier-Mojtabi *et al.* [12] used the Fourier-Galerkin spectral method to expand the three-dimensional temperature and fluid velocity, but only low orders of approximation were considered. In fact the main numerical results concerning the three-dimensional flows are due to Rao *et al.* [13] and Charrier-Mojtabi [14]. Rao *et al.* [13] used the Galerkin method and developed the temperature and the potential velocity vector into a truncated Fourier series. They described steady three-dimensional flows with orders of approximation up to $10\times 13\times 5$ in the r, ϕ and z directions, respectively. In their study, only one value of the radii ratio, namely R = 2, was considered for a longitudinal aspect ratio, A = 2, and Rayleigh numbers varying between 60 and 150. More recently, Charrier-Mojtabi [14] have used a spectral method which is based based on the mixed Fourier-Chebyshev approximation, with a formulation in terms of pressure and temperature to solve both the two- and three-dimensional Darcy-Boussinesq equations. The hysteresis loop, mentioned by Barbosa *et al.* [15] and [16] was not obtained in this study.

Aboubi *et al.* [17] and [18] reported studies of natural convection in an annular porous layer, with a centrifugal field, or in a horizontal annulus filled with an anisotropic porous medium while, in the same geometry, Kimura and Pop [19] have studied the non-Darcian effects on conjugate natural convection. In the same configuration, but for a fluid medium, some investigations have been made by Cheddadi *et al.* [20] for two-dimensional flows, and by Rao *et al.* [21] and Vafai *et al.* [22] for three-dimensional flows. The numerical technique used for all these works was a finite-difference method.

Experimental studies using the Christiansen effect to visualize the thermal two-dimensional fields have been carried out by Cloupeau *et al.* [23] for the description of two-dimensional flows with a cell of longitudinal aspect ratio of A = 0.5 and a radii ratio of R = 2. They only observed unicellular flows. Charrier-Mojtabi *et al.* [8], with a cell of dimensionless parameters A = 0.5 and R = 2, using the Christiansen effect, have also observed unicellular steady structures and have shown experimentally the physical existence of two-dimensional bicellular steady structures for Rayleigh numbers higher than 65 (± 5), with one small counterrotating cell at the top of the annulus. For Ra* higher than 65, either a two-dimensional unicellular flow or a two-dimensional bicellular flow can be observed for the same value of Ra* and each of them is stable.

In an eccentric annulus, Bau *et al.* [24] have obtained numerically, for the same set of dimensionless parameters (for example, R = 2, eccentricity of 0.6, Ra* = 200), either a unicellular steady flow or a multicellular convective structure with two small counterrotating cells at the top of the annulus, but this multicell structure was unstable.

Since higher heat transport rates are associated with multicellular flows, it seems important to establish which type of flow is physically realizable. Both linear and non-linear stability analyses of the Rayleigh-Benard problem have been the subject of numerous publications. In an annular layer, and due to the horizontal temperature gradient, a conduction regime cannot occur for a non-zero temperature difference. Due to the difficulty of obtaining the equation giving the monocellular basic flow in this configuration only a few stability studies are available.

Stability analyses have been performed by Caltagirone [1] and Himasekhar and Bau [6]. In Caltagirone [1], the transition between two-dimensional unicellular flow and three-dimensional flows with two-dimensional axisymmetrical perturbations depending on r and z, the radial and axial coordinates, respectively, were investigated. Based on physical considerations, this stability analysis was limited to the upper part of the annular layer and the axisymmetrical, two-dimensional disturbances considered were for the temperature and the velocity components in the radial and axial directions, respectively. Himasekhar and Bau [6], have considered the transition between the two-dimensional unicellular flow and two-dimensional multicellular flows using two-dimensional perturbation analysis based on the Galerkin method and a power series expansion of the Rayleigh number up to order 60 to approximate the basic two-dimensional unicellular steady-state solution. These studies by Caltagirone [1] and by Himasekhar and Bau [6] have contributed to a first understanding of convection in an annular porous medium.

The experimental studies conducted by Caltagirone [1] and by Charrier-Mojtabi *et al.* [8] have shown the existence of various convective regimes. The stability analyses of two-dimensional unicellular flow, as reported in the literature see for example Caltagirone [1] and Himasekhar and Bau [6] do not allow us to predict which regime (a two-or three-dimensional flow) will set in for a given triplet of values Ra, R and A.

To define the critical conditions for the onset of instabilities, we consider :

(i) A synthesis of the two-dimensional and three-dimensional numerical simulations using a collocation-Chebyshev method. This method gives an accurate description of the two-dimensional multicellular flows for a large range of Rayleigh numbers and radii ratios. The three-dimensional spiral flows are described in the vicinity of the transition with the two-dimensional unicellular flow. Bifurcation points between two-dimensional unicellular flow and two-dimensional multicellular flows or three-dimensional flows are also determined numerically. However, this procedure needs long computation times and only concerns discrete values of either R and A.

(ii) An extension of the previous study developed by Mojtabi *et al.* [7], who used an analytical method to construct an accurate solution to the equations governing two-dimensional natural convection for narrow gaps, is presented. The analytical solution, using a regular perturbation expansion was developed up to order 6. In the present study, we extend their earlier work, to a solution using an asymptotic development up to order 15, in the parameter $\varepsilon = (r_o - r_i)/r_i$. The solution is obtained with a symbolic

algebra code and convergence of the resulting series for the Nusselt number is analyzed. The domain of the convergence of this series is further increased by using the Padé process of acceleration which allows us to obtain an improved estimation of the Nusselt number for unicellular two-dimensional flows in a horizontal porous annulus.
(iii) A linear stability analysis of a steady, two-dimensional basic fluid flow and a non-linear stability analysis, using the energy method, as extended by Joseph [25] and used by Mojtabi *et al.* [26] in a fluid annulus, is presented. We consider the basic fluid flow resulting from the numerical scheme and the one obtained by the perturbation method and we examine the validity of the latter method.

PROBLEM FORMULATION

A porous annular layer which is bounded by two horizontal, concentric cylinders of axial length L is considered. Both the inner and outer cylinders are maintained at different but constant temperatures, T_i and T_o, respectively. The porous medium of porosity ε and permeability K, is saturated with an incompressible Newtonian fluid of kinematic viscosity υ. The saturated porous medium is equivalent to an artificial isotropic fluid of heat capacity $(\rho c)^* = \varepsilon(\rho c)_f + (1-\varepsilon)(\rho c)_s$ and the thermal conductivity λ^*, where $(\rho c)_s$ is the matrix heat capacity. This hypothesis has been extensively studied by Combarnous and Bories [27].

In addition to the Boussinesq approximation, several classical assumptions are made in order to simplify the formulation, namely,
(i) Darcy's law is assumed to be valid.
(ii) The inertia and viscous dissipation terms are neglected.
(iii) The fluid is assumed to be in thermal equilibrium within the porous matrix.
The range of validity of the above assumptions was analysed in detail in Aniri and Vafai [28] and it can be seen from their work that the above assumptions can be used for a large range of practical applications. The conservation of mass, momentum and energy are given as :

$$\nabla . \mathbf{V} = 0 \tag{1}$$

$$(\rho c)^* \frac{\partial T}{\partial t} - \lambda^* \nabla^2 T + (\rho c)_f \mathbf{V} . \nabla T = 0 \tag{2}$$

$$\varepsilon^{-1} \rho \frac{\partial \mathbf{V}}{\partial t} + \nabla p - \rho \mathbf{g} + \mu K^{-1} \mathbf{V} = 0 \tag{3}$$

where $\mathbf{g}$ is the gravitational acceleration, $\mathbf{V} = u\,\mathbf{e}_1 + v\,\mathbf{e}_2 + w\,\mathbf{e}_3$ is the fluid velocity, p is the pressure and T is the temperature, $\mathbf{e}_1$, $\mathbf{e}_2$ and $\mathbf{e}_3$ are the unit vectors of the cylindrical orthogonal system of axes r, ϕ and z, respectively, ϕ is referenced with respect to the upward vertical radius, see figure 1.
A linear state equation is assumed, namely,

$$\rho = \rho_1 \left[1 - \beta(T - T_1)\right] \tag{4}$$

where ρ is the fluid density at temperature T, ρ_1 is the fluid density at temperature T_1 and β is the coefficient of thermal expansion.
The boundary and initial conditions are :

$$T = T_i \;,\; \mathbf{n}.\mathbf{V} = 0 \quad \text{for } r = r_i \quad \forall t > 0 \tag{5}$$
$$T = T_o \;,\; \mathbf{n}.\mathbf{V} = 0 \quad \text{for } r = r_o \quad \forall t > 0 \tag{6}$$
$$V_\phi = \frac{\partial T}{\partial \phi} = 0, \quad \text{at } \phi = 0, \pi \quad V_z = \frac{\partial T}{\partial z} = 0, \quad \text{at } z = 0, L \tag{7}$$

where $\mathbf{n} = \mathbf{e}_1$ is the outward normal unit vector to the cylinders.

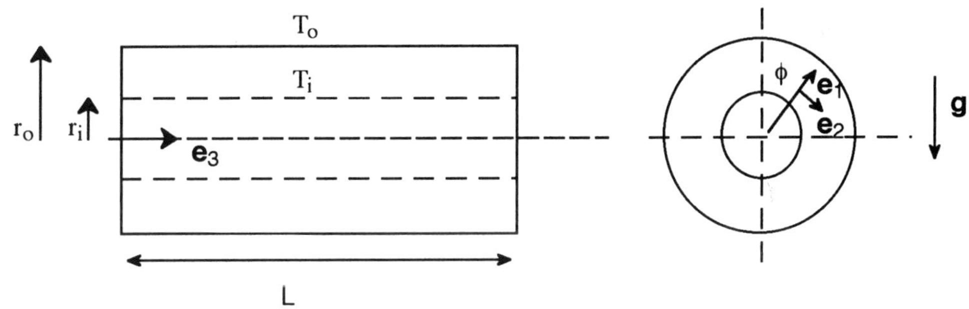

Figure 1 The porous layer and the coordinates system employed

Equations (1)-(7) are now put in dimensionless form with the following reference parameters : r_i for length, $(\rho c)^* r_i^2 / \lambda^*$ for time, $\Delta T = T_i - T_o$ for temperature, $\lambda^* / (\rho c)_f\, r_i$ for velocity and $\lambda^* \mu / K (\rho c)_f$ for pressure and, for simplicity, the same letters will denote both the physical and dimensionless parameters. The transformed system of equations (1)-(3) are now given by :

$$\nabla . \mathbf{V} = 0 \tag{8}$$
$$\frac{\partial T}{\partial t} - \nabla^2 T + \mathbf{V} . \nabla T = 0 \tag{9}$$
$$\varepsilon^{-1} P_r^{-1^*} M Da \frac{\partial \mathbf{V}}{\partial t} + \nabla p + Ra^* \mathbf{k} T + \mathbf{V} = 0 \tag{10}$$

where : $\mathbf{k} = -\cos\phi\, \mathbf{e}_1 + \sin\phi\, \mathbf{e}_2$, $Ra^* = g\beta\, (\rho c)_f\, \Delta T\, K r_i \,/\, (\lambda^* \nu)$, $P_r^* = (\rho c)_f \nu / \lambda^*$, $M = (\rho c)_f \,/\, (\rho c)^*$ and $Da = K / r_i^2$ are, respectively, the direction of the gravity field, the Rayleigh number, the Prandtl number, the ratio of the heat capacities and the

Darcy number which represents the pore fineness. In general, Da is very small $\left(\cong 10^{-4} - 10^{-6}\right)$, so the term containing Da may be neglected.

The boundary conditions (5)-(7), written in dimensionless form, are given by :

$$T = 1, \quad \mathbf{n}.\mathbf{V} = 0 \quad \text{for } r = 1 \quad \forall t > 0 \tag{11}$$

$$T = 0, \quad \mathbf{n}.\mathbf{V} = 0 \quad \text{for } r = R = r_o / r_i \quad \forall t > 0 \tag{12}$$

$$v = \frac{\partial T}{\partial \phi} = 0, \quad \text{at } \phi = 0, \pi \quad w = \frac{\partial T}{\partial z} = 0, \quad \text{at } z = 0, L / r_i \tag{13}$$

Equations (8)-(13), with Da = 0, will be taken as the starting point in our study. It appears, see Caltagirone [1] and Himasekhar and Bau [6] that, for small values of the Rayleigh number, the unicellular steady two-dimensional regime sets in with a vertical symmetry plane. The reference state, defined by the two-dimensional temperature and velocity fields, $T_o(r, \phi)$ and $\mathbf{V}_o(r, \phi)$, respectively, is obtained using a perturbation method, a finite-difference scheme and a spectral method.

NUMERICAL SOLUTION

A conformal transformation is used to move from polar (r, ϕ, Z) to cartesian coordinates (X = ln r , Y = ϕ , Z), the initial domain [1, R] × [0, π] × [0, A] is transformed into a cartesian domain: [0, ln R] × [0, π] × [0, A] .

Taking the divergence of Darcy's law, equation (10), then the velocity term is eliminated and the following transformed pressure and temperature equations are given by :

$$\frac{\partial^2 P}{\partial X^2} + \frac{\partial^2 P}{\partial Y^2} + \exp(2X)\frac{\partial^2 P}{\partial Z^2} = Ra^* \exp(X)[\cos(Y)\frac{\partial T}{\partial X} - \sin(Y)\frac{\partial T}{\partial Y}] \tag{14}$$

$$\frac{\partial^2 T}{\partial X^2} + \frac{\partial^2 T}{\partial Y^2} + \exp(2X)\frac{\partial^2 T}{\partial Z^2} = Ra^* \exp(X) T [\cos(Y)\frac{\partial T}{\partial X} - \sin(Y)\frac{\partial T}{\partial Y}] - [\frac{\partial P}{\partial X}\frac{\partial T}{\partial X} + \frac{\partial P}{\partial Y}\frac{\partial T}{\partial Y} + \exp(2X)\frac{\partial P}{\partial Z}\frac{\partial T}{\partial Z}] \tag{15}$$

with the boundary conditions :

$$X = 0: \quad T = 1, \quad \left.\frac{\partial P}{\partial X}\right|_{X=0} = Ra^* \cos(Y)$$

$$X = \ln R = \frac{1}{\alpha}: \quad T = 0, \quad \left.\frac{\partial P}{\partial X}\right|_{X=\ln R} = 0 \tag{16}$$

$$Y = 0, \pi: \quad \frac{\partial P}{\partial Y} = 0, \quad \frac{\partial T}{\partial Y} = 0; \quad Z = 0, A: \quad \frac{\partial P}{\partial Z} = 0, \quad \frac{\partial T}{\partial Z} = 0.$$

The set of equations (14)-(15), and the boundary conditions (16), are also valid for the two-dimensional steady state if all the derivatives $\frac{\partial}{\partial Z}$ and $\frac{\partial^2}{\partial Z^2}$ are set identically to zero.

The conformal mapping into a cartesian domain allows the easier use of the spectral method to obtain the numerical solution of the problem. However, it should be noted that for the three-dimensional case, the presence of the term exp (2X) in the Laplacian operator, see equations (14)-(15), may introduce some numerical instabilities and reduce the efficiency of this transformation.

Numerical Procedure: A Mixed Fourier-Chebyshev Approximation

In this section, the mixed Fourier-Chebyshev approximation is only presented for the three-dimensional case but for the two-dimensional case a similar development, using a stream function and temperature formulation, has been employed by Charrier-Mojtabi *et al.* [8].

To solve the equations (14-15), subject to the boundary conditions (16), the pressure P and the temperature T are expanded in terms of Fourier series in both the azimuthal and axial periodic directions and as Chebyshev polynomials of orders N, K in the confined radial direction as follows :

$$P_{N,K} = \sum_{n=0}^{N}\sum_{k=0}^{K} f_{nk}(X)\cos(nY)\cos(k\pi BZ)$$

$$T_{N,K} = 1 - \alpha X + \sum_{n=0}^{N}\sum_{k=0}^{K} g_{nk}(X)\cos(nY)\cos(k\pi BZ) \qquad (17)$$

where $B = (r_o - r_i)/L$. Expansions (17) are inseted into equations (14) and (15), which are then appropriately projected on to the Fourier basis functions (this procedure correspond to the identification of the coefficients of sin and cos after the substitution).

The functions f_{pk} and g_{pk} are thus solutions of the second-order differential system :

$$\frac{\partial^2 f_{pk}(x)}{\partial x^2} - \left(\frac{p^2}{4\alpha^2} + \frac{k^2\pi^2B^2e^{(x+1)/\alpha}}{4\alpha^2}\right)f_{pk}(x) = A_{pk} \qquad (18)$$

$$\frac{\partial^2 g_{pk}(x)}{\partial x^2} - \left(\frac{p^2}{4\alpha^2} + \frac{k^2\pi^2B^2e^{(x+1)/\alpha}}{4\alpha^2}\right)g_{pk}(x) = B_{pk} \qquad (19)$$

where $x = 2\alpha X - 1$; $x \in [-1,+1]$. The expressions for the terms A_{pk} and B_{pk} and the boundary conditions for the functions $f_{pk}(x)$ and $g_{pk}(x)$ are given in the appendix of Charrier-Mojtabi[14].

Equations (18) and (19) are discretized, using the collocation-Chebyshev method, see Canuto *et al.* [29], with the Gauss-Lobatto points : $\{ x_i = \cos(i\pi / (M-1)) , i = 0....M-1\}$. Near the boundaries $x = -1$ (i.e. $r = 1$) and $x = +1$ (i.e. $r = R$) this denser grid system ensures an accurate description of the radial boundary layers.

The functions $f_{pk}(x)$ and $g_{pk}(x)$ are expanded into a truncated series of Chebyshev polynomials as follows :

$$f_{pk}(x) = \sum_{m=1}^{M} \xi_{mpk} T_{m-1}(x) \quad \text{and} \quad g_{pk}(x) = \sum_{m=1}^{M} \theta_{mpk} T_{m-1}(x)$$

where $T_k(x) = \cos[k \arccos(x)]$. With the collocation-Chebyshev method, all the computations are made in the physical space. Thus the unknowns are not the spectral coefficients ξ_{mpk} and θ_{mpk} , but the values of the functions $f_{pk}(x)$ and $g_{pk}(x)$ at the collocation points $\{ x_i = \cos(i\pi /(M-1)), i= 0....M-1\}$.

The linearized version of equations (17) and (19) are solved using a diagonalization procedure of the Chebyshev second-order derivative operator, D^{*2}_{CL}, suitably modified to take into account the boundary conditions for the pressure (Neumann conditions) and the homogeneous boundary conditions for the temperature (Dirichlet conditions), see Haldenwang *et al.* [30]. The diagonalization itself is performed once and for all in a preprocessing storage step and for more details concerning this procedure (time integration, convergence criterion, etc.) the reader should consult Charrier-Mojtabi [14].

Multicellular Two-dimensional Flow Analysis

Comparison with the results obtained using a Ψ-T formulation. The results obtained with P-T formulation were first compared to those obtained using the stream function and the temperature (Ψ-T) formulation, see Charrier -Mojtabi *et al.* [8]. In both cases, different flow structures and isotherms may appear for the same values of Ra* and R, depending on the initial conditions introduced in the computations. For all the cases which were investigated in this chapter, namely $Ra^* \in [10, 350]$ and $R = 2^{1/4}, 2^{1/2}$ and 2, very good agreement was found between the results obtained using the two formulations.

Figures 2a and 2c show the streamlines and isotherms and figures 2b and 2d the pressure fields and the isotherms for the case Ra* = 120 and R = 2. For this configuration, with the approximation $M\times N=30\times 30$, a two-dimensional unicellular, figures 2a, 2b, or bicellular flow, figures 2c, 2d can be obtained. The flow configuration consists of one or two cells on half the domain. If the approximations higher than $M\times N=20\times 20$ is sufficient to describe the two-dimensional unicellular or bicellular flows, an approximation up to 45× 135 are necessary to obtain an accurate description of the two-dimensional multicellular flows.

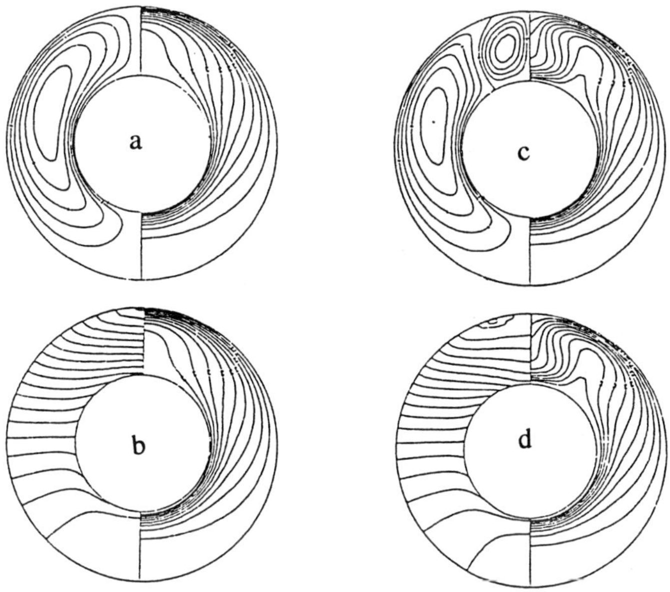

Figure 2 Streamlines and isotherms ; Pressure field and isotherms (Unicellular (a, b) and bicellular (c, d) flows) for Ra* = 120 and R = 2

It can be observed that the pressure field is very weakly affected by the flow structure modification and a similar behaviour is observed for the natural convection in a horizontal annulus filled with a fluid, see Cheddadi *et al.* [18].

Bifurcation phenomena for R = 2 and R = $2^{1/2}$. The numerical determination of the bifurcation points between the two-dimensional unicellular flow and multicellular flows requires an accurate description of the basic fluid flow. Therefore higher orders of approximation, up to M = 45 collocation points in the radial direction and N = 135 Fourier modes in the azimuthal direction have been considered.

The process of determinating the bifurcation point is discussed in detail in Charrier-Mojtabi [14], where it was found that the critical value of the Rayleigh number, corresponding to the transition from two-dimensional unicellular to two-dimensional bicellular flow, for R = 2, is $60.5 < Ra_c^* < 61.5$. This value, was found, using the order of approximation of 30×95, is lower than the value obtained numerically by Rao *et al.* [5] using the Galerkin method at low approximation

(Ra_c*= 65.5 ± 0.5) but it is close to the value suggested by Himasekhar and Bau [6], namely $Ra_c^* \approx 62$.

For the case $R = 2^{1/2}$, the critical Rayleigh number obtained for the transition between the two-dimensional unicellular and bicellular flows, using an order of approximation of 30×95, is $111.5 < Ra_c^* < 112$, whereas Himasekhar and Bau [6] found that $Ra_c^* \approx 112$. Further, our numerical analysis has show the existence of two stable solutions for $Ra^* > Ra_c^*$, for both R = 2 and $R = 2^{1/2}$ and the results are is in good agreement with those of Himasekhar and Bau [6] who used a linear stability technique. They showed that for R = 2 and $R = 2^{1/2}$, the loss of uniqueness occurs without the loss in stability of the unicellular flow and that these two branches of the solution lose their stability via a Hopf bifurcation for high Rayleigh numbers. On the contrary, for $R = 2^{1/4}$ and $2^{1/8}$, these authors have shown that the transition between the two-dimensional unicellular flow and two-dimensional multicellular flow occurs via a "perfect bifurcation", i.e. one solution branch loses stability whilst the another one gains it.

We find also for R = 2 that a two-dimensional tricellular flow appears for the lower-order approximations but disappears for the higher-order approximations whilst the two-dimensional bicellular flow persists even for higher-order approximations. These numerical results appear to agree well with the experimental visualizations of the thermal field using the Christiansen effect, see Charrier-Mojtabi *et al.* [8]. It should be noted that these experimental results concerning the bicellular flow is only observed in the case of a small cell length, namely A < 1/2, and our stability analysis has confirmed this result.

Analysis of the three-dimensional flows. The interest in this new two-dimensional P-T formulation is that it may be naturally extended to three dimensions. However, the three-dimensional study was limited to low orders of approximation, up to $M \times N \times K = (16)^3$ and this limitation is due to the presence of the term exp(x) in the equations (18) and (19). The explicit treatment of the terms $\exp(x) f_{pk}(x)$ and $\exp(x) g_{pk}(x)$ penalizes this procedure for higher orders of approximation. The same two values of R have been considered, R = 2 and $R = 2^{1/2}$, while the longitudinal aspect ratio, A, was varied from 0.5 to 2. Computations were conducted with Rayleigh numbers close to the critical value of the transition between the two-dimensional unicellular flow and the three-dimensional flow. For these values of Ra*, the flow is still mainly two-dimensional, except in the upper part of the layer and the approximation $M \times N \times K = (16)^3$ is sufficient to describe the three-dimensional effects, see Charrier-Mojtabi [14]. It can be seen that, for Rayleigh numbers close to the bifurcation point, the change from $M \times N \times K = 16 \times 12 \times 12$ to $16 \times 16 \times 16$ induces less than a 0.6 % change in the mean Nusselt number.

Transition between the two-dimensional and the three-dimensional convection regimes. We find numerically that the transition between the two-dimensional unicellular flow and the three-dimensional flow, for A = 2 and R = 2 corresponds to $55 < Ra_c^* < 60$ and this result is in reasonable agreement with the experimental results of Caltagirone[1], who obtained Ra_c^* exp. = 65 ±4 for R = 2 and A = 20.

As might be expected, the three-dimensional spiral flow produces a larger heat transfer rate than does the two-dimensional unicellular, but the difference is rather small. It can also be seen that the heat transfer rate for the three-dimensional flow is smaller than that for the two-dimensional bicellular flow and these results agree with the numerical results reported by Rao *et al.* [14]. For the three-dimensional case, good agreement is found with the experimental results of Caltagirone [1] for R = 2 and A = 20, and this is due to the periodicity of the flow along the axial direction.

Stability analyses performed by Caltagirone [1] and Himasekhar and Bau [6] do not allow us to forecast which flow will occur. One of the purposes of this analysis, for the case of narrow gaps ($R << 2^{1/2}$), is to determine a criterion for which a two-dimensional multicellular structure can appear before a three-dimensional structure appears.

ANALYTICAL SOLUTION

To obtain an analytical solution of the steady two-dimensional state, we use the dimensionless form of the governing equations formulated in terms of the stream function $\psi(r,\phi)$ and the temperature $T(r, \phi)$ namely,

$$\nabla^2 \psi = Ra_e^* [\sin(\phi)\frac{\partial T}{\partial r} + \frac{\cos(\phi)}{r}\frac{\partial T}{\partial \phi}]$$
$$\nabla^2 T = \frac{1}{r}(\frac{\partial \psi}{\partial \phi}\frac{\partial T}{\partial r} - \frac{\partial \psi}{\partial r}\frac{\partial T}{\partial \phi}) \tag{20}$$

with the boundary conditions :

$$\begin{aligned} r = 1: &\quad T = 1, \quad \psi = 0 \quad \forall \phi \\ r = R: &\quad T = 0, \quad \psi = 0 \quad \forall \phi \\ \phi = 0, \pi: &\quad \psi = 0, \quad \frac{\partial T}{\partial \phi} = 0 \quad \forall r \end{aligned} \tag{21}$$

In order to express equations (20) in terms of the relative gap with $\varepsilon = R-1$, the new coordinate x, given by $x = r-1/\varepsilon$, is used so that the domain shifts from $\frac{1}{\varepsilon} \le r \le 1 + \frac{1}{\varepsilon}$ to $0 \le x \le 1$ and ϕ remains unchanged. The resulting equations with $\xi = 1 + x\varepsilon$ are given by :

$$\xi^2\frac{\partial^2\psi}{\partial x^2}+\varepsilon\xi\frac{\partial\psi}{\partial x}+\varepsilon^2\frac{\partial^2\psi}{\partial\phi^2}=Ra_e^*[\,\xi^2\sin(\phi)\frac{\partial T}{\partial x}+\varepsilon\xi\cos(\phi)\frac{\partial T}{\partial\phi}]$$

$$\xi^2\frac{\partial^2 T}{\partial x^2}+\varepsilon\xi\frac{\partial T}{\partial x}+\varepsilon^2\frac{\partial^2 T}{\partial\phi^2}=\varepsilon\xi(\frac{\partial T}{\partial x}\frac{\partial\psi}{\partial\phi}-\frac{\partial T}{\partial\phi}\frac{\partial\psi}{\partial x}) \quad (22)$$

and the boundary conditions are given by :

$x=0$	$\Psi=0$	and	$T=1$	$\forall\,\phi$	(23)
$x=1$	$\Psi=0$	and	$T=0$	$\forall\,\phi$	(24)
$\phi=0,\pi$	$\Psi=0$	and	$\partial T/\partial\phi=0$	$\forall\,x$	(25)

The global Nusselt number Nu^*_g, which corresponds to the ratio of the total heat transfer to the conductive heat transfer, is given by :

$$Nu_g^*=-\frac{\ln(1+\varepsilon)}{\pi\varepsilon}\int_0^\pi\left.\frac{\partial T}{\partial x}\right|_{x=0}d\phi=-\frac{(1+\varepsilon)\ln(1+\varepsilon)}{\pi\varepsilon}\int_0^\pi\left.\frac{\partial T}{\partial x}\right|_{x=1}d\phi \quad (26)$$

The solution is periodic in the azimuthal direction, but not in the radial direction, and therefore the temperature T and the stream function Ψ are expanded as follows :

$$\Psi=\sum_{n=1}^{N}\sum_{i=0}^{I}\varepsilon^i f_n^i(x)\sin(n\phi),\quad T=\sum_{n=0}^{N}\sum_{i=0}^{I}\varepsilon^i g_n^i(x)\cos(n\phi) \quad (27)$$

These expansions are inserted into equations (22), and like terms in sin and cos and the coefficients of like powers of ε are equated and we obtain the set of second-order linear differential equations:

$$\frac{\partial^2 f_p^i(x)}{\partial x^2}=F(f_k^j,f_k'^j,g_m^n,g_m'^n,x)=F_p^i(x)$$

$$\frac{\partial^2 g_p^i(x)}{\partial x^2}=G(g_k^j,g_k'^j,f_m^n,f_m'^n,x)=G_p^i(x) \quad (28)$$

with j and n < i, and k and m < p, and $F_p^i(x)$ and $G_p^i(x)$ are polynomial functions.

The analytical solution of equations (28), to any desired order, is given by :

$$f_p^i(x)=\overline{F}_p^i(x)-x\overline{F}_p^i(x=1)\quad\text{and}\quad g_p^i(x)=\overline{G}_p^i(x)-x\overline{G}_p^i(x=1)$$

$$\text{with}\quad \overline{F}_p^i(x)=\int_0^x\!\!\int_0^y F_p^i(u)\,du\,dy\quad\text{and}\quad \overline{G}_p^i(x)=\int_0^x\!\!\int_0^y G_p^i(u)\,du\,dy \quad (29)$$

The degree of the polynomial expressions $f_p^i(x)$ and $g_p^i(x)$ increases as the order of approximation, I, increases.

For engineering applications, we are often concerned with the effect of the fluid motion on the heat transfer between the two cylinders and this can be evaluated by calculating the mean Nusselt number averaged over the inner cylinder surface namely,

$$Nu_g^* = -\ln(1+\varepsilon)\sum_{i=0}^{\infty}\varepsilon^{i-1}\left.\frac{dg_0^i}{dx}\right|_{x=0} \tag{30}$$

To find the functions $f_p^i(x)$ and $g_p^i(x)$ it is necessary to use symbolic language computations and in this study the Maple program has been employed. Once a solution for the temperature and the stream function is obtained up to order 15 in ε (the limit being imposed by the total memory of the computer), a similar expression is derived for the Nusselt number in terms of series in powers of ε. Therefore, the series representation of the Nusselt is given by :

$$\begin{aligned}
Nu_g^* &= 1 + 4.2163\times10^{-4}Ra_e^{*2}(\varepsilon^2-\varepsilon^3) + (2.4361\times10^{-4} - 1.6782\times10^{-7}Ra_e^{*2})Ra_e^{*2}\varepsilon^4 \\
&-(0.6559\times10^{-4} - 3.3563\times10^{-7}Ra_e^{*2})Ra_e^{*2}\varepsilon^5 - (0.5291\times10^{-4} + 1.5842\times10^{-7}Ra_e^{*2} \\
&-9.7372\times10^{-11}Ra_e^{*4})Ra_e^{*2}\varepsilon^6 + (1.1188\times10^{-4} - 3.6383\times10^{-7}Ra_e^{*2} - 2.9212\times \\
&10^{-10}Ra_e^{*4})Ra_e^{*2}\varepsilon^7 - (1.2949\times10^{-4} - 7.9426\times10^{-7}Ra_e^{*2} - 6.2682\times10^{-11}Ra_e^{*4} + \\
&6.5510\times10^{-14}Ra_e^{*6})Ra_e^{*2}\varepsilon^8 + (1.2388\times10^{-4} - 6.9602\times10^{-7}Ra_e^{*2} + 1.1128\times10^{-9} \\
&Ra_e^{*4} + 2.6204\times10^{-13}Ra_e^{*6})Ra_e^{*2}\varepsilon^9 - (1.0797\times10^{-4} - 6.4318\times10^{-8}Ra_e^{*2} + 2.1138 \\
&\times10^{-9}Ra_e^{*4} + 6.7203\times10^{-14}Ra_e^{*6} - 4.7865\times10^{-17}Ra_e^{*8})Ra_e^{*2}\varepsilon^{10} + (0.8933\times10^{-4} \\
&+6.7355\times10^{-7}Ra_e^{*2} + 1.8080\times10^{-10}Ra_e^{*4} - 2.3013\times10^{-12}Ra_e^{*6} - 2.3932\times10^{-16} \\
&\times Ra_e^{*8})Ra_e^{*2}\varepsilon^{11} + (-0.7180\times10^{-4} - 1.0255\times10^{-6}Ra_e^{*2} + 4.8716\times10^{-9}Ra_e^{*4} + \\
&4.2901\times10^{-12}Ra_e^{*6} - 2.3349\times10^{-16}Ra_e^{*8} - 3.6968\times10^{-20}Ra_e^{*10})Ra_e^{*2}\varepsilon^{12} + \\
&(5.6885\times10^{-5} + 8.0161\times10^{-7}Ra_e^{*2} - 6.4359\times10^{-9}Ra_e^{*4} + 3.8977\times10^{-12}Ra_e^{*6} + \\
&4.0335\times10^{-15}Ra_e^{*8} + 2.2181\times10^{-19}Ra_e^{*10})Ra_e^{*2}\varepsilon^{13} + (4.4878\times10^{-5} - 1.9984\times \\
&10^{-7}Ra_e^{*2} - 3.8791\times10^{-9}Ra_e^{*4} - 2.2877\times10^{-11}Ra_e^{*6} - 7.4128\times10^{-15}Ra_e^{*8} + \\
&4.3903\times10^{-19}Ra_e^{*10} + 3.0010\times10^{-23}Ra_e^{*12})Ra_e^{*2}\varepsilon^{14} + o(\varepsilon^{15})
\end{aligned}$$

where Ra_e^* is the Rayleigh number based on $(r_o\text{-}r_i)$. The domain of the convergence of this series is further increased by using the Padé process of acceleration which allows us to obtain an improved estimation of the Nusselt number.

LINEAR STABILITY

In contrast with Caltagirone [1], who neglected the role of the fluid velocity, we keep the velocity $\mathbf{V}_0$ in the perturbation equations. Introducing the temperature, fluid velocity and pressure perturbations, namely $\theta\,Ra^{-1}$ with $Ra = Ra^{*1/2}$, $\mathbf{v}$ and p, respectively, into the system of equations (8)-(10), we obtain, after some manipulations, the following governing equations :

$$\nabla . \mathbf{v} = 0 \tag{31}$$

$$\frac{\partial\theta}{\partial t} = \nabla^2\theta - \mathbf{V}_0 . \nabla\theta - Ra\,\mathbf{v} . \nabla T_o \tag{32}$$

$$0 = Ra\,\Im(\theta) - \nabla_r^2\,\mathbf{v} \tag{33}$$

where ∇_r^2 is the $\mathbf{e}_1$ component of the Laplacian vector and :

$$\begin{aligned}\Im(\theta) = &\left[\frac{\cos(\phi)}{r}\frac{\partial\theta}{\partial r} + \frac{\sin(\phi)}{r}\frac{\partial^2\theta}{\partial r\partial\phi} + \frac{\cos(\phi)}{r^2}\frac{\partial^2\theta}{\partial\phi^2} - \frac{\sin(\phi)}{r^2}\frac{\partial\theta}{\partial\phi} + \cos(\phi)\frac{\partial^2\theta}{\partial z^2}\right]\mathbf{e}_1 \\ &-\left[\sin(\phi)\frac{\partial^2\theta}{\partial r^2} + \frac{\cos(\phi)}{r}\frac{\partial^2\theta}{\partial r\partial\phi} - \frac{\cos(\phi)}{r^2}\frac{\partial\theta}{\partial\phi} + \sin(\phi)\frac{\partial^2\theta}{\partial z^2}\right]\mathbf{e}_2 \\ &-\left[\cos(\phi)\frac{\partial^2\theta}{\partial r\partial z} - \frac{\sin(\phi)}{r}\frac{\partial^2\theta}{\partial\phi\partial z}\right]\mathbf{e}_3\end{aligned} \tag{34}$$

Physical considerations, based on the observation of secondary phenomena occurring in the upper part of the annulus, lead us to simplify equations (32) and (33) by assuming the temperature and the fluid velocity perturbations to be two-dimensional. Thus, only the r and z velocity components will be considered, namely

$$\mathbf{v} = u(r,z)\,\mathbf{e}_1 + w(r,z)\,\mathbf{e}_3\,. \tag{35}$$

If we develop perturbations as periodic functions parallel to the cylinder axis, i.e. $\theta = \theta(r)\exp(i\,\alpha\,z)$ and $u = u(r)\exp(i\,\alpha\,z)$, and taking into account the previous physical approximations, we obtain :

$$\frac{\partial\theta}{\partial t} - (D^2 + \frac{1}{r}D - \alpha^2 - U_0\,D)\theta + Ra\frac{\partial T_O}{\partial r}u = 0 \tag{36}$$

$$Ra\,(\frac{D}{r} - \alpha^2)\theta - (D^2 + \frac{1}{r}D - \alpha^2 - \frac{1}{r^2})u = 0 \tag{37}$$

where U_0 and u are the $\mathbf{e}_1$ components of $\mathbf{V}_0$ and $\mathbf{v}$, respectively, α is the perturbation wavenumber $(\alpha = \frac{m\pi r_i}{L})$, m is the number of cells developed over the axial length L of the cylinders or the preferred modes when the two-dimensional basic fluid flow loses its stability in the annular horizontal layer of the axial length L and D stands for d/dr. The Galerkin method transforms equations (36) and (37) into a system of ordinary differential equations, and the perturbations are represented by a set of linearly independent functions which satisfy the boundary conditions :

$$\theta(r) = \sum_{k=1}^{N} a_k(t)\,\Theta_k(r), \quad u(r) = \sum_{k=1}^{N} b_k(t)\,U_k(r) \tag{38}$$

with boundary conditions :

$$\Theta_k(0) = \Theta_k(R) = 0 \quad \text{and} \quad U_k(0) = U_k(R) = 0\,. \tag{39}$$

The chosen trial functions, which satisfy the boundary conditions, are given by :

$$\Theta_k = U_k = \left[(r-1)(R-r)\right]^k / (R-1)^{2k}\,.$$

The critical Rayleigh number and the critical wavenumber have been determined with N = 1, 2 and 3 with good convergence (relative error between the approximation N = 2 and N = 3 is always less than 2×10^{-4}). The results are reproduced in figures 3 and 4 and Table 1. The critical Rayleigh number $Ra_L = Ra_L^*(R-1)$ and critical wavenumber $Sc_L = \alpha(R-1)$, both based on the annular layer thickness, obtained by the numerical model are plotted against the radius ratio $R-1$. This is done for both reference states (determined by the perturbation method and the numerical method). When R tends to unity, i.e. when the upper part of the annulus approaches the plane horizontal layer, Ra_L and Sc_L tends to $4\pi^2$ and π, respectively. This behaviour is similar to the onset of convection in the horizontal porous layer heated from below and represents the classical result see Combarnous and Bories [27].

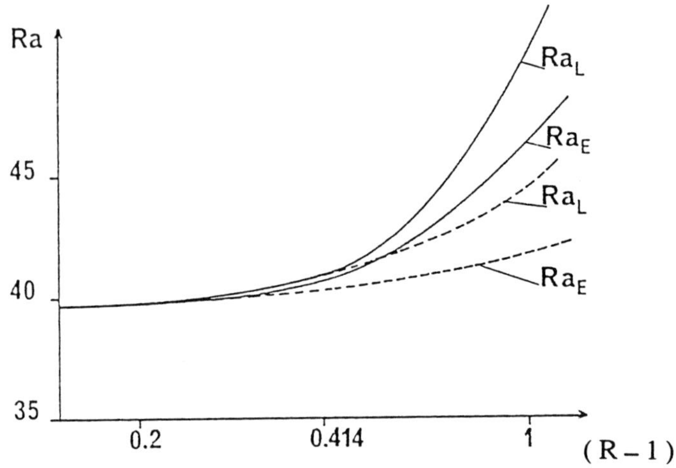

Figure 3 The critical Rayleigh number obtained by the linear and the energy theory, results are from the numerical scheme (——) and the perturbation method (-----)

Table 1 The critical Rayleigh number and wavenumber obtained by the linear theory for R = 1.2, $2^{1/2}$ and 2 using the Galerkin approximation

R	N = 1		N = 2		N = 3	
	Ra_c	S_c	Ra_c	S_c	Ra_c	S_c
Linear theory (numerical scheme)						
1.2	40.323	3.1688	39.819	3.1484	39.819	3.1486
$\sqrt{2}$	41.283	3.1854	40.833	3.1651	40.834	3.1653
2	50.151	3.2676	50.355	3.2543	50.345	3.2564
Linear theory (power-series expansion)						
1.2	40.403	3.1716	39.895	3.1514	39.895	3.1512
$\sqrt{2}$	41.385	3.1947	40.912	3.1752	40.912	3.1752
2	44.918	3.2930	44.625	3.2790	44.632	3.2790

As R increases, the difference between the results obtained by numerical scheme and those obtained by power - series expansion increases and this is due to the perturbation method power series development as a function of the Rayleigh number. The stability result obtained by the perturbation method is valid only for small values of the Rayleigh number.

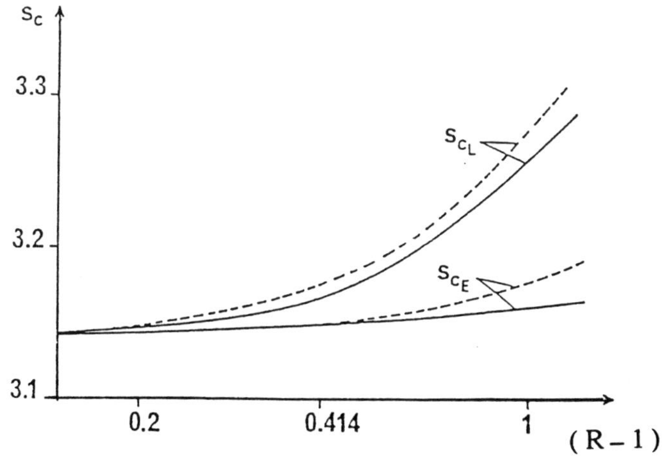

Figure 4 The critical wavenumber obtained by the linear and the energy theory, results are from the numerical scheme(———) and the perturbation method.(-------)

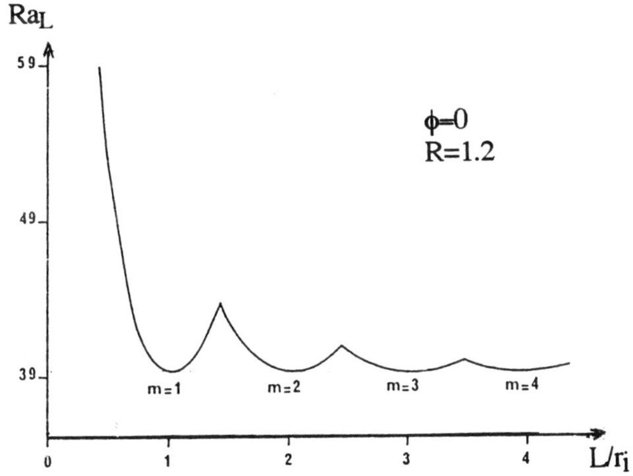

Figure 5 The variation of the critical Rayleigh number with L/r_i for R = 1.2 and for a number of cells chosen such that Ra is a minimum

Using the Galerkin method, our linear stability analysis shows for any value of ϕ, that near $\phi = 0$ the critical Rayleigh number and wavenumber vary slightly from $4\pi^2$ and π, respectively. The variations of Ra_L and Sc_L, as a function of ϕ, which varies between 0 and π, are also determined and the critical Rayleigh number Ra_L and wavenumber S_{cL} as a function of angular coordinate for R = 1.2 are plotted in figure 6. When the flow is studied in the whole annular layer, Ra_L is approximately the double of the minimum critical Rayleigh number obtained for $\phi = 0$, see figure 6. Thus the assumption that the flow stability can only be investigated at the top of the annulus is justified.

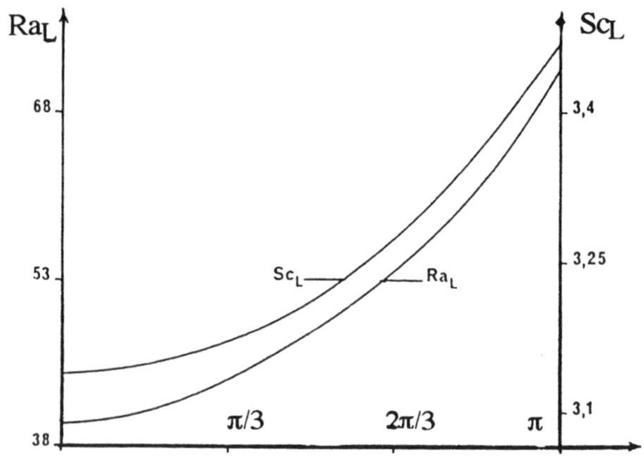

Figure 6 Linear critical Rayleigh number and wavenumber as a function of the angular coordinate for R=1.2

ENERGETIC STABILITY

The energy method was used to determine where, in the Ra^*-R plane, the basic flow, T_0 (r, ϕ) and $\mathbf{V}_0$ (r, ϕ) is steady regardless of the perturbation amplitude. In this section we use energy stability theory, cf. Joseph [25], to examine the stability of the basic flow to finite amplitude disturbances and this method enables us to find a lower bound for the value of Ra^* below which disturbances in some sense die away.

Considering equations (7) – (9) with Da = 0, and on introducing the temperature, fluid velocity and pressure perturbations (θRa^{-1} with $Ra = Ra^{*1/2}$, **v** and p, respectively), we obtain :

$$\nabla . \mathbf{v} = 0 \tag{40}$$

$$\frac{\partial\theta}{\partial t} - \nabla^2\theta + \mathbf{V}_o . \nabla\theta + \mathbf{v} . \nabla\theta + Ra\, \mathbf{v} . \nabla T_o = 0 \tag{41}$$

$$\mathbf{v} + Ra\, \mathbf{k}\, \theta + \nabla p = 0 \tag{42}$$

Multiplying equation (41) by θ and equation (42) by **v** and then integrating both the resulting equations over the whole volume Ω, with $< f >$ standing for $\int_\Omega f\, d\omega$ and using the divergence and Green's theorems, we obtain :

$$\frac{d}{dt} < \frac{1}{2}\theta^2 > = - < |\nabla \theta|^2 > - Ra < v . \nabla T_o \theta > \tag{43}$$

$$0 = Ra < \mathbf{v} . \mathbf{k} \theta > - < |\mathbf{v}|^2 > \tag{44}$$

We now consider a linear combination of these two equations and, with $\hat{\theta} = \lambda^{1/2}\theta$, we thus obtain :

$$\frac{d}{dt} < \frac{\hat{\theta}^2}{2} > = - Ra < (\frac{k}{\lambda^{1/2}} + \lambda^{1/2}\nabla T_o).v \hat{\theta} > - < |\nabla\hat{\theta}|^2 + |v|^2 > \tag{45}$$

with $\varepsilon = < \frac{\hat{\theta}^2}{2} >$ being the perturbation energy and thus we may write equation (45) in the form :

$$\frac{d\varepsilon}{dt} = - < |\nabla\hat{\theta}|^2 + |v|^2 > [1 - Ra \frac{< (\frac{k}{\lambda^{1/2}} + \lambda^{1/2}\nabla T_o).v\hat{\theta} >}{< |\nabla\hat{\theta}|^2 + |v|^2 >}] \tag{46}$$

The stability of the basic flow is dependent on the inequality proposed by Serrin [25], i.e. if $\hat{\theta}$ and $\mathbf{v}$ are functions such that $\hat{\theta}_{|s} = 0$, $\mathbf{v}_{|s} = 0$ and $\nabla . \mathbf{v} = 0$, then there exists a constant σ such that :

$$\frac{\sigma^2}{2} < \hat{\theta}^2 > \leq < \nabla\hat{\theta} \cdot \nabla\hat{\theta} > \tag{47}$$

On taking:

$$D = < |\nabla\hat{\theta}|^2 + |\mathbf{v}|^2 > = 1 \tag{48}$$

$$R_\lambda^{-1} = \max_H [- < (\frac{k}{\lambda^{1/2}} + \lambda^{1/2}\nabla T_o).v\hat{\theta} >] \tag{49}$$

where $Ra_E = \max_\lambda(R_\lambda)$, H is a functional space such that $\nabla . \mathbf{v} = 0$ and $\hat{\theta}|_s = 0$, and $\mathbf{v}|_s = 0$. Under these conditions we deduce the Euler – Lagrange equations for this variational problem, with $Ra_{cE}^{*1/2} = Ra_E$, as follows :

$$\nabla . \mathbf{v} = 0 \tag{50}$$

$$\frac{R_\lambda}{2}(\lambda^{1/2}\nabla T_0 + \frac{1}{\lambda^{1/2}}\mathbf{k}).\mathbf{v} - \nabla^2\hat{\theta} = 0 \tag{51}$$

$$\frac{R_\lambda}{2}(\lambda^{1/2}\nabla T_0 + \frac{1}{\lambda^{1/2}}\mathbf{k})\hat{\theta} + \nabla p + \mathbf{v} = 0 \tag{52}$$

Taking the curl of equation (52) twice and, using the properties of the solution T_0, the condition (48) and the assumption $\phi = 0$, which was justified previously (using the linear stability analysis), we obtain :

$$\frac{R_\lambda}{2}(\lambda^{1/2}\frac{\partial T_0}{\partial r} + \frac{1}{\lambda^{1/2}})\, u = \nabla^2\hat{\theta} \tag{53}$$

$$\frac{R_\lambda}{2}[\lambda^{1/2}(\frac{\partial T_0}{\partial r}\frac{\partial^2\hat{\theta}}{\partial z^2} - \frac{1}{r^2}\frac{\partial\hat{\theta}}{\partial r}\frac{\partial^2 T_0}{\partial \phi^2}) - \frac{1}{\lambda^{1/2}}(\frac{1}{r}\frac{\partial\hat{\theta}}{\partial r} + \frac{\partial^2\hat{\theta}}{\partial z^2})] = -\nabla^2{}_r\mathbf{v} \tag{54}$$

where $\nabla^2{}_r\mathbf{v}$ is the $\mathbf{e}_1$ component of the Laplacian of the vector $\mathbf{v}$ ($\nabla_r{}^2\mathbf{v} = \frac{\partial^2 u}{\partial z^2} + \frac{\partial^2 u}{\partial r^2} + \frac{1}{r}\frac{\partial u}{\partial r} - \frac{u}{r^2} + \frac{1}{r^2}\frac{\partial^2 u}{\partial\phi^2} - \frac{2}{r^2}\frac{\partial v}{\partial\phi}$). The temperature and radial velocity perturbations ($\hat{\theta}$ and u) are developed : $\hat{\theta} = \theta(r)\exp(isz)$ and $u = u(r)\exp(isz)$ and they are represented by a set of trial functions which satisfy the boundary conditions. Applying the Galerkin method to the system (53–54), we obtain :

$$\sum_{k=1}^{N} a_k[\int_1^R \theta_m(L(\theta_k) + \frac{\theta_k}{r^2})r dr] - \sum_{k=1}^{N} b_k \int_1^R \frac{R_\lambda}{2}[\lambda^{1/2}(\frac{\partial T_0}{\partial r} + \frac{1}{\lambda^{1/2}})u_k\,\theta_m\, r dr] = 0 \tag{55}$$

$$\sum_{k=1}^{N} a_k\{\int_1^R \frac{R_\lambda}{2}[\lambda^{1/2}(\alpha^2\theta_k\frac{\partial T_0}{\partial r} + \frac{D\theta_k}{r^2}\frac{\partial^2 T_0}{\partial\phi^2}) - \frac{1}{\lambda^{1/2}}(\frac{D\theta_k}{r} - \alpha^2\theta_k)]u_k r dr\} - \sum_{k=1}^{N} b_k[\int_1^R L(u_k)u_m\, r dr] = 0 \tag{56}$$

where α is the wavenumber, $D = \frac{\partial}{\partial r}$ and $L = \frac{\partial^2}{\partial r^2} + \frac{1}{r}\frac{\partial}{\partial r} + \frac{1}{r^2}\frac{\partial^2}{\partial\phi^2} + \frac{\partial^2}{\partial z^2} - \frac{1}{r^2}$. The trial functions in r used are the same as the trial functions used previously (in the linear stability analysis).

Equations (55) and (56) can admit non-zero solutions only if the Nth order determinant formed is zero but this condition is satisfied only for particular values of R_λ, the smallest eigenvalue being the desired critical value. For a given R, there exists a critical Rayleigh number Ra_E for which stability is assured and is defined by :

$$Ra_E = \min_{\alpha}(\max_{\lambda}(R_\lambda))$$

We have thus determined from equations (55) and (56) the global stability curve and deduced the critical values Ra_E and sc_E for each of the values of R and several degrees of approximation see table 2. Figures 3 and 4 show how sc_E and Ra_E depend on R and Again both reference states defined by the perturbation method and by numerical model were considered.

Table 2 The critical Rayleigh number and wavenumber obtained by the energy theory for R = 1.2, $2^{1/2}$ and 2 using the Galerkin approximation

R	N = 1		N = 2		N = 3	
	Ra_c	S_c	Ra_c	S_c	Ra_c	S_c
Energy method (numerical scheme)						
1.2	40.263	3.1649	39.758	3.1442	39.758	3.1442
$\sqrt{2}$	41.031	3.1710	40.572	3.1505	40.572	3.1505
2	46.677	3.1775	46.522	3.1610	46.500	3.1621
Energy method (power-series expansion)						
1.2	40.263	3.1649	39.748	3.1444	39.748	3.1443
$\sqrt{2}$	40.884	3.1724	40.381	3.1519	40.381	3.1519
2	42.348	3.2002	41.835	3.1794	41.835	3.1794

CONCLUSIONS

The linear stability theory, defined by assuming infinitely small perturbations, gives a sufficient instability criterion but finite-amplitude solutions can exist for values of Ra less than the critical value found for infinitesimal disturbances (linear stability theory). In the non-linear theory, where the energy perturbation is considered, a necessary condition for the asymptotic stability has been obtained. It was found, for Rayleigh numbers $Ra \in [Ra_E, Ra_L]$ and for non-infinitesimal perturbations, that nothing can be concluded about the stability of the system. For $Ra < Ra_E$ the basic flow is globally stable for any pertubation, regardless of the amplitude. For $Ra > Ra_L$, the two-dimensional basic flow is linearly unstable and always reaches a three-dimensional, or a new two-dimensional, flow depending on the aspect ratio of the cylinders R and the axial aspect ratio of the cell A. The results obtained clearly show that Ra_E and Sc_E tend respectively to $Ra_L = 4\pi^2$ and $Sc_L = \pi$ when the upper part of the annulus approaches the plane horizontal layer. Thus subcritical instability is not possible when R tends to 1 which is a classical result for the onset of convection in a horizontal porous layer which is heated from below. However, as the radius ratio R increases, the difference between these two critical values increases. It should be noted that the critical wavenumbers follow qualitatively the same pattern the as the Rayleigh number, i.e. $S_{cL} \geq S_{cE}$.

Bifurcation points between two-dimensional unicellular flows and two-dimensional multicellular or three-dimensional flows have been numerically determined. For $R = 2^{1/2}$ and $A \geq 1$, the three-dimensional numerical results show that only the two-dimensional unicellular flow exists if $Ra^* \leq Ra^*_{c1}$, where Ra^*_{c1} is localized between 90 and 100 for $R = 2^{1/2}$ and A = 2. The stability analysis performed with the two-dimensional axisymmetric perturbations depending on r and z to predict the transition towards three-dimensional flows are in good agreement with the numerical and experimental results, while the two-dimensional simulations or stability analysis using two-dimensional disturbances, localised in the basic flow plane, show that the two-dimensional unicellular and bicellular flow are stable for $Ra^* \geq Ra^*_{c2}$ ($Ra^*_{c2} \approx 112$ for $R = 2^{1/2}$). These two fluid flows have been observed experimentally in an annular cavity with A = 0.5. The stability analysis with an axisymmetric perturbations show that when A decreases from 1 to 0, the linear critical Rayleigh number increases. For small values of the axial extension of the cell A, it is possible to determine a criterion corresponding to the onset of two-dimensional multicellular flows before a three-dimensional steady flow is established.

For $R << 2^{1/2}$, the basic solution corresponding to the two-dimensional unicellular flow is investigated by the regular perturbation method in powers of ε = R–1, or the Rayleigh number. The influence of the parameter ε is then analyzed and this permits us to give a development of the critical Rayleigh number and wavenumber, depending on the small parameter ε, Charrier-Mojtabi [32] and Charrier-Mojtabi et *al.* [33]. For the limiting case of $\varepsilon \rightarrow 0$, we find the well-known result $Ra^*_c = 4\pi^2$ for an annular cell of infinite extension.

REFERENCES

1. J.P. Caltagirone, Thermoconvective instabilities in a porous medium bounded by two concentric horizontal cylinders, *J. Fluid Mech.* **76**, 337-362 (1976).
2. P.J. Burns and C.L. Tien, Natural convection in porous medium bounded by concentric spheres and horizontal cylinders, *Int. J. Heat Mass Transfer*, **22**, 929-939 (1979).
3. A. Mojtabi , D. Ouazar and M.C. Charrier-Mojtabi , An efficient finite element code for 2D steady state in a porous annulus, *Proc. Int. Conf. Numerical Methods for Thermal Problems,* Montreal, **5**, 644-654 (1987).
4. M.C. Charrier-Mojtabi and J.P. Caltagirone, Numerical simulation of natural convection in an annular porous layer by spectral method, *Proc. 1st Int. Conf. of Numerical Methods for Non-Linear Problems*, Swansea, 821-828 (1980).
5. Y.F. Rao , K. Fukuda and S. Hasegawa S., Steady and transient analysis of natural convection in a horizontal porous annulus with Galerkin method, *Journal of Heat Transfer,* **109**, 919-927 (1987).
6. K. Himasekhar and H. H. Bau, Two-dimensional bifurcation phenomena in thermal convection in horizontal concentric annuli containing saturated porous media, *J. Fluid* Mech. **187**, 267-300 (1988).

7. A. Mojtabi and M.C. Charrier-Mojtabi, Analytical solution of steady natural convection in an annular porous medium evaluted with a symbolic algebra code, *J. Heat Transfer* **114**, 1065-1067 (1992).
8. M.C. Charrier-Mojtabi, A. Mojtabi, M. Azaiez and G. Labrosse, Numerical and experimental study of multicellular free convection flows in an annular porous layer, *Int. J. Heat Mass Transfer*, **34**, 3061-3074 (1991).
9. M.C. Charrier-Mojtabi, A. Mojtabi, H. Khallouf, A pseudo-spectral method for natural convection in an annular porous layer: A pressure and temperature formulation, National Heat Transfer Conference, *ASME, HTD* **194**, 75-81 San Diego, August (1992).
10. K. Himasekhar and H.H. Bau, Large Rayleigh number convection in a horizontal eccentric annulus containing saturated porous media, *Int. J. Heat Mass Transfer,* **29**, 703-712 (1986).
11. K. Fukuda,Y. Takuta, S. Hasegawa, S. Shimomura and K. Sanokawa K, Three-dimensional natural convection in a porous medium between concentric inclined cylinders, A.S.M.E, HTD **8**, 97-103, Orlando ,U.S.A. (1980).
12. M.C. Charrier-Mojtabi , A. Mojtabi and J.P. Caltagirone, Three dimensional convection in a horizontal porous layer, *Euromech* **138**, 75-77, Karlsruhe (1981).
13. Y.F. Rao, K. Fukuda and S. Hasegawa, A numerical study of three dimensional natural convection in a horizontal porous annulus with Galerkin method, *Int. J. Heat* Mass Transfer **31**, 695-707 (1988).
14. M.C. Charrier-Mojtabi, Numerical solution of two and three dimensional free convection flows in horizontal porous annulus using pressure and temperature formulation *Int. J. Heat Mass Tranfer* **40,** 1521-1533 (1997).
15. J.P. Barbosa , and E. Saatdjian, Natural convection in porous horizontal cylindrical annulus, *J. Heat Transfer* **116**, 621-626 (1994).
16. J.P. Barbosa , and E. Saatdjian, Natural convection in porous cylindrical annuli. Int. J. Numer. Methods Heat Fluid Flow, **5**, 3-17 (1995).
17. K. Aboubi, L. Robillard and E Bilgen, Convective heat transfer in an annular porous layer with centrifugal force field. Numer. Heat Transfer A, **28**, 375-388 (1995).
18. K. Aboubi, L. Robillard and E Bilgen, Natural convection in horizontal annulus filled with an anisotropic porous medium. Proc. ASME/JSME Thermal Engineering Joint Conf. **3**, 415-422 (1995).
19. S. Kimura and I. Pop, Non-Darcian effects on conjugate natural convection between horizontal concentric cylinders filled with a porous medium. Fluid Dyn. Res., **7**, 241-253 (1991).
20. A. Cheddadi, J.P. Caltagirone, A. Mojtabi and K. Vafai, Free Two-dimensional convective bifurcation in a horizontal annulus, *J. Heat Transfer* **114**, 99-106 (1992).
21. Y.F. Rao, K. Fukuda and S. Hasegawa , A numerical study of three dimensional natural convection in a horizontal porous annulus with Galerkin method, *Int. J. Heat Mass Transfer* **31**, 695-707 (1988).

22. K. Vafai and J. Ettefagh, An investigation of transient thre-dimensional buoyancy-driven flow and heat tranfer in a closed horizontal annulus, *Int. J. Heat Mass Tranfer* **34**, 2555-2570 (1991).
23. M. Cloupeau and S. Klarsfeld, Visualization of thermal fields in saturated porous media by Christiansen effect, *Applied Optics,* **12**, 198-204 (1973).
24. H.H.Bau, G. McBlane and I. Sarferstein, Numerical simulation of thermal convection in an eccentric annulus containing porous media, *ASME 83 WA/HT* 34 (1983).
25. D.D. Joseph Stability of fluid motions I and II Springer, Berlin. Tracts in Net. Phil. **27** and **28** (1976).
26. A. Mojtabi and J.P. Caltagirone , Energy Stability of a natural convective flow in a horizontal annular space , *Phys. Fluids,* **22**, 1208-1209 (1979).
27. M. Combarnous and S. Bories Hydrothermal convection in saturated porous media, Advance in hydroscience, Academic Press, 231-307 (1975).
28. A. Aniri and K. Vafai Analysis of dispersion effects and non thermal equilibrium, non-Darcian, variable porosity incompressible flow through porous medium. *Int. J. Heat Mass Tranfer,* **37,** 939-954 (1994).
29. C. Canuto , M.Y. Hussaini .,T.A. Zang and A. Quarteroni , *Spectral Methods in fluids Dynamics*. Springer, Berlin (1988).
30. P. Haldenwang, G. Labrosse , S. Abboudi and M. Deville, Chebyshev 3D spectral and 2D pseudospectral solvers for the Helmholtz equation, *J. Comp. Phys.,* **55**, 277-290 (1984).
31. M.C. Charrier-Mojtabi and A. Mojtabi, Numerical simulation of three-dimensional free convection in a horizontal porous annulus, *Proc. 10th Int. Heat Transfer Conference, Brighton,U.K.*, **2**, 319-324 (1994).
32. M.C. Charrier-Mojtabi, Etude numérique théorique et expérimentale des écoulements thermoconvectifs bidimensionnels et tridimensionnels en couche annulaire poreuse horizontale, Thèse de Doctorat d'Etat , University of Bordeaux I (1993).
33. M.C. Charrier-Mojtabi and A. Mojtabi, Stabilité des écoulements de convection naturelle en espace annulaire poreux horizontal, *C.R. Acad. Sci. Paris,* t. **320**, Série IIb, 177-183 (1995).

A UNIFIED TREATMENT OF DARCY-FORCHHEIMER BOUNDARY-LAYER FLOWS

A. NAKAYAMA

Department of Mechanical Engineering, Shizuoka University
3-5-1 Johoku, Hamamatsu, 432 JAPAN

INTRODUCTION

In the study of mixed convection within non-Darcy porous media, Nakayama and Pop [1] considered the Forchheimer-extended Darcy law. They argued that it is the total velocity magnitude over the heat transfer surface that virtually determines the convective heat transfer from the surface, and they chose the slip velocity at the wall as a velocity scale to form a *modified* Peclet number.

The unified treatment, based on this modified Peclet number, enabled them to extract *all* possible similarity solutions for free, forced and mixed convection within Darcy and non-Darcy porous media. It can also be seen that there exists three limiting flow regimes, namely, the forced convection, the Darcy free convection and the Forchheimer free convection regimes. Appropriate dimensionless groups for identifying these regimes are found to be the *micro-scale* Reynolds and Grashof numbers based on the square root of the permeability. Upon taking these dimensionless numbers as the variables, a regime map can be constructed to show the three asymptotic convection regimes, and also another three intermediate convection regimes, namely, the Darcy mixed convection, the Darcy-Forchheimer free convection and the Forchheimer mixed convection regimes.

In this chapter, we shall revisit the problems of non-Darcian convective heat transfer, using the Darcy-Forchheimer boundary-layer equations, using a similar approach to that described by Nakayama and Pop [1].

MAGNITUDE ANALYSIS OF THE BOUNDARY-LAYER-EQUATIONS FOR POROUS MEDIA

The classical boundary-layer concept can readily be exploited for the governing equations of mass, momentum and energy conservations, which were generalized by Vafai and Tien [2] for convective flows in porous media. Let us consider the case of steady, forced convection from an isothermal flat plate which is embedded in a fluid-saturated porous medium, as shown in Figure 1. The boundary-layer equations are given by

$$\frac{\partial u}{\partial x}+\frac{\partial v}{\partial y}=0 \tag{1}$$

$$\frac{\rho}{\varepsilon^{+2}}\left(u\frac{\partial u}{\partial x}+v\frac{\partial u}{\partial y}\right)=-\frac{dp}{dx}+\frac{\mu}{\varepsilon^{+}}\frac{\partial^2 u}{\partial y^2}-\frac{\mu}{K}u-\frac{C\rho}{K^{1/2}}u^2 \tag{2}$$

$$u\frac{\partial T}{\partial x}+v\frac{\partial T}{\partial y}=\alpha_e\frac{\partial^2 T}{\partial y^2} \tag{3}$$

with the boundary conditions:

$$y=0: \quad u=v=0,\ T=T_w \tag{4a}$$

$$y\to\infty: \quad u=u_e,\ T=T_e \tag{4b}$$

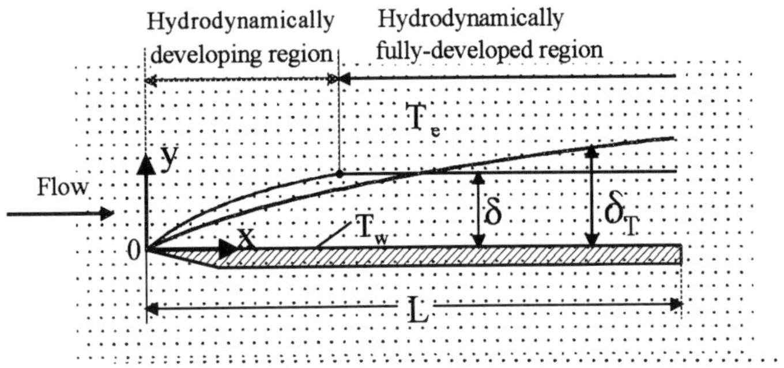

Figure 1: Development of boundary-layer

where u and v are the Darcian velocities in the x and y directions, respectively, while p and T are the local pressure and temperature; ρ and μ are the density and viscosity of the fluid; K and ε^+ are the permeability and porosity, respectively, of the porous medium; α_e is the effective thermal diffusivity of the fluid-saturated porous medium, accounting for both molecular diffusion and mechanical dispersion; C is the Forchheimer constant. Let us write equation (2) at the outer-edge of the boundary-layer, as

$$-\frac{dp}{dx}=\frac{\mu}{K}u_e+\frac{C\rho}{K^{1/2}}u_e^2 \tag{5}$$

Unlike in clear fluid flows over a flat plate, the pressure decreases in the streamwise direction linearly (with respect to x) due to the Darcy and Forchheimer resistances caused by the solid matrix. In the vicinity of the leading edge, the convective inertia terms on the left hand side of equation (2) may become comparable in magnitude to the Darcy and Forchheimer terms, since the velocity gradient is steep there. Let us estimate the extent of the region $0 < x < x_c$, where the convective inertia terms are appreciable, such that

$$\frac{\rho}{\varepsilon^{+2}}\left(u\frac{\partial u}{\partial x}+v\frac{\partial u}{\partial y}\right)\sim\frac{\rho}{\varepsilon^{+2}}\frac{u_e^2}{x_c}\sim\frac{\mu}{K}u_e+\frac{C\rho}{K^{1/2}}u_e^2$$

and thus, we may define the hydrodynamically developing entry length by

$$x_c \equiv \frac{\rho\, u_e^2}{\varepsilon^{+2}\left(\frac{\mu}{K}u_e+\frac{C\rho}{K^{1/2}}u_e^2\right)} \tag{6}$$

Usually, K is very small, so that the convective inertia terms are appreciable only in the immediate vicinity of the leading edge, where the boundary-layer theory itself fails anyway. For this reason, the convective inertia terms are often dropped from the equation in the case of natural porous materials.

Nakayama et al. [3] estimated the viscous boundary-layer thickness δ for the upstream region $0 < x < x_c$ and for the downstream region $x > x_c$, respectively. In the upstream region, the convective inertia predominates over the total of the Darcy and Forchheimer resistances, such that

$$\underset{\text{Boundary friction}}{\frac{\mu}{\varepsilon^{+}}\frac{u_e}{\delta^2}} \sim \underset{\text{Convective inertia}}{\frac{\rho}{\varepsilon^{+2}}\frac{u_e^2}{x}}$$

Thus, we have

$$\delta \sim \left(\frac{\varepsilon^{+}\nu x}{u_e}\right)^{1/2} = \frac{x}{Re_x^{1/2}} \quad \text{for} \quad 0 < x < x_c \tag{7}$$

where

$$Re_x = \frac{\left(u_e/\varepsilon^{+}\right)x}{\nu} \tag{8}$$

is the *macro-scale* Reynolds number, which is based on the intrinsic average velocity.

On the other hand, in the downstream region, the Darcy and Forchheimer resistances dominate over the convective inertia, such that

$$\frac{\mu}{\varepsilon^{+}}\frac{u_e}{\delta^2} \quad \sim \quad \frac{\mu}{K}u_e + \frac{C\rho}{K^{1/2}}u_e^2$$

Boundary friction — Darcy and Forchheimer resistances

Thus, we have

$$\delta \sim \left(\frac{K/\varepsilon^{+}}{1+Re_K}\right)^{1/2} \quad \text{for} \quad x > x_c \tag{9}$$

where

$$Re_K = \frac{CK^{1/2}u_e}{\nu} \sim \frac{\text{Forchheimer drag}}{\text{Darcy drag}} \tag{10}$$

is the *micro-scale* Reynolds number, which is based on the square root of the permeability and the Darcian velocity. It may be concluded that the viscous boundary-layer thickness increases , since $\delta \propto \sqrt{x}$ from the leading edge (as in clear fluids) and approaches its upper bound given by equation (7), in the vicinity $x \approx x_c$. Similarly, we estimate that the thermal boundary-layer thickness $\delta_T(> \delta)$ as

$$u_e \frac{T_w - T_e}{x} \sim \alpha_e \frac{T_w - T_e}{\delta_T^2}$$

Convection — Conduction

Thus, we have

$$\delta_T \sim \left(\frac{\alpha_e x}{u_e}\right)^{1/2} = \frac{x}{Pe_x^{1/2}} \tag{11}$$

where

$$Pe_x = \frac{u_e x}{\alpha_e} \tag{12}$$

is the local *Peclet number*. Hence, the thermal thickness δ_T grows as $\delta_T \propto \sqrt{x}$, and this is illustrated in Figure 1. Therefore, the local Nusselt number for the case of pure forced convection may be estimated as

$$Nu_x \equiv \frac{q_w x}{k_e(T_w - T_e)} \sim \frac{x}{\delta_T} \sim Pe_x^{1/2} \tag{13}$$

A similar order of magnitude analysis can be made for the case of pure free convection. Setting the *x*-coordinate vertically upwards for the case of Darcy free convective flow over a vertical flat plate, we expect

$$\frac{\mu}{K} u_{max} \quad \sim \quad \rho\, g\beta\,(T_w - T_e)$$

Darcy resistance Buoyancy force

Thus, we estimate

$$u_{max} \sim \frac{Kg\beta\,(T_w - T_e)}{\nu}$$

Replacing u_e in equations (10) and (12) by u_{max}, the *micro-scale* Grashof number (based on the square root of the permeability) and the *macro-scale* Rayleigh number appropriate for free convective flows in porous media (i.e. the Darcy-Rayleigh number) can be formed as

$$Gr_K = \frac{CK^{3/2} g\beta\,(T_w - T_e)}{\nu^2} \sim \frac{\text{Buoyancy - induced Forchheimer drag}}{\text{Buoyancy - induced Darcy drag}} \tag{14}$$

$$Ra_x = \frac{Kg\beta\,(T_w - T_e)x}{\alpha_e \nu} \tag{15}$$

repectively. Thus, replacing Pe_x in equation (13) by Ra_x, the local Nusselt number for the case of free convection can be estimated as

$$Nu_x \sim Ra_x^{1/2} \tag{16}$$

DARCY-FORCHHEIMER BOUNDARY-LAYER EQUATIONS

We shall examine the full momentum boundary-layer equation (2), and seek further possible simplifications of the equation. Let us consider again the case of steady forced convection, as illustrated in Figure 1, and estimate the hydrodynamically developing entry length x_c according to equation (6).

For typical momentum boundary layers in commercial low-density Foametals of $K = 10^{-7} m^2$, $\varepsilon^+ = 0.98$ and $u_e = 0.1m$, we approximately have x_c = 0.5 mm for air flow and x_c = 2 mm for water flow. Thus, even in high porosity media, the effect of convective inertia is appreciable only within a short distance from the leading edge. Since a similar argument can be made for free and mixed convection situations, we may conclude that, for most practical situations, except for flows in highly permeable media , the convective inertia can be left out of equation (12) which becomes

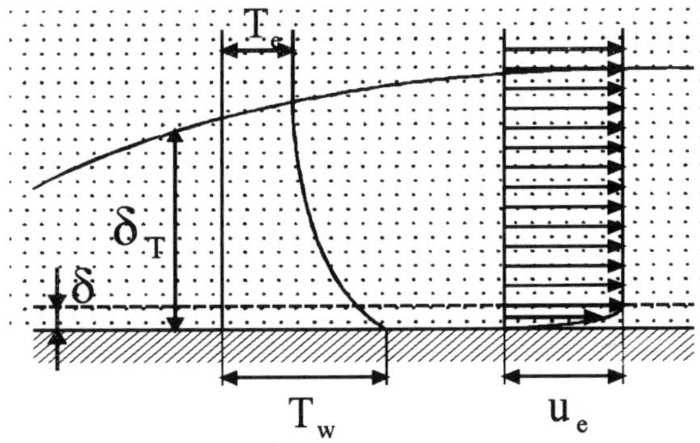

Figure 2: Velocity and temperature distributions in a porous medium

$$-\frac{1}{\rho}\frac{dp}{dx}+\frac{\nu}{\varepsilon^+}\frac{\partial^2 u}{\partial y^2}-\frac{\nu}{K}u-\frac{C}{K^{1/2}}u^2=0 \tag{17}$$

Thus, we obtain the boundary-layer expression of the heuristic Brinkman-Forchheimer-extended Darcy law. Furthermore, we note that, for most porous materials, the viscous boundary-layer thickness δ is quite small as we estimate its order from equation (9), which, for the air flow in the aforementioned high porosity material, gives $\delta \approx 0.3\text{mm}$. In fact, the viscous boundary-layer is so thin that the boundary viscous effects on the gross quantities, such as flow rate and pressure drop, are indeed insignificant and can be neglected in most practical situations. Thus, over almost the entire flat surface, the viscous boundary-layer is confined to a thin layer which is very close to the wall, that, across the thermal boundary-layer, the temperature distribution is essentially free from the boundary viscous effects, as illustrated in Figure. 2. The situation is analogous to the low Prandtl number situation of clear fluid flows without a porous matrix. The foregoing observations prompt us to drop the Brinkman term from equation (17), and discard the no-slip condition at the wall so that the flow may slip over the wall.

The final set of Darcy-Forchheimer boundary-layer equations, which includes the buoyancy force term, which is valid for both plane and axisymmetric bodies of arbitrary shape, becomes

$$\frac{1}{r^*}\frac{\partial\left(r^* u\right)}{\partial x}+\frac{\partial v}{\partial y}=0 \tag{18}$$

$$\frac{\nu}{K}u+\frac{C}{K^{1/2}}u^2=\frac{\nu}{K}u_e+\frac{C}{K^{1/2}}u_e^2+g_x\,\beta\left(T-T_e\right) \tag{19}$$

$$u\frac{\partial T}{\partial x}+v\frac{\partial T}{\partial y}=\alpha_e\frac{\partial^2 T}{\partial y^2} \tag{20}$$

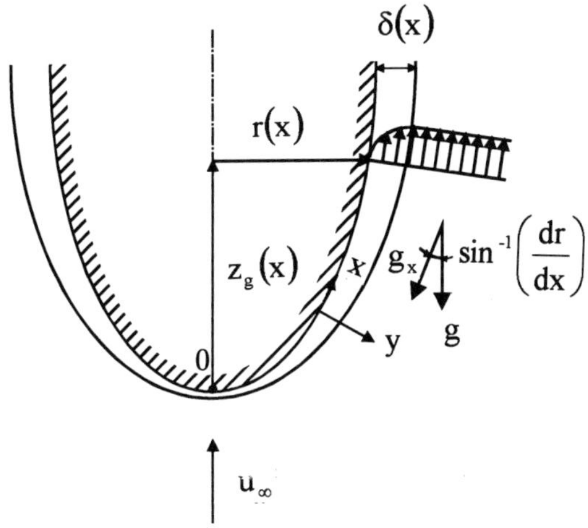

Figure 3: Boundary-layer coordinates

with the boundary conditions:

$$y=0: \qquad v=0, \quad T=T_w(x) \tag{21a}$$

$$y\to\infty: \qquad u=u_e(x) \quad (\text{ or } \quad T=T_e) \tag{21b}$$

where

$$r^*=\begin{cases}1 & :\text{plane body}\\ r(x) & :\text{axisymmetric body}\end{cases} \tag{22}$$

and

$$g_x=g\frac{dz_g}{dx}=g\left\{1-\left(\frac{dr}{dx}\right)^2\right\}^{1/2} \tag{23}$$

g_x is the tangential component of the gravitational acceleration vector, whereas z_g is the elevation measured from the front stagnation point, as indicated in Figure 3. For the case of axisymmetric bodies, it is assumed that the body radius $r(x)$ is quite large relative to the boundary-layer thickness, so that the transverse radial curvature effect is negligible. Further, we should note that the pressure gradient term is eliminated using the external flow relationship:

$$-\frac{1}{\rho}\frac{d\left(p+\rho g z_g\right)}{d x}=\frac{\nu}{K}u_e+\frac{C}{K^{1/2}}u_e^2 \tag{24}$$

The above set of Darcy-Forchheimer boundary-layer equations holds for forced, free and mixed convection from plane and axisymmetric bodies of arbitrary shape in most practical fluid-saturated porous media, except in media with extremely high permeability. In what follows, we shall examine this set of Darcy-Forchheimer boundary-layer equations to classify possible flow and heat transfer regimes and construct a useful flow regime map, with which we can consult, as we attack a particular convection problem associated with porous media.

MODIFIED PECLET NUMBER AND FLOW REGIME MAP

Most previous studies on mixed convection correlate the local Nusselt number in terms of either Pe_x (to treat mostly the forced flow dominated cases), or Ra_x (to treat mostly the buoyancy force dominated cases). However, any mixed convection analysis which employs either Pe_x or Ra_x inevitably suffers from a singularity under a certain physical limiting condition. For example, if Pe_x is used for the scaling, a singularity will appear as $Ra_x / Pe_x \to \infty$. In what follows, we shall seek what seems to be the most appropriate dimensionless number to correlate the Nusselt number for all convective flow situations, without encountering any singularities in either of the physical limiting conditions. Mixed convection over plane and axisymmetric bodies of arbitrary shape will be considered, using the Darcy-Forchheimer boundary-layer equations under the assumptions of uniform porosity and negligible thermal dispersion. The Darcy-Forchheimer momentum equation (19) may readily be solved for the velocity u as

$$u=\frac{\nu}{2CK^{1/2}}\left[\left\{\left(1+2Re_K\right)^2+4Gr_K\left(\frac{T-T_e}{T_w-T_e}\right)\right\}^{1/2}-1\right] \tag{25a}$$

where the micro-scale Reynolds and Grashof numbers are defined using the local values of u_e, g_x and $\left(T_w-T_e\right)$, as

$$Re_K(x)=CK^{1/2}u_e(x)/\nu \tag{25b}$$

and

$$Gr_K(x)=CK^{3/2}g_x(x)\beta\left\{T_w(x)-T_e\right\}/\nu^2 \tag{25c}$$

The external velocity $u_e(x)$ can be determined from potential flow theory.

Substituting the boundary condition (21a) into expression (25), we find the slip velocity at the wall u_w as

$$u_w = \frac{\nu}{2CK^{1/2}}\left[\left\{(1+2\mathrm{Re}_K)^2 + 4\mathrm{Gr}_K\right\}^{1/2} - 1\right] \tag{26}$$

which indicates that the velocity field is established as the result of both an external flow and the buoyancy force. Nakayama and Pop [1] argued that it is the total velocity magnitude over the heat transfer surface that virtually determines the convective heat transfer from the heated surface, and chose the slip velocity at the wall as the velocity scale, and they proposed a new dimensionless number, namely, the *modified* Peclet number as

$$\mathrm{Pe}_x^* = \frac{u_w x}{\alpha_e} = \mathrm{Pe}_x \frac{\left\{(1+2\mathrm{Re}_K)^2 + 4\mathrm{Gr}_K\right\}^{1/2} - 1}{2\mathrm{Re}_K} \tag{27}$$

On the other hand, the energy equation (20), may be scaled as

$$u_w \frac{\Delta T_w}{x} \sim \alpha_e \frac{\Delta T_w}{\delta_T^2} \tag{28a}$$

Hence, we may expect that the general expression for the local Nusselt number Nu_x, which is valid for all convection modes, to be given in terms of the modified Peclet number Pe_x^* as

$$\mathrm{Nu}_x \sim \frac{x}{\delta_T} \sim \mathrm{Pe}_x^{*1/2} \tag{28b}$$

Under the physical limiting conditions, as described below in terms of Re_K and Gr_K, the modified Peclet number expression (27) reduces to the corresonding simple expressions. Expanding the terms in the numerator of the last expression in equation (27), we find

Regime I (*Forced convection regime*):

$$\mathrm{Nu}_x^2 \sim \mathrm{Pe}_x^* = \mathrm{Pe}_x \quad \text{for} \quad \mathrm{Re}_K + \mathrm{Re}_K^2 >> \mathrm{Gr}_K \tag{29a}$$

Regime II (*Darcy free convection regime*):

$$\mathrm{Nu}_x^2 \sim \mathrm{Pe}_x^* = \mathrm{Ra}_x \quad \text{for} \quad \mathrm{Re}_K << \mathrm{Gr}_K << 1 \tag{29b}$$

Regime III (*Forchheimer free convection regime*):

$$Nu_x^2 \sim Pe_x^* = Ra_x^* \quad \text{for} \quad Re_K + Re_K^2 << Gr_K \quad \text{and} \quad Gr_K >> 1 \tag{29c}$$

where

$$Ra_x^* = \frac{K^{1/4}\{g_x\beta\,(T_w - T_e)\}^{1/2}x}{C^{1/2}\alpha_e} = \frac{Ra_x}{Gr_K^{1/2}} \tag{30}$$

is another dimensionless number appropriate for the Forchheimer free convection regime.

The foregoing three flow regimes, namely, the *forced convection regime*, the *Darcy free convection regime* and the *Forchheimer free convection regime*, are identified by I, II and III in Figure 4. In the forced convection regime, the effect of the buoyancy force is negligible, whereas it predominates over the effect of forced flow, in both the Darcy and Forchheimer free convection regimes. Moreover, in the Darcy free convection regime, the Forchheimer drag due to the porous inertia is negligible, whereas it dominates over the Darcy drag in the Forchheimer free convection regime. Another three intermediate regimes (as indicated in Figure 4), which connect the foregoing three regimes, namely, the *Darcy mixed convection regime* (IV), the *Darcy-Forchheimer free convection regime* (V) and the *Forchheimer mixed convection regime* (VI), may be identified as follows:

Regime IV (*Darcy mixed convection regime*):

$$Nu_x^2 \sim Pe_x^* = Pe_x + Ra_x \quad \text{for} \quad Re_K \sim Gr_K << 1 \tag{31a}$$

Regime V (*Darcy-Forchheimer free convection regime*):

$$Nu_x^2 \sim Pe_x^* = Ra_x \frac{(1+4Gr_K)^{1/2} - 1}{2Gr_K} \quad \text{for} \quad Gr_K \sim 1 \quad \text{and} \quad Re_K << 1 \tag{31b}$$

Regime VI (*Forchheimer mixed convection regime*):

$$Nu_x^2 \sim Pe_x^* = \left(Pe_x^2 + Ra_x^{*2}\right)^{1/2} \quad \text{for} \quad Gr_K \sim Re_K^2 >> 1 \tag{31c}$$

Thus, the modified Peclet number Pe_x^* reduces to the conventional dimensionless numbers under the corresponding physical limiting conditions given in terms of the micro-scale dimensionless numbers, Re_K and Gr_K. The flow regime map, shown in

Figure 4, enables us to find what convection mode prevails for a given set of the micro-scale dimensionless numbers. In the following sections, we shall consider each flow regime and investigate possible similarity solutions.

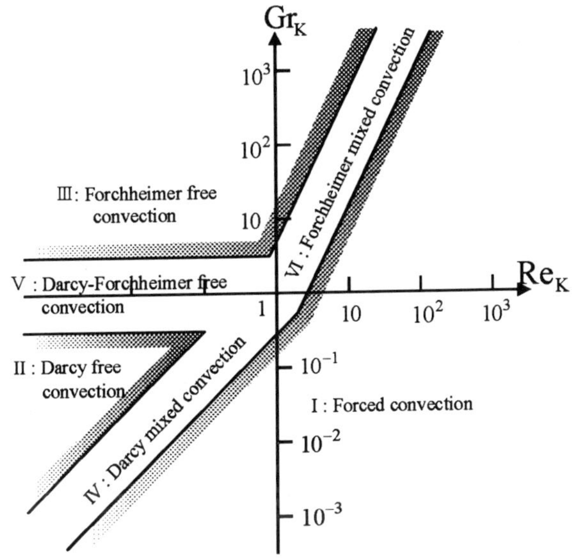

Figure 4: Flow regime map

It is interesting to note that we have introduced three macro-scale dimensionless numbers, namely, Pe_x, Ra_x and Ra_x^*, and two micro-scale dimensionless numbers namely, Re_K and Gr_K. However, only three among these five dimensionless numbers are independent, since we have the following inter-relations between the macro- and micro-dimensionless numbers:

$$Ra_x / Pe_x = Gr_K / Re_K \tag{32a}$$

and

$$Ra_x / Ra_x^* = Gr_K^{1/2} \tag{32b}$$

UNIFIED TRANSFORMATIONS FOR DARCY-FORCHHEIMER BOUNDARY-LAYER EQUATIONS

Having established the modified Peclet number, we propose general transformations, which are valid for all plane and axisymmetric bodies, as follows:

$$f(x,\eta) = \frac{\psi}{\alpha_e r^* (Pe_x^* I)^{1/2}} \tag{33}$$

$$\theta(x,\eta) = \frac{T - T_e}{T_w - T_e} \tag{34}$$

and

$$\eta = \frac{y}{x}\left(\frac{Pe_x^*}{I}\right)^{1/2} \tag{35}$$

where

$$I = \frac{\int_0^x (T_w - T_e)^2 u_w r^{*2}\, dx}{(T_w - T_e)^2 u_w r^{*2} x} \tag{36}$$

is a function which adjusts the scale in the y-direction according to a given body geometry, and its surface temperature distribution. Note that I = 1 for the case of an isothermal flat plate.

After some manipulations using these transformed variables, the momentum and energy equations reduce to

$$f = \frac{\int_0^\eta \left\{(1+2Re_K)^2 + 4Gr_K\theta\right\}^{1/2} d\eta - \eta}{\left\{(1+2Re_K)^2 + 4Gr_K\right\}^{1/2} - 1} \tag{37}$$

and

$$\theta'' + \left(\frac{1}{2} - m_T I\right) f\theta' - m_T I \theta f' = I x \left(f' \frac{\partial \theta}{\partial x} - \theta' \frac{\partial f}{\partial x}\right) \tag{38}$$

where

$$m_T(x) \equiv \frac{d\ln(T_w - T_e)}{d\ln x} = \frac{x}{(T_w - T_e)} \frac{dT_w}{dx} \tag{39}$$

The transformed boundary conditions are

$$\eta = 0 \quad : \quad \theta = 1 \tag{40a}$$

$$\eta \to \infty : \quad \theta = 0 \tag{40b}$$

The Darcian velocities are given by

$$u = u_w f' \tag{41a}$$

and

$$v = \frac{\alpha_e}{x}\left(\frac{Pe_x^*}{I}\right)^{1/2}\left\{\left(m_T I - \frac{1}{2}\right)f + \left(\frac{1}{2} - m_T I - I\frac{d\ln u_w r^*}{d\ln x}\right)\eta f' - I x\frac{\partial f}{\partial x}\right\} \tag{41b}$$

The resulting set of equations (37) and (38) subject to boundary conditions (40a) and (40b) include all possible solutions to the free, forced and mixed convection problems of Darcy-Forchheimer flows. Once the temperature distribution θ is known by solving the set of transformed equations, the local Nusselt number, which is our primary concern, may readily be evaluated from

$$Nu_x = \frac{q_w x}{(T_w - T_e)k_e} = -\theta'(x,0)\left(\frac{Pe_x^*}{I}\right)^{1/2} \tag{42}$$

In general, the transformed equations (37) and (38) must be solved using one of the many boundary-layer solution methodologies, such as the local similarity and non-similarity solution methodologies. However, there are certain cases, in which the partial differential equation (38), along with (37), reduces to an ordinary differential equation. In what follows, we shall seek such similarity solutions, focusing on each convective flow regime.

FORCED CONVECTION REGIME (Regime I : $Re_K + Re_K^{\ 2} >> Gr_K$)

Noting that $Pe_x^* = Pe_x$ since $u = u_e$, for this pure forced convection regime, we find

$$f = \eta \tag{43a}$$

and the product $m_T I$, which is the only parameter appearing on the left-hand side of equation (38), reduces to

$$m_T I = \frac{d\ln(T_w - T_e)}{d\ln x}\frac{\int_0^x (T_w - T_e)^2 u_e r^{*2}\,dx}{(T_w - T_e)^2 u_e r^{*2} x} \tag{43b}$$

The above equation indicates that similarity solutions are possible when the wall temperature varies according to

$$(T_w - T_e) \propto \xi_1^{\lambda_1} \tag{44a}$$

where

$$\xi_1 \equiv \int_0^x u_e\, r^{*2}\,dx \tag{44b}$$

and thus we have

$$m_T I = \frac{\lambda_1}{1+2\lambda_1} \tag{44c}$$

For this particular case, the energy equation reduces to the following ordinary differential equation:

$$\theta'' + \frac{1}{2(1+2\lambda_1)}\eta\theta' - \frac{\lambda_1}{1+2\lambda_1}\theta = 0 \tag{45}$$

which has to be solved subject to the boundary condition (40). The equation can readily be integrated for a given value of λ_1, using any standard numerical integration scheme.

For the case of a semi-infinite flat plate, we note that $\xi_1 = u_e x \propto x$, and hence, $\lambda_1 = m_T = \text{constant}$. Thus, the above energy equation is identical to what we would obtain for the case of forced convection from a flat plate with its wall temperature following the power function of the distance from the leading edge as

$$(T_w - T_e) \propto x^{\lambda_1}$$

Therefore, the numerical values of $\theta'(0\,;\,\lambda_1)$ furnished for the specific case of a flat plate are directly applicable for all cases of plane and axisymmetric bodies of arbitrary shape. We can readily translate these flat plate results to calculate the local Nusselt number from a body of arbitrary shape with its wall temperature distribution given by the above power function, using the useful expression:

$$Nu_x = -(1+2\lambda_1)^{1/2}\theta'(0;\lambda_1)\left(\frac{d\ln\xi_1}{d\ln x}\right)^{1/2} Pe_x^{1/2} \tag{46}$$

The above expression is deduced from expression (42), noting the relationship, $1/I = (1+2\lambda_1)(d\ln\xi_1 / d\ln x)$. The geometric effect on Nu_x is accounted in terms of the parameter ξ_1 defined by equation (43), which readily yields

$$\xi_1 \propto x^{1+m} \quad \left(\text{i.e. } \lambda_1 = \frac{m_T}{1+m}\right) \qquad : \text{wedge} \tag{47a}$$

$$\xi_1 \propto x^{3+m} \quad \left(\text{i.e. } \lambda_1 = \frac{m_T}{3+m}\right) \qquad : \text{cone} \tag{47b}$$

$$\xi_1 \propto \{1 - \cos(x/R)\} \qquad : \text{horizontal circular cylinder} \tag{47c}$$

$$\xi_1 \propto \left\{\cos^3(x/R) - 3\cos(x/R) + 2\right\} \qquad : \text{sphere} \qquad (47d)$$

where the potential flow solutions, namely, $u_e \propto x^m$ for a wedge and a cone, and $u_e \propto \sin(x/R)$ for a cylinder and a sphere have been assumed. Thus, the function, $(d\ln\xi_1 / d\ln x)$, which is required for the evaluation of Nu_x is given by

$$\frac{d\ln\xi_1}{d\ln x} = 1 + m \qquad : \text{wedge} \qquad (47e)$$

$$\frac{d\ln\xi_1}{d\ln x} = 3 + m \qquad : \text{cone} \qquad (47f)$$

$$\frac{d\ln\xi_1}{d\ln x} = \frac{(x/R)\sin(x/R)}{1-\cos(x/R)} \qquad : \text{horizontal circular cylinder} \qquad (47g)$$

$$\frac{d\ln\xi_1}{d\ln x} = \frac{3(x/R)\sin^3(x/R)}{\cos^3(x/R) - 3\cos(x/R) + 2} \qquad : \text{sphere} \qquad (47h)$$

We should note that the following wedge-cone inter-relationship holds:

$$\left(Nu_x / Pe_x^{1/2}\right)_{cone} = 3^{1/2}\left(Nu_x / Pe_x^{1/2}\right)_{wedge}\Bigg|_{\substack{m_{wedge}=m_{cone}/3 \\ m_{Twedge}=m_{Tcone}/3}} \qquad (48)$$

The numerical integration results for $\theta'(0;\lambda_1)$, as obtained by Nakayama and Koyama [3], are tabulated in Table 1 in terms of $Nu_x / Pe_x^{1/2}(d\ln\xi_1/d\ln x)^{1/2} = -(1+2\lambda_1)\theta'(0;\lambda_1)$ for all two-dimenisional bodies of arbitrary shape. Especially for the case of an isothermal wall, we have $-\theta'(0;0) = 1/\sqrt{\pi}$. Thus, we obtain a general expression for Nu_x, which is valid for all plane and axisymmetric bodies with an isothermal wall, as

$$Nu_x = \frac{1}{\pi^{1/2}}\left(\frac{d\ln\xi_1}{d\ln x}\right)^{1/2} Pe_x^{\ 1/2} \qquad : \text{isothermal wall} \qquad (49)$$

The local surface heat flux variation over an isothermal sphere and an isothermal circular cylinder may be found in Nakayama and Koyama [4]. An average Nusselt number expression can be obtained by numerically integrating expression (46) over a heat transfer surface, for a given function, $(\mathrm{dln}\xi_1 / \mathrm{dlnx})$. For example, we have

$$\mathrm{Nu} \equiv \frac{(q_w)_{av}(2R)}{(T_w - T_e)k_e} = 1.02 \left(\frac{u_\infty 2R}{\alpha_e}\right)^{1/2} \quad : \text{Isothermal horizontal circular cylinder} \quad (50)$$

Table 1: Heat transfer results on pure forced convection

λ_1	$-\theta'(0;\lambda_1)$	$\mathrm{Nu_x}/(\mathrm{Pe_x dln}\xi_1/\mathrm{dlnx})^{1/2}$
-0.5	0	0
-0.3	0.447	0.283
-0.1	0.539	0.482
0	0.564	0.564
0.1	0.583	0.639
0.2	0.598	0.707
1/3	0.613	0.791
0.5	0.627	0.886
0.8	0.644	1.038
1.0	0.651	1.128
1.5	0.665	1.329
2.0	0.673	1.504

DARCY FREE CONVECTION REGIME (Regime II: $\mathrm{Re_K} << \mathrm{Gr_K} << 1$)

In this free convection regime, the Forchheimer term is negligible, as compared with the Darcy term, such that we have $\mathrm{Pe}_x^* = \mathrm{Ra}_x$, and equation (37) yields

$$f = \int_0^\eta \theta \, d\eta \qquad (51a)$$

The product $m_T I$, as defined by equations (36) and (39), is given by

$$m_T I = \frac{\mathrm{dln}(T_w - T_e)}{\mathrm{dlnx}} \frac{\int_0^x (T_w - T_e)^3 g_x r^{*2} \, dx}{(T_w - T_e)^3 g_x r^{*2} x} \qquad (51a)$$

Thus, similarity solutions exist when the wall temperature varies as

$$(T_w - T_e) \propto \xi_2^{\lambda_2}$$

where

$$\xi_2 \equiv \int_0^x g_x r^{*2}\, dx \tag{52}$$

such that

$$m_T I = \frac{\lambda_2}{1+3\lambda_2} \tag{53}$$

and the energy equation reduces to

$$\theta'' + \frac{1+\lambda_2}{2(1+3\lambda_2)}\theta' \int_0^\eta \theta\, d\eta - \frac{\lambda_2}{1+3\lambda_2}\theta^2 = 0 \tag{54}$$

which is identical to what Cheng and Minkowycz [5] considered for the case of Darcy free convection from a vertical flat plate subject to the wall temperature following the power function of x with an exponent λ_2. Thus, the numerical results provided by Cheng and Minkowycz [5] are directly available and these results for a flat plate may readily be translated to those for an arbitrary body, by use of the general expression as

$$Nu_x = -(1+3\lambda_2)^{1/2}\theta'(0;\lambda_2)\left(\frac{d\ln\xi_2}{d\ln x}\right)^{1/2} Ra_x^{1/2} \tag{55}$$

The numerical values of $Nu_x / Ra_x^{1/2}(d\ln\xi_2 / d\ln x)^{1/2} = -(1+3\lambda_2)^{1/2}\theta'(0;\lambda_2)$ and $-\theta'(0;\lambda_2)$ are tabulated in Table 2. It is interesting to note that $q_w = 0$ everywhere when $\lambda_2 = -1/3$ and, moreover, for the case of isothermal bodies, we have

$$Nu_x = 0.444\left(\frac{d\ln\xi_2}{d\ln x}\right)^{1/2} Ra_x^{1/2} \quad : \text{isothermal wall} \tag{56}$$

The foregoing equations are valid for all plane and axisymmetric bodies, and ξ_2 may readily be evaluated from expression (52) for any particular geometry of interest. We should note that the function $(d\ln\xi_2 / d\ln x)$ assumes the value of unity for a vertical (or inclined) plate, and three for a vertical cone with its apex pointing downward, such that

$$\left(Nu_x / Ra_x^{1/2}\right)_{cone} = 3^{1/2}\left(Nu_x / Ra_x^{1/2}\right)_{plate}\Big|_{m_{T_{plate}} = m_{T_{cone}}/3} \tag{57}$$

Table 2: Heat transfer results on Darcy free convection

λ_2	$-\theta'(0;\lambda_2)$	$Nu_x/(Ra_x d\ln\xi_2/d\ln x)^{1/2}$
-1/3	0	0
-0.3	0.233	0.0735
-0.2	0.370	0.234
-0.1	0.419	0.350
0	0.444	0.444
1/9	0.461	0.532
0.2	0.470	0.594
1/3	0.479	0.678
0.4	0.483	0.717
0.5	0.488	0.771
0.6	0.491	0.822
0.8	0.497	0.916
1.0	0.501	1.001
1.5	0.506	1.187
2.0	0.510	1.349

whereas the function becomes identical to $(d\ln\xi_1/d\ln x)$ for a sphere and a circular cylinder, as listed in expressions (47g) and (47h). The heat flux variations over an isothermal sphere, and an isothermal circular cylinder, may be found in Nakayama and Koyama [6]. The average Nusselt number expressions may be sometimes more useful than the local ones. Therefore, we carry out a numerical integration over the surface of the cylinder surface, and the average Nusselt number, which is based on the diameter for the case of an isothermal horizontal circular cylinder, $Nu = 2R(q_w)_{av}/(T_w - T_e)k_e$, can be found to be

$$Nu = 0.565\left(\frac{Kg\beta(T_w - T_e)2R}{\alpha_e \nu}\right)^{1/2} \quad : \text{ circular cylinder } (T_w = \text{constant}) \tag{58}$$

It can readily be shown that the wall heat flux q_w, along a vertical plate (or a cone pointing vertically downward) remains constant when $m_T = 1/3$. Noting that $(T_w - T_e) = q_w x / Nu_x k_e$, the unknown wall temperature for a given constant heat flux q_w, may be evaluated from

$$Nu_x = \left\{-2^{1/2}\theta'\left(0;\frac{1}{3}\right)\right\}^{2/3}\left(\frac{Kg\beta\, q_w x^2}{\alpha_e k_e \nu}\right)^{1/3} = 0.772\left(\frac{Kg\beta\, q_w x^2}{\alpha_e k_e \nu}\right)^{1/3}$$

$$: \text{ vertical plate } (q_w = \text{constant}) \tag{59a}$$

$$\mathrm{Nu}_x = \left\{-2\theta'\left(0;\frac{1}{9}\right)\right\}^{2/3}\left(\frac{Kg_x\beta\, q_w x^2}{\alpha_e k_e v}\right)^{1/3} = 0.947\left(\frac{Kg_x\beta\, q_w x^2}{\alpha_e k_e v}\right)^{1/3}$$

: vertical cone (q_w = constant) (59b)

FORCHHEIMER FREE CONVECTION REGIME

(**Regime III:** $\mathrm{Re}_K + \mathrm{Re}_K^{\ 2} << \mathrm{Gr}_K$ and $\mathrm{Gr}_K >> 1$)

In this free convection regime, the Forchheimer term describing the porous inertial contribution predominates over the Darcy term, such that $\mathrm{Pe}_x^* = \mathrm{Ra}_x^*$, and equations (36), (37) and (39) give

$$f = \int_0^\eta \theta^{1/2}\, d\eta \tag{60a}$$

$$m_T I = \frac{d\ln(T_w - T_e)}{d\ln x}\frac{\int_0^x (T_w - T_e)^{5/2} g_x^{\ 1/2} r^{*2}\, dx}{(T_w - T_e)^{5/2} g_x^{\ 1/2} r^{*2} x} \tag{60b}$$

Thus, similarity solutions exist when the wall temperature varies as

$$(T_w - T_e) \propto \xi_3^{\ \lambda_3}$$

where

$$\xi_3 \equiv \int_0^x g_x^{\ 1/2} r^{*2}\, dx \tag{61}$$

such that

$$m_T I = \frac{\lambda_3}{1 + \frac{5}{2}\lambda_3} \tag{62}$$

and the energy equation reduces to

$$\theta'' + \frac{2+\lambda_3}{2(2+5\lambda_3)}\theta'\int_0^\eta \theta^{1/2}\, d\eta - \frac{2\lambda_3}{2+5\lambda_3}\theta^{3/2} = 0 \tag{63}$$

The general expression for a body of arbitrary shape, with its wall temperature following the above power function, is given by

$$\mathrm{Nu}_x = -\left(1 + \frac{5}{2}\lambda_3\right)^{1/2}\theta'(0;\lambda_3)\left(\frac{d\ln\xi_3}{d\ln x}\right)^{1/2}\mathrm{Ra}_x^{*\,1/2} \tag{64}$$

The numerical values of $Nu_x / Ra_x^{*1/2} (d\ln\xi_3 / d\ln x)^{1/2} = -\{1 + (5/2)\lambda_3\}^{1/2} \theta '(0\,;\lambda_3)$ and $-\theta '(0\,;\lambda_3)$ are tabulated in Table 3 , following Nakayama et al. [7]. In particular, for the case of isothermal bodies, we have

$$Nu_x = 0.494\left(\frac{d\ln\xi_3}{d\ln x}\right)^{1/2} Ra_x^{*1/2} \quad : \text{isothermal wall} \tag{65}$$

where the function $(d\ln\xi_3 / d\ln x)$ assumes the value of unity for a vertical plate, and three for a vertically downward pointing cone, just as in the case of Darcy free convection. Thus, the inter-relationship between the plate and cone can be found as

$$\left(Nu_x / Ra_x^{*1/2}\right)_{cone} = 3^{1/2}\left(Nu_x / Ra_x^{*1/2}\right)_{plate}\Bigg|_{m_{T_{plate}} = m_{T_{cone}}/3} \tag{66}$$

Moreover, we note that

$$\frac{d\ln\xi_3}{d\ln x} = \frac{(x/R)\sin^{1/2}(x/R)}{\int_0^{x/R} \sin^{1/2}(x/R)\, d(x/R)} \quad : \text{horizontal circular cylinder} \tag{67a}$$

and

$$\frac{d\ln\xi_3}{d\ln x} = \frac{(x/R)\sin^{5/2}(x/R)}{\int_0^{x/R} \sin^{5/2}(x/R)\, d(x/R)} \quad : \text{sphere} \tag{67b}$$

Performing a numerical integration, the average Nusselt number based on the diameter for the case of an isothermal horizontal circular cylinder, $Nu = 2R(q_w)_{av} / (T_w - T_e)k_e$, may be found to be

$$Nu = 0.689\left(\frac{K^{1/2} g\beta\, (T_w - T_e)(2R)^2}{C\alpha_e^2}\right)^{1/4} \quad : \text{horizontal circular cylinder} \quad (T_w = \text{constant}) \tag{68}$$

The wall heat flux q_w along a vertical plate (or a vertically downward pointing cone) remains constant when $m_T = 2/5$. The unknown wall temperature, for a given constant heat flux q_w, may be evaluated from the following local Nusselt number expression:

$$Nu_x = \left\{-2^{1/2}\theta'\left(0;\frac{2}{5}\right)\right\}^{4/5}\left(\frac{K^{1/2}g\beta\, q_w x^3}{C\alpha_e^{\,2}k_e}\right)^{1/5} = 0.806\left(\frac{K^{1/2}g\beta\, q_w x^3}{C\alpha_e^{\,2}k_e}\right)^{1/5}$$

: vertical plate (q_w = constant) (69a)

$$Nu_x = \left\{-2\theta'\left(0;\frac{2}{15}\right)\right\}^{4/5}\left(\frac{K^{1/2}g_x\beta\, q_w x^3}{C\alpha_e^{\,2}k_e}\right)^{1/5} = 1.024\left(\frac{K^{1/2}g_x\beta\, q_w x^3}{C\alpha_e^{\,2}k_e}\right)^{1/5}$$

: vertical cone (q_w = constant) (69b)

Table 3: Heat transfer results on Forchheimer free convection

λ_3	$-\theta'(0;\lambda_3)$	$Nu_x/(Ra_x^* d\ln\xi_3/d\ln x)^{1/2}$
-0.4	0	0
-0.3	0.348	0.174
-0.2	0.427	0.302
-0.1	0.469	0.406
0	0.494	0.494
2/15	0.515	0.595
0.2	0.523	0.641
0.3	0.533	0.704
0.4	0.540	0.763
0.5	0.546	0.819
0.8	0.558	0.966
1.0	0.563	1.054
1.5	0.572	1.247
2.0	0.577	1.414

INTERMEDIATE FLOW REGIMES

Let us consider the three intermediate flow regimes, namely, the Darcy mixed convection , the Darcy-Forchheimer free convection, and the Forchheimer mixed convection regimes, identified as Regimes IV, V and VI, respectively, in Figure 4.

Darcy mixed convection regime (Regime IV: $Re_K \sim Gr_K << 1$)

Cheng [8] investigated this mixed convection regime, in which expression (27) reduces to $Pe_x^* = Pe_x + Ra_x$. The study on this regime reveals that only a limited number of similarity solutions are possible because of the very restrictive conditions, see Nakayama and Koyama [4]. Nakayama and Koyama [4] found similarity solutions for isothermal spheres and cylinders, and vertical wedges and cones with the surface temperature varying with the same power index as that of the external velocity, namely $m_T = m$. For these cases, all terms on the right hand side of equation (38) vanish, and the resulting ordinary differential equation can be numerically integrated in the usual fashion. Subsequently, Nakayama and Pop [1] correlated the numerical results, and proposed the following approximate formula, which is valid for isothermal spheres, horizontal cylinders and vertical plates:

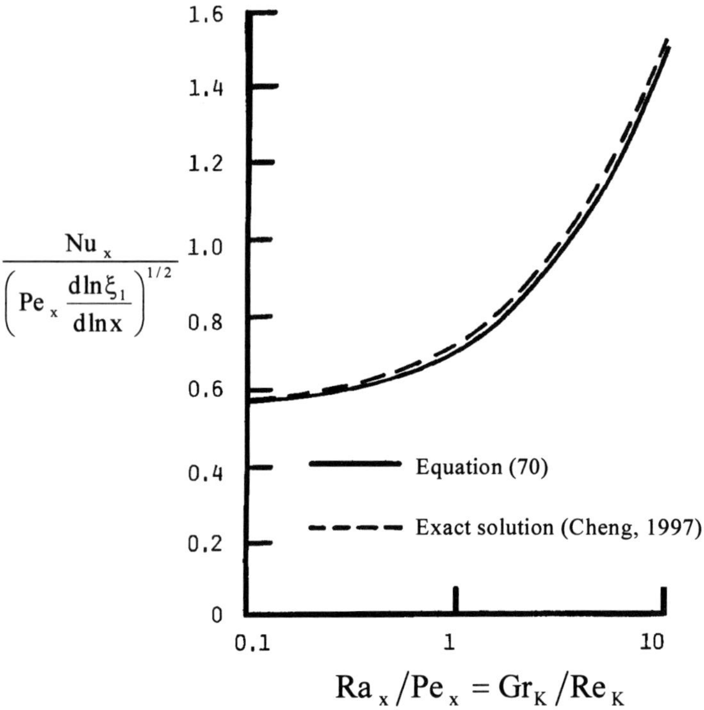

Figure 5: Nusselt number for Darcy mixed convection from an isothermal surface

$$Nu_x = \frac{1}{\pi^{1/2}} \left(\frac{d\ln\xi_1}{d\ln x} \right)^{1/2} \left(Pe_x + 0.62 Ra_x \right)^{1/2} \quad : \text{isothermal wall} \qquad (70)$$

We should note that $(d\ln\xi_1 / d\ln x) = (d\ln\xi_2 / d\ln x)$, since T_w = constant and $g_x \propto u_e$ should hold in this case. The approximate formula (70), which asymptotically reduces to both the forced convection formula (49) and the Darcy free convection formula (56), is compared against Cheng's exact solution in Figure 5, and it is observed that there is a good agreement between the two solutions.

Darcy-Forchheimer free convection regime (Regime V: $Gr_K \sim 1$ and $Re_K << 1$)

Plumb and Huenefeld [9] were the first to find a similarity solution for an isothermal flat plate in the Darcy-Forchheimer free convection regime. Another class of similarity solutions were found by Nakayama, et al. [10] for curved surfaces, in which the wall temperature decreases in the streamwise direction. For the isothermal case, they modified the Bejan-Poulikakos' formula [11], based on Oseen's linearized solution, to approximate closely the numerical integration results, and they proposed the following approximate formula which are valid for isothermal vertical plates and cones:

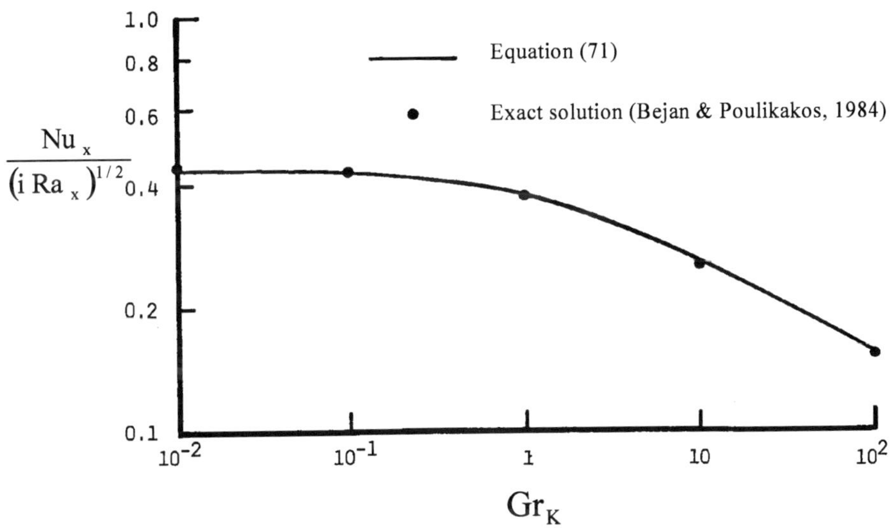

Figure 6: Nusselt number for Darcy-Forchheimer free convection from an isothermal surface

$$Nu_x = \frac{\left[16Gr_K^2 - \left\{(1+4Gr_K)^{1/2} - 1\right\}^2\right]^{1/2}}{8Gr_K}\left\{iRa_x \frac{(1+4Gr_K)^{1/2} - 1}{2Gr_K}\right\}^{1/2} \qquad (71)$$

where

$$i = \begin{cases} 1 & : \text{vertical plate} \\ 3 & : \text{vertical cone pointing downward} \end{cases} \qquad (72)$$

The expression (71) reduces to $Nu_x = 0.433(i\,Ra_x)^{1/2}$ in the limit of Darcy free convection, and to $Nu_x = 0.500(i\,Ra_x^*)^{1/2}$ for the limit of Forchheimer free convection.. The formula (71) is shown along with the exact solution, Bejan and Poulikakos [11], in Figure 6, where we observe excellent agreement between the formula and exact solution.

Forchheimer mixed convection regime (Regime VI: $Gr_K \sim Re_K^{\,2} >> 1$)

Nakayama and Pop [1] pointed out that only a limited number of similarity solutions exist in this mixed convection regime, such as the stagnation flow around a sphere or a horizontal circular cylinder with a sinusoidal wall temperature distribution, and a vertical wedge or a cone pointing vertically downward with its wall temperature varying as $(T_w - T_e) \propto x^{2m}$, where *m* is the power index of the external flow velocity. In particular, for the case of an isothermal vertical flat plate, we may use the approximate formula given by Nakayama and Pop [1] as follows:

$$Nu_x = \frac{1}{\pi^{1/2}}\left(Pe_x^2 + 0.59 Ra_x^{*2}\right)^{1/4} \quad \text{: isothermal vertical flat plate} \quad (73)$$

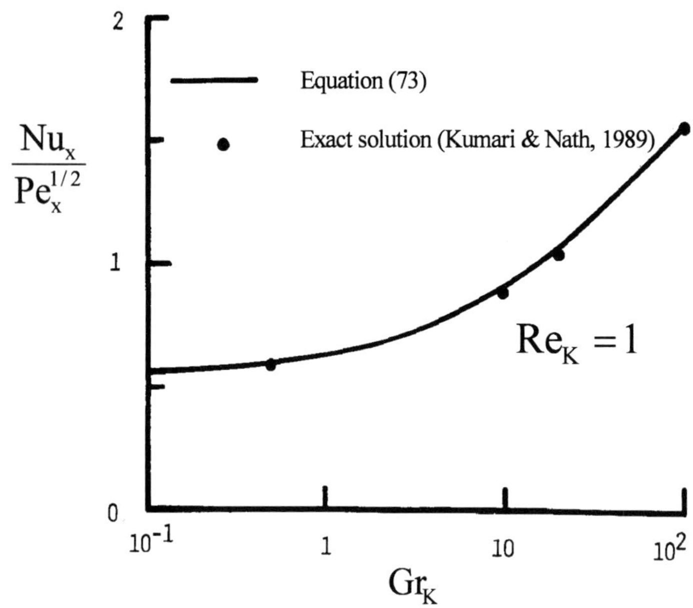

Figure 7: Nusselt number for Forchheimer mixed convection from an isothermal surface

which asymptotically reduces to the forced convection formula (49), and to the Forchheimer free convection formula (65). The validity of the above approximate formula can be confirmed in Figure 7, which compares the formula (73) with the exact values given by Kumari and Nath [12].

CONCLUDING REMARKS

For the case of an isothermal flat plate, the foregoing three Nusselt number expressions for the intermediate flow regimes, overlap each other, as illustrated in Figure 8. These three expressions are quite useful for practical estimations of heat transfer rates. We should consult with the flow regime map and the corresponding asymptotic expressions provided above, before performing numerical integrations for a number of sets of the flow parameters, since the flow regime map and asymptotic expressions may save a significant amount of computer time which would otherwise be necessary to perform such a parametric study.

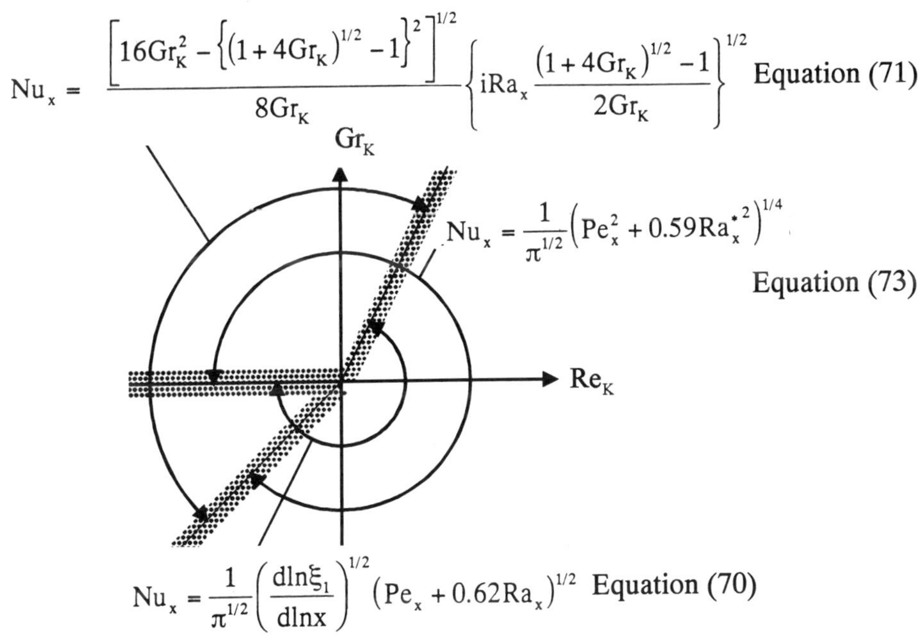

Figure 8: Heat transfer formulae for an isothermal flat plate

References

1. A. Nakayama and I. Pop, A unified similarity transformation for free, forced and mixed convection in Darcy and non-Darcy porous media, Int. J. Heat Mass Transfer **34**, 357-367 (1991).
2. K. Vafai and C. L. Tien, Boundary and inertia effects on flow and heat transfer in porous media, Int. J. Heat Mass Transfer **24**, 195-203 (1981).
3. A. Nakayama, T. Kokudai and H. Koyama, Non-Darcian boundary layer flow and forced convective heat transfer over a flat plate in a fluid-saturated porous medium, ASME J. Heat Transfer **112**, 157-162 (1990).
4. A. Nakayama and H. Koyama, General similarity transformation for combined free and forced convection flows within a fluid-saturated porous medium, Trans. ASME J. Heat Transfer **109**, 1041-1045 (1987).
5. P. Cheng and W. J. Minkowycz, Free convection about a vertical plate embedded in a saturated porous medium with application to heat transfer from a dike, J. Geophys. Res **82**, 2040-2044 (1977).
6. A. Nakayama and H. Koyama, Free convective heat transfer over a non-isothermal body of arbitrary shape embedded in a fluid-saturated porous medium, Trans. ASME J. Heat Transfer **109**, 125-130 (1987).
7. A. Nakayama, A., T. Kokudai and H. Koyama, Forchheimer free convection over a non-isothermal body of arbitrary shape in a saturated porous medium, Trans. ASME J. Heat Transfer **112**, 511-515 (1990).
8. P. Cheng, Combined free and forced convection flow about inclined surfaces in porous media, Int. J. Heat Mass Transfer **20**, 807-814 (1977).
9. O. A. Plumb, and J. C. Huenefeld, Non-Darcy natural convection from heated surfaces in saturated porous media, Int. J. Heat Mass Transfer **24**, 765-768 (1981).
10. A. Nakayama, H. Koyama and F. Kuwahara, Similarity solution for non-Darcy free convection from a non-isothermal curved surface in a fluid-saturated porous medium, Trans. ASME J. Heat Transfer **111**, 807-811 (1989).
11. A. Bejan and D. Poulikakos, The non-Darcy regime for vertical boundary layer natural convection in a porous medium, Int. J. Heat Mass Transfer **27**, 717-722 (1984).
12. M. Kumari and G. Nath, Non-Darcy mixed convection boundary layer flow on a vertical cylinder in a saturated porous medium, Int. J. Heat Mass Transfer **32**, 183-187 (1989).

TRANSIENT CONVECTION HEAT TRANSFER IN A POROUS MEDIUM: EXTERNAL FLOWS

I. POP[1], D.B. INGHAM[2] and J.H. MERKIN[2]

[1]Faculty of Mathematics, University of Cluj R-3400 Cluj, CP253, Romania
[2]Department of Applied Mathematical Studies, University of Leeds, Leeds LS2 9JT, UK

INTRODUCTION

Interest in the study of the phenomena of convective flow through a fluid-saturated porous medium stems both from fundamental considerations, such as the development of better insights into the nature of the underlying physical processes, and from practical considerations, such as the fact that these idealized configurations serve as a launching pad for modelling the analogous transfer processes in more realistic systems. On the other hand, such idealized geometries also provide a testing ground for checking the validity of the theoretical analysis. Consequently, considerable research effort has been expended on understanding the convective transfer processes between porous media and objects of different shapes embedded in these media.

The practical interest in convective heat transfer in a porous medium is expanding rapidly, due to the wide range of applications in contemporary technologies. These important applications include such areas as geothermal energy utilization, thermal energy storage and recoverable systems, petroleum reservoirs, insulation of high temperature gas-solid reaction vessels, chemical catalytic convectors, storage of grain, fruits and vegetables, pollutant dispersion in aquifers, industrial and agricultural water distribution, buried electrical cables, combustion in situ in underground reservoirs for the enhancement of oil recovery, ceramic radiant porous burners used by industrial firms as efficient heat transfer devices and the reduction of hazardous combustion products using catalytic porous beds. These problems have attracted the attention of engineers and scientists from very diversified disciplines, such as applied mathematics, mechanical, chemical, civil, nuclear and bio-engineering, food science and geothermal physics.

It is readily recognized that a wealth of information is now available on convective heat transfer phenomena for fluid-saturated porous media under most conditions of practical interest. Consequently, satisfactory means have evolved for the estimation of the velocity and temperature fields, as well as for the drag, heat, and mass transfer coefficients in applications involving porous media. Several excellent review articles, books and monographs which summarize the state-of-the-art are available in the open literature and these testify to the maturity of this field, e.g. Combarnous and Bories [1], Cheng [2,3], Bejan [4], Ene and Polisevski [5], Tien and Vafai [6], Nield and Bejan [7], Nakayama [8], Kaviany [9] and Kimura *et al.* [10].

Although the earliest reference to fluid flowing through a porous media dates back to Darcy [11], its importance and ramifications in process design and operation have been recognized only during the last two to three decades. Consequently, a wealth of information on the mechanism of convective heat transfer in a porous medium in a variety of processes and unit operations involving a range of geometric configurations has occurred over recent years. However, the bulk of the available information pertains to what one might call the engineering analysis, although considerable progress has also been made in developing better insights into the underlying physical processes. Indeed, several momentum transport models were proposed and used during the last few years, such as (i) the Forchheimer-extended Darcy model, (ii) the Brinkman-Forchheimer-extended Darcy model and (iii) the general equation model, which is the same as model (ii) with an additional convection inertial term included. With the widespread availability and power of computers, most of these models have been thoroughly investigated, see Nakayama [8], but very few comparisons have been made with experimental data, the main reason for this being the difficulty in gathering meaningful experimental data.

A cursory inspection of the existing references on convective flow in porous media reveals that, in general, steady-state phenomena have been extensively studied, whereas unsteady phenomena have received relatively much less attention. The mechanism causing an unsteady flow may either act at the boundaries, and this may be through a change in one or more of the dependent variables, or it may be present within the fluid volume. Examples of the former include unsteadiness resulting from the movement of the system boundaries relative to the fluid and changing the upstream or the inlet conditions. Unsteadiness of the latter type results from changing the body forces, wall and internal energy generation rates, or the pressure gradients. More complex unsteadiness may include several of these effects simultaneously. In the present review article, we are concerned only with external flows, i.e. convective flows over surfaces which are embedded in porous media, which are heated or cooled in an unsteady manner. Such a review will not only result in the consolidation of the currently available information, but also facilitate the identification of the areas meriting further work. The analogous information pertaining to a viscous fluid has been reviewed recently by Pop *et al.* [12].

BASIC EQUATIONS

As remarked earlier, the present chapter sets out to provide a review of the existing literature on the role of time dependent fluid flow characteristics on the convective interphase transport between a fluid-saturated porous medium and solid surfaces which are immersed in the porous medium. To simplify the mathematical analysis, the following assumptions are made: (i) the flow is sufficiently slow such that the convective fluid and porous medium are in local thermodynamical equlibrium; (ii) the fluid and the porous medium are homogeneous and isotropic; (iii) the pressure and the temperature are such that the fluid remains in the liquid phase; (iv) the porous medium is fully saturated; (v) Darcy's law is valid in thermally active systems; (vi) the Boussinesq approximation is valid, i.e. the fluid density ρ decreases linearly with the fluid temperature T^* according to $\rho = \rho_0[1-\beta(T^*-T_0)]$, where ρ_0 denotes the fluid density at a reference temperature T_0; and (vii) all the other fluid and porous medium properties are assumed constant. With these assumptions, the basic equations of convective flow in a porous medium are given by, see Nield and Bejan [7],

$$\nabla^*.\mathbf{V}^* = 0 \tag{1}$$

$$\frac{\mu}{\kappa}\mathbf{V}^* = \rho_0\mathbf{g}[1-\beta(T^*-T_0)] - \nabla^* p^* \tag{2}$$

$$\sigma\frac{\partial T^*}{\partial t^*} + (\mathbf{V}^*.\nabla^*)T^* = \alpha\nabla^{*2}T^* \tag{3}$$

where $\mathbf{V}^*$ is the mean velocity vector, or the Darcy velocity, t^* denotes the time, $\mathbf{g}$ is the gravitational acceleration , p^* is the pressure, κ is the permeability, μ is the viscosity of the fluid, and σ and α are defined by $\sigma = (\rho C_p)m/(\rho_0 C_p)_f$ and $\alpha = k/(\rho_0 C_p)_f$, respectively.

A comprehensive set of boundary-layer equations, which are valid in the Darcy-Boussinesq approximation, is given in the next section. The equations are comprehensive in that they include subset equations obtained by using different scalings according to whether the plates are inclined, vertical or horizontal. The specific scaling used to arrive at the asymptotically correct boundary-layer equations depends also on whether it is the temperature of the plate, or the heat flux at the plate, that is suddenly changed at time $t^* = 0$ from the ambient conditions.

FLAT PLATES

The problem of convective heat transfer caused by a heated or cooled flat plate which is placed adjacent to a semi-infinite porous medium provides one of the most basic scenarios in the area of heat transfer, and is thus of considerable theoretical and practical interest. In this section an overview of the work performed on unsteady boundary-layer flow near a flat plate embedded in a porous medium is presented.

Inclined plates: similarity solutions

Consider a fluid-saturated porous medium which is bounded by a flat impermeable heated plate inclined at an angle ε to the horizontal, with x^* and y^* being the coordinates along and normal to the plate, respectively. We specify that the temperatures of the plate and the ambient medium are $T_w^*(x^*, t^*)$ and $T_\infty^*(x^*, t^*)$, respectively, where $T_w^* > T_\infty^*$. It can be readily shown that equations (1)–(3) can be written in non-dimensional form as, see for example Johnson and Cheng [13],

$$\frac{\partial^2 \psi}{\partial x^2} + \frac{\partial^2 \psi}{\partial y^2} = Ra\left(\frac{\partial \theta}{\partial y}\sin(\varepsilon) - \frac{\partial \theta}{\partial x}\cos(\varepsilon)\right) \tag{4}$$

$$\frac{\partial \theta}{\partial t} + \frac{\partial \psi}{\partial y}\frac{\partial \theta}{\partial x} - \frac{\partial \psi}{\partial x}\frac{\partial \theta}{\partial y} = \frac{\partial^2 \theta}{\partial x^2} + \frac{\partial^2 \theta}{\partial y^2} \tag{5}$$

where ψ is the stream function defined in the usual way, namely $u = \partial\psi/\partial y$ and $v = -\partial\psi/\partial x$, and $Ra = g\kappa\beta\Delta T L/\nu\alpha$ is the Rayleigh number. The non-dimensional variables are defined as

$$x = x^*/L \ , \ y = y^*/L \ , \ t = t^*\alpha/(\sigma L^2) \ , \ \psi = \psi^*/\alpha \ , \ \theta = (T^* - T_0)/\Delta T$$

where ΔT is a reference temperature.

If we assume that the convection takes place in a thin layer adjacent to the heated surface such that $\partial/\partial x \ll \partial/\partial y$ and if we take $\Delta T = g\kappa\beta L/\nu\alpha$, equations (4) and (5) then become

$$\frac{\partial^2 \psi}{\partial y^2} = \frac{\partial \theta}{\partial y}\sin(\varepsilon) - \frac{\partial \theta}{\partial x}\cos(\varepsilon) \tag{6}$$

$$\frac{\partial \theta}{\partial t} + \frac{\partial \psi}{\partial y}\frac{\partial \theta}{\partial x} - \frac{\partial \psi}{\partial x}\frac{\partial \theta}{\partial y} = \frac{\partial^2 \theta}{\partial y^2} \tag{7}$$

Johnson and Cheng [13] were the first to derive the necessary and sufficient conditions under which similarity solutions exist for equations (6) and (7) when they are subject to the general boundary conditions

$$\psi = 0 \ , \ \theta = \theta_w(x,t) \qquad \text{at} \quad y = 0 \tag{8a}$$

$$\frac{\partial \psi}{\partial y} = 0 \ , \ \theta = \theta_\infty(x,t) \qquad \text{as} \quad y \to \infty \tag{8b}$$

where the condition $\partial\psi/\partial y = 0$ as $y \to \infty$ implies that any variation in $\theta_\infty(x,t)$ does not induce fluid motions, i.e. a stable environment is assumed. Johnson and Cheng [13] showed that only very specific solutions exist for unsteady free convection about an inclined flat plate in a porous medium. A summary of all the physical realizable solutions is presented in Table 1 of the paper by Johnson and Cheng [13].

Vertical plates with a sudden increase in the temperature

Consider now a vertically heated plate ($\varepsilon = \pi/2$) which is embedded in a fluid-saturated porous medium which is at a uniform temperature T_∞. We assume that at $t^* = 0$ the temperature of the plate, or the heat flux at the plate, is suddenly increased to the constant value T_w^*, or q_w, respectively. In this case, the unsteady boundary-layer equations can be reduced to a single equation, which can be written in non-dimensional form as, see Cheng and Pop [14],

$$\frac{\partial}{\partial \tau}\int_0^\infty \theta \, dy + \frac{\partial}{\partial X}\int_0^\infty \theta^2 \, dy = -\left(\frac{\partial \theta}{\partial y}\right)_{y=0} \tag{9}$$

This equation is to be solved subject to the boundary and initial conditions

$$\theta(X, y, 0) = 0 \tag{10a}$$

$$\theta(X, 0, \tau) = 1 \qquad \text{or} \qquad \frac{\partial \theta}{\partial y}(X, 0, \tau) = -1 \tag{10b}$$

$$\theta(X, \infty, \tau) = 0 \tag{10c}$$

in which the non-dimensional variables are defined as

$$\tau = \alpha t^*/(\sigma L^2)\ ,\ y = y^*/L\ ,\ X = x^{*2}/(L^2 Ra_x)\ ,\ \theta = (T^* - T_\infty)/\Delta T$$

where $Ra_x = g\kappa\beta\Delta T x^*/\nu\alpha$ is the local Rayleigh number and $\Delta T = T_w^* - T_\infty$ for the case of a constant wall temperature and $\Delta T = q_w L/k$ for the case of a constant heat flux at the plate, respectively.

Cheng and Pop [14] solved equation (9) assuming a temperature profile of the form

$$\theta = \mathrm{erfc}(\eta) \tag{11}$$

for the plate with a step increase in the wall temperature and

$$\theta = \frac{\sqrt{\pi}}{2}\,\delta\ \mathrm{erfc}(\eta) \tag{12}$$

for the plate with a step increase in the heat flux at the wall, where erfc is the complementary error function and

$$\eta = y/\,\delta \tag{13}$$

with $\delta(x, t)$ being the boundary-layer thickness.

Substituting equations (11), (13) into equation (9) and after integrating yields

$$\frac{\partial \delta}{\partial \tau} + A\,\frac{\partial \delta}{\partial X} = \frac{2}{\delta} \tag{14}$$

subject to the initial and boundary conditions

$$\tau = 0\ : \qquad \delta(X,0) = 0 \tag{15a}$$

$$\tau \geq 0\ : \qquad \delta(X,\tau) = 0 \qquad \text{at} \quad X = 0 \tag{15b}$$

where

$$A = 2 - \sqrt{2} = 0.5857$$

Equation (14) is a first-order, linear, partial differential equation of the hyperbolic type and it has been solved exactly by the method of characteristics and approximately by the Kármán-Pohlhausen integral-method. The method of characteristics gives

$$\delta = 2\sqrt{\tau} \tag{16}$$

if the initial condition (15a) is used, and

$$\delta = 2\sqrt{X/A} = 2.61\sqrt{X} \tag{17}$$

if the boundary condition (15b) is imposed. Note that the expression for δ changes from equation (16) to equation (17) along the limiting characteristic

$$\tau = X/A = 1.707X \tag{18}$$

After δ has been determined, the surface heat flux given by

$$q_w = -k\left(\frac{\partial T^*}{\partial y^*}\right)_{y^*=0} = \frac{2k\Delta T}{\delta\sqrt{\pi}}$$

can be calculated. As shown in Figure 1, equation (18) is a straight line which divides the $x - \tau$ plane into two regions: the lower region for $\tau < \tau_s$ with δ given by equation (16) and the upper region for $\tau \geq \tau_s$ with δ given by equation (17), where τ_s is the time taken to reach steady-state flow given by

$$\tau_s = \left(\frac{\sigma}{A}\right) x^2/\ Ra_x \tag{19}$$

Therefore, we have

$$\theta = \operatorname{erfc}(y/2\sqrt{\tau}) \quad , \quad q_w = \frac{k\Delta T}{L\sqrt{\pi\tau}} \tag{20}$$

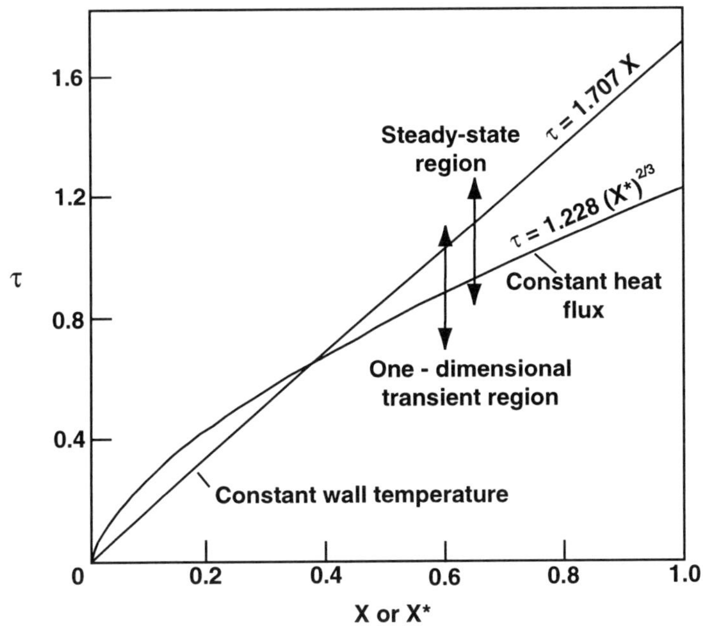

Figure 1 The limiting characteristics for the cases of constant wall temperature and constant heat flux, where $X^* = (x^*/L)^3(1/Ra_x^*)$ and $Ra_x^* = g\kappa\beta q_w {x^*}^2/(\nu\alpha k)$

for $\tau < \tau_s$, and

$$\theta = \text{erfc}(0.3826\xi) \quad , \quad q_w = \frac{k\Delta T}{x^*}\sqrt{\frac{ARa_x}{\pi}} \tag{21}$$

for $\tau \geq \tau_s$, where $\xi = y(Ra_x)^{1/2}/x$. Using equation (21), we can express the local Nusselt number Nu_x as

$$\frac{Nu_x}{(Ra_x)^{1/2}} = 0.4318 \tag{22}$$

where $Nu_x = q_w x^*/k\triangle T$.

It is worth mentioning that equation (20) is identical to the exact solution for the one-dimensional heat conduction equation in a semi-infinite porous medium with an initial temperature T_∞ when its bounding surface is suddenly raised to a temperature T_w^*. Thus, during the initial stage when the leading edge effect is not felt, the solution for the temperature, or for the velocity field, as given by equation (20) is independent of x^*. We therefore identify the lower region ($\tau \geq \tau_s$) in Figure 1 as the transient one-dimensional conduction region. On the other hand, equation (21) is valid in the upper region ($\tau \geq \tau_s$). Furthermore, we notice that the constant

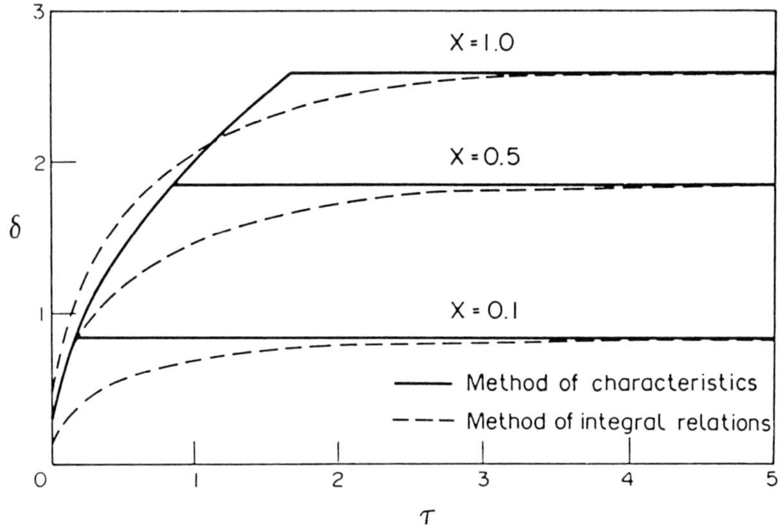

Figure 2 The growth of the boundary-layer thickness for the case of a step increase in wall temperature

0.43 in equation (22) is about 1.2% lower than the exact value of 0.444, based on the similarity solution found by Cheng and Minkonwycz [15] for the steady free convection flow along an isothermal plate which is embedded in a porous medium. This shows that they are in very good agreement.

Equation (14) has been also solved by Cheng and Pop [14] using the method of integral relations. The variation of δ with τ is shown in Figure 2.

The transient free convection boundary-layer solution adjacent to a horizontal plate ($\varepsilon = 0$) in a porous medium was also reported by Pop and Cheng [16] using the present method. We mention here that the problem of suddenly cooling vertical and horizontal surfaces in porous media has been studied by Ingham *et al.* [17, 18].

Ingham and Brown [19] were the first to present detailed solutions for the problem of free convection from a vertical flat plate embedded in a porous medium when the temperature of the plate is suddenly raised from T_∞ to a value which is proportional to $x^{*\lambda}$, where λ is a prescribed constant. Scaling the stream function ψ with $Ra^{1/2}$, the y-coordinate with $Ra^{1/2}$ and time t with Ra, reduces the full planar Darcian momentum and energy equations (4) and (5) with $\varepsilon = \pi/2$, in the limit $Ra \to \infty$, to the following boundary-layer system, see Ingham and Brown [19],

$$\frac{\partial^2 \psi}{\partial y^2} = \frac{\partial \theta}{\partial y} \tag{23}$$

$$\frac{\partial \theta}{\partial t} + \frac{\partial \psi}{\partial y}\frac{\partial \theta}{\partial x} - \frac{\partial \psi}{\partial x}\frac{\partial \theta}{\partial y} = \frac{\partial^2 \theta}{\partial y^2} \tag{24}$$

subject to the initial and boundary conditions

$$t < 0 : \qquad \psi = \theta = 0 \qquad \text{all} \quad x,\, y \tag{25a}$$

$$t \geq 0 : \qquad \psi = 0, \quad \theta_w = x^\lambda \qquad \text{on} \quad y = 0,\, x \geq 0 \tag{25b}$$

$$\frac{\partial \psi}{\partial y} = 0, \quad \theta = 0 \qquad \text{as} \quad y \to \infty,\, x \geq 0 \tag{25c}$$

To reduce the number of independent variables in equations (23) and (24) from three to two, the following new variables are introduced

$$\psi = x^{(1+\lambda)/2} f(\eta, \tau) \quad , \quad \eta = y x^{(\lambda-1)/2} \quad , \quad \tau = t x^{\lambda - 1} \tag{26}$$

On using equations (23) and (26), the temperature θ is given by

$$\theta = x^\lambda \frac{\partial f}{\partial \eta} \tag{27}$$

Then, substituting equations (26) and (27) into equation (24) gives

$$\left[1 - (1-\lambda)\tau \frac{\partial f}{\partial \eta}\right] \frac{\partial^2 f}{\partial \eta \partial \tau} + \left[(1-\lambda)\tau \frac{\partial f}{\partial \tau} - \frac{1+\lambda}{2} f\right] \frac{\partial^2 f}{\partial \eta^2} + \lambda \left(\frac{\partial f}{\partial \eta}\right)^2 = \frac{\partial^3 f}{\partial \eta^3} \tag{28}$$

with boundary conditions (25) becoming

$$\tau = 0 : \qquad f = 0 \qquad \text{for all} \quad \eta \tag{29a}$$

$$\tau \geq 0 : \qquad f = 0 \quad , \quad \frac{\partial f}{\partial \eta} = 1 \qquad \text{at} \quad \eta = 0 \tag{29b}$$

$$\frac{\partial f}{\partial \eta} \to 0 \qquad \text{as} \quad \eta \to \infty \tag{29c}$$

The variables η and τ are the appropriate ones to use for studying the final decay to the steady state solution, and by setting $\partial/\partial\tau \equiv 0$ in equation (28) leads to the steady-state boundary-layer equation

$$f''' + \frac{1+\lambda}{2} f f'' - \lambda f'^2 = 0 \tag{30a}$$

subject to the boundary conditions

$$f(0) = 0 \quad , \quad f'(0) = 1 \quad , \quad f'(\infty) = 0 \tag{30b}$$

where primes denote differentiation with respect to η. However, in the initial period of the flow development, the boundary-layer grows as though the wall was infinitely long with the effect of the finite leading edge (at $x = 0$) being felt only at later times. Consequently, more suitable independent variables to use for small values of τ are

$$\xi = \eta/(2\tau^{1/2}) \quad \text{and} \quad \tau \tag{31a}$$

with the dependent variable

$$f = 2\tau^{1/2}F(\xi,\tau) \tag{31b}$$

Using the transformation (31), equation (28) becomes

$$\frac{\partial^3 F}{\partial \xi^3} + \left[2\xi + 4\tau\lambda F - 4(1-\lambda)\tau^2\frac{\partial F}{\partial \tau}\right]\frac{\partial^2 F}{\partial \xi^2} - 4\tau\lambda\left(\frac{\partial F}{\partial \xi}\right)^2$$

$$-4\tau\left[1-(1-\lambda)\tau\frac{\partial F}{\partial \tau}\right]\frac{\partial^2 F}{\partial \xi \partial \tau} = 0 \tag{32}$$

while the initial and boundary conditions (29) remain essentially unchanged.

From equation (30) we obtain, after a little algebra,

$$2(2\lambda+1)\int_0^\infty f f'^2 \, d\eta = 1 \tag{33}$$

where $f \geq 0$ for all $0 \leq \eta < \infty$. From equation (33) it follows directly that, for equation (30) to have a solution, we require

$$\lambda > -1/2$$

with the solution becoming singular as $\lambda \to -1/2$.

For the numerical solutions of equations (28) and (32) there are two cases to consider, namely $\lambda \geq 1$ and $-1/2 < \lambda < 1$, respectively. The details are described fully by Ingham and Brown [19] and are not repeated here. The rate of approach to the steady-state solution (asymptotic solution) has been found to be of the form

$$f(\eta,\tau) = f_0(\eta) + \tau^\lambda f_1(\eta) + h.o.t., \qquad \lambda \neq 1 \tag{34}$$

where $f_0(\eta)$ is the solution of equation (30) and $\gamma = -2 - 3/(\lambda - 1)$.

Figures 3 and 4 show the variation of the non-dimensional temperature θ as a function of η at various values of τ and $\lambda = 1, 2$ and 4 (Figure 3), and $\lambda = 0.25, 0, -0.2$ and -0.425 (Figure 4), respectively. It is seen from these figures that the steady-state boundary-layer solution, as given by equation (34), is achieved for large τ, with the rate of approach to the steady state increasing as λ is increased. As λ approaches its lower limit of $\lambda = -1/2$, the steady state takes increasingly longer to be attained. This observation is further confirmed in Figure 5, which shows the variation of $(\partial\theta/\partial\eta)_{\eta=0}$ with τ for some values of λ.

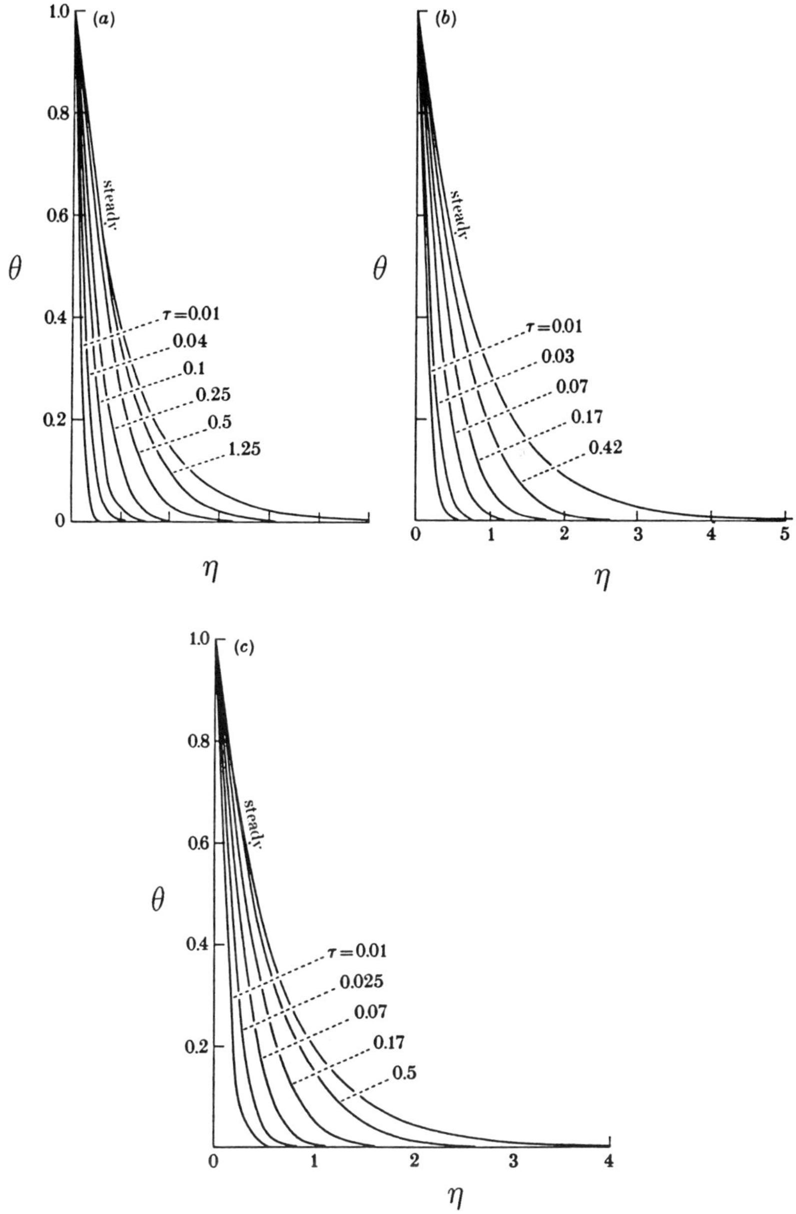

Figure 3 The non-dimensional temperature distribution as a function of η at various values of τ: (a) $\lambda = 1$, (b) $\lambda = 2$ and (c) $\lambda = 4$

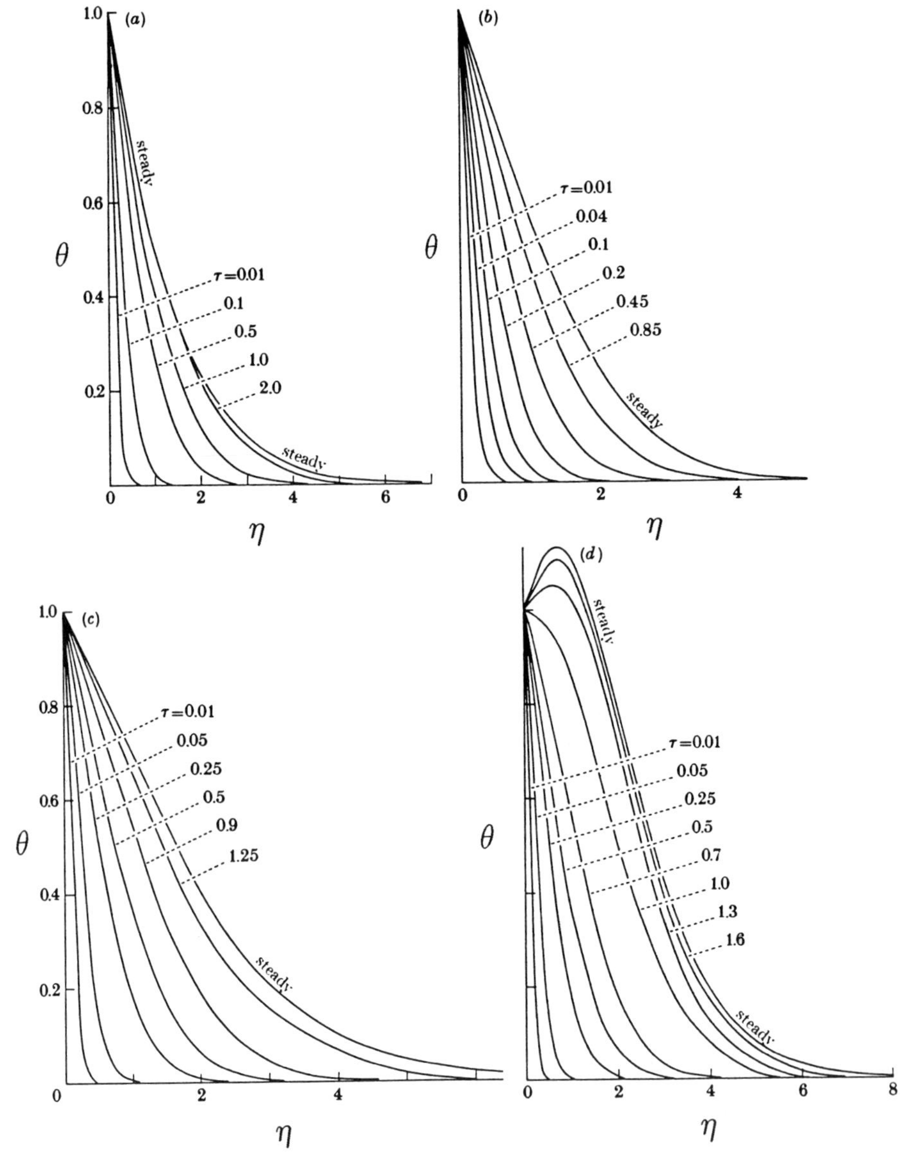

Figure 4 The variation of the non-dimensional temperature as a function of η at various values of τ: (a) $\lambda = 0.25$, (b) $\lambda = 0$, (c) $\lambda = -0.2$ and (d) $\lambda = -0.425$

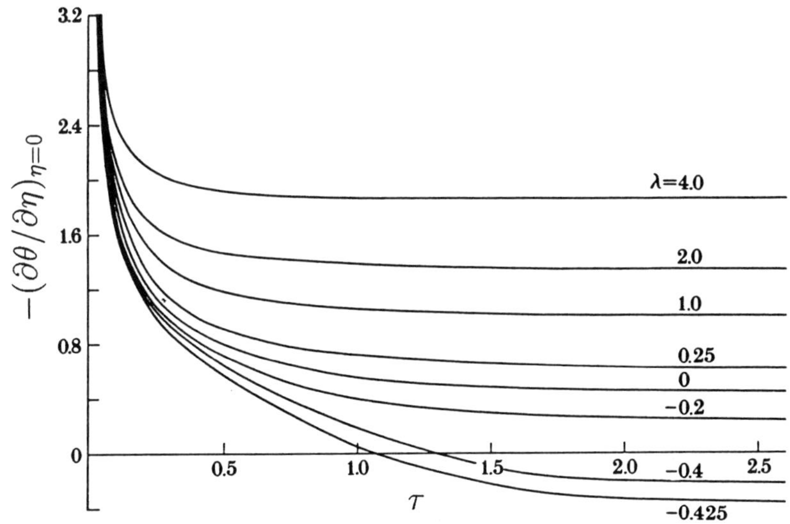

Figure 5 The variation of the local Nusselt number with τ for various values of λ

Vertical plates with a sudden increase in the heat flux

Merkin and Zhang [20] have presented solutions for the boundary-layer flow generated on a vertical surface ($\varepsilon = \pi/2$) in a porous medium in the case when the wall is heated at a rate q_w which is proportional to x^λ for time $t \geq 0$. In this case the scalings are $Ra^{1/3}$, $Ra^{-1/3}$, $Ra^{1/3}$ and $Ra^{2/3}$ for ψ, θ, y and t, respectively, where now $Ra = g\kappa\beta q_w L^2/(k\mu\alpha)$. Thus, in the limit $Ra \to \infty$, equations (4) and (5) reduce to equations (23) and (24), respectively. The variables ψ, θ, η and τ are now of the form

$$\psi = x^{(2+\lambda)/3} f(\eta,\tau) \quad , \quad \theta = x^{(1+2\lambda)/3}\frac{\partial f}{\partial \eta} \tag{35a}$$

$$\eta = y x^{(\lambda-1)/3} \quad , \quad \tau = t x^{2(\lambda-1)/3} \tag{35b}$$

so that equations (23) and (24) become

$$\left[1 - \frac{2}{3}(1-\lambda)\tau\frac{\partial f}{\partial \eta}\right]\frac{\partial^2 f}{\partial \eta \partial \tau} - \left[\left(\frac{2+\lambda}{3}\right) f - \frac{2}{3}(1-\lambda)\tau\frac{\partial f}{\partial \tau}\right]\frac{\partial^2 f}{\partial \eta^2}$$
$$+ \left(\frac{1+2\lambda}{3}\right)\left(\frac{\partial f}{\partial \eta}\right)^2 = \frac{\partial^3 f}{\partial \eta^3} \tag{36}$$

with the initial and boundary conditions

$$\tau < 0 : \qquad f = 0 \qquad \text{all} \quad \eta \tag{37a}$$

$$\tau \geq 0 : \qquad f = 0 \quad , \quad \frac{\partial^2 f}{\partial \eta^2} = -1 \qquad \text{on} \quad \eta = 0 \tag{37b}$$

$$\frac{\partial f}{\partial \eta} \to 0 \qquad \text{as} \quad \eta \to \infty \tag{37c}$$

By taking $\partial/\partial\tau \equiv 0$ in equation (36) leads to the steady-state boundary-layer equation

$$f''' + \frac{2+\lambda}{3} f f'' - \frac{1+2\lambda}{3} f'^2 = 0 \tag{38a}$$

along with the boundary conditions

$$f(0) = 0 \quad , \quad f''(0) = -1 \quad , \quad f'(\infty) = 0 \tag{38b}$$

From equation (38) it follows that

$$(1+\lambda) \int_0^\infty f'^2 \, d\eta = 1$$

which shows that, for equation (38) to have a solution, we require

$$\lambda > -1$$

with the solution becoming singular as $\lambda \to -1$.

However, in the initial stages of the growth of the boundary-layer, the most appropriate variables to use are

$$f = \tau F(\xi, \tau) \quad , \quad \xi = \eta/\tau^{1/2} \quad \text{and} \quad \tau \tag{39}$$

On substituting equation (39) into equation (36), we obtain

$$\left[\tau - \frac{2}{3}(1-\lambda)\tau^{5/2}\right] \frac{\partial^2 F}{\partial \xi \partial \tau} - \left[\frac{1}{2}\xi + \lambda \xi^{3/2} F - \frac{2}{3}(1-\lambda)\tau^{5/2} \frac{\partial F}{\partial \tau}\right] \frac{\partial^2 F}{\partial \xi^2}$$

$$+ \frac{1}{2}\frac{\partial F}{\partial \xi} + \lambda \tau^{3/2} \left(\frac{\partial F}{\partial \xi}\right)^2 = \frac{\partial^3 F}{\partial \xi^3} \tag{40}$$

while the initial and boundary conditions (37) remain essentially unchanged. At $\tau = 0$ the equation (40) becomes, with $F(\xi, 0) = F_0(\xi)$,

$$F_0''' + \frac{1}{2}\xi F_0'' - \frac{1}{2} F_0' = 0 \tag{41a}$$

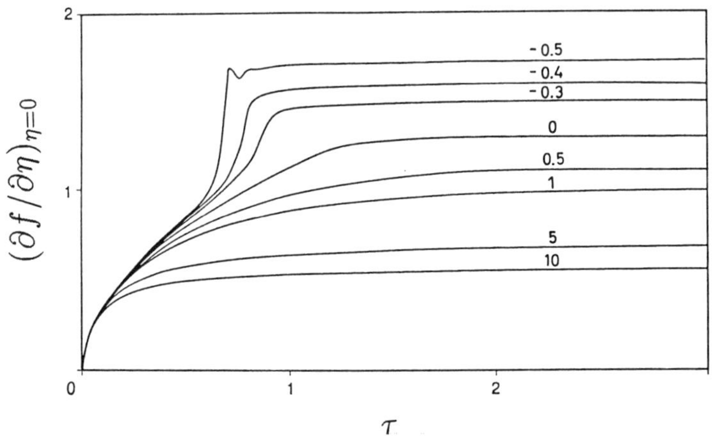

Figure 6 A graph of $(\partial f/\partial \eta)_{\eta=0}$ against τ for a range of values of λ as obtained from the numerical solution of equations (36) and (40)

for all values of λ, and has to be solved subject to

$$F_0(0) = 0 \quad , \quad F_0''(0) = -1 \quad , \quad F_0'(\infty) = 0 \tag{41b}$$

The solution of equation (41) is

$$F_0' = \frac{2}{\sqrt{\pi}} \exp(-\xi^2/4) - 3\sqrt{2}\,\mathrm{erfc}(\xi/\sqrt{2}) \tag{42}$$

From equations (34) and (40) it then follows, for $\tau \ll 1$, that

$$\left(\frac{\partial f}{\partial \eta}\right)_{\eta=0} = \frac{2}{\sqrt{\pi}}\,\tau^{1/2} + \ldots \tag{43}$$

Equations (36) and (40) were integrated numerically by Merkin and Zhang [20] in the range $-1 < \lambda < 1$ and $\lambda > 1$, using the same method as that proposed by Ingham and Brown [19]. Some of the results obtained are presented in Figure 6, where a plot of $(\partial f/\partial \eta)_{\eta=0}$ against τ for a range of values of λ is given. This figure shows a similar behaviour for $\tau \ll 1$ for all values of λ, as predicted by equation (43), and also by equation (34), where now γ is given by $\gamma = -3(\lambda+1)/2(\lambda-1)$, for τ large. The rate of approach to the asymptotic solution is seen to decrease slightly as the value of λ is decreased. This can also be seen in Figure 7, where plots of $\partial f/\partial \eta$ against η at various values of τ for $\lambda = 5.0$ and 1.0 are given. Figure 8 illustrates the variation of the velocity field $\partial f/\partial \eta$ against η at various values of τ

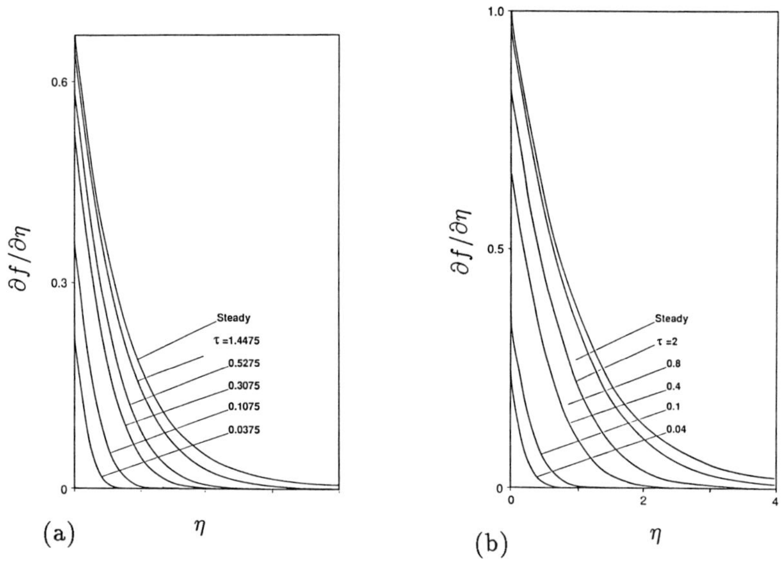

Figure 7 Graphs of $\partial f/\partial \eta$ against η for (a) $\lambda = 5$ and (b) $\lambda = 1$, given at various values of τ, as obtained from the numerical solution of equations (36) and (40)

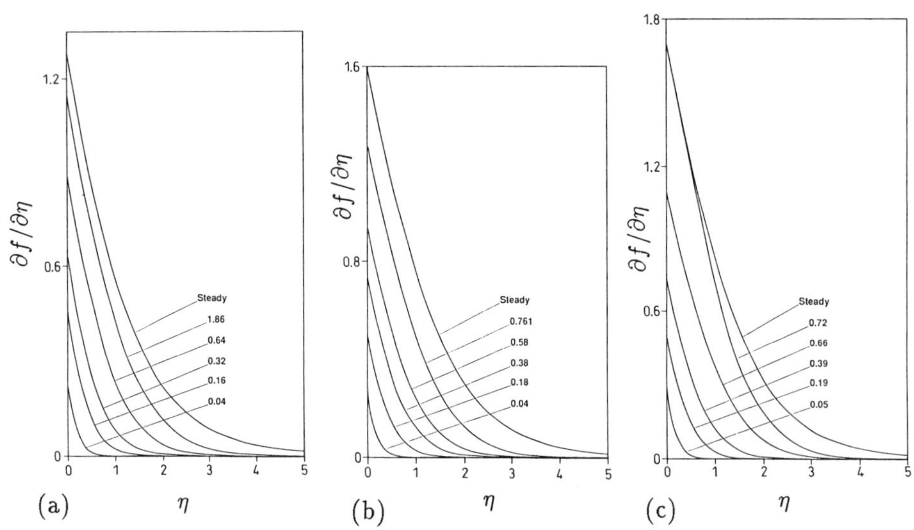

Figure 8 Graphs of $\partial f/\partial \eta$ against η for (a) $\lambda = 0$, (b) $\lambda = -0.4$ and (c) $\lambda = -0.5$, given at various values of τ, as obtained from the numerical solution of equations (36) and (40)

for $\lambda = 0, -0.4$ and -0.5. It is seen from the first two graphs that the velocity field has a steady progression from the small time solution to the steady-state (asymptotic) solution. However, for $\lambda = -0.5$ we can see that $\partial f/\partial \eta$ has the same values for a range of η both at $\tau = 0.72$ and in the steady-state solution, suggesting an "overshoot" in $\partial f/\partial \eta$ in this case.

A number of approximate solution techniques are also available in the open literature for the unsteady convective flow in the vicinity of a vertical flat plate which is embedded in a porous medium under different conditions as described by Raptis [21], Ingham and Pop [22], Haq and Mulligan [23], Pop and Herwig [24] and Rao and Pop [25].

Vertical plates which are suddenly cooled

Consider a vertical flat plate in a porous medium and assume that the general transient arises from a sudden change in the level of energy input flux on the surface of the plate, i.e. a steady input heat flux q_2'' is changed at $t^* = 0$ to a new steady level q_2'' and is maintained at this value for $t^* > 0$. Using the appropriate variables as defined in Harris *et al.* [26, 27], equations (23) and (24) can be reduced to the following form,

$$\frac{\partial^3 f}{\partial \eta^3} + \left(-1 + 2\tau \frac{\partial f}{\partial \eta}\right) \frac{\partial^2 f}{\partial \eta \partial \tau} + \left(2f - 2\tau \frac{\partial f}{\partial \tau}\right) \frac{\partial^2 f}{\partial \eta^2} - \left(\frac{\partial f}{\partial \eta}\right)^2 = 0 \qquad (44)$$

which is to be solved with the boundary conditions

$$f(0,\tau) = 0 \quad , \quad \frac{\partial^2 f}{\partial \eta^2}(0,\tau) = -\frac{q_2''}{q_1''} \quad , \quad \frac{\partial f}{\partial \eta}(\infty,\tau) = 0 \qquad (45)$$

At $\tau = 0$ the flow is steady and hence $f(\eta, 0) = f_0(\eta)$, say, so that, from equation (44), $f_0(\eta)$ satisfes the equation

$$f_0''' + 2f_0 f_0'' - f_0'^2 = 0 \qquad (46a)$$

subject to the boundary conditions

$$f_0(0) = 0 \quad , \quad f_0''(0) = -1 \quad , \quad f_0'(\infty) = 0 \qquad (46b)$$

For $\tau \ll 1$ there exists an inner boundary-layer which is described by equation (44) and outside this layer the flow remains the initial boundary-layer profile as given by f_0 in equation (46). Since the appropriate length scale for small τ is $\tau^{1/2}$, we introduce the variables

$$f(\eta, \tau) = \tau F(\xi, \tau) \quad , \quad \xi = \eta / 2\tau^{1/2} \qquad (47)$$

Equation (44) thus becomes

$$\frac{1}{8}\frac{\partial^3 F}{\partial \xi^3} - \frac{1}{4}\frac{\partial F}{\partial \xi} + \frac{1}{2}\left(-\tau + \tau^{5/2}\right)\frac{\partial^2 F}{\partial \xi^2} + \frac{1}{2}\left(\frac{\xi}{2} - \tau^{5/2}\frac{\partial F}{\partial \tau}\right)\frac{\partial^2 F}{\partial \xi^2} = 0 \tag{48}$$

which is subject to the boundary conditions, from conditions (45),

$$F(0,\tau) = 0 \quad , \quad \frac{\partial^2 F}{\partial \xi^2}(0,\tau) = -4\left(\frac{q_2''}{q_1''}\right) \tag{49}$$

The solution for $\partial F/\partial \xi$ in this inner region is given by

$$\frac{\partial F}{\partial \xi} = 2a\tau^{-1/2} - 4\xi + 4(1-R)\left[\xi \mathrm{erfc}(\xi) - \frac{1}{\sqrt{\pi}}\exp(-\xi^2)\right]$$

$$+2b^2\xi^2\tau^{1/2} + \frac{2b}{\sqrt{\pi}}(1-R)\tau\exp(-\xi^2) + \mathbf{O}(\tau^{3/2}) \tag{50}$$

where $R = q_2''/q_1''$ and $b = f_0'(0) = 0.8987$. Thus, the resulting velocity, or temperature function, for small times, is given by

$$\frac{\partial f}{\partial \eta} = \frac{\partial f_0}{\partial \eta} + (1-R)\,\eta\,\mathrm{erfc}\left(\frac{\eta}{2\tau^{1/2}}\right)$$

$$-\frac{1}{\sqrt{\pi}}(1-R)\left(2\tau^{1/2} - a\tau^{3/2}\right)\exp(-\eta^2/4\tau) + \mathbf{O}(\tau^2) \tag{51}$$

Another quantity of interest is the non-dimensional plate temperature which is given by

$$\theta_w(\tau) = \left(\frac{\partial f}{\partial \eta}\right)_{\eta=0} = b + \frac{1}{\sqrt{\pi}}(1-R)(-2\tau^{1/2} + b\tau^{3/2}) + \mathbf{O}(\tau^2) \tag{52}$$

for $\tau \ll 1$.

The transient effects, which are described by the small time solutions (51) and (52), continue to penetrate outwards into the boundary-layer to evolve into a new steady-state flow. In order to match these small and large time solutions, a numerical solution of the full boundary-layer equation (44) has to be used. To do this we notice that equation (44) is parabolic and therefore it can be integrated numerically using a step-by-step method.

This marching method enables the solution $f_0(\eta)$ at time $\tau = 0$ to proceed in time and gives a complete solution for $\tau \leq \tau^*$, where τ^* is the maximum value of τ reached in the numerical scheme, which is less than τ satisfying $2\tau\frac{\partial f}{\partial \eta}(0,\tau) = 1$. At large times the solution for $\theta(\eta,\tau)$ is known to approach the steady state profile $\theta(\eta,\infty)$ associated with the heat flux q_2'' at the plate. The solution of the ordinary differential equation governing this steady-state flow can be obtained as a similarity solution of equation (46), from which it can be shown that

$$\theta(\eta,\infty) = R^{2/3} f_0'(\eta R^{1/2}) \tag{53}$$

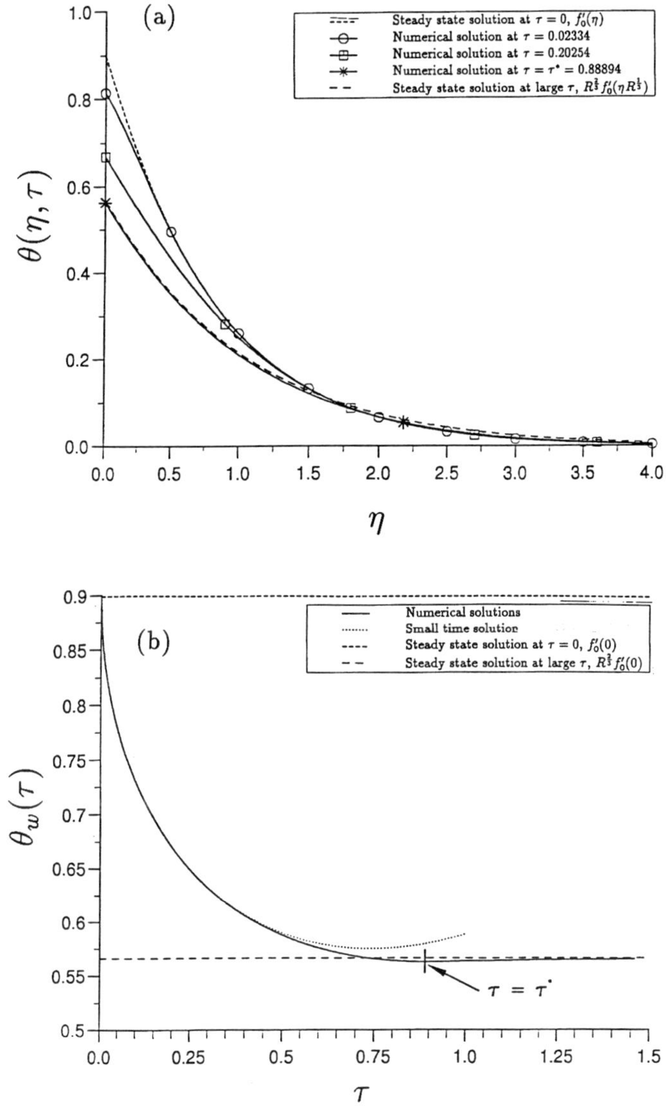

Figure 9 (a) The variation of the non-dimensional temperature $\theta(\eta, \tau)$ as a function of η at various values of τ and the steady state solutions at $\tau = 0$ and $\tau = \infty$. (b) The variation of the wall temperature $\theta_w(\tau)$ as a function of τ, the small time solution and the steady state solutions at $\tau = 0$ and $\tau = \infty$. $R = 0.5$

The numerical details of solving equation (44) are fully described by Harris *et al.* [27, 28]. The variation of the temperature profiles $\theta(\eta,\tau)$ and also the wall temperature $\theta_w(\tau)$ at various values of τ for $R = 0.5$ is shown in Figure 9. By plotting the steady-state profiles, as predicted by equation (46) and the appropriate similarity solution (53), the transition from $\tau = 0$ to $\tau = \infty$ is clearly illustrated in Figure 9. This figure also demonstrates that, initially, the effects of the change in the surface heat flux at the plate are not felt near the outer edge of the boundary-layer. Furthermore, it is found that the transient solution develops close to the small time solution (52) and is graphically almost identical when $\tau < 0.4$ and $\tau < 0.15$ for $R = 0.5$ and 2, respectively. At large values of time, the temperature at the wall, $\theta_w(\tau)$, approaches the steady-state solution (53), but overshoots slightly and the numerical method breaks down near the local maximum, or minimum, of $\theta_w(\tau)$ for $R > 1$ or $R < 1$, respectively. The temperature profile $\theta(\eta,\tau^*)$ displayed in Figure 9 shows that a local turning point in $\theta(\eta,\tau)$ only occurs near to the wall and for large values of η the temperature is always bounded by the two steady-state values at $\tau = 0$ and $\tau = \infty$, respectively.

The main objective of the papers reviewed above was to study the transient convective flow adjacent to a flat plate in a porous medium in which the plate is considered infinitesimally thin and, therefore, the heat conduction within the plate is neglected. However, there are many problems of practical importance where the thermal conditions at the plate depend strongly on the conductive heat transfer within the solid plate. In these situations the convective flow must be studied as a mixed problem, termed a conjugate problem, where there is an interaction between the convective flow in the fluid and the heat conduction within the solid plate. Until recently, there have been only two papers by Vynnycky and Kimura [29] and Shu and Pop [30] on the transient conjugate free convection due to a vertical plate of finite extent adjacent to a semi-infinite porous medium and much work remains to be done on this subject.

It is worth mentioning to this end that certain results exist for the transient non-Darcy free and forced convection heat transfer from a flat plate embedded in a fluid-saturated porous medium, see Chen *et al.* [31], and Jang *et al.* [32].

HORIZONTAL CYLINDERS

One of the earliest analyses on the unsteady heat transfer from a horizontal circular cylinder which is immersed in a forced convection Darcy flow was carried out by Sano [33]. He reported asymptotic solutions which were valid at large and small values of the Peclet numbers in the case when the unsteady temperature field is produced by a step change in the wall temperature. As regards the transient free convection from a horizontal cylinder in a porous medium, detailed analytical and numerical work has been performed by Ingham *et al.* [34], Pop *et al.* [35, 36], Tyvand [37], Sundfor and Tyvand [38] and Bradean *et al.* [39].

We consider a horizontal circular cylinder of radius a embedded in a porous medium which is at a constant temperature T_∞. At time $t^* = 0$, the temperature of the cylinder is suddenly increased to the constant value $T_w(> T_\infty)$ and is maintained at this temperature thereafter. We take cylindrical polar coordinates (r^*, ϕ, z^*) with the axis of the cylinder r^*=0 and $\phi = 0$ as the downward vertical, and assume all quantities are independent of z^*. If we first introduce the dimensionless variables

$$t = t^* \alpha Ra/(\sigma a^2) \quad , \quad r = r^*/\, a \quad , \quad \mathbf{V} = \mathbf{V}^* a/(\alpha Ra)$$

$$p = p^*/(\alpha Ra) \quad , \quad \theta = (T^* - T_\infty)/\Delta T$$

and then eliminate the pressure p, equations (1)–(3) can be written as, see Ingham it et al. [34],

$$\frac{\partial^2 \psi}{\partial \xi^2} + \frac{\partial^2 \psi}{\partial \phi^2} + e^{\xi} \left(\frac{\partial \theta}{\partial \xi} \sin(\phi) + \frac{\partial \theta}{\partial \phi} \cos(\phi) \right) = 0 \tag{54}$$

$$\frac{\partial \theta}{\partial t} + e^{-2\xi} \left(\frac{\partial \psi}{\partial \phi}\frac{\partial \theta}{\partial \xi} - \frac{\partial \psi}{\partial \xi}\frac{\partial \theta}{\partial \phi} \right) = \frac{e^{-2\xi}}{Ra} \left(\frac{\partial^2 \theta}{\partial \xi^2} + \frac{\partial^2 \theta}{\partial \phi^2} \right) \tag{55}$$

where $\xi = \ln r$ and the velocity components are written in terms of the stream function ψ as

$$u = e^{-\xi}\, \frac{\partial \psi}{\partial \phi} \quad , \quad v = -e^{-\xi}\, \frac{\partial \psi}{\partial \xi} \tag{56}$$

If the symmetry condition at $\phi = \pi$ is enforced, the initial and boundary conditions are

$$t < 0 : \qquad \psi = \theta = 0 \qquad (\xi \geq 0) \tag{57a}$$

$$t \geq 0 : \qquad \psi = 0 \quad , \quad \theta = 1 \qquad \text{on} \quad \xi = 0 \tag{57b}$$

$$\frac{\partial \psi}{\partial \xi} = 0 \quad , \quad \theta = 0 \qquad \text{as} \quad \xi \to \infty \tag{57c}$$

$$\psi = \frac{\partial \theta}{\partial \phi} = 0 \qquad \text{on} \quad \phi = 0 \quad \text{and} \quad \phi = \pi \ (\xi \geq 0) \tag{57d}$$

When the cylinder is impulsively heated, a boundary-layer of thickness proportional to $(t/Ra)^{1/2}$ is formed initially on the surface of the cylinder. This suggests the introduction of a new variable η defined as $\xi = s\eta$ and a scaling for ψ with s into equations (54) and (55), where $s = 2(t/Ra)^{1/2}$. The unsteady boundary-layer equations for this problem are then obtained by setting $s = 0$. After some algebra, we finally obtain,

$$\frac{\partial \psi}{\partial \eta} = -\theta \sin(\phi) \tag{58}$$

$$\frac{\partial^2 \theta}{\partial \eta^2} + 2\eta \frac{\partial \theta}{\partial \eta} - 4t \frac{\partial \theta}{\partial t} = 4t \left(\frac{\partial \psi}{\partial \phi}\frac{\partial \theta}{\partial \eta} - \frac{\partial \psi}{\partial \eta}\frac{\partial \theta}{\partial \phi} \right) \tag{59}$$

and the initial and boundary conditions (57) remain essentially unchanged.

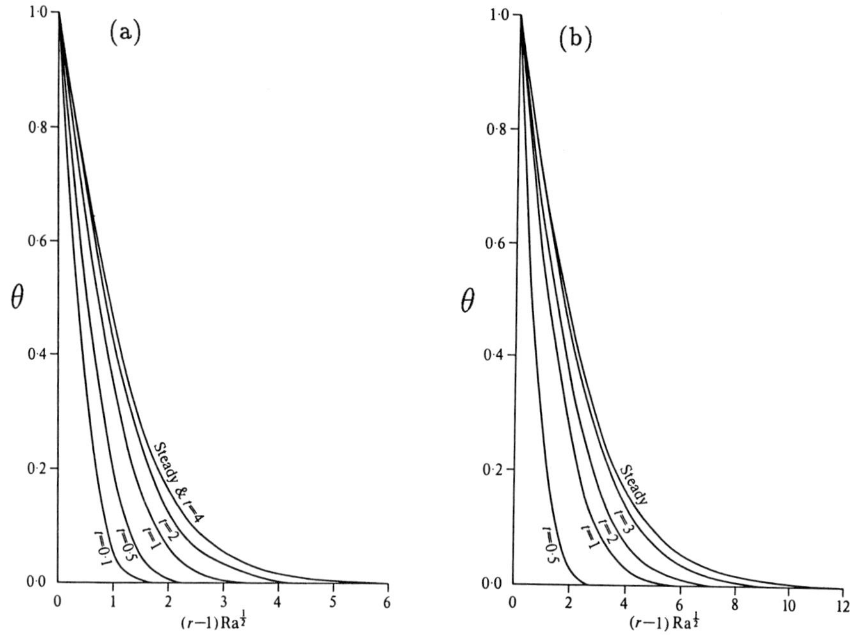

Figure 10 The temperature at various times as a function of the radial distance at (a) $\phi = 0^o$ and (b) $\phi = 180^o$

Equations (58) and (59) were first solved analytically for small values of time and numerical solutions obtained using a standard finite-difference method. The nature of the boundary-layer collision at the top of the cylinder is further discussed. The numerical work strongly suggests that no singularity is generated at a finite value of time and this is contrary to the results obtained by several authors when dealing with the similar problem in a viscous (non-porous) fluid.

Figure 10 shows the temperature variation with the radial distance at $\phi = 0^o$ (bottom stagnation point) and 180° (top stagnation point). Also the variation of the heat transfer coefficient Q given by

$$Q = -\frac{a(Ra)^{-1/2}}{\Delta T}\left(\frac{\partial T^*}{\partial r^*}\right)_{r^*=a} = -(Ra)^{-1/2}\left(\frac{\partial \theta}{\partial r}\right)_{r=1} \tag{60}$$

as a function of ϕ at various times is depicted in Figure 11. Also shown is the steady-state profile as obtained by Merkin [40]. It is seen that the unsteady results approach the steady-state solution at large values of time.

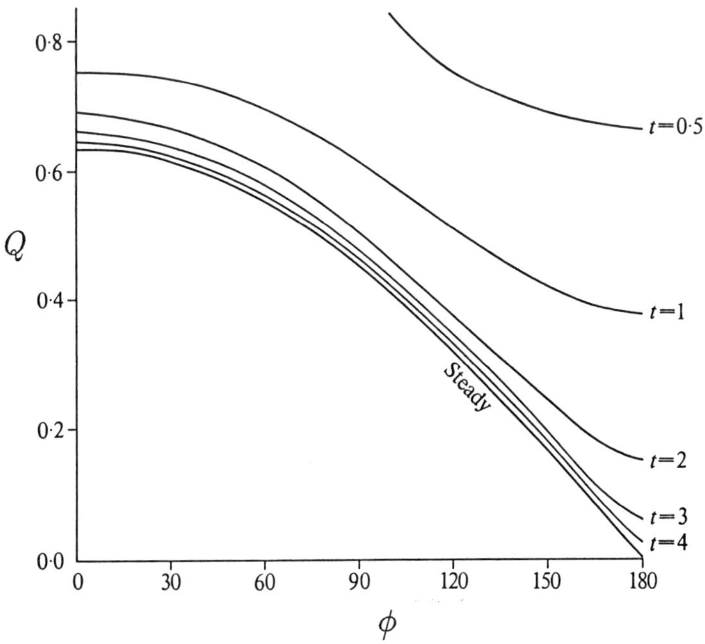

Figure 11 The variation of the heat transfer coefficient Q as a function of ϕ at various times

Very little has been reported on the transient convection in a porous medium which surrounds a vertical circular cylinder with a sudden change in the wall temperature since the study of Kimura [41].

Before concluding this subsection it is worthwhile to recount briefly some studies concerning other transient processes in a porous medium. Recently, Nguyen *et al.* [42] reported results for the transient free convection from a horizontal circular cylinder which is embedded in a porous medium in which the flow is chemically reactant. Also, the only pertinent paper on unsteady free convection boundary-layer flow in a porous medium near a general three-dimensional stagnation point on an isothermal surface is due to Pop and Merkin [43].

SPHERES

In contrast to the cylinder problem, the corresponding studies concerning the transient convective heat transfer from a sphere which is immersed in a porous medium are only of recent origin. Sano and Okihara [44] and Sano [45] have studied the transient case using asymptotic theory in terms of small values of the Rayleigh number. On the other hand, Nguyen and Paik [46] have investigated the transient nature of the transport phenomena under conditions where boundary-layer theory fails to capture the correct physics of the mixed convection associated with a sphere in a porous medium saturated with water using Chebyshev-Legendre spectral method. This flow situation is known to exist in water, especially near the freezing point. Very recently, Yan *et al.* [47] have presented detailed numerical results for the transient free convection from a sphere which is buried in a porous medium for $0.01 \leq Ra \leq 200$ using a finite-difference method. Both the cases of constant wall temperature and constant heat flux at the wall were considered.

CONCLUDING REMARKS

The bulk of the literature available on transient free and mixed convection flow from solid surfaces which are embedded in a fluid-saturated porous medium has been throughly reviewed in this chapter. In particular, consideration has been given to the cases of plates, cylinders and spheres. However, most studies are based on the use of the classical Darcy law model and theoretical developments based on more realistic fluid models are required when studying unsteady flows in porous media.

It is perhaps fair to state that the experimental activity lags far behind the corresponding theoretical developments in the general area of boundary-layer flows in a porous medium. However, the available theoretical and experimental results do indicate the availability of satisfactory means for estimating the convective transfer coefficients, under free and mixed convection regimes, from plates, cylinders and spheres. More work which covers a wider range of geometric surfaces is required to generalize the conclusions presented for the configurations analised in this chapter.

Several problem areas which are important in physical applications, as well as in theory, still persist. Turbulent flows and chemically reactive flows in a porous medium are examples of such areas which must be investigated in the near future. The limited available results on steady-state, convective heat transfer in a porous medium which is saturated with saline water in which the density extremum might occur or saturated with a non-Newtonian fluid can be extended to include time-dependent conditions. The analogous phenomena in viscoelastic fluids remain virtually unexplored thus far. Even scarcer are the results on unsteady convective heat transfer in a porous medium which is saturated with a micropolar fluid.

Thus, considerable scope exists for theory as well as careful experimentation in this branch of porous media technology. Due to the numerous applications of porous media where unsteady flows arise, increasing effort will surely be spent in understanding the complicated non-steady effects in the coming years. It certainly merits much more attention than it has received in the past.

REFERENCES

1. M.A. Combarnous and S.A. Bories, Hydrothermal convection in saturated porous media, *Adv. Hydroscience* **10**, 231-237 (1975).

2. P. Cheng, Heat transfer in geothermal systems, *Adv. Heat Transfer* **14**, 1-105 (1978).

3. P. Cheng, Natural convection in a porous medium: external flows. In: Natural Convection: Fundamentals and Applications (eds. S. Kakac, W. Aung and R. Viskanta), Hemisphere, Washington, DC, 475-513 (1985).

4. A. Bejan, Convective heat transfer in porous media. In: Handbook of Single-phase Convective Heat Transfer (eds. S. Kakac, R.K. Shah and W. Aung), Chapter 16, Wiley, New York (1987).

5. H.I. Ene and D. Polisevski, Thermal Flow in Porous Media, D. Reidel, Dordrecht, The Netherlands (1987).

6. C.L. Tien and K. Vafai, Convective and radiative heat transfer in porous media, *Adv. Appl. Mech.* **27**, 225-281 (1990).

7. D.A. Nield and A. Bejan, Convection in Porous Media, Springer, New York (1992).

8. A. Nakayama, PC-Aided Numerical Heat Transfer and Convective Flow, CRC Press, Tokyo (1995).

9. M. Kaviany, Principles of Heat Transfer in Porous Media, 2nd edition, Springer, New York (1995).

10. S. Kimura, T. Kiwata, A. Okajima and I. Pop, Conjugate natural convection in porous media, *Adv. Water Resources* **20**, 111-126 (1997).

11. H. Darcy, Les Fontaines Publiques de la Ville de Dijon, Victor Dalmont, Paris (1856).

12. I. Pop, D.B. Ingham and J.H. Merkin, Transient convective heat transfer in external flow. To appear in: Time-Dependent Nonlinear Convection (ed. P. Tyvand), Computational Mechanics Publications, Southampton (1997).

13. C.H. Johnson and P. Cheng, Possible similarity solutions for free convection boundary layers adjacent to flat plates in porous media, *Int. J. Heat Mass Transfer* **21**, 709-718 (1978).

14. P. Cheng and I. Pop, Transient free convection about a vertical flat plate imbedded in a porous medium, *Int. J. Engng. Sci.* **22**, 253-264 (1984).

15. P. Cheng and W.J. Minkowycz, Free convection about a vertical flat plate embedded in a porous medium with application to heat transfer from a dike, *J. Geophys. Res.* **82**, 2040-2044 (1977).

16. I. Pop and P. Cheng, The growth of a thermal layer in a porous medium adjacent to a suddenly heated semi-infinite horizontal surface, *Int. J. Heat Mass Transfer* **26**, 1574-1576 (1983).

17. D.B. Ingham, J.H. Merkin and I. Pop, Flow past a suddenly cooled vertical flat surface in a saturated porous medium, *Int. J. Heat Mass Transfer* **25**, 1916-1919

(1982).

18. D.B. Ingham, J.H. Merkin and I. Pop, Flow past a suddenly cooled horizontal flat surface in a saturated porous medium, *Acta Mechanica* **56**, 205-217 (1985).

19. D.B. Ingham and S.N. Brown, Flow past a suddenly heated vertical plate in a porous medium, *Proc. Roy. Soc. London* **A 403**, 51-80 (1986).

20. J.H. Merkin and G. Zhang, The boundary-layer flow past a suddenly heated vertical surface in a saturated porous medium, *Warme- und Stoffubertr.* **27**, 299-304 (1992).

21. A. Raptis, Unsteady free convective flow through a porous medium, *Int. J. Engng. Sci.* **21**, 345-348 (1983).

22. D.B. Ingham and I. Pop, The sudden melting of a thin vertical flat plate in a Darcian free convection flow, *Acta Mechanica* **67**, 77-93 (1988).

23. S. Haq and J.C. Mulligan, Transient free convection about a vertical flat plate embedded in a saturated porous medium , *Num. Heat Transfer* **A 18**, 227-242 (1990).

24. I. Pop and H. Herwig, Transient mass transfer from an isothermal vertical flat plate embedded in a porous medium, *Int. Comm. Heat Mass Transfer* **17**, 813-821 (1990).

25. K.N. Rao and I. Pop, Transient free convection in a fluid saturated porous medium with temperature dependent viscosity, *Int. Comm. Heat Mass Transfer* **21**, 573-581 (1994).

26. S.D. Harris, D.B. Ingham and I. Pop, Transient free convection on a vertical plate subjected to a change in surface heat flux in porous media, *Fluid Dynamics Research* **18**, 313-324 (1996).

27. S.D. Harris, D.B. Ingham and I. Pop, Free convection from a vertical plate in a porous medium subjected to a sudden change in surface heat flux, *Transport in Porous Media* **26**, 207-226 (1997).

28. S.D. Harris, D.B. Ingham and I. Pop, Free convection from a vertical plate in a porous media subjected to a sudden change in surface temperature, *Int. Comm. Heat Mass Transfer* **24**, 543-552 (1997).

29. M. Vynnycky and S. Kimura, Transient conjugate free convection due to a vertical plate in a porous medium, *Int. J. Heat Mass Transfer* **38**, 219-231 (1995).

30. J.-J. Shu and I. Pop, Transient conjugate free convection along a vertical surface in porous media. *To be published in Int. J. Engng. Sci.* (1997).

31. C.K. Chen, C.I. Hung and H.C. Horng, Transient natural convection on a vertical flat plate embedded in a high-porosity medium, *ASME J. Energy Res. Tech.* **109**, 112-118 (1987).

32. J.Y. Jang, D.J. Tzeng and H.J. Shaw, Transient free convection with mass transfer on a vertical plate embedded in a high porosity medium, *Num. Heat Transfer* **A 20**, 1-18 (1991).

33. T. Sano, Unsteady heat transfer from a circular cylinder immersed in a Darcy flow, *J. Engng. Math.* **14**, 177-190 (1980).

34. D.B. Ingham, J.H. Merkin and I. Pop, The collision of free-convection boundary layers on a horizontal cylinder embedded in a porous medium, *Quart. J. Mech. Appl. Math.* **36**, 313-335 (1983).

35. I. Pop, D.B. Ingham and P. Cheng, Transient free convection about a horizontal circular cylinder in a porous medium, *Fluid Dynamics Research* **12**, 295-305 (1993).

36. I. Pop, D.B. Ingham and R. Bradean, Transient free convection about a horizontal circular cylinder in a porous medium with constant surface flux heating, *Acta Mechanica* **119**, 79-91 (1996).

37. P.A. Tyvand, First-order transient free convection about a horizontal cylinder embedded in a porous medium, *Fluid Dynamics Research* **15**, 277-294 (1995).

38. H.O. Sundfor and P.A. Tyvand, Transient free convection in a horizontal porous cylinder with a sudden change in wall temperature. In: Waves and Nonlinear Processes in Hydrodynamics (eds. J. Grue, B. Gjevik and J.E. Weber), Kluwer, Holland, 291-302 (1996).

39. R. Bradean, D.B. Ingham, P.J. Heggs and I. Pop, Unsteady free convection adjacent to an impulsively heated horizontal circular cylinder in a porous medium. *To be published in Num. Heat Transfer* **A** (1997).

40. J.H. Merkin, Free convection boundary layers on axisymmetric and two-dimensional bodies of arbitrary shape in a saturated porous medium, *Int. J. Heat Mass Transfer* **22**, 1461-1462 (1979).

41. S. Kimura, Transient forced and natural convection heat transfer about a vertical cylinder in a porous medium, *Int. J. Heat Mass Transfer* **32**, 617-620 (1989).

42. H.D. Nguyen, S. Paik, R.W. Douglas and I. Pop, Unsteady non-Darcy reaction-driven flow from an isotropic cylinder in porous media, *Chem. Engng. Sci.* **51**, 4963-4977 (1996).

43. I. Pop and J.H. Merkin, Three-dimensional Darcian free convection near a stagnation point on an isothermal surface, *Transport in Porous Media* **2**, 357-362 (1987).

44. T. Sano and R. Okihara, Natural convection around a sphere immersed in a porous medium at small Rayleigh numbers, *Fluid Dynamics Research* **13**, 39-44 (1994).

45. T. Sano, Unsteady forced and natural convection around a sphere immersed in a porous medium, *J. Engng. Math.* **30**, 515-525 (1996).

46. H.D. Nguyen and S. Paik, Unsteady mixed convection from a sphere in water-saturated porous media with variable surface temperature/heat flux, *Int. J. Heat Mass Transfer* **37**, 1783-1793 (1994).

47. B. Yan, I. Pop and D.B. Ingham, A numerical study of unsteady free convection from a sphere in a porous medium, *Int. J. Heat Mass Transfer* **40**, 893-903 (1997).

THERMAL BOUNDARY–LAYER INSTABILITIES IN POROUS MEDIA: A CRITICAL REVIEW

D. A. S. REES

Department of Mechanical Engineering, University of Bath,
Claverton Down, Bath, BA2 7AY, UK

INTRODUCTION

The aims of this review are (i) to provide a comprehensive survey of the present knowledge of how thermal boundary layers in porous media are destabilised, (ii) to give a critical account of the present state–of–the–art, and (iii) to recommend strategies for the future development of this important topic. At present there are, to the author's knowledge, fewer than thirty publications which deal with the instability of thermal boundary layers in porous media, only two of which are concerned with the nonlinear development of instability. Therefore this topic can only be regarded as being in its infancy. By contrast, the study of the instability of the well–known Blasius boundary–layer flow of an isothermal clear fluid over a flat plate left the realms of linearised theory based on the parallel–flow approximation many decades ago. Weakly nonlinear theories have been developed, as have nonparallel analyses, and it is now possible to employ the techniques of Direct Numerical Simulation to determine in detail the temporal evolution of primary and secondary instabilities en route to fully developed turbulence. Thus our present topic is lagging well behind in its development, especially when it is realised that fully numerical simulations in porous media flows are no more difficult to obtain than are Navier–Stokes simulations, and are, in fact, easier should boundary effects in the form of the Brinkman terms be neglected. It is hoped, therefore, that the present review will serve as a stimulus for greatly increased work in this field.

We will also emphasize the sometimes subtle roles played by the various approximations which are used routinely in this topic, namely, the boundary–layer approximation and the parallel–flow approximation. The former serves to reduce the full, spatially elliptic, equations of motion to either a set of ordinary differential equations (thereby obtaining a self–similar flow), or, at worst, a parabolic set of partial differential equations (thereby obtaining a nonsimilar flow which is computed using a suitable marching scheme). The latter serves to simplify the linearised disturbance equations to ordinary differential form and therefore these solutions are relatively easy to obtain compared with the full partial differential form. Both approximations are discussed and we show that care must be taken in their use. It is also hoped that bringing these issues to the fore will allow the subject to progress in a more informed and effective manner.

In this review we will attempt to maintain a consistent notation throughout. Upper–case letters will usually refer to dimensional variables, and lower–case to dimensionless variables. The w–subscript refers to conditions at the heated surface, while the ∞–subscript refers to ambient conditions. The variable η is the similarity or pseudo–similarity variable, and ξ and Ra_x are scaled streamwise coordinates; precise definitions of these will depend on the flow configuration in question. Coordinates used as subscripts denote derivatives with respect to that coordinate, while primes denote derivatives with respect to η. Occasionally these conventions will differ from those in some of the papers reviewed.

The references follow the usual numerical style with the following modification. References [1] to [25] comprise all the known work on the title topic; these are ordered according to their appearance. The remainder are concerned with those papers which have a strong bearing on the review, such as those dealing with basic boundary–layer flows in porous media, or those on the instability of clear fluids.

THE NATURE OF THE BOUNDARY–LAYER APPROXIMATION

There are three methods commonly used to obtain the equations for thermal boundary–layer flows in porous media, and these will be illustrated using the near–horizontal free convection boundary–layer flow from a heated surface. Assuming the simplest possible situation, namely, that Darcy's law is valid, that the Boussinesq approximation applies, that the flow is steady, and that the medium is homogeneous, nondeformable and isotropic, the governing equations are,

$$\frac{\partial U}{\partial X} + \frac{\partial V}{\partial Y} + \frac{\partial W}{\partial Z} = 0, \tag{1a}$$

$$U = -\frac{K}{\mu}\Big[\frac{\partial P}{\partial X} - \rho_\infty g\beta(T - T_\infty)\sin\phi\Big], \tag{1b}$$

$$V = -\frac{K}{\mu}\Big[\frac{\partial P}{\partial Y} - \rho_\infty g\beta(T - T_\infty)\cos\phi\Big], \tag{1c}$$

$$W = -\frac{K}{\mu}\frac{\partial P}{\partial Z}, \tag{1c}$$

$$U\frac{\partial T}{\partial X} + V\frac{\partial T}{\partial Y} + W\frac{\partial T}{\partial Z} = \alpha\Big[\frac{\partial^2 T}{\partial X^2} + \frac{\partial^2 T}{\partial Y^2} + \frac{\partial^2 T}{\partial Z^2}\Big]. \tag{1e}$$

Here X, Y and Z are the dimensional streamwise, cross–stream and spanwise coordinates, respectively, U, V and W are the corresponding fluid flux velocities, P is the pressure and T the temperature. The material parameters, K, μ, g, β and α have their usual meanings, T_∞ is the ambient temperature of the medium, and ρ_∞ is the corresponding fluid density. The inclination above the horizontal of the upward–facing heated surface is ϕ. The heated surface has a uniform surface temperature distribution: $T_w = T_\infty + \Delta T$ for $X > 0$, where ΔT is the temperature difference between the wall and the ambient medium. The resulting two–dimensional flow may be studied by introducing a streamfunction in the form,

$$U = \frac{\partial \Psi}{\partial Y}, \qquad V = -\frac{\partial \Psi}{\partial X}, \qquad W = 0. \tag{2}$$

If, in addition, it is assumed that there are no Z–variations, equations (1) become,

$$\frac{\partial^2\Psi}{\partial X^2} + \frac{\partial^2\Psi}{\partial Y^2} = \frac{\rho_\infty g\beta K}{\mu\alpha}\left[\frac{\partial T}{\partial Y}\sin\phi - \frac{\partial T}{\partial X}\cos\phi\right] \tag{3a}$$

$$\frac{\partial^2 T}{\partial X^2} + \frac{\partial^2 T}{\partial Y^2} = \frac{\partial\Psi}{\partial Y}\frac{\partial T}{\partial X} - \frac{\partial\Psi}{\partial X}\frac{\partial T}{\partial Y}. \tag{3b}$$

The heated surface is assumed to be semi–infinite in extent and therefore there is no external physical lengthscale which may be used to nondimensionalise the equations. The three methods of analysing the boundary–layer flow treat this absence in different ways.

Using the local Rayleigh number — method 1

This is the most frequently used method for studying boundary–layer flows. It proceeds initially by assuming that $X \gg Y$, that is, the boundary–layer thickness is much smaller than the distance from the leading edge. It follows that the X–derivative diffusion terms on the left hand sides of equations (3a) and (3b) may be neglected in favour of the Y–derivatives. On following this procedure, equations (3) become parabolic and may be transformed into a more suitable form using the set of transformations,

$$\psi = \alpha\, Ra_x^{1/3} f(\xi,\eta), \qquad T = T_\infty + \Delta T\,\theta(\xi,\eta), \tag{4a, b}$$

where

$$\eta = \frac{Y}{X} Ra_x^{1/3}, \qquad \xi = Ra_x^{1/3}\tan\phi, \qquad Ra_x = \frac{\rho g\beta K\Delta T\cos\phi}{\mu\alpha}X. \tag{4c, d, e}$$

Finally, f and θ may easily be shown to satisfy the pair of equations,

$$f'' = \xi\theta' + \tfrac{2}{3}\eta\theta' - \tfrac{1}{3}\xi\theta_\xi, \qquad \theta'' + \tfrac{1}{3}f\theta' = \tfrac{1}{3}\xi(f'\theta_\xi - \theta' f_\xi), \tag{5a, b}$$

subject to the boundary conditions

$$f(0) = 0, \qquad \theta(0) = 1 \qquad \text{and} \qquad f',\ \theta \to 0 \quad \text{as} \quad \eta \to \infty. \tag{5c, d, e, f}$$

Although the present notation is slightly different, the above analysis follows very closely that of Jang and Chang [3] who considered the more general case of a power–law surface temperature distribution.

When $\phi = 0$ the heated surface is horizontal, and, according to Jang and Chang [5] this corresponds to the $\xi = 0$ solution of equations (5), which was first presented by Chang and Cheng [26]; such a flow is termed self–similar. When ϕ is positive, equations (5) are solved using a marching scheme, such as the Keller–box method (see Keller and Cebeci [27]). The flow and temperature profiles change as ξ increases; such a flow is termed non–similar.

The problem of the absence of a physical lengthscale has been circumvented by the definitions contained in equation (4) where Ra_x plays the role of a stretched streamwise coordinate and η is the pseudo–similarity variable. It is interesting to note that the governing partial differential equations have not been nondimensionalized using this formulation, at least not in the usual way. It is straightforward to show that $X \gg Y$ using the definition of η when $Ra_x \gg 1$, and therefore a very large local Rayleigh number is a necessary condition for the boundary–layer approximation to be valid. Most workers in the general field of boundary–layer flows use such an approach, but it is sufficient to cite the very first two papers dealing with the basic flows in porous media, Cheng and Chang [26] and Cheng and Minkowycz [28], and to mention that most of the linear stability papers cited in [1]–[25] also do so. Stability analyses undertaken using this formulation of the basic flow result in *finite* values for Ra_x which correspond to neutral stability.

Using a fictitious lengthscale — method 2

An alternative approach is to define a fictitious lengthscale denoted by L. The idea here is that L is used to define a global (rather than a local) Rayleigh number. Therefore if attention is fixed at some station downstream of the leading edge then this could be defined as the dimensional lengthscale and therefore the equivalent nondimensional distance from the leading edge is precisely unity. Further, when the Rayleigh number is much greater than 1, then the boundary–layer is fully developed.

Beginning with the full dimensional equations (3) the following scalings are defined:

$$(X,Y) = L(x,y), \qquad T = T_\infty + \Delta T\ \theta, \qquad \Psi = \alpha\psi, \tag{6}$$

where Ψ is defined in equation (2), and therefore equations (3) become

$$\frac{\partial^2\psi}{\partial x^2} + \frac{\partial^2\psi}{\partial y^2} = Ra\left[\frac{\partial\theta}{\partial y}\sin\phi - \frac{\partial\theta}{\partial x}\cos\phi\right], \tag{7a}$$

$$\frac{\partial^2 T}{\partial x^2} + \frac{\partial^2 T}{\partial y^2} = \frac{\partial\psi}{\partial y}\frac{\partial\theta}{\partial x} - \frac{\partial\psi}{\partial x}\frac{\partial\theta}{\partial y}. \tag{7b}$$

Here

$$Ra = \frac{\rho_\infty g\beta KL\Delta T}{\mu\alpha} \tag{8}$$

is the Rayleigh number (or Darcy–Rayleigh number) based on the lengthscale L. Given the definition of L as the streamwise lengthscale of interest, x should be set to be $O(1)$ as $Ra \to \infty$ in a formal asymptotic analysis.

As interest is focused on how inclinations of the surface from the horizontal modify the horizontal boundary–layer flow, a scale analysis is undertaken which balances the magnitude of the y–derivative term on the left hand side of equation (7a) with the x–derivative term on the right. Such an analysis also balances the y–diffusion term in equation (7b) with the right hand side terms. The resulting scalings follow: $\psi = O(Ra^{1/3})$ and $y = O(Ra^{-1/3})$. Thus, when $Ra \to \infty$ then $y \ll x$ in magnitude and the x–derivative diffusion terms in equation (7) may be formally neglected. Indeed, within this asymptotic framework the size of the neglected diffusion terms are easily seen to be $O(Ra^{-2/3})$ relative to the retained terms. The governing parabolic equations may now be obtained using the substitution,

$$\psi = Ra^{1/3}x^{1/3}f(\xi,\eta), \qquad \theta = \theta(\xi,\eta), \tag{9a,b}$$

where η and ξ now have the following definitions,

$$\eta = \frac{Ra^{1/3}y}{x^{2/3}}, \qquad \xi = x^{1/3}, \tag{10a, b}$$

and hence

$$f'' = Ra^{1/3}\xi\theta' \sin\phi + [\tfrac{2}{3}\eta\theta' - \tfrac{1}{3}\xi\theta_\xi]\cos\phi, \qquad \theta'' + \tfrac{1}{3}f\theta' = \tfrac{1}{3}\xi[f'\theta_\xi - \theta' f_\xi], \tag{11a, b}$$

are to be solved subject to boundary conditions (5c–f).

The presence of the Ra term in equation (11a) signals a difficulty as this term is asymptotically larger than the remaining terms. However, this difficulty may be avoided by introducing the scaling $\phi = O(Ra^{-1/3})$, which is equivalent to having a heated surface which is very nearly horizontal. If $\phi = Ra^{-1/3}$ is set in (11a), then the governing equations become,

$$f'' = \xi\theta' + \tfrac{2}{3}\eta\theta' - \tfrac{1}{3}\xi\theta_\xi, \qquad \theta'' + \tfrac{1}{3}f\theta' = \tfrac{1}{3}\xi[f'\theta_\xi - \theta' f_\xi], \tag{12a, b}$$

which are identical to equations (5a,b). Other numerical multiples of $Ra^{-1/3}$ will result in equations identical to equations (12) if the definitions of f, ξ and η are modified slightly; see Rees and Riley [29] or Ingham et al. [30] for details.

With this method an artificial lengthscale has been introduced. The chief advantage of such a device is to allow the boundary–layer to be studied using rigorous methods of asymptotic methods. Here, the Rayleigh number, based on L, is taken as the asymptotically large parameter. Definitive statements can be made about the size of neglected terms, something which is done with more difficulty when using method 1, and often is not done. The above–quoted papers, Rees and Riley [29] and Ingham et al. [30], are typical examples of the use of this method.

Using a lengthscale suggested by the physical parameters — method 3

The definition of Ra_x in expression (4e) may be written in the form,

$$Ra_x = \frac{\rho g\beta K\Delta T\cos\phi}{\mu\alpha}X \equiv X\cos\phi/L \equiv x\cos\phi, \tag{13}$$

and it may be shown that this quantity is dimensionless. Setting $Ra_x = x\cos\phi$ can therefore be thought of as defining a lengthscale, L, in terms of the properties of the fluid and the medium. Thus if L, which is given by

$$L = \frac{\mu\alpha}{\rho g\beta K\Delta T}, \tag{14}$$

is chosen as a lengthscale, then the value of Ra defined for method 2 is precisely unity, and the value of Ra_x used in method 1 is precisely $x\cos\phi$, where x is nondimensional. The above analysis of method 2 carries through in the same way, except that $Ra = 1$ must be substituted into equations (9) to (11). It is difficult to translate the method 2 criterion for the asymptotic magnitude of ϕ into a satisfactory form for study, but method 3 yields no problems when studying the horizontal ($\phi = 0$) self–similar flow, or other self–similar flows.

Analyses undertaken using this method are asymptotic analyses with x as the asymptotically large parameter. Neglected terms in the full elliptic equations have well–defined orders of magnitude, but these are now expressed in terms of x. A typical paper which uses this approach is the high–order, boundary–layer analysis of Riley and Rees [31]. Stability analyses result in *finite* values of x beyond which the boundary–layer is deemed unstable.

Discussion

All three of the above methods have been used to derive the boundary–layer equations for convection in a porous medium, but all three suffer from disadvantages which will now be discussed.

The first method is not often used as an asymptotic method, but rather as an approximation to the governing equations where streamwise diffusion has been neglected. Many of the users of this method, especially those working outside the field of porous medium convection, typically invoke what is termed the boundary–layer approximation by neglecting such derivatives, and then proceed with a set of substitutions analogous to expressions (4). In the interests of balance, it must be said that the higher–order analyses of Chang and Cheng [32] and Cheng and Hsu [33] show that method 1 may be formulated in terms of a rigorous asymptotic analysis. However, method 1 can very easily lead to incorrect results if not applied correctly, and this is especially true for nonsimilar flows. An example of this is the above (method 1) basic flow analysis taken from Jang and Chang [5] which is concerned with vortex instabilities of the inclined thermal boundary–layer flow. Although streamwise diffusion was neglected in favour of cross–stream diffusion, due to the thinness of the boundary–layer, both the temperature terms on the right hand side of equation (3a) were retained. There are good formal reasons for doing this when using method 2, and the inclination of the surface must then be sufficiently small. But there is no mention in the text of Jang and Chang [5] of restrictions on the inclination angle, and it is claimed there that the stability analysis is valid for a wide range of inclinations. However, the balancing of the two temperature terms at $O(1)$ inclinations implies necessarily that both the streamwise and cross–stream diffusion terms must be of the same order of magnitude, and hence that the flow is elliptic. Of course, this negates the original assumption that the boundary–layer approximation is valid and renders their results incorrect.

Method 2 does not suffer from the danger of misapplication as it is a rigorous asymptotic analysis. But, as will be seen below, there are difficulties in presenting a similarly rigorous stability analysis within the same framework. This may be illustrated best by considering briefly the vortex stability of a generally inclined heated surface with $0 < \phi < \frac{\pi}{2}$. If the nondimensionalisation (6) is used, with appropriate scalings for Z, U, V, W and P, then equations (1) become

$$\frac{\partial u}{\partial x} + \frac{\partial v}{\partial y} + \frac{\partial w}{\partial z} = 0, \tag{15a}$$

$$u = -\frac{\partial p}{\partial x} + Ra\,\theta \sin\phi, \qquad v = -\frac{\partial p}{\partial y} + Ra\,\theta\cos\phi, \qquad w = -\frac{\partial p}{\partial z}, \tag{15b, c, d}$$

$$u\frac{\partial \theta}{\partial x} + v\frac{\partial \theta}{\partial y} + w\frac{\partial \theta}{\partial z} = \frac{\partial^2 \theta}{\partial x^2} + \frac{\partial^2 \theta}{\partial y^2} + \frac{\partial^2 \theta}{\partial z^2}. \tag{15e}$$

The basic flow is given by $\psi = Ra^{1/2}x^{1/2}f(\eta)$, $\theta = \theta(\eta)$ where $u = \psi_y$, $v = -\psi_x$, $w = 0$ and η is now defined according to

$$\eta = Ra^{1/2}y/x^{1/2}. \tag{16}$$

The functions $f(\eta)$ and $\theta(\eta)$ satisfy the equations,

$$f' = g\sin\phi, \qquad g'' + \tfrac{1}{2}fg' = 0, \qquad f(0) = 0, \quad g(0) = 1, \qquad g \to 0 \quad \text{as} \quad \eta \to \infty. \tag{17}$$

Denoting the basic flow by a "B" superscript, the linearised equations for perturbations to this flow are

$$\frac{\partial u}{\partial x} + \frac{\partial v}{\partial y} + \frac{\partial w}{\partial z} = 0, \tag{18a}$$

$$u = -\frac{\partial p}{\partial x} + \theta\sin\phi, \qquad v = -\frac{\partial p}{\partial y} + \theta\cos\phi, \qquad w = -\frac{\partial p}{\partial z}, \tag{18b, c, d}$$

$$u\frac{\partial \theta^B}{\partial x} + v\frac{\partial \theta^B}{\partial y} + u^B\frac{\partial \theta}{\partial x} + v^B\frac{\partial \theta}{\partial y} = \frac{\partial^2 \theta}{\partial x^2} + \frac{\partial^2 \theta}{\partial y^2} + \frac{\partial^2 \theta}{\partial z^2}. \tag{18e}$$

Assuming that $z = O(y) = O(Ra^{-1/2})$ when $x = O(1)$, see expression (16), and taking θ to be $O(1)$, since this is a homogeneous system, we now proceed to find scalings for the other variables. Clearly equation (18a) indicates that that the magnitude of u is at most $O(v\,Ra^{1/2})$ where the magnitude of v is to be found. As w and p_z balance in equation (18d), then so do v and p_y in equation (18c). However, from equation (18c), v must be at least of magnitude $O(Ra)$, given the size of the temperature term. In turn, this means that the magnitude of the $v\theta_y^B$ term in equation (18e) is $O(Ra^{3/2})$, which is asymptotically larger than all the other terms in that equation. Apart from the streamwise diffusion term, which is $O(1)$, the others are $O(Ra)$ in size. Thus this formulation would seem to lead to a contradiction as no single term can be asymptotically larger than all the others in an equation. There is a resolution to this problem which involves introducing a thin sublayer within the main boundary–layer, but this will be discussed later.

Why, then, does method 1 seem to succeed in giving what look like good stability analyses when the rigorous method 2 (when applied naively) quite obviously does not? The straightforward answer to this is that method 1 computes a value for Ra_x and therefore the *necessary* requirement for the validity of the boundary–layer approximation, that $Ra_x \gg 1$, is over–ruled. Indeed, the strict imbalance in the sizes of the terms in equation (18e) is ignored in Jang and Chang [5], where the term in their analysis which is equivalent to the present troublesome term is found to be multiplied explicitly by $Ra_x^{1/2}$ (see equation (42) in [5]). Thus, if method 1 is treated as a rigorous asymptotic analysis as $Ra_x \to \infty$, then presentation of results in the form of finite values of Ra_x means that the method is inconsistent.

Stability analyses using method 3 are prone to the same difficulties as method 1. The use of a coordinate, x, as the arbitrarily large parameter makes it difficult to formulate this type of analysis, but finite values of x are still computed. In method 3 it is easier to 'hide' the fact that x is asymptotically large. For self-similar flows these values will be the same as those of $Ra_x \cos\phi$ using method 1. However, method 3 is the best way to attempt numerical simulations of the full nonlinear equations where the boundary-layer approximation has not been invoked.

So far, then, the picture that has been painted is a somewhat depressing one. None of the methods used for deriving both the boundary-layer and linearised perturbation equations fare particularly well. Method 2 is the most reliable method for computing the basic flow, but a straightforward linear stability analysis, as expounded above, is seemingly inconsistent. Methods 1 and 3 can give good results for the basic flow, but care must be taken over their implementation. However, in stability analyses, orders of magnitude often are not consistently applied with method 1, and method 3 can only be used with ease when the basic flow is self-similar.

An alternative and quite reasonable view of the method 1 stability analysis is that Ra_x should not be treated formally as a large parameter, but rather that the large Ra_x basic flow is only being used as an approximation to the flow at a finite value of Ra_x. This means that there is no inconsistency in computing a finite value of Ra_x in such a stability analysis. The accuracy of the resulting stability criterion then depends on (a) how good an approximation the boundary-layer solution is to the exact solution, and (b) the consequences of neglecting streamwise diffusion in the disturbance equations. With this in mind, the next step is to review the method 1 stability analyses which are contained in the literature.

LINEAR ANALYSES USING THE LOCAL RAYLEIGH NUMBER

The pioneering work in this field is that of Hsu et al. [1], who studied the vortex instability of convection induced by horizontal heated surfaces with a power-law wall temperature distribution. The basic flow is self-similar (see Cheng and Chang [26]) and the authors follow the method 1 analysis. Their detailed analysis will not be repeated here, but Hsu et al. make mention of the "bottling effect", a term coined by Haaland and Sparrow [34], which is used to allow some terms in the disturbance equations to be neglected. The bottling effect causes the disturbances to be confined to the boundary-layer because there is a net inflow to the boundary-layer caused by fluid entrainment. The streamwise derivative in the disturbance continuity equation is neglected, which allows a disturbance streamfunction to be defined; this is termed the assumption of local similarity. An important feature of the analysis is that the disturbance functions are assumed to display the same functional dependence on x as their basic flow counterparts. Thus both the basic and disturbance temperature fields are taken to be functions of the similarity variable multiplied by X^m. Solutions for various exponents, m, are presented in terms of neutral curves of Ra_x against a scaled wavenumber, k, where $k = aX\, Ra_x^{-1/3}$, and where all disturbances are taken to be proportional to $\exp(iaZ)$, where a is the (constant) wavenumber. These curves are shown in Figures 1 for $0 \leq m \leq 2$. Hsu et al. [1] also show that the minimum value for Ra_x and the corresponding value of k both vary almost linearly with m; this is reproduced in Figure 2.

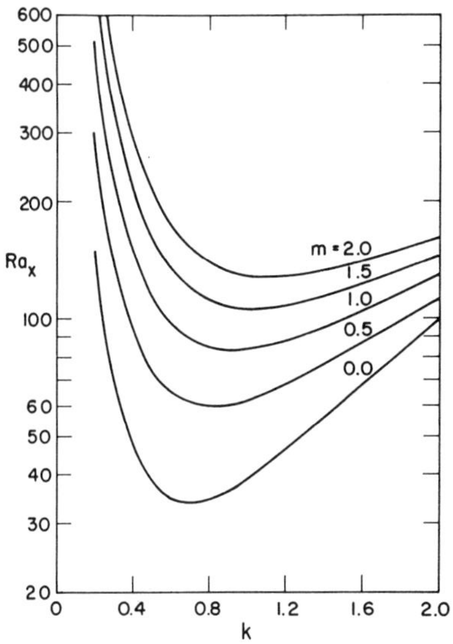

Figure 1 Rayleigh numbers as a function of spanwise wavenumber at neutral stability. This is figure 2 from Hsu et al. [1]

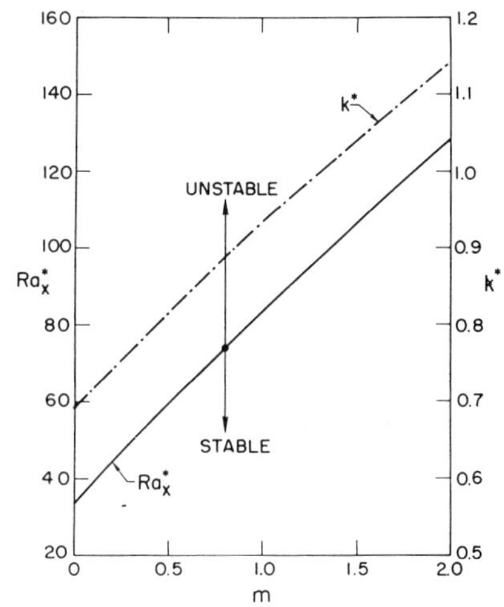

Figure 2 Critical Rayleigh numbers and associated wavenumbers as a function of m. This is figure 3 from Hsu et al. [1]

Typical disturbance streamlines and isotherms corresponding to a cross–section of the vortex system are shown in Figure 3 and confirm that the disturbance is confined to the boundary–layer.

The analysis of Hsu et al. [1] was then applied to inclined surfaces by Hsu and Cheng [2] who followed a method 1 version of the analysis contained between equations (13) and (18). They found that the two parameters, Ra_x and ϕ may be combined into a single nondimensional group, an important result, and therefore stability criteria were presented in the form of curves of $Ra_x \cot^2 \phi$ against $k = ax\, Ra_x^{-1/2}$, where a is again the vortex wavenumber. These curves are shown in Figure 4 and the nearly linear dependence of the minimising values of $Ra_x \cot^2 \phi$ and k with m are shown in Figure 5. Note that Hsu and Cheng use α_0 as the inclination angle from the vertical and hence $\phi = \pi - \alpha_0$ and $Ra_x \cot^2 \phi = Ra_x \tan^2 \alpha_0$. Clearly these results imply that the vertical thermal boundary–layer is stable, for the critical value of Ra_x becomes asymptotically large as $\phi \to \pi/2$.

Hsu and Cheng [3] then extended the work of Hsu et al. [1] to mixed convection. They found that a free stream velocity proportional to $X^{(2m-1)/3}$ when the surface temperature is proportional to X^m yields a self–similar basic flow. Any other free–stream velocity gives nonsimilarity. The basic flow can be described in terms of m and a mixed convection parameter, M, given by $M = Ra_x/Pe_x^{3/2}$, where

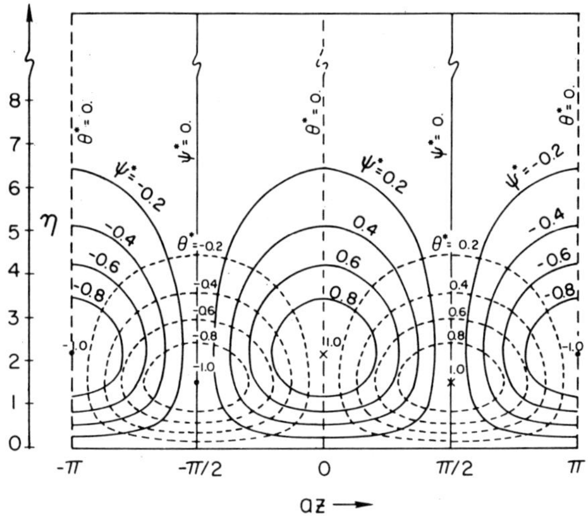

Figure 3 Secondary flow streamlines (solid lines) and isotherms (dashed lines) at the onset of vortex instability. This is figure 2 from Hsu et al. [1]

$Pe_x = u_\infty X/\alpha$ is a local Peclet number, and $u_\infty(X)$ is the free stream velocity. The stability characteristics are given in terms of the variation of Pe_x with wavenumber for chosen values of M and m. It is found that decreasing values of M, which correspond to a more strongly forced convection, cause the critical value of Pe_x to increase and the critical wavenumber to decrease. This can be explained by noting that the thinning effect on the basic flow of increasing the free–stream velocity causes the local Rayleigh number based on the local boundary–layer thickness to decrease.

A similar extension to the inclined analysis of Hsu and Cheng [2] was undertaken by the same authors in Hsu and Cheng [4]. In this case the mixed convection parameter is $M = Ra_x/Pe_x$, and the free stream velocity is proportional to X^m in order to retain self–similarity. The qualitative behaviour of the neutral curves is the same as in Hsu and Cheng [4] when the forced flow is in the same direction as that of free convection. But when the forced convection opposes buoyancy forces, the critical wavenumber increases as M decreases. Once more (see Hsu and Cheng [2]) the parameters Pe_x and ϕ collapse into a single dimensionless group for presentation of the stability analysis.

The analysis of Hsu and Cheng [2] was re–examined by Jang and Chang [5], who claimed that the former results were not valid as the normal component of the buoyancy force had been neglected. Jang and Chang's formulation of the basic flow is almost identical to that contained in the subsection describing method 1, and this was shown in the first paragraph of the following discussion subsection to be in error. Their basic flow corresponds to a near–horizontal heated surface, rather than a generally inclined one, and therefore their stability results are based on an incorrect basic flow.

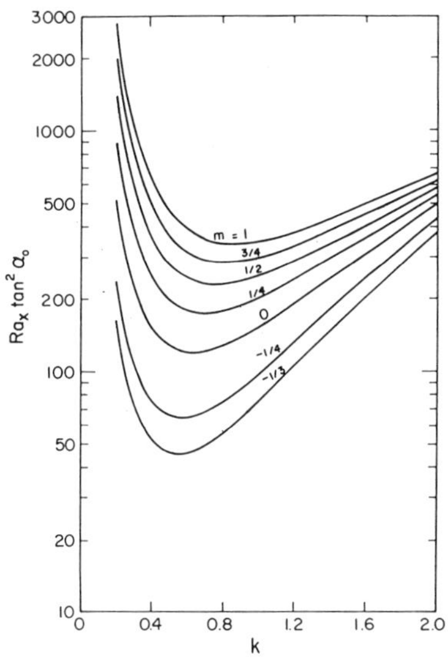

Figure 4 Values of $Ra_x \tan^2 \alpha_0$ as a function of dimensionless spanwise wavenumber, k. This is figure 2 from Hsu and Cheng [2] and is reproduced by permission of A.S.M.E.

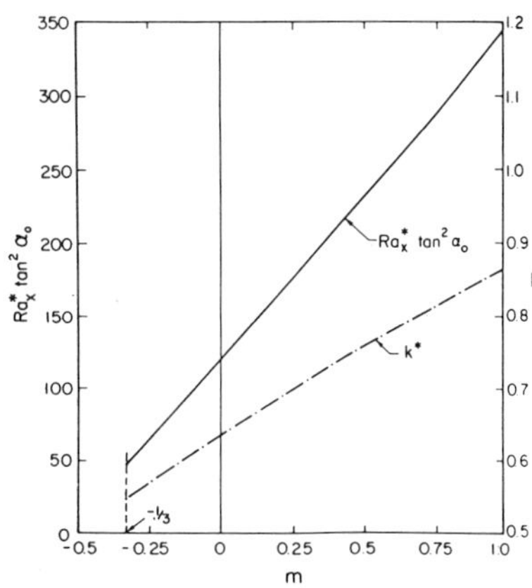

Figure 5 The critical values of $Ra_x \tan^2 \alpha_0$ and wavenumber as functions of m. This is figure 3 from Hsu and Cheng [2] and is reproduced by permission of A.S.M.E.

After these first forays into the field of thermal boundary–layer instabilities in porous media, there appeared a long series of papers by Jang, Chang and their co–workers ([6]–[18]) which have detailed the effects of various extensions to the problems considered in [1] to [4]. Each one will be reviewed briefly below, concentrating first on those for which the basic flow is self–similar.

Jang and Chang [6] considered how mass–transfer effects modify the conclusions of Hsu et al. [1]. Jang and Chang, as in all their other papers, assume the principle of local similarity and neglect streamwise derivatives in the perturbation equations. Thus they follow closely the methodology used in Hsu et al. [1]. Attention is restricted to the case of an isothermal and isosolutal surface, since the presence of mass transfer effects already give rise to two extra parameters, namely, the Lewis number, Le, and a buoyancy ratio, N. The case of $N = 0$ corresponds to having no mass transfer effects, and positive values of N to when mass transfer acts in the same sense as buoyancy effects. Critical values of Ra_x are found to vary markedly with changes in both Le and N. It is found that Ra_x decreases when either Le or N increase. Ra_x is independent of Le when $N = 0$, a result which is intuitive. It is worth noting that critical values of Ra_x become very low (less than 10) for large values of N and Le. Detailed variations of Ra_x are shown in Figure 6.

Maximum density effects on the stability of inclined free convection flow were studied in Jang and Chang [7]. They considered the stability of water–saturated media near the maximum density point at $T_m = 4°$C. The critical value of

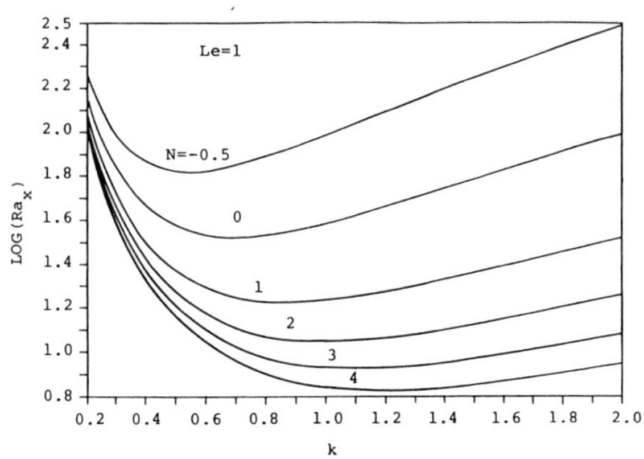

Figure 6 Neutral stability curves for selected values of N with $Le = 1$. This is figure 7a from Jang and Chang [6]

$Ra_x \cot^2 \phi$ (in our notation) decreases as the absolute value of $(T_m - T_\infty)/(T_w - T_\infty)$ increases. Again it is possible for these values to become less than 10 under some circumstances. Another paper dealing with this problem is by Jang and Chang [8] and it aims to correct the analysis of Jang and Chang [7] in the same way that Jang and Chang [5] attempted to correct that of Hsu and Cheng [2]. As the analysis of Jang and Chang [8] follows very closely that of Jang and Chang [5], save that maximum density effects are included, it is clear that the basic flow presented in Jang and Chang [8] is also invalid, and therefore the stability results are unreliable.

Jang and Leu [9] and Leu and Jang [10] considered how variations with temperature of the viscosity, μ, affect the stability of the horizontal free convective boundary–layer. The former paper is concerned with a uniformly heated surface, whereas the latter, unusually in this review, is concerned with a uniform heat flux surface. The authors assume that the basic flow is self–similar and deduce that an isothermal surface is the only case for which the assumption is true. On assuming an exponential form for the viscosity/temperature relation, the stability analysis shows that the critical value of Ra_x increases as μ_w/μ_∞ increases. However, that Nield [35] questions the general conclusion of Leu and Jang [10] because the fact that the Rayleigh number definition is not unique necessarily implies that the occurrence of destabilisation or stabilisation depends on the precise definition of Ra_x.

Variable porosity, permeability and thermal diffusivity effects on horizontal free convection were presented by Jang and Chen [11]. Motivated by the fact that porosity changes often occur near a solid boundary (see Vafai [36]), termed the channelling effect, the authors fitted an exponential relation to expressions for these properties. A power–law surface temperature distribution was taken. Self–similarity was ensured by assuming that the variation in permeability, given by $K(Y) = K_\infty(1 + 3e^{-Y/\gamma})$, is made a function of η; this is done by choosing the "constant", γ, to be precisely $X/Ra_x^{1/3}$. Although reference is made to other papers citing support for this mathematical device, it is difficult to see how a porous medium could have a region of decreased permeability the thickness of which is an exact proportion of the local boundary–layer thickness. If K is kept as a function of Y, a physically reasonable assumption, then the basic flow is nonsimilar and the stability analysis would then need modification.

A study of the combined effects of fluid inertia (Forchheimer form–drag) and thermal dispersion on horizontal free convection was undertaken by Jang and Chen [12]. The basic flow is self–similar only when the power–law exponent of the surface temperature is precisely 0.5. In addition, a uniform external flow allows mixed convection effects also to be considered without loss of self–similarity. Thermal dispersion is found to stabilise the flow, and this becomes increasingly pronounced as the free stream velocity increases. Inertia, on the other hand, destabilises the flow by increasing the thickness of the basic boundary–layer flow.

The remaining papers by Jang and co–workers use nonsimilar flows as the basic flow and therefore the methodology of computing neutral values of Ra_x has to change slightly. Generally the nonsimilar basic flow is computed using the Keller–box method [27], and the value of Ra_x is found by solving the ordinary differential disturbance equations. Basic flow solutions at values of Ra_x lying between the streamwise gridpoints of the Keller–box computation are interpolated to find the neutral curves. It is interesting to note that the Keller–box method itself may be modified to compute eigenvalues such as growth rates (see Lewis et al. [37] and Shu and Wilks [38]), but such an extension of the basic numerical method has not yet been used in stability calculations.

Jang and Lie [13] considered mixed convection flow from horizontal and inclined power–law heated surfaces; in this regard the problem is identical to that of Hsu and Cheng [3]. Again these authors considered that Hsu and Cheng [3] were in error by neglecting the normal component of the buoyancy force. In this case Jang and Lie [13] retain the generally inclined pseudo–similarity variable, $\eta = Ra_x^{1/2}Y/X^{1/2}$. They also assume that both the X and Y temperature derivatives are of the same order of magnitude at $O(1)$ inclination angles. Again, the flow must be elliptic under these conditions and hence the boundary–layer approximation is either invalid, or their flow corresponds only to asymptotically small inclinations from the horizontal.

Inertia (Forchheimer drag) effects on horizontal free convection were considered in Chang and Jang [14] where the surface temperature varies as X^m. In general the flow is nonsimilar, but when $m = \frac{1}{2}$ it becomes self–similar. The authors present a thorough study of the basic flow, as this had not previously appeared in the literature. We note that an analysis for the isothermal case has only recently been published, see Rees [39]. Generally it is found that the presence of inertia destabilises the flow. A secondary purpose of his paper was to investigate the effect of the presence of those terms with first degree streamwise derivatives which originate in the advective terms of the full disturbance equations. When present, they tend to stabilise the flow relative to when they are absent. The additional effects of including the advective inertia terms and the Brinkman terms were investigated by Chang and Jang [15]. Much of this paper is taken up with details of the basic flow as that had not been investigated previously. There still remains scope for further analysis in this area. Indeed, even the effect on the basic flow of having the Brinkman term as the only extension to Darcy's law has not yet been studied in detail — unpublished work by the present author indicates a developing two–layer structure within the main boundary–layer at large distances from the leading edge, which is similar to the vertical boundary–layer flow considered by Kim and Vafai [40]. Returning to Chang and Jang [15], the stability analysis depends on a large number of independent parameters and therefore it is difficult to condense the detailed conclusions. The further inclusion of variable porosity, permeability and thermal diffusivity effects, but with no advective inertia terms may be found in Jang and Chen [16].

Horizontal mixed convection flow from a surface with power–law heating, inertia (Forchheimer drag) and boundary (Brinkman) effects are studied in Lie and Jang [17]. Boundary effects are again found to stabilise the flow, as are inertia effects (in contradistinction to the results of Chang and Jang [14]). The addition effect of uniform suction or blowing at the heated surface was considered by Jang et al. [18]. The main conclusion is that suction stabilises the flow, while blowing destabilises it, and, again, this may be related to the local boundary–layer thickness.

OTHER LINEARISED ANALYSES

All the papers reviewed above consider vortex disturbances. In the analogous field of clear–fluid convection it is well–known that the most dangerous disturbance changes from streamwise vortices to two–dimensional travelling waves as the inclination of the uniformly heated surface becomes more nearly vertical, see Lloyd and Sparrow [41]. It is possible that waves are more dangerous, i.e. have a lower critical Ra_x, in porous medium convection. Of all the modifications to Darcy's law considered by Jang, Chang and co–workers, it is the Brinkman term, which adds a 'viscous' term to Darcy's law, which is most likely to affect the qualitative nature of the most dangerous disturbance. It is possible that a remark in Hsu et al. [1], stating that experimental evidence favours vortices, has caused effort to be concentrated almost entirely on vortices, while neglecting waves. However, the paper by Rees and Bassom [19] concludes that waves grow beyond $Ra_x = 28.90$ (in the notation of [1]) compared with $Ra_x = 33.47$ for vortices, see Hsu et al. [1] — this is for an isothermal horizontal surface. Behind this rather surprising and counter–intuitive result lies the question: what is the role played by the assumed x–dependence of the disturbances? At present the answer is unknown, but it is extremely likely that these critical values will depend on the chosen form of the x–dependence.

A paper by Storesletten and Rees [20] dealing with instability from uniformly heated horizontal or inclined surfaces is concerned with the accuracy of the basic flow. This concern was motivated by the facts that, high–order analyses of the basic flow, such as that of Riley and Rees [31] for horizontal convection, proceeds in powers of $x^{-1/3}$, and the critical value of x for vortices is 33.47, according to Hsu et al. [1]. Thus successive terms in the high–order analysis do not vary greatly in their magnitude at the point of incipient instability, and therefore the leading–order boundary–layer flow is inaccurate there. The aim of Storesletten and Rees [20] was to investigate the effect on stability of including further terms in the representation of the basic flow. As in other high–order analyses, the porous medium is assumed to bounded by two semi–infinite surfaces, the second of which is either insulated or at the ambient temperature, and the included angle between the surfaces is termed the wedge angle. The main conclusion of Storesletten and Rees [20] is that stability results for generally inclined surfaces are very reliable when the heated surface is near the vertical, but that they become increasingly unreliable as the surface tends

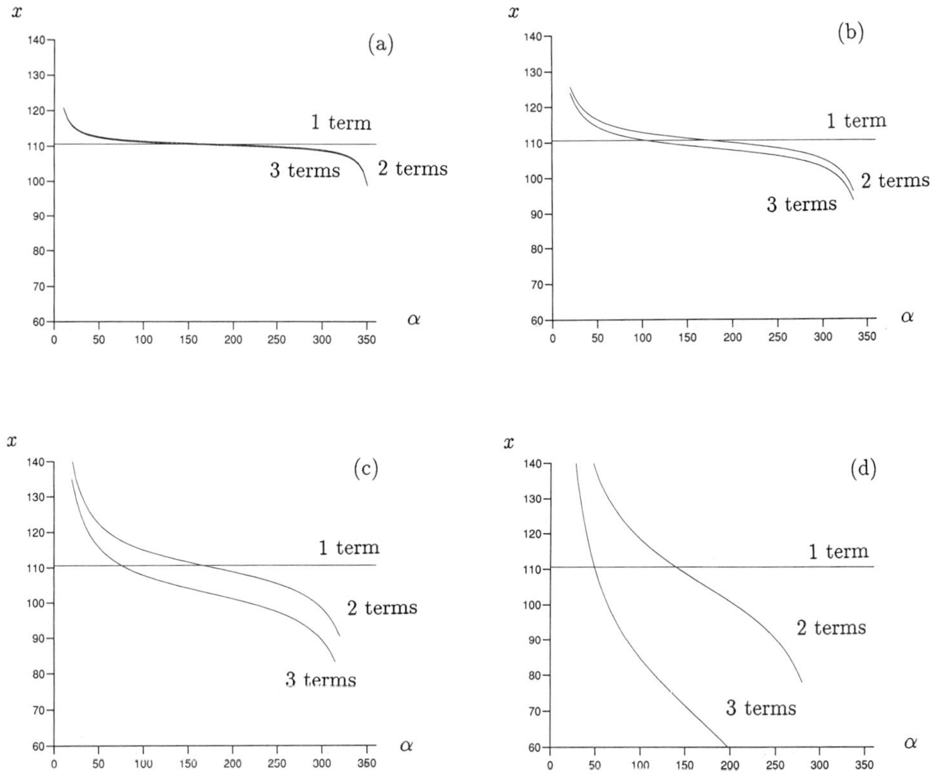

Figure 7 Variation of the critical distance, x, with wedge angle, α, and the number of terms in the asymptotic series for the basic flow. (a) $\phi = 85°$; (b) $\phi = 75°$; (c) $\phi = 60°$; (d) $\phi = 30°$. This is figure 2a–d from Storesletten and Rees [20]

towards the horizontal; see Figure 7.

When the surface is horizontal the similarity variable is different, and the analysis has to be reworked. The critical value of x depends very strongly on the wedge angle between the surfaces; see Figure 8. In fact, when the wedge angle is less than about 62° the wavespeed of travelling waves is negative, implying that the instability travels upstream. This unphysical result shows that the basic flow obtained even with a high-order analysis (which goes only as far as the third term because of the appearance of eigensolutions) is still insufficiently accurate for horizontal convection.

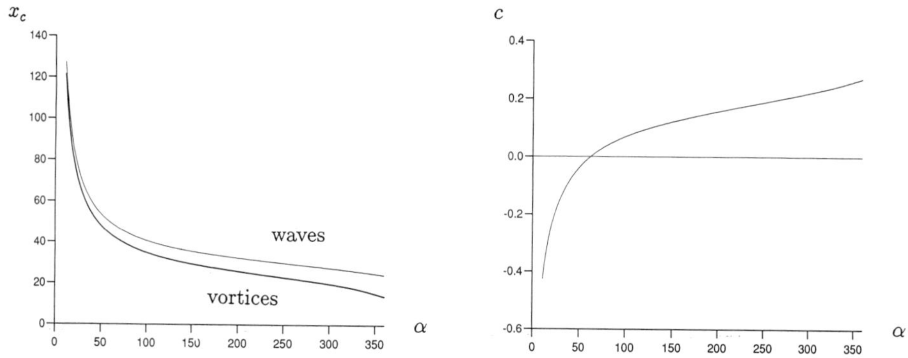

Figure 8 Variation with wedge angle, α, of the critical value of x for waves and vortices, x_c, and the wavespeed, c. This is figure 4 and figure 5 from Storesletten and Rees [20]

THE NATURE OF LINEAR STABILITY THEORY

Without exception, papers [1]–[18] use what Jang and Chang [6] term the principle of local similarity to develop their linearised stability theories. In practice streamwise derivatives in the diffusion terms are neglected in the perturbation equations. Then the vortex perturbations are assumed to take the form, $x^m \mathcal{F}(\eta)e^{iaz}$, and the disturbance equations reduce to an ordinary differential eigenvalue problem for the critical distance, x or Ra_x, depending on the formulation. This procedure is also known as the parallel–flow approximation. In this section we will explore what it means for linear stability theory to include the effects of non–parallelism of the basic flow. To do this we consider the analogous problem of convection in layers of finite thickness heated from below. In particular, we will consider the Bénard–like problem of Walton [42] which deals with a layer where the gap between the surfaces increases linearly, but very gradually. This nonuniformity is accounted for by using a weakly nonlinear theory and it models the increasing thickness of most boundary–layer flows together with a nonuniform/nonparallel basic flow. A similar analysis for convection in a porous layer results in the same canonical equation. This qualitative analysis is novel, and has not appeared elsewhere in the literature.

The weakly nonlinear analysis of Walton [42] is assumed and not reproduced here. The analysis results in an equation governing the amplitude, A, of vortex disturbances to the basic flow:

$$A_\tau = XA - A_{XXXX} - A^3. \tag{19}$$

Here, X is a 'slow' lengthscale (much greater than the vortex wavelength) and its value in the term XA represents the increase in the depth of the layer. The slow timescale is denoted τ. In this discussion the definition of X has nothing to do with that used in the rest of the paper. Equation (19) governs how weakly nonlinear convection evolves in time, and the A_{XXXX} term represents how disturbances diffuse in the streamwise direction, this is a nonparallel term. The Rayleigh number does not appear in this equation but its definition is based on the layer depth at $X = 0$ and its value is precisely $4\pi^2$, the value corresponding to the onset of convection in a uniform layer; see Lapwood [43]. The A^3 term in equation (19) arises from nonlinear terms in the governing equations, and it restrains the unbounded growth of linearly unstable convection.

Linear theory will neglect the nonlinear term, but parallel analyses also neglect the X–derivative term. Thus the equivalent of papers [1]–[19] in Walton's problem is the equation,

$$A_\tau = XA, \tag{20}$$

which has a solution $A = A_0(X)e^{X\tau}$, where $A_0(X)$ is the disturbance profile at $\tau = 0$. Clearly $X = 0$ corresponds to neutral stability in this context, exponential decay occurs when $X < 0$ and growth when $X > 0$. Thus a well–defined neutral position has been found, qualitatively the same as in [1]–[19].

Consider now the effect of nonparallelism by including the diffusion term. For simplicity we will assume that we are at neutral stability and therefore the time derivative term is absent. Now equation (19) reduces to

$$0 = XA - A_{XXXX}, \tag{21}$$

which is a fourth–order version of Airy's equation. We will assume that solutions are sought in the region $-\infty < X < \infty$, and for physical reasons we require the solution to decay as $X \to -\infty$, but the appropriate solution also decays as $X \to \infty$. A straightforward asymptotic analysis of equation (21) for large negative X shows that A may take any one of the four functional forms,

$$A \sim (-X)^{-5/6} \exp\Big[\frac{4}{5\sqrt{2}}(\pm 1 \pm i)(-X)^{5/4}\Big], \tag{22a}$$

whereas, for large positive X,

$$A \sim X^{-5/6} \exp\Big[\pm(4i/5)X^{5/4}\Big], \qquad A \sim X^{-5/6} \exp\Big[\pm(4/5)X^{5/4}\Big], \tag{22b}$$

with these forms being valid up to a multiplicative factor. It is possible to show that there is a unique function which exhibits decay as $X \to \pm\infty$, and we note that this decay is superexponential for negative X but is algebraic (with a decreasing period of oscillation) for positive X; in this regard it is similar to the Airy function, $Ai(-X)$. Thus the neutral mode cannot be considered in terms of a location beyond which instability grows, but rather it is a distinct mode whose profile must be computed rather than assumed in advance.

When temporal growth is included, equation (19) becomes

$$A_\tau = XA - A_{XXXX}, \tag{23}$$

which has the solution,

$$A = e^{\lambda\tau} B(X - \lambda), \tag{24}$$

where $B(X)$ satisfies equation (21), and λ has an arbitrary real value. As this solution is valid for any value of λ we have now a continuous set of modes, indistinguishable by profile, but distinguished by their growth rate, or, equivalently, by their location in space. Assuming that it is possible to extend Fourier analysis to cover solutions of equation (21), then the general solution may be written in terms of an initial disturbance profile using a fourth–order–Airy version of a Fourier integral.

If now a sidewall is placed at some finite, negative value of X, thereby restricting the layer to be semi–infinite, the above continuous set of modes becomes a discrete set since the sidewall boundary conditions confer a restriction on which of the original set of modes can be chosen. Such a qualitative change in the nature of the modal set does not occur in the semi–infinite *uniform* Bénard problem with a sidewall, see Cross et al. [44], but once the layer ceases to be uniform, then such a qualitative change takes place. Thus the problem is characterised by a set of modes with discrete values of the exponential growth rate.

How do these qualitative results from the model problem carry over to the instability of boundary–layer flows? First, we note that the basic flow is nonparallel and the depth of the layer increases as X increases, thereby increasing the Rayleigh number based on the local layer depth. Second, it is necessary to be convinced that the presence of a sidewall in the model problem is a good analogue of the leading edge in boundary–layer flows. To see this, consider the inclined uniformly heated surface, the equations for which are given in equations (15). This analysis will follow method 3 and therefore we set $Ra = 1$ in equations (15). Additionally, equations (15) will be extended by assuming that the flow may be unsteady. For illustration purposes we assume that the flow is two–dimensional and therefore let $u = \psi_y$, $v = -\psi_x$ and $w = 0$. Equations (15) reduce to

$$\frac{\partial^2\psi}{\partial x^2}+\frac{\partial^2\psi}{\partial y^2}=\frac{\partial\theta}{\partial y}\sin\phi-\frac{\partial\theta}{\partial x}\cos\phi, \qquad \frac{\partial\theta}{\partial t}+\frac{\partial\psi}{\partial y}\frac{\partial\theta}{\partial x}-\frac{\partial\psi}{\partial x}\frac{\partial\theta}{\partial y}=\frac{\partial^2\theta}{\partial x^2}+\frac{\partial^2\theta}{\partial y^2}, \quad (25a,b)$$

where t is nondimensional time. Assume that the porous medium is bounded by a flat surface at $y = 0$, where $\theta = 1$ on $x > 0$ and $\theta_y = 0$ on $x < 0$. As $y \to \infty$ then $\theta \to 0$. When this surface is vertical, the introduction of parabolic coordinates (a Schwartz–Christoffel transformation) renders these full equations soluble precisely in terms of the Cheng and Minkowycz [28] leading order boundary–layer solution. If the parabolic coordinates are defined according to

$$x=\frac{\xi^2-\eta^2}{4}, \qquad y=\frac{\xi\eta}{2}, \tag{26}$$

then equations (25) are transformed to

$$\frac{\partial^2\psi}{\partial\xi^2}+\frac{\partial^2\psi}{\partial\eta^2}=\tfrac{1}{2}\Big(\eta\frac{\partial\theta}{\partial\xi}+\xi\frac{\partial\theta}{\partial\eta}\Big)\sin\phi-\tfrac{1}{2}\Big(\xi\frac{\partial\theta}{\partial\xi}-\eta\frac{\partial\theta}{\partial\eta}\Big)\cos\phi, \tag{27a}$$

$$\frac{(\xi^2+\eta^2)}{4}\frac{\partial\theta}{\partial t}+\frac{\partial\psi}{\partial\eta}\frac{\partial\theta}{\partial\xi}-\frac{\partial\psi}{\partial\xi}\frac{\partial\theta}{\partial\eta}=\frac{\partial^2\theta}{\partial\xi^2}+\frac{\partial^2\theta}{\partial\eta^2}. \tag{27b}$$

The flow domain within this coordinate system is $0 \leq \xi < \infty$ and $0 \leq \eta < \infty$. Therefore $\xi = 0$, which is the insulated surface, $y = 0$, $x < 0$, may be seen to correspond to the sidewall in the model problem.

There is one difference between this boundary–layer problem and the model problem. The former has a boundary–layer thickness in terms of either y or η which varies over $O(1)$ distances in terms of x or ξ, whereas the thickness of the layer in the model problem varies very slowly. However, this is only a quantitative difference, the advantage of slow variation in the model problem is that much more analytical progress can be made. The quick variation in the boundary–layer thickness of the present problem means that a weakly nonlinear analysis cannot be undertaken, but it is possible to derive from equations (27) a linearised set of equations which are equivalent to equation (23). If the steady basic flow satisfying equations (27) is denoted by (ψ^B, θ^B), then equations (27), after linearisation about this solution becomes,

$$\frac{\partial^2 \psi}{\partial \xi^2} + \frac{\partial^2 \psi}{\partial \eta^2} = \tfrac{1}{2}\Big(\eta \frac{\partial \theta}{\partial \xi} + \xi \frac{\partial \theta}{\partial \eta}\Big) \sin\phi - \tfrac{1}{2}\Big(\xi \frac{\partial \theta}{\partial \xi} - \eta \frac{\partial \theta}{\partial \eta}\Big) \cos\phi, \tag{28a}$$

$$\frac{(\xi^2+\eta^2)}{4}\frac{\partial \theta}{\partial t} + \frac{\partial \psi^B}{\partial \eta}\frac{\partial \theta}{\partial \xi} + \frac{\partial \psi}{\partial \eta}\frac{\partial \theta^B}{\partial \xi} - \frac{\partial \psi^B}{\partial \xi}\frac{\partial \theta}{\partial \eta} - \frac{\partial \psi}{\partial \xi}\frac{\partial \theta^B}{\partial \eta} = \frac{\partial^2 \theta}{\partial \xi^2} + \frac{\partial^2 \theta}{\partial \eta^2}. \tag{28b}$$

This system is fourth–order in ξ, equivalent to X in equation (19), and, when ψ and θ are set proportional to $e^{\lambda t}$, there results a partial differential eigenvalue problem for the growth rate λ. Apart from an additional cross–stream dependence on η, equations (28) are identical in form to equation (23). Although no numerical solutions of equation (28) yet exist, it is clear, given the analogy, that the full linearised problem incorporating nonparallel and streamwise diffusion effects will yield an infinite set of modes each with their associated growth rate. Moreover, this set of growth rates will form a discrete set, rather than a continuous set. Thus the full linearised equations, like the model problem, cannot give a critical distance from the leading edge beyond which disturbances grow. This paradigm shift may be difficult to accept, but it is the inevitable consequence of the presence of streamwise diffusion in the linearised equations.

ANALYSES BASED ON THE EXACT SOLUTION

The remaining papers in this review do not use the boundary–layer approximation to the steady basic flow as a basis for stability analyses. Rather, the exact solutions of Rees and Bassom [45] were used. In all cases in this section the Rayleigh number is scaled out of the full governing equations, as per method 3.

The vertical surface

The simplest possible case of Darcy flow induced by a uniformly heated vertical surface was treated in Rees [21] and Lewis et al. [22]. Both papers use the parabolic coordinate system defined by expressions (26), and equations (28) were solved with $\phi = \pi/2$. In Rees [21], direct numerical simulations using the full equations were presented, and it was concluded that even large perturbations to the basic steady

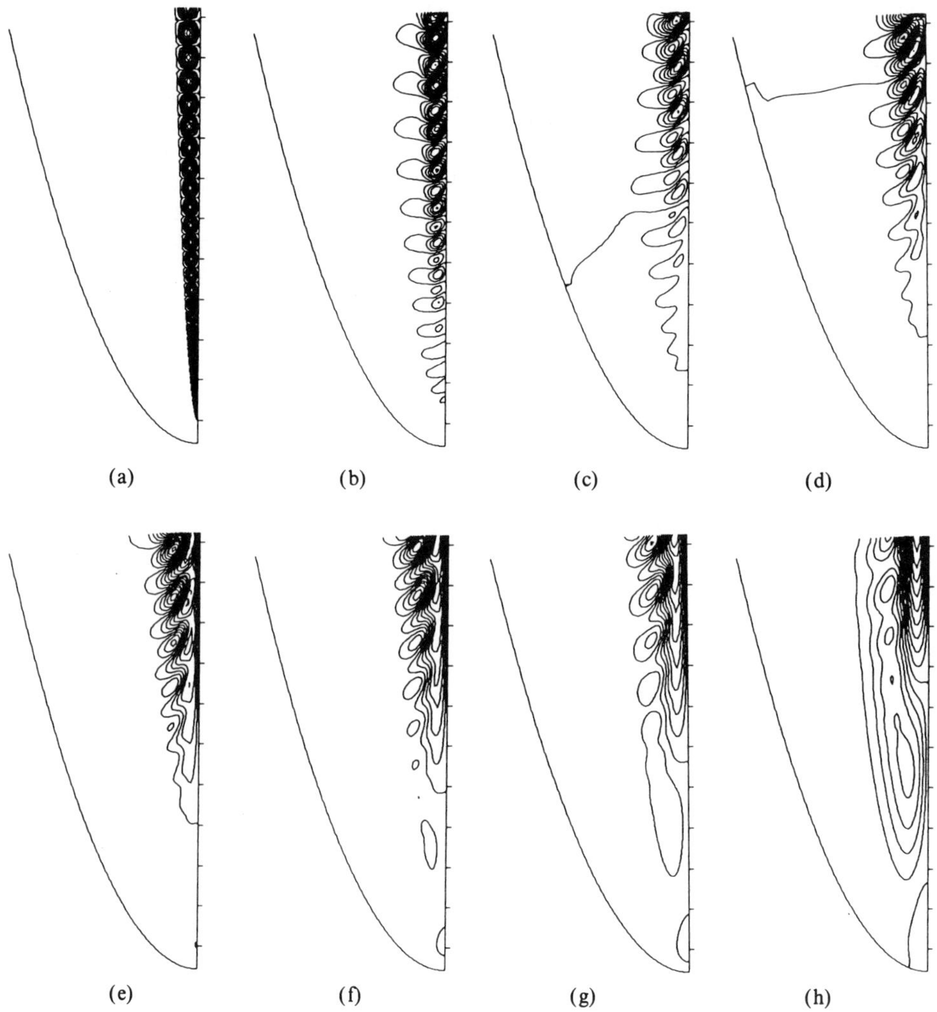

Figure 9 Perturbation isotherms for the disturbance shown in (a). Isotherms are plotted at intervals of $|\delta\theta|_{\max}/10$ for each frame. Frame (a) has $|\delta\theta|_{\max} = 1$, and frame (h) has $|\delta\theta|_{\max} = 1.147 \times 10^{-3}$. (a) $t = 0$; (b) $t = 100$; (c) $t = 200$; (d) $t = 300$; (e) $t = 400$; (f) $t = 500$; (g) $t = 600$; (h) $t = 850$. Reproduced by permission of Birkhäuser–Verlag. This is figure 5 from Lewis et al. [21]

flow decay eventually; a typical solution in terms of perturbation isotherms is presented in Figure 9.

If an analogy is drawn between this flow and that of a vertical channel with a horizontal temperature drop (the Darcy–Bénard problem rotated through 90°), then such a result is not surprising for Gill [46] demonstrated analytically that the layer problem is linearly stable for all values of the Rayleigh number. However, the result of Rees [21] cannot be considered definitive because a numerical simulation of this sort must, necessarily, involve only a finite computational domain. This observation motivated the asymptotic analysis of Lewis et al. [22]. In this paper the authors first re–examined Gill's work to determine the growth rates of wave disturbances in the limit of large Rayleigh numbers and showed that the exponential growth rate is given by

$$\lambda = i\alpha Ra - 0.56(1 + i\sqrt{3})(\alpha Ra)^{2/3} + \dots, \tag{29}$$

as $Ra \to \infty$, where α is the wavenumber. As the leading order real term is negative, this confirms and quantifies Gill's result. The disturbance is found to split into a two–layer structure: one region is of thickness $O(Ra^{-1/3})$ immediately next to one of the walls, and the other fills the rest of the channel. Equation (29) applies to modes concentrated near the cold wall. For the vertical boundary–layer, the same analysis yields

$$\lambda = -\frac{2i\alpha}{\xi} - 1.46(1 - i\sqrt{3})\left(\frac{\alpha}{\xi}\right)^{4/3} + \dots. \tag{30}$$

From this we see that the magnitude of the real part of λ decays as ξ increases, but that it remains negative. Thus the boundary–layer becomes less stable further downstream, but it always remains stable. Again, in the large–ξ limit the disturbance develops a two layer structure with the modes concentrated mainly in a thin layer of thickness $\eta = O(\xi^{-1/3})$.

The horizontal surface

When the heated surface is horizontal, instability is expected *a priori* since the steady basic boundary–layer flow is unstably stratified. According to Hsu et al. [1], the point of incipient instability is very close to the leading edge. We have already seen how the inaccuracy caused by a poor representation of the basic flow affects stability criteria, and it is very likely that the principle of local similarity used by many authors is similarly affected. Thus there are only two possible avenues of study for the horizontal problem. The first is to develop an asymptotic analysis which is valid far downstream of the leading edge, and the other is to solve the full governing equations. The former is undertaken in Bassom and Rees [23], and first steps towards the latter in Rees [24] and Rees and Bassom [25].

In Bassom and Rees [23] the analysis was carried out in the coordinate system defined by

$$x = \xi(\xi^2 - 3\eta^2)/27, \qquad y = \eta(3\xi^2 - \eta^2)/27, \tag{31}$$

which is again a Schwartz–Christoffel transformation; see [45]. The stability results of Hsu et al. [1] were recomputed in this coordinate system in order to determine the neutral curve in terms of ξ against the vortex wavenumber. This curve gives slightly different values from that of Hsu et al. [1] because the exact solution is used as the basic profile. The thrust of the paper is to determine the asymptotic structure of disturbances at large values of ξ (or x). This is a long and involved analysis, and the reader is referred to that paper for details, but again disturbances are found to inhabit a thin region (in terms of η) immediately adjacent to the heated surface. In this case it is shown that $\eta = O(\xi^{-1/3})$ is the thickness of this sublayer.

Numerical simulations of the full, unsteady two–dimensional equations of motion,

$$\psi_{\xi\xi} + \psi_{\eta\eta} = \frac{1}{9}\Big[2\xi\eta\theta_\eta + (\eta^2 - \xi^2)\theta_\xi\Big], \qquad \theta_t = \frac{81}{(\xi^2+\eta^2)^2}\Big[\theta_{\xi\xi} + \theta_{\eta\eta} + \psi_\xi\theta_\eta - \psi_\eta\theta_\xi\Big], \tag{32}$$

are presented in Rees [24] and in much more detail in Rees and Bassom [25]. The central implications of this study are (i) localised disturbances placed well beyond the theoretical point of neutral stability propagate both upstream and downstream, showing that ellipticity in an inherent part of the evolution of disturbances; (ii) developing waves may be ejected from the boundary–layer to form plumes; (iii) the strength of such eruptions can be such that the additional entrainment of fluid they induce causes substantial thinning of the boundary–layer even very close to the leading edge; (iv) cell merging occurs; (v) the flow is essentially chaotic, though the flow may, for short periods of time and at locations sufficiently close to the leading edge, remain roughly periodic in time before events downstream affect the flow. An example of a simulation from Rees [24] is shown in Figure 10, where cell–merging and the creation of recirculating regions may be seen. This figure shows that waves grow in strength while advecting downstream and that these waves seem to maintain a roughly constant wavelength, at least when they are fairly close to the leading edge. As other figures in Rees and Bassom [25] indicate, the further away from the leading edge, the more chaotic the behaviour of the flow seems to be. Given that waves seem not to propagate upstream closer to the leading edge than about $x = 80$ (c.f. the critical value of roughly 28 given in Rees and Bassom [19]) it is clear that these numerical simulations also support the misgivings about the accuracy of the basic flow used in linearised analyses using the parallel–flow approximation. Clearly Rees and Bassom [25] only represents the first faltering steps in what remains to be a large research programme, but it does indicate that there is much exciting work still to be done. Of particular interest are vortex simulations, but since the flow will then be three–dimensional, the numerical effort will be substantially greater.

CONCLUSION

In this review an exhaustive survey has been presented of papers dealing with the instability of thermal boundary layers in porous media. In addition, a large amount of space has necessarily been devoted to various theoretical issues which have a very strong bearing on the topic. The methods used to analyse not only the basic flow, but also their instability, have been subject to careful scrutiny.

Many works have appeared which use linearised analyses based on the parallel–flow approximation. Although a large number of extensions to Darcy's law have been studied, the overall method itself has been called to account. The analysis of Storesletten and Rees [20], in particular, shows that the inaccuracy of the leading order boundary–layer solution has a strong effect on the stability criteria. It also shows that when the heated surface is nearly vertical, then such results may be regarded as reliable. Detailed asymptotic analyses valid for large distances from the leading have been undertaken, and these account for nonparallel effects. For the vertical thermal boundary–layer such results are very useful in determining that the flow is stable. For the horizontal boundary–layer it is doubtful whether the results

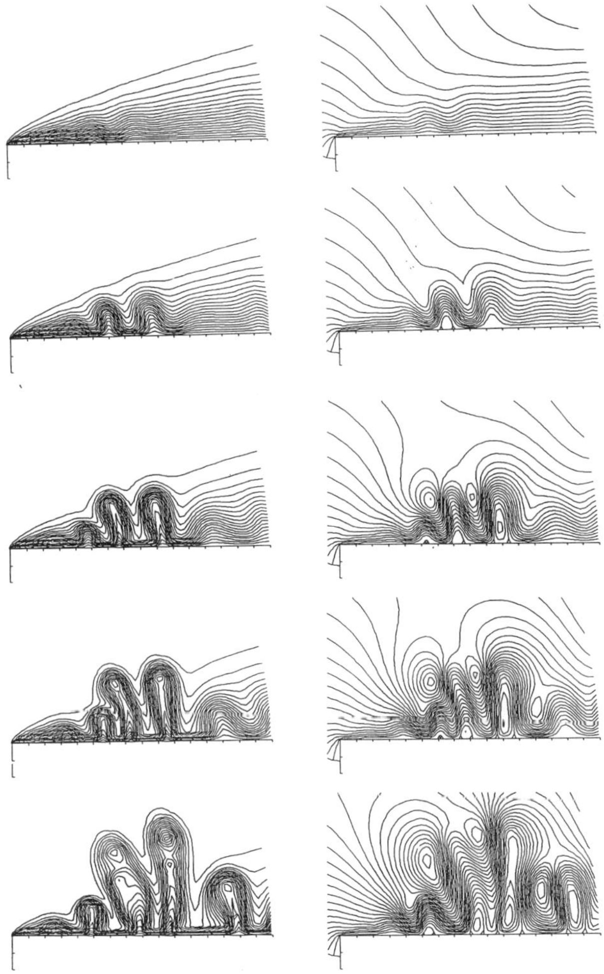

Figure 10 Instantaneous isotherms (left) and streamlines (right) at successive points in time. Tickmarks on the axes are at intervals of 250. Reproduced by permission of Springer–Verlag. This is figure 1 from Rees [24]

of the analysis are of practical importance, since instabilities enter the strongly nonlinear regime even at stations fairly close to the leading edge. Solutions of the full linearised equations are not yet available; these would enable us to determine modal growth rates and hence the detailed effect of ellipticity (or streamwise diffusion, or nonparallelism). No weakly nonlinear analyses have appeared in the literature, although it may be possible to apply the techniques of Chen et al. [47] to vortex disturbances in near-vertical boundary layers.

Only a few numerical simulations of the full governing equations have been undertaken, and those have concentrated only on two–dimensional flows. At present nothing is known with any assurance about the form of the preferred mode of instability when the heated surface is close to the horizontal, even for a relatively simple flow such as that given by Darcy's law. Therefore is very likely that the results even of Rees and Bassom [25] may not be definitive; this conjecture can only be tested by solving the full three–dimensional equations. Another numerical approach which may prove informative but with a much reduced computational time is the Parabolised Stability Equations method. This method has been used to great effect in the computation of the nonlinear development of instabilities in the Blasius boundary–layer, see Bertolotti et al. [48], and many other boundary–layer flows. A very recent review of this methodology is given in Herbert [49].

Clearly the work which remains in order to give a good understanding of instability mechanisms and their subsequent temporal evolution is substantial. When other effects, such as those considered by Jang, Chang and co–workers, are included in the analysis, it could very well be the case that qualitative results, such as the form of the most dangerous disturbance, may change. Indeed this is quite likely to be the case for Darcy–Brinkman flow, as the resulting equations are much closer to the familiar Navier–Stokes equations than are those for Darcy flow, and as the vertical thermal boundary–layer flow of a clear fluid is destabilised by waves, rather than by vortex disturbances.

Acknowledgements

The author would like to express his indebtedness to his co–workers, Prof. D. S. Riley, Prof. L. Storesletten, Dr. A. P. Bassom and Miss Samantha Lewis for their part in the work described here. The author is also grateful to Prof. D. A. Nield for bringing references Leu and Jang [10] and Nield [35] to his attention.

REFERENCES

1. C.T. Hsu, P. Cheng, and G.M. Homsy, Instability of free convection flow over a horizontal impermeable surface in a porous medium, *Int. J. Heat Mass Transf.* **21**, 1221–1228 (1978).
2. C.T. Hsu and P. Cheng, Vortex instability in buoyancy induced flow over inclined heated surfaces in porous media, *Trans. A.S.M.E. J. Heat Transf.* **101**, 660–665 (1979).
3. C.T. Hsu and P. Cheng, Vortex instability of mixed convection flow in a semi–infinite porous medium bounded by a horizontal surface, *Int. J. Heat Mass Transf.* **23**, 789–798 (1980).
4. C.T. Hsu and P. Cheng, The onset of longitudinal vortices in mixed convective flow over an inclined surface in a porous medium, *Trans. A.S.M.E. J. Heat Transf.* **102**, 544–549 (1980).
5. J.Y. Jang and W.J. Chang, Vortex instability of buoyancy–induced inclined boundary–layer flow in a saturated porous–medium, *Int. J. Heat Mass Transf.* **31**, 759–767 (1988).
6. J.Y. Jang and W.J. Chang, The flow and vortex instability of horizontal natural–convection in a porous–medium resulting from combined heat and mass buoyancy effects, *Int. J. Heat Mass Transf.* **31**, 769–777 (1988).

7. J.Y. Jang and W.J. Chang, Vortex instability of inclined buoyant layer in porous media saturated with cold water, *Int. Comm. Heat Mass Transf.* **14**, 405–416 (1987).
8. J.Y. Jang and W.J. Chang, Maximum density effects on vortex instability of horizontal and inclined buoyancy–induced flows in porous media, *Trans. A.S.M.E. J. Heat Transf.* **111**, 572–574 (1989).
9. J.Y. Jang and J.S. Leu, Variable viscosity effects on the vortex instability of free–convection boundary–layer flow over a horizontal surface in a porous–medium, *Int. J. Heat Mass Transf.* **36**, 1287–1294 (1993).
10. J.S. Leu and J.Y. Jang, Variable viscosity effects on the vortex instability of the convective boundary layer flow over a horizontal uniform heat flux surface in a saturated porous medium, *Proc. 6th Int. Symp. on Transport Phenomena and Thermal Engineering, Seoul, Korea,* **1**, 203–208 (1993).
11. J.Y. Jang and J.L. Chen, Variable porosity effect on vortex instability of a horizontal mixed convection flow in a saturated porous–medium, *Int. J. Heat Mass Transf.* **36**, 1573–1582 (1993).
12. J.Y. Jang and J.L. Chen, Thermal dispersion and inertia effects on vortex instability of a horizontal mixed convection flow in a saturated porous–medium, *Int. J. Heat Mass Transf.* **36**, 383–389 (1993).
13. J.Y. Jang and K.N. Lie, Vortex instability of mixed convection flow over horizontal and inclined surfaces in a porous–medium, *Int. J. Heat Mass Transf.* **35**, 2077–2085 (1992).
14. W.J. Chang and J.Y. Jang, Inertia effects on vortex instability of a horizontal natural–convection flow in a saturated porous–medium, *Int. J. Heat Mass Transf.* **32**, 541–550 (1989).
15. W.J. Chang and J.Y. Jang, Non–Darcian effects on vortex instability of a horizontal natural–convection flow in a porous–medium, *Int. J. Heat Mass Transf.* **32**, 529–539 (1989).
16. J.Y. Jang and J.L. Chen, Variable porosity and thermal dispersion effects on vortex instability of a horizontal natural–convection flow in a saturated porous–medium, *Wärme–Stoffübertrag.* **29**, 153–160 (1994).
17. K.N. Lie and J.Y. Jang, Boundary and inertia effects on vortex instability of a horizontal mixed convection flow in a porous–medium, *Num. Heat Transf. Part A – Applications* **23**, 361–378 (1993).
18. Jang, J. Y., K.N. Lie and J.L. Chen, The influence of surface mass flux on vortex instability of a horizontal mixed convection flow in a saturated porous–medium, *Int. J. Heat Mass Transf.* **38**, 3305–3311 (1995).
19. D.A.S. Rees and A.P. Bassom, The linear wave instability of boundary–layer flow–induced by a horizontal heated surface in porous media, *Int. Comm. Heat Mass Transf.* **21**, 143–150 (1994).
20. L. Storesletten and D.A.S. Rees, The influence of higher–order effects on the linear instability of thermal boundary layer flow in porous media, Submitted to *Int. J. Heat Mass Transf.*
21. S. Lewis, A.P. Bassom and D.A.S. Rees, The stability of vertical thermal boundary layer flow in a porous medium, *Eur. J. Mech. B: Fluids* **14**, 395–408 (1995).
22. D.A.S. Rees, Nonlinear wave stability of vertical thermal boundary layer flow in a porous medium, *Journal of Applied Mathematics and Physics (Z.A.M.P.)* **44**, 306–313 (1993).

23. A.P. Bassom and D.A.S. Rees, The linear vortex instability of flow–induced by a horizontal heated surface in a porous–medium, *Quart. J. Mech. Appl. Math.* **48**, 1–19 (1995).

24. D.A.S. Rees, Numerical simulation of thermal boundary layer instabilities in porous media, *Proc. I.U.T.A.M. Symposium on "Nonlinear instabilities of nonparallel flows", Clarkson University, New York* 242–247 (1993).

25. D.A.S. Rees and A.P. Bassom, The nonlinear nonparallel wave instability of boundary–layer flow–induced by a horizontal heated surface in porous media, *J. Fluid Mech.* **253**, 267–295 (1993).

26. P. Cheng and I.–D. Chang, On buoyancy-induced flows in a saturated porous medium adjacent to horizontal heated surfaces, *Int. J. Heat Mass Transf.* **19**, 1267–1272 (1976).

27. H.B. Keller and T. Cebeci, Accurate numerical methods for boundary layer flows 1. Two dimensional flows, *Proc. Int. Conf. Numerical Methods in Fluid Dynamics. Lecture Notes in Physics* Springer–New York (1971).

28. P. Cheng and W.J. Minkowycz, Free convection about a vertical flat plate imbedded in a porous medium with application to heat transfer from a dike, *J. Geophys. Res.* **82**, 2040–2044 (1977).

29. D.A.S. Rees and D. S. Riley, Free convection above a near–horizontal semi–infinite heated surface embedded in a saturated porous medium, *Int. J. Heat Mass Transf.* **28**, 183–190 (1985).

30. D.B. Ingham, J.H. Merkin and I. Pop, Natural convection from a semi–infinite flat plate inclined at a small angle to the horizontal in a saturated porous medium, *Acta Mechanica* **57**, 183–202 (1985).

31. D.S. Riley and D.A.S. Rees, Non–Darcy natural convection from arbitrarily inclined heated surfaces in saturated porous media, *Quart. J. Mech. Appl. Math.* **38**, 277–295 (1985).

32. I.–D. Chang and P. Cheng, Matched asymptotic expansions for free convection about an impermeable horizontal surface in a porous medium, *Int. J. Heat Mass Transf.* **26**, 163–174 (1983).

33. P. Cheng and C.T. Hsu, Higher order approximations for Darcian free convection about a semi–infinite vertical flat plate, *Trans. A.S.M.E. J. Heat Transf.* **106**, 143–151 (1984).

34. S.E. Haaland and E.M. Sparrow, Vortex instability of natural convection flow on inclined surfaces, *Int. J. Heat Mass Transf.* **16**, 2355–2367 (1973).

35. D.A. Nield, Estimation of an effective Rayleigh number for convection in a vertically inhomogeneous porous medium or clear fluid, *Int. J. Heat Fluid Flow* **15**, 337–340 (1994).

36. K. Vafai, Convective flow and heat transfer in variable–porosity media, *J. Fluid Mech.* **147**, 233–259 (1984).

37. S. Lewis, D.A.S. Rees and A.P. Bassom, High wavenumber convection in tall porous containers heated from below, To appear in *Quart. J. Mech. Appl. Math.*

38. J.J. Shu and G. Wilks, Mixed convection laminar–film condensation on a semi–infinite vertical plate, *J. Fluid Mech.* **300**, 207–229 (1995).

39. D.A.S. Rees, The effect of inertia on free convection from a horizontal surface embedded in a porous medium, *Int. J. Heat Mass Transf.* **39**, 3425–3430 (1996).

40. S.J. Kim and K. Vafai, Analysis of natural convection about a vertical surface embedded in a porous medium, *Int. J. Heat Mass Transf.* **32**, 665–677 (1989).

41. J.R. Lloyd and E.M. Sparrow, On the instability of natural convection flow on inclined plates, *J. Fluid Mech.* **42**, 465–470 (1970).

42. I.C. Walton, On the onset of Rayleigh–Bénard convection in a fluid layer of slowly varying depth, *Stud. Appl. Math.* **67**, 199–216 (1982).

43. E.R. Lapwood, Convection of a fluid in a porous medium, *Proc. Camb. Phil. Soc.* **44**, 508–521 (1948).

44. M.C. Cross, P.G. Daniels, P.C. Hohenberg and E.D. Siggia, Phase–winding solutions in a finite container above the convective threshold, *J. Fluid Mech.* **127**, 155–183 (1983).

45. D.A.S. Rees and A.P. Bassom, Some exact solutions for free convective flows over heated semi–infinite surfaces in porous media, *Int. J. Heat Mass Transf.* **34**, 1564–1567 (1991).

46. A.E. Gill, A proof that convection in a porous medium is stable, *J. Fluid Mech.* **35**, 545–547 (1969).

47. C.C. Chen, A. Labhari, H.–C. Chang, and R.E. Kelly, Spanwise pairing of finite–amplitude longitudinal vortex rolls in inclined free–convection boundary layers, *J. Fluid Mech.* **231**, 73–111 (1991).

48. F.P. Bertolotti, Th. Herbert and P.R. Spalart, Linear and nonlinear stability of the Blasius boundary layer, *J. Fluid Mech.* **242**, 441–474 (1992).

49. Th. Herbert, Parabolized stability equations, *Ann. Rev. Fluid Mech.* **29**, 245–283 (1997).

EFFECTS OF ANISOTROPY ON CONVECTIVE FLOW THROUGH POROUS MEDIA

L. STORESLETTEN

Department of Mathematics, Agder College, Kristiansand, Norway

INTRODUCTION

Thermally driven convection in fluid-saturated porous media is of considerable interest in a variety of geophysical and technological problems. There are important applications in geothermal energy utilization, oil reservoir modelling, building thermal insulation, nuclear waste disposals and chemical catalytic converters, to name but a few.

So far, the substantial part of theoretical and experimental investigations on convective flow in porous media has dealt with *isotropic* materials. However, in many practical situations the porous materials are *anisotropic* in their mechanical and thermal properties. Anisotropy is generally a consequence of preferential orientation or asymmetric geometry of grain or fibres. An example of such a medium is loft insulation which ususally has lower permeability across the insulating layer than it has in the perpendicular directions. Another important example is geological systems with anisotropic sediments and rocks. In his book, Bear [1] discusses anisotropy in natural materials as soils and geological formations. Sediments are commonly deposited in such a way that their permeability in the horizontal directions (unless tilting of the formation occurs) is greater than in other directions. In most stratified materials the resistance to the flow is smaller (i.e. permeability greater) along the planes of deposition than across them.

This paper presents a review of the research on convective flow through anisotropic porous media. Emphasis is placed on the effects of anisotropy on the marginal stability limit at the onset of convection, the flow structure and the heat transfer rate. The topics covered are convection in horizontal layers, inclined layers, enclosures filled with a porous medium, convective boundary-layer flow, hydrodynamic dispersion, double-diffusive convection, and anisotropy in multilayered media.

There are two shorter reviews by McKibbin ([2] ,[3]) on this theme, dealing with layered and anisotropic media, and a judicious selection of papers is also quoted in the book by Nield and Bejan [4] . Since then, however, a number of significant papers have been published on this subject.

We have assumed that the reader is conversant with the basic theory and results of convection in isotropic porous media.

GOVERNING EQUATIONS

The majority of theoretical studies on thermally driven flows in porous media is based on Darcy's law and the Oberbeck-Boussinesq approximation, confer Nield and Bejan [4] . For a homogeneous *anisotropic* medium these equations are

$$\mu\, \mathbf{v} + \mathcal{K} \cdot (\nabla p - \rho\, \mathbf{g}) = 0 \tag{1}$$

$$\nabla \cdot \mathbf{v} = 0 \tag{2}$$

$$c \frac{\partial T}{\partial t} + \mathbf{v} \cdot \nabla T = \nabla \cdot (\mathcal{D} \cdot \nabla T) \tag{3}$$

$$\rho = \rho_0 \left(1 - \gamma\, (T - T_0)\right). \tag{4}$$

Here equation (1) is the momentum equation (Darcy's law), (2) is the continuity equation, (3) the heat equation and (4) the equation of state. Moreover, t is the time, μ is the viscosity, $\mathbf{v}$ the seepage velocity, p the pressure, ρ the density, ρ_0 the density of the fluid at temperature T_0, T the temperature, $\mathbf{g}$ the acceleration due to gravity, c the heat capacity ratio, γ the thermal expansion coefficient, $\mathcal{K}$ the permeability tensor and $\mathcal{D}$ the thermal diffusivity tensor.

HORIZONTAL POROUS LAYERS

The first studies on convective flow in *anisotropic* porous media appeared in the middle of the seventies and concentrated on the porous media analogue of the Bénard problem. Castinel and Combarnous [5] found the criterion for the onset of convection in a *horizontal layer* with anisotropic permeability. They reported experimental results which agree fairly well with their theoretical predictions. Epherre [6] extended the stability analysis by including anisotropy in the thermal diffusivity. It was shown that anisotropy in the mechanical and thermal properties affects the marginal stability criterion as well as the preferred width of the convection cells. The investigations were restricted to two-dimensional anisotropy, i.e. the layer was horizontally isotropic. However, a few years later Kvernvold and Tyvand [7] performed a theoretical analysis on linear and nonlinear convection in a horizontal layer where a three-dimensional anisotropy was included. Let us now concentrate our attention on the model studied by Kvernvold and Tyvand [7] , deduce the linear stability analysis and report some of the main results.

We consider a fluid-saturated porous layer which is bounded above and below by two infinite and impermeable horizontal planes. The upper and lower boundaries are separated by a distance h and are at constant temperatures T_0 and $T_0 + \Delta T$, respectively. Here the characteristic temperature difference ΔT is positive, which means that the layer is heated from below. The medium is assumed to have coinciding principal axes of permeability and thermal diffusivity. One of these axes is directed vertically upwards, in the z direction, and the x and y axes are defined by the directions of the other two principal axes. The quantities K_1, K_2, K_3 and D_1, D_2, D_3 are the components of permeability and diffusivity in the x, y and z directions, respectively. The dimensionless anisotropy tensors are then written as:

$$\mathcal{K} = \xi_1 \mathbf{i}\mathbf{i} + \xi_2 \mathbf{j}\mathbf{j} + \mathbf{k}\mathbf{k} \tag{5}$$

$$D = \eta_1 \mathbf{i}\mathbf{i} + \eta_2 \mathbf{j}\mathbf{j} + \mathbf{k}\mathbf{k} \tag{6}$$

where

$$\xi_1 = K_1/K_3, \qquad \xi_2 = K_2/K_3, \qquad \eta_1 = D_1/D_3, \qquad \eta_2 = D_2/D_3. \tag{7}$$

When the temperature difference ΔT is sufficiently small there exists a motionless conduction state $\mathbf{v} = \mathbf{0}, p = p_s, \rho = \rho_s$ and $T = T_s$ where

$$T_s = T_0 + \frac{\Delta T}{h}(h - z). \tag{8}$$

To investigate the linear stability of the basic flow and the steady motions occurring at the onset of convection, we let

$$\mathbf{v} = (\tilde{u}, \tilde{v}, \tilde{w}), \qquad p = p_s + \tilde{p}, \qquad \rho = \rho_s + \tilde{\rho}, \qquad T = T_s + \tilde{\theta}. \tag{9}$$

Here $(\tilde{u}, \tilde{v}, \tilde{w})$, $\tilde{p}, \tilde{\rho}$ and $\tilde{\theta}$ denote the perturbations of velocity, pressure, density and temperature, respectively. Substituting these expressions into equations (1)-(4) and neglecting the nonlinear term $\mathbf{v} \cdot \nabla\tilde{\theta}$, we obtain the linearized perturbation equations. Moreover, the variables are made dimensionless with scales of length, time, velocity, temperature and pressure given by h, $ch^2/D_3, D_3/h, \Delta T, \mu D_3/K_3$, respectively. By eliminating $\tilde{p}, \tilde{u}, \tilde{v}$ and $\tilde{\rho}$ from the perturbation equation and for convenience deleting the tilda, we obtain the following governing equations:

$$\left(\xi_1 \frac{\partial^2}{\partial x^2} + \xi_2 \frac{\partial^2}{\partial y^2} + \frac{\partial^2}{\partial z^2}\right) w = Ra \left(\xi_1 \frac{\partial^2}{\partial x^2} + \xi_2 \frac{\partial^2}{\partial y^2}\right) \theta \tag{10}$$

$$\frac{\partial \theta}{\partial t} - w = \left(\eta_1 \frac{\partial^2}{\partial x^2} + \eta_2 \frac{\partial^2}{\partial y^2} + \frac{\partial^2}{\partial z^2}\right) \theta \tag{11}$$

where the Rayleigh number Ra is given by

$$Ra = \frac{K_3\, g\, \gamma\, \Delta T\, h}{D_3\, \nu} \tag{12}$$

and $\nu = \mu/\rho_0$ is the kinematic viscosity. Impermeable and perfectly heat-conducting boundaries yield the condition

$$\theta = w = 0 \quad \text{at} \quad z = 0, 1. \tag{13}$$

The preferred mode of disturbances which satisfies the boundary conditions (13) are given by

$$\begin{aligned} w &= A\, \sin(\pi\, z) \exp\left(i\,(kx + ly) + \sigma\, t\right) \\ \theta &= B\, \sin(\pi\, z)\, \exp\left(i\,(kx + ly) + \sigma\, t\right) \end{aligned} \tag{14}$$

where k and l are the wave numbers in the x and y directions, respectively, σ is the growth rate, and A and B are constants. Since the boundary value problem (10)-(13) is self adjoint, σ is real and neutral stability is given by $\sigma = 0$. The Rayleigh number for the onset of convection is found to be

$$Ra = \frac{\xi_1 k^2 + \xi_2 l^2 + \pi^2}{\xi_1 k^2 + \xi_2 l^2} \left(\eta_1 k^2 + \eta_2 l^2 + \pi^2\right). \tag{15}$$

Minimizing equation (15) with respect to k and l yields the critical Rayleigh number

$$Ra_c = \pi^2 \left(\min \left\{ \left(\frac{\eta_1}{\xi_1}\right)^{\frac{1}{2}}, \left(\frac{\eta_2}{\xi_2}\right)^{\frac{1}{2}} \right\} + 1 \right)^2. \tag{16}$$

Concerning the critical wave numbers k_c and l_c there are three cases:

Case A: $\eta_1/\xi_1 < \eta_2/\xi_2$ which implies

$$k_c = \pi\, (\xi_1 \eta_1)^{-\frac{1}{4}}, \; l_c = 0 \tag{17}$$

which gives convection rolls aligned in the y direction.

Case B: $\eta_1/\xi_1 > \eta_2/\xi_2$ which implies

$$k_c = 0, \; l_c = \pi\, (\xi_2 \eta_2)^{-\frac{1}{4}} \tag{18}$$

which gives convections rolls aligned in the x direction.

Case C: $\eta_1/\xi_1 = \eta_2/\xi_2$. Here the orientation of the rolls is undetermined. The critical wave number vector

$$\boldsymbol{\alpha} = k_c\,\mathbf{i} + l_c\,\mathbf{j} \tag{19}$$

is constrained by the relation

$$(\xi_1\eta_1)^{1/2}\,k_c^2 + (\xi_2\eta_2)^{1/2}\,l_c^2 = \pi^2. \tag{20}$$

In the case of horizontal isotropy, i.e. $\xi_1 = \xi_2 = \xi$ and $\eta_1 = \eta_2 = \eta$, the results reduce to

$$Ra_c = \pi^2\left[1 + \left(\frac{\eta}{\xi}\right)^{1/2}\right]^2, \qquad \alpha_c = \left(k_c^2 + l_c^2\right)^{1/2} = \pi\,(\xi\eta)^{-1/4} \tag{21}$$

which were first obtained by Epherre [6] . The effects of anisotropy on the onset of convection are easily seen from equations (21). Let us keep K_3 and D_3 (vertical permeability and diffusivity) fixed, and vary the corresponding horizontal quantities. Then an increased horizontal permeability reduces the critical Rayleigh number and increases the preferred cell width. Increasing the horizontal diffusivity will lead to the opposite effect.

Kvernvold and Tyvand [7] also investigated the supercritical, steady two-dimensional motion based on the nonlinear equations. Regions of stable wave numbers and Rayleigh numbers were found. It was shown that the Nusselt number Nu and the stability regions depend on the anisotropy parameters only through the ratios ξ_1/η_1 and ξ_2/η_2. The velocity and temperature fields, however, depend on the appropriate values of ξ_1, ξ_2, η_1 and η_2. They also showed that, if Nu is displayed as a function of Ra/Ra_c, all curves for different values of ξ_1/η_1 start out with the same slope, namely 2.0. Some of the results are related to insulation techniques, and this aspect is discussed separately in their work.

The inclusion of three-dimensional anisotropy introduced by Kvernvold and Tyvand [7] did not lead to any essential new flow patterns at convection onset compared to the isotropic case. But this is true only if one of the principal axes of the anisotropic medium is normal to the layer. Such a restriction was maintained in almost all work in the field up to the beginning of the nineties. Tyvand and Storesletten [8] seem to be the first to have studied natural convection in an anisotropic medium where none of the principal axes are vertical. They considered Bénard convection in a horizontal porous layer with anisotropic permeability, transversely isotropic, whereas the orientation of the longitudinal principal axis is arbitrary. The permeability tensor then is

$$\mathcal{K} = K_L\,\mathbf{i}'\mathbf{i}' + K_T\left(\mathbf{j}'\mathbf{j}' + \mathbf{k}'\mathbf{k}'\right). \tag{22}$$

Here K_L and K_T denote the longitudinal and transverse components of the permeability, respectively, and the longitudinal axis (with unit vector $\mathbf{i}'$) makes an angle β with the horizontal direction. Furthermore, the Rayleigh number is defined by $Ra = K_T\, g\, \gamma\, h\, \Delta T/\,(D\, \nu)$, where D is the isotropic thermal diffusivity. The linear stability and the steady flow pattern at the onset of convection were examined. Two different types of convection cells (rolls) were found, both two-dimensional: If the transverse permeability is larger than the longitudinal permeability $(K_T > K_L)$, the planes of motion are tilted, but the lateral cell walls are vertical as usual. For the converse case $(K_T < K_L)$, the flow occurs in vertical planes, but the lateral cell walls are tilted as well as curved. The preference for these different flow patterns is explained as a preference for flow directions with as large tangential permeability as possible. This preference also gives rise to a tendency of concentration of flow along the cell boundaries when the anisotropy increases, in the case when the cell walls are tilted. In the former case $(K_T > K_L)$, the critical Rayleigh number Ra_c and the corresponding critical wave number m_c are

$$Ra_c = \pi^2 \left[1 + \left(\frac{\xi}{\xi \cos^2\beta + \sin^2\beta}\right)^{1/2}\right]^2, \quad m_c = \pi \left(\cos^2\beta + \xi^{-1}\sin^2\beta\right)^{\frac{1}{4}} \tag{23}$$

where $\xi = K_T/K_L$. The tilt angle α, i.e. the angle which the planes of motion make with the vertical direction, is given by

$$\tan\alpha = \frac{(1-\xi)\tan\beta}{\xi + \tan^2\beta}. \tag{24}$$

It follows that α tends to zero if $\xi \to 1$ or if $\beta \to 0^o$ or 90°. For fixed ξ, the plane of motion obtains its maximum tilt when

$$\tan\beta = \xi^{1/2} \tag{25}$$

and the corresponding maximum value is

$$|\alpha|_{\max} = \arctan\left(\frac{\xi - 1}{2\xi^{1/2}}\right). \tag{26}$$

For example, in the case $\xi = 2$, $|\alpha|_{\max} = 19.47^o$ is obtained at $\beta = 54.73^o$. The tilt angle α is purely mechanically determined, as it is derived from a component of Darcy's law without any buoyancy terms. As α is completely decoupled from thermal effects, equation (24) is obviously not restricted to thermally driven flows and Tyvand [9] found the same formula as (24) for the tilt angle of free surface flow cells in a porous medium with the same type of anisotropy as assumed in the present paper.

In the latter case $(K_T < K_L)$, the critical Rayleigh number is calculated numerically. For given ξ, the number Ra_c attains its maximum for $\beta = 0^o$ and $\beta = 90^o$. The minimum value of Ra_c, which depends on β, is calculated as well. These results have applications in insulation techniques, as described by Kvernvold and Tyvand [7]. The case when $K_T < K_L$

may be interpreted as a medium composed of parallel fibres. The results show that there is nothing to be gained by an oblique orientation of the fibres, since the critical Rayleigh number is always reduced compared with a perpendicular or parallel orientation of fibres *vs.* boundaries. The case $K_T > K_L$ may be interpreted as a medium composed of perforated parallel plates, and the work conforms the plausible conjecture that the maximum Rayleigh number, i.e. optimal insulating properties, is obtained when these perforated plates are horizontal.

Storesletten [10] has studied the analogous problem for a horizontal layer with anisotropy in the thermal diffusivity. The diffusivity tensor is then given by

$$\mathcal{D} = D_L \mathbf{i}' \, \mathbf{i}' + D_T \left(\mathbf{j}' \, \mathbf{j}' + \mathbf{k}' \mathbf{k}' \right) \tag{27}$$

where D_L and D_T are the longitudinal and transverse components of the diffusivity, and the longitudinal principal axis (with unit vector $\mathbf{i}'$) makes an angle β with the horizontal direction. Two different types of convection cells (rolls) were found at the onset of convection, both two-dimensional. If the longitudinal diffusivity is larger than the transverse $(D_T < D_L)$, the convection cells are rectangular with vertical planes of motion like the isotropic case. At this point, there is an esential difference between the present case and the analogous case for a porous medium with anisotropic permeability, where the planes of motion were tilted (Tyvand and Storesletten [8]). For the converse case, $D_T > D_L$, the planes of motion are still vertical, whereas the lateral cell walls are tilted as well as curved, see Figure 1. The streamlines are computed at $\beta = 40.1^o$ for the cases $\eta = 1.00$, 2.00, 4.00 and 8.00. The preference for these different flow patterns is explained as a preference for flow directions with as small a tangential diffusivity along the streamlines as possible.

In the first case, $D_T < D_L$, the critical Rayleigh number Ra_c and the corresponding wave number m_c are found to be

$$Ra_c = \pi^2 \left[1 + \left(\cos^2 \beta + \eta^{-1} \sin^2 \beta \right)^{1/2} \right]^2, \quad m_c = \pi \left(\cos^2 \beta + \eta^{-1} \sin^2 \beta \right)^{1/4} \tag{28}$$

where $\eta = D_T / D_L$. It follows that $Ra_c \to 4\pi^2$ as $\eta \to 1$. For fixed η, Ra_c obtains its minimum value of $4\pi^2$ for $\beta = 0^o$ and its maximum value of $\pi^2 \left(1 + \xi^{-1/2} \right)$ for $\beta = 90^o$. In the second case, $D_T > D_L$, the critical Rayleigh number is determined numerically. For η fixed, Ra_c attains its maximum $4\pi^2$ for $\beta = 0^o$ and $\beta = 90^o$. The minimum values depending on β are calculated.

A two dimensional linear analysis of the decay of a disturbed free surface in an anisotropic porous layer was made by Tyvand [9] . The layer rests on an impermeable horizontal bottom and has an arbitrary two-dimensional anisotropy in the permeability, given by equation (22). The tilt of the streamlines is studied and for infinite depth these streamlines are straight lines making an angle α with the vertical, given by equation (24).

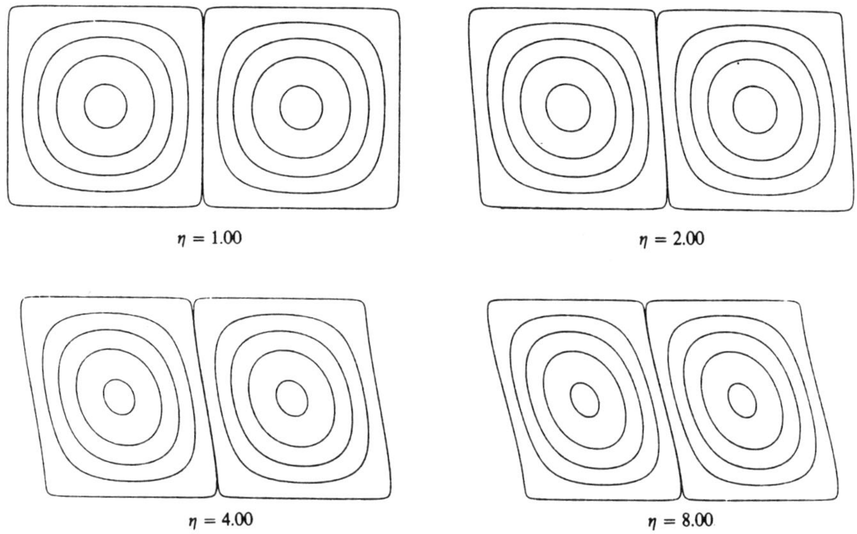

Figure 1 Computed streamlines at $\beta = 40.1^o$ for the cases $\eta = 1.00$, 2.00, 4.00 and 8.00

The formula for the decay rate σ is found to be

$$\sigma = \frac{(K_L K_T)^{1/2}\,\rho g}{\mu} \tanh\left[\frac{(K_L K_T)^{1/2}\,k\,d}{K_L \sin^2\beta + K_T \cos^2\beta}\right] \tag{29}$$

where k is the number of a horizontal Fourier component and d is the depth.

The effects of anisotropy and surface boundary conditions on convection in a horizontal porous layer were studied by McKibbin [11] . The layer is heated from below by an isothermal impermeable boundary and with an upper surface that is isothermal but which has a general boundary condition of the form $p + \lambda\ \partial p/\partial n\ = constant$. Here p is the pressure and λ is a parameter taking the limiting value 0 for a constant pressure boundary and ∞ for an impermeable boundary. It was found that there is always some recirculation of fluid within the porous layer for finite λ. In the case $\lambda = 0$ there is a stagnation point in the interior of the layer as well as on the surface. The effects of anisotropy in permeability and thermal conductivity, expressed in terms of the parameters $\xi = K_H/K_V$ and $\eta = k_H/k_V$, respectively, are dependent on the ratio ξ/η only (the subscripts H and V refer to the horizontal and vertical directions). The recirculation diminishes as $\xi/\eta \to 0$ and there is full recirculation of the fluid within the layer as $\xi/\eta \to \infty$.

Chen, Chen and Pearlstein [12] presented a numerical study of the conditions for the onset of convection in an anisotropic porous layer underlaying a fluid layer. They consider horizontal layers of infinite extent, where the top of the fluid layer and the bottom of the porous layer are bounded by rigid walls maintained at different constant temperatures, which are low at the top and high at the bottom. Darcy's law is used in the porous medium and the Beavers-Joseph condition is applied at the interface between the two layers. It is found that the effects of anisotropy on the onset of convection are most profound for small values of the depth ratio ζ (ratio of the fluid layer thickness to porous layer thickness), since in this case, the onset of convection corresponds to significant motion in both layers. For larger values of ζ, the onset of motion is increasingly confined to the fluid layer, with the transport of heat through the porous layer occuring primarily by conduction. Thus, the influence of an anisotropic permeability on the stability conditions for larger ζ is less significant than the effects of an anisotropic thermal diffusivity.

A similar study, primarily an extension of the analysis of Chen et al. [12] , was made by Chen and Hsu [13] . They performed a linear analysis for the convective instability in a system consisting of a fluid layer overlying a porous layer, in which both the permeability and the diffusivity are vertically anisotropic and inhomogeneaous. A wide-range parametric study is carried out, in terms of the depth ratio ζ , the permeability ratio ξ (ratio of horizontal to vertical permeability), thermal diffusivity ratio η (ratio of horizontal to vertical diffusivity), and the inhomogeneity parameters. For $\zeta \leq 0.1$, the critical Rayleigh number Ra_c for the porous medium is found to be essentially an explicit function of ξ/η and the corresponding critical wave number a_c is found to be a function of $(\xi\eta)^{-1/4}$. Since the porous layer dominates the two-layer system by convection, the anisotropic and inhomogeneous effects are significant. For $\zeta \geq 0.2$, however, no explicit function for either Ra_c and a_c is obtained and the anisotropic and inhomogeneous effects are not significant because the onset of convection is largely confined to the fluid layer. Generally, the vertically increasing permeability leads to a less stable state with smaller critical wavelength; and the vertically increasing thermal diffusivity leads to a more stable state with larger critical wavelength.

The theoretical studies above deal with models based on Darcy's law. However, Qin and Kaloni [14] presented an analysis of the onset of convection in a horizontal porous layer with anisotropic permeability, where they used Brinkman's equation (instead of Darcy's law). Such a model could be used to describe thermally driven flows in sparcely packed fibrous insulation materials. For both free and rigid boundaries, the critical Rayleigh number and wave number at marginal stability were calculated. In both cases it was found that there exist ranges for which the stability criterion is intermediate between the asymptotic limit cases of Darcy's law (porosity tends to zero) and ordinary viscous fluid (porosity tends to infinity). For stress-free boundaries, there is a specific direction along which rolls are preferred. The alignment depends on the magnitude of the permeability ratios ξ_1 and ξ_2, defined by equation (7). For rigid boundaries, the cases of insulating boundary and conducting boundary were studied by two different numerical techniques. In both cases, critical Rayleigh numbers were found.

Straughan and Walker [15] developed a linear instability analysis and a nonlinear energy stability analysis for convection in an anisotropic porous medium. They take the configuration of Tyvand and Storesletten [8] but allow a quadratic temperature dependence for the density. This is a physically important change since many geological problems involve convection with water near the freezing point, and in this situation penetrative convection may develop. Tyvand and Storesletten [8] found that at onset of convection the growth rate σ is real and so convection arises via a stationary convection mechanism. This is the case also for penetrative convection with a quadratic density model when the medium is *isotropic*. However, Straughan and Walker [15] find that any departure from isotropy yields a complex growth rate and so the convection process always begins via an oscillatory convection mechanism. Finally they briefly examine an internal heat source model for penetrative convection in an anisotropic medium.

INCLINED POROUS LAYERS

There are a couple of papers which examine the effect of an anisotropic permeability on convective flow in inclined layers. Trew and McKibbin [16] studied two-dimensional convection in a system of porous layers for small fluid velocities. Thus, as a first approximation, heat transfer is assumed to occur only by diffusive processes and convective heat transfer is ignored. A mathematical model of the system is constructed and applied to various physical configurations, including non-symmetric multi-layered systems. It is shown that layers with high anisotropy ratios $(K_x >> K_z)$ increase the impact of that layer on the flow only slightly. Here K_x and K_z are the permeability components along and across the layer, respectively. In contrast, layers with a low permeability ratio $(K_x << K_z)$ cause that layer to have little effect on the flow, even at high aspect ratios. It was found that system asymmetry caused the greatest effect on the flow, even at smaller aspect ratios.

Storesletten and Tveitereid [17] have examined the onset of convection (three-dimensional disturbances) in an inclined porous layer with anisotropic permeability. The layer is bounded above and below by two infinite and impermeable heat conducting planes, making an angle α with respect to the horizontal. The boundaries are maintained at constant, different temperatures, the lower boundary being the hotter. At subcritical Rayleigh numbers, $Ra < Ra_c$, there exists a stable basic flow, and when Ra is sufficiently increased, instability occurs and a convective motion is set up. There exist two different types of convection rolls, depending on α and the anisotropy ratio $\xi = K_L/K_T$. Here K_L and K_T are the longitudinal and transverse components of the permeability, respectively, where the direction of the basic flow is denoted longitudinal. When $K_L < K_T$, Ra_c is found to be $4\pi^2$, independent of ξ and α. The motion occurring at the onset of convection corresponds to longitudinal rolls with axes parallel to the basic flow, like the isotropic case. When $K_L > K_T$, Ra_c is calculated numerically for various ξ and α. There exists a critical angle α_c where transverse horizontal convection rolls (TR) are preferred at the onset of convection when $\alpha < \alpha_c$, and longitudinal rolls (LR) are preferred when $\alpha > \alpha_c$, see Figure 2.

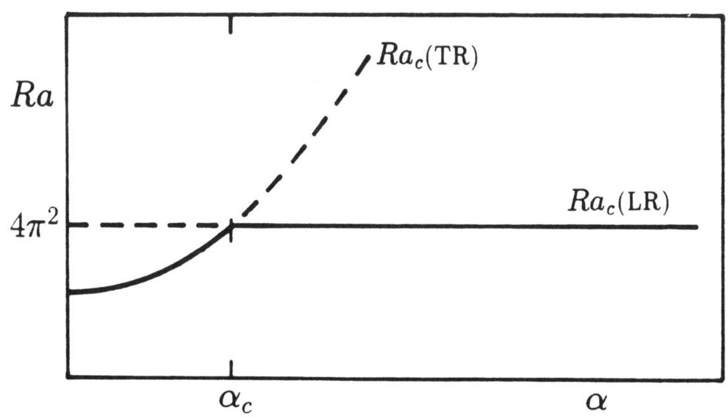

Figure 2 Sketch of the linear stability boundaries as a function of the tilt angle α. The conduction state is unstable with respect to transverse rolls for $Ra > Ra_c$ (TR), and with respect to longitudinal rolls for $Ra > Ra_c$ (LR). The critical tilt angle α_c denotes the intersection between the stability boundaries

ENCLOSURES FILLED WITH A POROUS MEDIUM

A number of papers on the subject has dealt with convective flow in enclosures filled with an anisotropic fluid-saturated medium. These studies are all restricted to two-dimensional convection, which seems to be consistent only for short cylinders, thin boxes or cavities. Beck [18] carried out a linear stability analysis of convection in a finite three-dimensional box of isotropic porous medium. His results show that near convection onset the fluid motion for thin boxes is always two-dimensional in the form of roll-cells. This fact was confirmed by Storesletten and Tveitereid [19] who considered convective flow in a porous horizontal circular cylinder in which the cylinder wall was non-uniformly heated to establish a linear temperature in the vertical direction, with the end sections insulated. For long cylinders ($L > 0.86$) a unique three-dimensional flow was determined at the onset of convection, while a two-dimensional convection occurs for short cylinders ($L < 0.86$), where L is the ratio between the length and radius of the cylinder. We assume similar results to be valid also for anisotropic media.

Convective heat transfer through porous materials in a vertical slot was examined analytically and numerically by Burns, Chow and Tien [20] . This paper seems to be the first to study the effect of anisotropy on convective flow on a porous medium confined in a two-dimensional rectangular enclosure. The vertical sidewalls are heated and cooled, respectively, and the horizontal boundaries are insulated or perfectly conducting. The analysis demonstrates the dependence of the Nusselt number on the Rayleigh number, aspect ratio, and anisotropic permeabilities. A simple analytic formula for calculating the heat transfer has been developed after obtaining a matching coefficient by comparison with numerical solutions. Both free and forced convection, simulating wall leakage in common building structures, are considered.The results may be useful in connection with thermal insulation of buildings.

Nilsen and Storesletten [21] analysed two-dimensional convection in horizontal rectangular channels. The channel walls, assumed to be impermeable and perfectly heat conducting, are nonuniformly heated to establish a linear temperature distribution in the vertical direction. It is shown that the critical Rayleigh number Ra_c and the flow structure at the onset of convection depend on the anisotropy aspect ratios ξ and η defined by

$$\xi = \frac{K_H}{K_V}\left(\frac{h}{a}\right)^2, \qquad \eta = \frac{D_H}{D_V}\left(\frac{h}{a}\right)^2 \tag{30}$$

where K_H, K_V and D_H, D_V are the horizontal and vertical components of the permeability and thermal diffusivity, respectively, and a and h are the horizontal and vertical dimensions of the channels. The main results are as follows:

(a) When $\xi = \eta$, which includes the isotropic situation,

$$Ra_c = 4\pi^2 (1 + \xi)\,. \tag{31}$$

At the onset of convection there exist two different steady flow patterns consisting of n and $n+1$ cells, respectively, where n depends on the aspect ratio a/h. For an isotropic medium, where $1/\xi = (a/h)^2$, the number of cells is given as follows:

$$\left(\frac{a}{h}\right)^2 \le 3 \quad \text{yields 2 and 3 cells} \tag{32}$$

$$(n-1)^2 - 1 \le \left(\frac{a}{h}\right)^2 \le n^2 - 1 \quad \text{yields } n \text{ and } n+1 \text{ cells.} \tag{33}$$

These results differ essentially from the well studied case with insulating sidewalls, see e.g. Sutton [22] and Beck [18] . This is not unexpected since conducting sidewalls change the physical conditions considerably. The heat exchange through the lateral walls increases the critical Rayleigh number and alters the cellular fomations at the onset of convection. A nonlinear stability analysis at moderate supercritical Rayleigh numbers shows that both of the flow structures are stable against two-dimensional perturbations. In addition, it turns out that any composition of these two flow patterns is a stable nonlinear solution.

(b) In the anisotropic case, $\xi \neq \eta$, there exists a unique steady flow pattern at the onset of convection consisting of n cells if $\xi < \eta$, and $n+1$ cells if $\xi > \eta$, where n depends on the product $\xi\eta$ in the following way:

$$(\xi\eta)^{-1/2} < 3 \text{ and } \begin{cases} \xi < \eta \text{ yields 2 cells} \\ \xi > \eta \text{ yields 3 cells} \end{cases} \tag{34}$$

$$(n-1)^2 - 1 < (\xi\eta)^{-1/2} < n^2 - 1 \text{ and } \begin{cases} \xi < \eta \text{ yields } n \text{ cells} \\ \xi > \eta \text{ yields } n+1 \text{ cells.} \end{cases} \tag{35}$$

The computed streamlines and isotherms are displayed in Figure 3 for $\xi = 0.5$, $\eta = 2$ and $\xi = 2$, $\eta = 0.5$. Nonlinear stability analysis shows that the unique solution is stable against

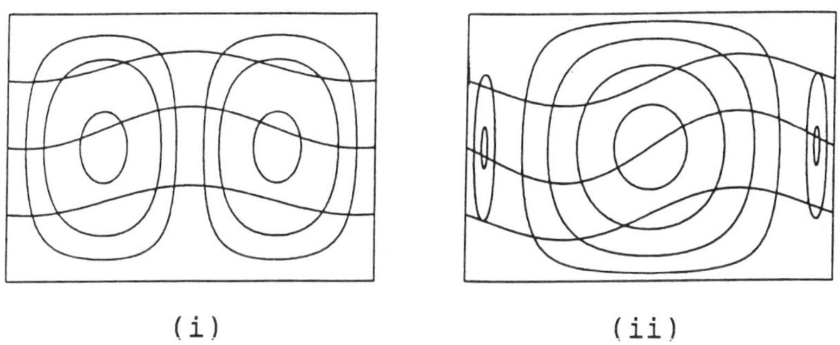

Figure 3 Computed streamlines and isotherms for the anisotropic cases when $(a/h)^2 = 2$ and (i) $\xi = 0.5$, $\eta = 2$, $Ra_c = 157.60$; (ii) $\xi = 2$, $\eta = 0.5$, $Ra_c = 39.39$

two-dimensional perturbations for all values of the parameters ξ, η.

A numerical study on natural convection in an enclosure filled with an anisotropic medium was carried out by Ni and Beckermann [23] . They considered a two-dimensional square cavity where the vertical sidewalls were held isothermally at different temperatures, while the horizontal top and bottom walls were insulated. The effect of mechanical and thermal anisotropy on the streamlines, isotherms and heat transport is studied. The permeability ratio and thermal diffusivity ratio are defined by

$$K^* = K_V/K_H \quad \text{and} \quad k^* = D_V/D_H \tag{36}$$

where the subscripts V and H refer to vertical and horizontal directions, respectively. The main results are reported as follows:

When compared to isotropic porous media, natural convection in the anisotropic case has the following physical characteristics:

(i) A high permeability ratio $(K^* > 1)$ caused channeling of the flow along the vertical

isothermal walls, a higher flow intensity in the enclosure and, consequently, a higher Nusselt number.
(ii) A low permeability ratio $(K^* < 1)$ causes channeling of the flow along the horizontal boundaries, thicker velocity boundary layers along the vertical walls, a lower flow intensity in the enclosure, and, hence, a smaller Nusselt number. All Nusselt numbers approach unity in the limit $K^* \to 0$.
(iii) A large thermal conductivity ratio $(k^* > 1)$ causes a higher flow intensity in the enclosure, the isotherms to be more vertical, the heat transfer to be more dominated by conduction, and, hence, a lower Nusselt number. All Nusselt numbers approach unity in the limit of $k^* \to \infty$.
(iv) A low thermal conductivity ratio $(k^* < 1)$ has very little influence on the flow and heat transfer patterns, and the Nusselt numbers for a given Ra and K^* are approximately the same as for $k^* = 1$.
(v) The permeability and thermal conductivity ratios generally have opposing effects on the Nusselt number.
(vi) The dependence of the flow and heat transfer on the Rayleigh number is quite similar to the isotropic case, while the absolute values of the Nusselt number change drastically for $K^* \neq 1$ and $k^* > 1$.

A similar problem was studied numerically by Zhang [24]. He examined convective heat transfer in a two-dimensional rectangular porous cavity, heated from the side. The medium is characterized by an anisotropic permeability with inclined principal axes with respect to the vertical, i.e. the tensor is given by equation (22). The streamlines and isotherms at various values of the anisotropy ratio are presented, as well as the influence of the anisotropy orientation (inclination). It turns out that the flow changes from a mild convection to a boundary layer regime as the orientation of the principal axes with higher permeability is changed from horizontal to vertical.

Zhang, Nguyen and Kahawita [25] considered Bénard convection in a two-dimensional rectangular cavity filled with a porous medium. The horizontal top and bottom walls are respectively cooled and heated, while the vertical walls are insulated. The porous medium is anisotropic in permeability with inclined principal axes. A linear analysis is used to find the marginal stability limit while a finite difference method is applied to numerically simulate the convection flows. When the inclination angle is zero (longitudinal direction is horizontal), the critical Rayleigh number is

$$Ra_c = 2\pi^2 (1 + R) \tag{37}$$

where the anisotropy ratio R in this case is given by $R = K_T/K_L$, confer equation (22). As expected, this result differs from the corresponding result (31) when the vertical walls were perfectly heat-conducting, studied by Nilsen and Storesletten [21]. Moreover, Zhang et al. [25] found that both the permeability ratio and inclination angle of the principal axes considerably modify the stability limit (Ra_c), the flow pattern and the heat transfer from that under isotropic conditions. Four solution branches have been discovered. The existence or the coexistence of the four branches depends strongly on the anisotropy and its orientation.

For a given permeability and on one solution branch, the maximum heat transfer rate is observed at an inclination angle where a minimum critical Rayleigh number is predicted by the linear stability analysis. Both the anisotropy and the inclination angle have the effect of enhancing the flow in the diagonal region of the cavity, and some symmetry properties of the flow structure observed in the isotropic case are destroyed.

The natural convection problem related to a two-dimensional rectangular cavity is extended in many directions and studied by several authors. Chang and Lin [26] made a numerical study to analyze the wall conduction effect. The two-dimensional porous cavity is formed by walls having finite conductance, and are assumed to be of the same material. The left and right of the outside walls are kept at constant but different temperatures, while the horizontal connecting walls are assumed to be insulated on the outside. The numerical results indicate that the ansisotropic permeability affects the flow field and heat transfer rate significantly. A critical value of the anisotropic thermal diffusivity ratio D^* may exist such that the Nusselt number reaches a minimum. This critical value decreases with an increase in the value of the anisotropic permeability ratio K^*, where $K^* = K_V/K_H$ and $D^* = D_V/D_H$. Moreover, wall conductance effect can lead to large changes in the Nusselt numbers.

Degan, Vasseur and Bilgen [27] considered the case where the vertical walls are heated and cooled, respectively, by a constant heat flux, while the horizontal walls are insulated. The principal axes of the anisotropic permeability are oblique to the gravity vector, whereas those of the thermal diffusivity coincide with the horizontal and vertical coordinate axes. An analytical solution, valid for flow in slender enclosures, is derived on the basis of a parallel flow approximation while a finite difference method is used to numerically simulate the fluid flow, temperature distribution and heat transfer ratio. It was shown that both the permeability ratio and inclination angle α of the principal axes have a strong influence on the convective flow, and the main results of Ni and Beckermann [23] (corresponding to $\alpha = 0$) are confirmed. Moreover, it is shown that a maximum (minimum) heat transfer within the cavity is obtained when the porous matrix is oriented in such a way that the principal axis with higher permeability is parallel (perpendicular) to the vertical direction. This result is applicable to insulation techniques relative to a vertical porous layer: The best possible insulation is reached when the principal axis with lower permeability is parallel to the vertical.

A closely related problem has recently been studied by Degan and Vasseur [28] . In the present case the sidewalls of the rectangular cavity are heated and cooled isothermally, while the horizontal walls still are insulated. The thermally isotropic, porous medium is anisotropic in permeability with principal axes oblique to the gravity vector. Both a numerical approach and an analytical solution valid in the boundary layer region, are used to investigate the problem. Again it was shown that the permeability ratio and inclination angle α of the principal axes have a strong influence on flow structure and heat transfer rate. Earlier results of a similar type reported by Degan et al. [27] are confirmed. In particular, it was shown that the heat transfer is at a maximum when the orientation of the principal axes with higher permeability is parallel to the gravity, while it is minimum when it is perpendicular to the gravity.

Kimura, Masuda and Hayashi [28] made a nonlinear analysis of natural convection in an anisotropic porous medium by employing a perturbation method for small Rayleigh numbers. They consider a two dimensional porous cavity heated from the side, while the top and bottom walls are insulated. The anisotropy ratios of permeability and thermal conductivity are considered in the range

$$0.1 \leq K_H/K_V \leq 100, \quad 0.01 \leq k_H/k_V \leq 100 \tag{38}$$

where K_H, K_V and k_H, k_V are the horizontal and vertical components of the permeability and conductivity, respectively. The effects of anisotropy of both permeability and thermal conductivity (diffusivity), and of the geometric aspect ratios on the flow pattern and the heat transfer are documented. In general, it is found that the geometric aspect ratio gives the same effect on the heat transfer and the convective pattern as the anisotropic properties of the porous medium. The results are presented in a number of figures showing how the streamlines and isotherms depend on the various parameters, and showing the average Nusselt number variation as a function of Ra.

A numerical study of two-dimensional convection in a vertical cylinder filled with an anisotropic porous medium was performed by Chang and Hsiao [30] . The top section and the circular cylinder wall are kept at a uniform high temperature, while the temperature of the bottom section is low. The paper investigates the effect of the anisotropic permeability and conductivity ratios, geometrical aspect ratio and Rayleigh number variation on the flow field and heat transfer. The numerical solutions show significant effects of these parameters. The major fraction of the total heat transfer comes from the lower sidewall, while the top wall seems to be insulated in all cases analysed. The heat transfer is more evident in the lower section of height as the aspect ratio A $(= height/radius)$ decreases. The heat transfer rate of the sidewall and bottom wall increases, either decreasing the values of anisotropic permeability ratio or increasing the values of anisotropic thermal conductivity ratio.

CONVECTIVE BOUNDARY-LAYER FLOW

The effect of anisotropy on thermal boundary-layer flow in porous media was first studied by Ene [31] . The convective flow is induced by a vertical, uniformly heated plate embedded in a fluid-saturated medium. A combined study of the effects of anisotropic permeability and diffusivity is presented by using the method of integral relations. The work considers the leading order boundary-layer flow, and it is assumed that the principal axes of the anisotropy tensors are free to rotate only in one direction. An analysis is made concerning the correct use of the thermal boundary-layer approximation in case of anisotropic medium. It turns out that the boundary-layer hypothesis is not valid if the permeability along the plate, K_x, is much greater than the permeability in the direction orthogonal to the plate, K_y. Furthermore, it is shown that if $K_y > K_x$, then there is an increase in the temperature field (compared to the case $K_y < K_x$).

A similar configuration was considered by Rees and Storesletten [32] , who examined the detailed effects of an anisotropic permeability. In contrast to Ene [31] , they allow the principal axes of the permeability to rotate freely, i.e. the axes may have an arbitrary orientation. A leading-order boundary-layer theory is presented. It is shown that there are two primary effects which are evident when considering this flow, namely, (i) a changed boundary-layer thickness (which has important rammifications for the rate of heat transferred into the porous medium) and (ii) an anisotropically induced drift in the spanwise direction. Conditions are found which determine when the drift is present or absent, and, if present, its strength.

A complementary study examining the effects of an anisotropic thermal diffusivity has recently been made by Storesletten and Rees [33] . Convection is induced by a generally inclined, uniformly heated surface embedded in a fluid-saturated medium. A third-order boundary-layer theory is presented in order to obtain accurate information on the effect of the anisotropy. It is shown that the thickness of the resulting leading order boundary-layer flow depends on the precise nature of the anisotropy. On the other hand, the anisotropic diffusivity does not induce a fluid drift in the spanwise direction, a result which is different from their earlier study [32] of the effects of an anisotropic permeability. It is found that the second-order temperature field does not contribute to the overall rate of heat transfer. Finally, they show that the third order correction to the leading order rate of heat transfer is given in terms of an explicit formula, depending on the diffusivity ratios, all three rotation angles of the principal axes, the angle of the inclination of the heated surface and the wedge angle.

HYDRODYNAMIC DISPERSION

Several papers have analyzed the influence of *hydrodynamic dispersion* on convection in anisotropic media. Tyvand [34] studied the onset of convection in a horizontal layer, heated from below, and bounded by impermeable and heat conducting boundaries. Horizontal isotropy with respect to permeability and thermal diffusivity is assumed. The dispersion, caused by a uniform basic flow, is considered in the limits of small and large Péclet numbers. The analysis shows that the onset of convection in both cases is independent of longitudinal dispersion, while dispersion in lateral directions has stabilizing effects, but the stabilizing effect is much stronger in case of large Péclet number. In both cases, the most unstable mode consists of stationary rolls aligned in the direction of the basic flow. The critical Rayleigh number and wave number at marginal stability are found analytically for both small and large Péclet numbers. When the uniform basic flow tends to zero the results coincide with those found by Epherre [6] , given in equation (21).

The effect of heat dispersion on steady finite amplitude convection in anisotropic porous layer was investigated by Tyvand [35] . The configuration is similar to that considered in his paper [34] three years earlier. The work is restricted to small Péclet numbers, and an amplitude expansion is applied to solve the nonlinear problem. The results indicate that the combination of anisotropy and dispersion effects may have a profound influence on the heat transfer through a porous layer. This influence is much stronger than the isolated effects of anisotropy and dispersion. This is due to the nonlinear character of the problem, by which superposition of effects is strongly violated.

A theoretical and experimental study on hydrodynamic dispersion in a saturated porous medium was made by Fattah and Hoopes [36]. A tensor model for the dispersion coefficient for flow in a homogeneous medium with anisotropic permeability was discussed. The tensor components were evaluated from experiments with an anisotropic homogeneous porous material consisting of thin, alternating layers of two types of sand, having different mean particle sizes. Tests were carried out for flow parallel, perpendicular, and inclined at 60° to the direction of the layers. The permeability and the longitudinal dispersion coefficient were found to be second-order tensors with equal and constant eccentricities and with major and minor principal axes oriented along and perpendicular to the directions of sand layers, respectively. The lateral dispersion coefficient was found to be a second-order tensor whose principal axes were orthogonal to those of the permeability tensor and whose eccentricity increased with increasing seepage velocity.

Howle and Georgiadis [37] performed a numerical study in order to examine the effect of anisotropic dispersion on heat transfer in a horizontal porous layer, heated from below. The momentum equation, including an inertial (Forchheimer) term, is coupled to a single energy equation which contains the full anisotropic tensor of effective thermal conductivity. A parametric study of two-dimensional steady cellular convection reveals the following: (i) Dispersion increases the net heat transfer after a Rayleigh number in the range 100-200. As the degree of anisotropy of the effective thermal conductivity is increased, the wall averaged Nusselt number is decreased. (ii) Using the available Rayleigh number - wave number variation data does not affect the divergence between simulation and experiment.

DOUBLE-DIFFUSIVE CONVECTION

In double-diffusive convection the gradients that provide the driving buoyancy force are induced by the combined effects of temperature and species concentration nonuniformities present in the fluid. The effects arise from the fact that heat diffuses more rapidly than a dissolved substance.

Tyvand [38] was the first to study double-diffusive convection in an anisotropic porous medium. He considered a horizontal layer which retains horizontal isotropy with respect to permeability, thermal diffusivity and solute diffusivity. It was shown that for porous media with thermally insulating solid matrices the stability diagram has the same shape as in the isotropic case. The critical wave number is constant and equal to that of the one-component case. For thermally conducting matrices, new features may occur, namely, the locus of the direct mode in the stability diagram may not be a straight line, and the corresponding wave number may be nonconstant. When the thermal diffusivity ratio η_T is less than the solute diffusivity ratio η_S, results indicate that such anisotropy tends to reduce the danger of pollution in the form of salt fingers penetrating into cold freshwater quifers from overlying warm saltwater. Here η_T and η_S are the quotients of the horizontal and vertical components.

The onset of double-diffusive convection in a rotating porous layer of infinite horizontal extent was investigated by Patil, Paravathy and Venkatakrishnan [39] . An anisotropic permeability is considered, with horizontal isotropy. Numerical computations indicate that

(i) for certain parameters a bottom-heavy arrangement destabilizes a rotating anisotropic porous layer, (ii) the lower the anisotropy parameters, the higher the range of bottom-heavy solute gradient for which there is destabilization, (iii) increase in the anisotropy parameter stabilizes the system, and (iv) for some values of the parameters rotation destabilizes the system, though in general, it has a stabilizing effect.

Double-diffusive convection in layered anisotropic porous media was studied numerically by H. D. Nguyen, S. Paik and R. W. Douglass [40] . A rectangular enclosure, consisting of two anisotropic porous layers with dissimilar hydraulic and transport properties, was considered. Darcy's law together with the energy and specie equations were solved numerically. Four different sets of boundary constraints were imposed on the system, including aiding diffusion, opposing diffusion and the two modes of crossed diffusion. The results show that each set of boundary conditions produces distinct flow, tempearture and concentration fields. The overall heat transfer rates may or may not be sensitive to the Rayleigh numbers, depending on the orientation of the boundary conditions of the temperature and concentration fields.

ANISOTROPY AND MULTILAYERED MEDIA

There is a correspondence between layering and anisotropy in porous media, i.e. a layered porous medium can, under certain conditions, be represented by an anisotropic model medium. This fact was demonstrated in a work by Marcus and Evenson [41] . By considering uniform flows, i.e. uniform flows in each layer, they showed that the set of scalar layer permeabilities of a multilayered medium is macroscopically equivalent to an average permeability tensor. Later it was noted by Wooding [42] that in a geothermal system with a ground structure composed of many successively laid strata of different permeability, the overall horizontal permeability may be up to ten times as large as the vertical component. He considered a porous layer in which the permeability is anisotropic and may also vary with depth. Criteria were found for onset of two- and three-dimensional convection, and both impermeable and free upper boundaries were treated. As expected the system is more unstable with free upper boundaries than with impermeable boundaries. But the difference becomes small when the horizontal permeability component K_H becomes much larger than the vertical component K_V, because then vertical flow is more difficult than horizontal flow.

The principle of large-scale anisotropy due to small scale layering was also applied in a work by McKibbin and Tyvand [43] . They study thermal convection in a bounded multilayered porous medium heated from below. The medium is periodically layered with respect to permeability and thermal conductivity. The onset of convection as well as slightly supercritical convection are investigated. The results indicate that anisotropic modelling proves useful even for small numbers of layers as long as the motion is 'large-scale convection'. However, modelling always fails for motion of 'local convection' type.

Time-dependent, two-dimensional convection in porous media with permeability fields that are anisotropic, layered or both, was studied numerically by Rosenberg and Spera [44] . This seems to be the first study incorporating both anisotropy and layering in the permeability field. Their work derives from an interest in 'oceanic crust' models which take into account layers of lava flows or other geological deposits forming a porous medium composed of layers of materials with different permeabilities and different anisotropies. Through numerical simulations they demonstrate that the dynamics of convection vary significantly with the spatial variation and degree of anisotropy of the permeability field. For a given Rayleigh number in a homogeneous anisotropic porous medium, the strength of the flow and heat transfer decrease as the anisotropy ratio $\xi = K_H/K_V$ decreases, i.e. when the medium becomes less permeable in the horizontal direction.

Rosenberg, Spera and Haymon [45] analyzed the relationship between flow and permeability field in seafloor hydrothermal systems. Using realistic permeability fields for young oceanic crust, they performed numerical simulations of finite amplitude, steady and unsteady convective fluid flow in layered and/or anisotropic porous media heated from below. Particular patterns of fluid flow and temperature in mid-ocean ridge hydrothermal systems were investigated. On the flanks of mid-ocean ridges, permeability measurements in deep-sea boreholes suggest that only the top few hundred metres of oceanic crust is permeable. Given this permeability structure, their models predict that convection occurs in the form of numerous cells with aspect ratios of order unity within this permeable layer. Such convection results in phenomena which are in agreement with field observations. Several models are studied, based on various realistic permeability structures. In their models, near-field effects dominate over far-field effects in systems with lateral variations in permeability, and no large-scale flow develops between widely spaced areas of contrasting permeability. The time to steady-state and and evolution of convective flow in porous media vary considerably with initial conditions. Their simulations suggest that the time to steady-state is relatively long and that it is possible that hydrothermal convection at the ridge axis never reaches a steady-state flow pattern. It is possible, however, that such systems may be quasi-steady for significant periods of time.

Acknowledgements

The author would like to express his indebtedness to his co-workers, Prof. M. Tveitereid, Dr. T. Nilsen, Prof. P.A. Tyvand and Dr. D.A.S. Rees for their part in the work described here. The author is also grateful to Prof. D.B. Ingham and Prof. I. Pop for their encouragement and assistance during the preparation of this review article and to Prof. D.A.Nield for reading the manuscript and making helpful corrections.

REFERENCES

1. J. Bear, *Dynamics of Fluids in Porous Media*, Elsevier, New York (1972).
2. R. McKibbin, Thermal convection in layered and anisotropic porous media: a review. In *Convective flows in Porous Media* (eds. R.A. Wooding and I. White), DSIR, Wellington, NZ, pp 113-127 (1985).

3. R. McKibbin, Convection and heat transfer in layered and anisotropic porous media. *Proceedings of the ICHMT International Seminar* on "Heat and mass transfer in porous media", Dubrovnik, Yugoslavia, May 20-24, 1991, pp 327-336 (1991).
4. D.A. Nield and A. Bejan, *Convection in Porous Media*, Springer-Verlag, Berlin (1992).
5. G. Castinel and M. Combarnous, Critère d'apparition de la convection naturelle dans une couche poreuse anisotrope horizontale, *C.R. Acad. Sc.*, **B. 278**, 701-704 (1974).
6. J. F. Epherre, Critère d'apparition de la convection naturelle dans une couche poreuse anisotrope, *Rev. Gén. Thermique* **168**, 949-950 (1975).
7. O. Kvernvold and P. A. Tyvand, Nonlinear thermal convection in anisotropic porous media, *J. Fluid Mech.* **90**, 609-624 (1979).
8. P. A. Tyvand and L. Storesletten, Onset of convection in an anisotropic porous medium with oblique principal axes, *J. Fluid Mech.* **226**, pp. 371-382 (1991).
9. P. A. Tyvand, Decay of a disturbed free surface in an anisotropic medium, *J. Hydrol.* **83**, 367-371 (1986).
10. L. Storesletten, Natural convection in a horizontal porous layer with anisotropic thermal diffusisvity, *Transport in Porous Media* **12**, 19-29 (1993).
11. R. McKibbin, Thermal convection in a porous layer: Effects of anisotropy and surface boundary conditions, *Transport in Porous Media* **1**, 271-292 (1986).
12. F. Chen, C.F. Chen and A.J. Pearlstein, Convective instability in superposed fluid and anisotropic porous layers, *Phys. Fluids A* **3**, 556-565 (1991).
13. F. Chen and L.H. Hsu, Onset of thermal convection in an anisotropic and inhomogeneous porous layer underlaying a fluid layer, *J. Appl. Phys* **69**, 6289-6301 (1991).
14. Y. Qin and P. N. Kaloni, Convective instabilities in anisotropic porous media, *Studies in Applied Mathematics* **91**, pp. 189-204 (1994).
15. B. Straughan and D. W. Walker, Anisotropic porous penetrative convection, *Proc. R. Soc. Lond.* A **452**, 97-115 (1996).
16. M. Trew and R. McKibbin, Convection in anisotropic inclined porous layers, *Transport in Porous Media* **17**, 271-283 (1994).
17. L. Storesletten and M. Tveitereid, Onset of convection in an inclined porous layer with anisotropic permeability, To appear in *J. Theo. Appl. Fluid Mech.* (1997).
18. L. J. Beck, Convection in a box of porous material saturated with fluid, *Phys. Fluids* **15**, 1377-1388 (1972).
19. L. Storesletten and M. Tveitereid, Natural convection in a horizontal porous cylinder, *Int. J. Heat Mass Transfer* **34**, 1959-1968 (1991).
20. P. J. Burns, L. C. Chow and C. L. Tien, Convection in a vertical slot filled with porous insulation, *Int. J. Heat Mass Transfer* **20**, 919-926 (1977).
21. T. Nilsen and L. Storesletten, An analytical study on natural convection in isotropic and anisotropic porous channels, *Trans ASME: J. Heat Transfer* **112**, 396-401 (1990).
22. F. M. Sutton, Onset on convection in a porous channel with net through flow, *Phys. Fluids* **13**, 1931-1934 (1970).
23. J. Ni and C. Beckermann, Natural convection in a vertical enclosure filled with anisotropic porous media, *Trans. ASME: J. Heat Transfer* **113**, 1033-1037 (1991).

24. X. Zhang, Convective heat transfer in a vertical porous layer with anisotropic permeability, *CANCAM* **2**, 579-580 (1993).
25. X. Zhang, T. H. Nguyen and R. Kahawita, Convective flow and heat transfer in an anisotropic porous layer with principal axes non-coincident with the gravity vector, *ASME HTD* **264**, 79-86 (1993).
26. W. J. Chang and H. C. Lin, Natural convection in a finite wall rectangular cavity filled with an anisotropic porous medium, *Int. J. Heat Mass Transfer* **37**, 303-312 (1994).
27. G. Degan, P. Vasseur and E. Bilgen, Convective heat transfer in a vertical anisotropic porous layer, *Int. J. Heat Mass Transfer* **38**, 1975-1987 (1995).
28. G. Degan and P. Vasseur, Natural convection in a vertical slot filled with an anisotropic porous medium with oblique principal axes, *Numerical Heat Transfer* **A30**, 397-412 (1996).
29. S. Kimura, Y. Masuda and K. Hayashi, Natural convection in an anisotropic porous medium heated from the side, *Heat Transfer Japanese Research* **22**, 139-153 (1993).
30. W. J. Chang and C. F. Hsiao, Natural convection in a vertical cylinder filled with anisotropic porous media, *Int J. Mass Heat Transfer* **36**, 3361-3367 (1993).
31. H. I. Ene, Effects of anisotropy on the free convection from a vertical plate embedded in a porous medium, *Transport in Porous Media* **6**, 183-194 (1991).
32. D. A. S. Rees and L. Storesletten, The effect of anisotropic permeability on free convective boundary layer flow in porous media, *Transport in Porous Media* **19**, 79-92 (1995).
33. L. Storesletten and D. A. S. Rees, An analytical study of free convective boundary layer flow in porous media: The effect of anisotropic diffusivity, To appear in *Transport in Porous Media* (1997).
34. P. A. Tyvand, Heat dispersion effect on thermal convection in anisotropic porous media, *J. Hydrol.* **34**, 335-342 (1977).
35. P. A. Tyvand, Influence of heat dispersion on steady convection in anisotropic porous media, *J. Hydrol.* **52**, 13-23 (1981).
36. Q. N. Fattah and J. A. Hoopes, Dispersion in anisotropic homogeneous porous media, *J. Hydraulic Engineering - ASCE* **111**, 810-827 (1985).
37. L. E. Howle and J. G. Georgiadis, Natural convection in porous media with anisotropic dispersive thermal conductivity, *Int. J. Heat Mass Transfer* **37**, 1081-1094 (1994).
38. P.A. Tyvand, Thermohaline instability in anisotropic porous media, *Water Resources Res.* **16**, 325-330 (1980).
39. P. R. Patil, C. P. Paravathy and K. S. Venkatakrishnan, Thermohaline instability in a rotating anisotropic porous medium, *Appl. Scientific Res.* **46**, 73-88 (1989).
40. H. D. Nguyen, S. Paik and R. W. Douglass, Study of double-diffusive convection in layered anisotropic porous media, *Numerical Heat Transfer*, **B26**, 489-505 (1994).
41. M. Marcus and D. E. Evenson, Directional permeability in anisotropic porous media, *Water Resources Contrib.* no. **31**, Hydrology Laboratory, University of California, Berkeley (1961).
42. R. A. Wooding, Large-scale geothermal field parameters and convection theory, *New Zealand Journal of Science* **21**, 219-228 (1978).

43. R. McKibbin and P. A. Tyvand, Anisotropic modelling of thermal convection in multi-layered porous media, *J. Fluid Mech.* **118**, 315-339 (1982).
44. N. D. Rosenberg and F. J. Spera, Role of anisotropic and/or layered permeability in hydrothermal convection, *Geophysical Research Letters* **17**, 235-238 (1990).
45. N. D. Rosenberg, F. J. Spera and R. M. Haymon, The relationship between flow and permeability field in seafloor hydrothermal systems, *Earth and Planetary Science Letters* **116**, 135-153 (1993).

FREE CONVECTION IN ROTATING POROUS MEDIA

PETER VADASZ

Department of Mechanical Engineering, University of Durban-Westville,
Private Bag X54001, Durban 4000, South Africa

INTRODUCTION

The study of flow in rotating porous media is motivated by its practical applications in engineering. Engineering applications include, among others, the food, chemical and materials processing industries and rotating machinery. Packed bed mechanically agitated vessels are used in the food processing and chemical engineering industries in batch processes. The packed bed consists of solid particles or fibres of material which form a rotating solid matrix, due to the mechanical agitation, while fluid flows through the pores. The role of the flow of fluid through these beds can vary from drying processes to extraction of soluble components from the solid particles. The molasses in centrifugal crystal separation processes in the sugar milling industry and the extraction of sodium alginate from kelp are just two examples of such processes. Modelling of flow and heat transfer in porous media is also applied in the design of heat pipes using porous wicks and includes effects of boiling in unsaturated porous medium, surface tension driven flow with heat transfer and condensation in unsaturated porous media. Plumb [1] presented a comprehensive review of the heat transfer in unsaturated porous media flow with particular applications to the heat pipe technology. When the heat pipe is used for cooling devices which are subject to rotation, the corresponding centrifugal and Coriolis effects become relevant as well. The problem of solidification of binary alloys includes also transport phenomena in rotating porous media as the dendritic mushy zone is regarded for all practical purposes as a porous medium . Other applications of the porous medium approach are discussed by Nield and Bejan [2] and Bejan [3] in comprehensive reviews of the fundamentals of heat convection in porous media. Bejan [3] mentions among the applications of heat transfer in porous media the process of cooling of winding structures in high-power density electric machines. When this applies to a rotor of an electric machine, say a generator (or motor), rotation effects become relevant as well.

With the emerging utilisation of the porous medium approach in non-traditional fields, including some applications in which the solid matrix is subjected to rotation, a thorough understanding of the flow and heat transfer in a rotating porous medium becomes essential.

Relatively limited research results are available for natural convection in rotating porous media, e.g. Rudraiah, Shivakumara and Friedrich [4], Patil and Vaidyanathan [5], Jou and Liaw [6, 7] and Palm and Tyvand [8]. Nield [9] while presenting a comprehensive review of the stability of convective flows in porous media finds also that the effect of rotation on convection in a porous medium attracted limited interest and the lack of experimental results is particularly noticed.

The main reason behind the lack of interest for this type of flow is probably the fact that **isothermal flow in homogeneous porous media following Darcy's law is irrotational** (Bear [10]) and hence the effect of rotation on this flow is not significant. However, for a heterogeneous medium with spatial dependent permeability, or for free convection in a non-isothermal homogeneous porous medium, the flow is not irrotational any more and hence the effects of rotation become significant. In some applications these effects can be small, e.g. when the porous media Ekman number is high.

The methodology adopted in this chapter consists of a presentation of the dimensionless equations governing the flow and transport phenomena in a rotating frame of reference, followed by a classification of convection problems. The available results to the different types of problems are then presented and finally recommendations are proposed regarding further research needed on topics which fall short of available or reliable results.

GOVERNING EQUATIONS

Darcy Model

The dimensionless continuity, Darcy, and energy equations governing the flow and heat transfer in rotating homogeneous porous media are presented in the form

$$\nabla \cdot \boldsymbol{q} = 0 \tag{1}$$

$$\boldsymbol{q} = -\left[\nabla p_r + Ra_g T \nabla\left(\hat{\boldsymbol{e}}_g \cdot \boldsymbol{X}\right) - Ra_\omega T \hat{\boldsymbol{e}}_\omega \times \left(\hat{\boldsymbol{e}}_\omega \times \boldsymbol{X}\right) + \frac{1}{Ek_\Delta} \hat{\boldsymbol{e}}_\omega \times \boldsymbol{q}\right] \tag{2}$$

$$\frac{\partial T}{\partial t} + \boldsymbol{q} \cdot \nabla T = \nabla^2 T + \dot{Q} \tag{3}$$

where provision was made in equation (3) for the possibility of internal heat generation $\dot{Q}$. For cases without heat generation the value of $\dot{Q}$ is set equal to zero. In equations (1)-(3) $\boldsymbol{q}$ is the dimensionless filtration velocity (Darcy's flux), p_r is the dimensionless reduced pressure generalised to include the constant components of the centrifugal as well as the gravity terms, T is the dimensionless temperature, $\hat{\boldsymbol{e}}_\omega$ and $\hat{\boldsymbol{e}}_g$ are unit vectors in the direction of the imposed angular velocity and gravity, respectively, and $\boldsymbol{X} = x\hat{\boldsymbol{e}}_x + y\hat{\boldsymbol{e}}_y + z\hat{\boldsymbol{e}}_z$ is the general position vector, where $\hat{\boldsymbol{e}}_x, \hat{\boldsymbol{e}}_y$ and $\hat{\boldsymbol{e}}_z$ are unit vectors in the x, y and z directions, respectively. The values l_c^2/α_{eo}, α_{eo}/l_c, and $\mu_o\alpha_{eo}/k_o$ have been used in equations (1)-(3) to scale the time, t_*, the filtration velocity, $\boldsymbol{q}_*$, and pressure, p_*, where the subscript $(\bullet)_*$ stands for identifying dimensional quantities, α_{eo} is the effective thermal diffusivity extended to account for the ratio between the heat capacity of the fluid and the effective heat capacity of the porous domain, l_c is a characteristic length scale, k_o is the permeability and μ_o is the dynamic viscosity. A linear approximation has been assumed for the relationship between the density and temperature in the form $\rho = 1 - \beta_T T$. Furthermore, by using the Boussinesq approximation (Boussinesq [11]) the density has been assumed constant everywhere, except when it appears in a body force term in Darcy's law, equation (2). In the density-temperature relationship above, the notation used was $\beta_T = \beta_* \Delta T_c$ where β_* is the thermal expansion coefficient and ΔT_c is a characteristic temperature difference used to scale the temperature in the equations. The dimensionless groups in equation (2) are the porous media gravity related Rayleigh number (Ra_g), the Rayleigh number related to the centrifugal body force (Ra_ω), and the porous media Ekman number Ek_Δ defined as

$$Ra_g = \frac{\beta_* \Delta T_c g_* l_c k_o}{\nu_o \alpha_{eo}} \quad ; \quad Ra_\omega = \frac{\beta_* \Delta T_c \omega_*^2 l_c^2 k_o}{\nu_o \alpha_{eo}} \quad ; \quad Ek_\Delta = \frac{\phi \nu_o}{2\omega_* k_o} \tag{4}$$

where ϕ is the porosity, ω_* is the angular velocity of rotation, g_* is the magnitude of the gravitational acceleration, ν_o is the kinematic viscosity, β_* is the thermal expansion coefficient and ΔT_c is a characteristic temperature difference. These dimensionless groups control the significance of the different phenomena. Therefore, the value of the Ekman number (Ek_Δ) controls the significance of the Coriolis effect and the ratio between the gravity related Rayleigh number (Ra_g) and the Rayleigh number related to the centrifugal body force (Ra_ω) controls the significance of gravity with respect to centrifugal forces as far as free convection is concerned. This ratio is $Ra_g/Ra_\omega = g_*/\omega_*^2 l_c$.

Equations (1)-(3) include three distinct mechanisms of coupling with two types of coupling being linear. First, the coupling between the pressure terms and the filtration velocity, i.e. a coupling between equation (1) and the three components of equation (2). The second linear type of coupling is due to the Coriolis acceleration acting on the filtration velocity components in the plane perpendicular to the direction of the angular velocity. The third type of coupling is non-linear as it involves the gravity, or the centrifugal buoyancy terms, in equation (2) and the energy equation (3). As $\boldsymbol{q}$ in equation (2) is dependent on temperature (T) due to the buoyancy terms, it follows that the convective term $\boldsymbol{q}\cdot\nabla T$ in the energy equation introduces the non-linearity. Therefore the coupling between the temperature, T, and the filtration velocity, $\boldsymbol{q}$, is the only source of non-linearity in the Darcy's formulation of fluid flow and heat transfer in rotating porous media.

Decoupling the equations is difficult without losing generality, although the linear types of coupling can be resolved. Resolving the Coriolis related coupling is particularly useful. In doing so a Cartesian coordinate system is used and without loss of generality one can choose $\hat{\boldsymbol{e}}_\omega = \hat{\boldsymbol{e}}_z$. A further choice is made which this time reduces the generality of the problem, stating that the gravity acceleration is assumed to be collinear with the z axis and directed downwards, i.e. $\hat{\boldsymbol{e}}_g = -\hat{\boldsymbol{e}}_z$. Some generality is lost as problems where the direction of rotation and gravity are not collinear will not be represented by the resulting equations. Nevertheless, it is of interest to demonstrate this particular choice of system as it represents a significant number of practical cases. Subject to this choice ($\hat{\boldsymbol{e}}_\omega = \hat{\boldsymbol{e}}_z$ and $\hat{\boldsymbol{e}}_g = -\hat{\boldsymbol{e}}_z$), equation (2) can be expressed in the following form (dropping the Δ from Ek_Δ and the r from p_r for convenience)

$$\left[1 + Ek^{-1}\hat{\boldsymbol{e}}_z \times\right]\boldsymbol{q} = -\left[\nabla p + Ra_\omega T\left(x\hat{\boldsymbol{e}}_x + y\hat{\boldsymbol{e}}_y\right) - Ra_g T\hat{\boldsymbol{e}}_z\right] \tag{5}$$

Classification of Rotating Flows in Porous Media

Rotating flows in porous media can be dealt with by classifying them first into three major categories, namely,

(i) *Isothermal flows in heterogeneous porous media subject to rotation.*
(ii) *Convective flows in non-isothermal homogeneous porous media subject to rotation.*
(iii) *Convective flows in non-isothermal heterogeneous porous media subject to rotation.*

This chapter is concerned with the second category. Convective flows in rotating porous media are further classified into three categories and each one of them can be separated into three cases. In order to justify this classification we proceed first by defining the phenomenon of free convection. Free convection is the phenomenon of fluid flow driven by density variations in a fluid subject to body forces. Therefore, there are two necessary conditions to be met in order to obtain convective flow; *(i) the existence of density variations within the fluid* and *(ii) the fluid must be subjected to body forces*. As density is in general a function of pressure p, temperature T, and solute concentration S (in the case of a binary mixture), i.e. $\rho \equiv \rho(p,T,S)$, and its variation with respect to pressure is much smaller than that with respect to temperature or concentration, it means that the convection can be driven either by temperature variations or by variations in solute concentration or by both. In the two latter cases

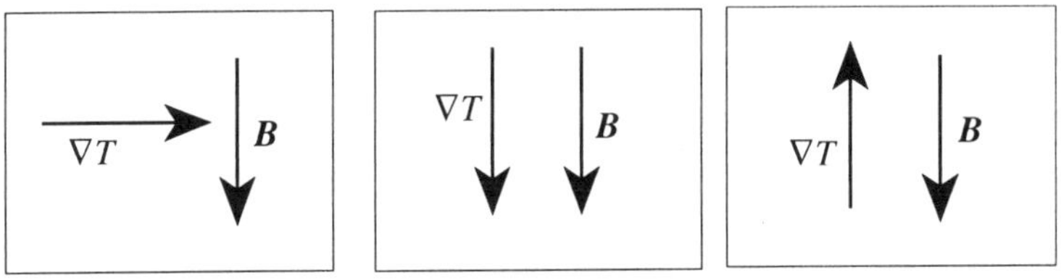

Unconditional Convection ***Conditional Convection*** ***No Convection***

(a) (b) (c)

Figure 1 The effect of the relative orientation of the temperature gradient with respect to the body force on the setup of convection

the relationship for the density should be extended to include S in the form $\rho = 1 - \beta_T T + \beta_S S$, the solute transport equation should be included in the model and equations (2) and (5) should be extended as well to account for the solute effect on the density. Gravity, centrifugal forces, electromagnetic forces (in the case of liquid metals subject to an electric field) are only examples of body forces which represent the second necessary condition for free convection to occur. Some of these body forces are constant, like gravity for example, others can vary linearly with the perpendicular distance from the axis of rotation like the centrifugal force. The lack of body forccs, for example under micro-gravity conditions pertaining in outer space, prevents the occurrence of free convection. The two conditions mentioned above are indeed necessary for free convection to occur, however they are not sufficient. The relative orientation of the density gradient with respect to the body force is an important factor for providing the sufficient conditions for convection to occur. This is shown graphically in Figure 1 for the particular case of thermal convection, where $\boldsymbol{B}$ represents the body force and $\nabla\rho = -\beta_T \nabla T$ is the direction of the density gradient. Given this basic introduction on the causes for the setup of convection the following classification of convective flows is introduced: *(i) Convection due to thermal buoyancy, (ii) Convection due to thermo-solutal buoyancy.* A third category related to convection due to solutal buoyancy alone could have been introduced but it wouldn't bring any significant contribution to *(i)* above as the results obtained for thermal buoyancy alone can be easily converted to the third case by simple analogy. This chapter focuses on reviewing the first category, while the convection due to thermo-solutal buoyancy certainly deserves a special and separate attention.

Three separate cases for each of the above categories may be considered, depending on the driving body force, i.e. *the convection driven by gravity, convection driven by the centrifugal force* and *the convection driven by both gravity and centrifugal force.*

ONSET OF FREE CONVECTION DUE TO THERMAL BUOYANCY CAUSED BY CENTRIFUGAL BODY FORCES

General Background

Considering Darcy's regime for a homogeneous porous medium subject to a centrifugal body force and neglecting gravity ($Ra_g/Ra_\omega << 1$), equations (1), (2) and (3) with $Ra_g = 0$ represent the mathematical model for this case. As the objective in the first instance is to establish the convective flow under small rotation rates, $Ek >> 1$ and as a first approximation the Coriolis effect can be neglected, i.e. $Ek \to \infty$. Following these conditions, the governing equations when heat generation is absent (i.e. $\dot{Q} = 0$) becomes

$$\nabla \cdot \boldsymbol{q} = 0 \tag{6}$$

$$\boldsymbol{q} = -\left[\nabla p - Ra_\omega T \hat{\boldsymbol{e}}_\omega \times \left(\hat{\boldsymbol{e}}_\omega \times \boldsymbol{X}\right)\right] \tag{7}$$

$$\frac{\partial T}{\partial t} + \boldsymbol{q} \cdot \nabla T = \nabla^2 T \tag{8}$$

The three cases corresponding to the relative orientation of the temperature gradient with respect to the centrifugal body force as presented in Figure 1 are considered. Case 1(a) in Figure 1 corresponding to a temperature gradient which is perpendicular to the direction of the centrifugal body force leads to unconditional convection and the solution representing this convection pattern is the objective of the investigation. Cases 1(b) and 1(c) in Figure 1, corresponding to temperature gradients collinear with the centrifugal body force, represent a stability problem. The objective is then to establish the stability condition, as well as the convection pattern, when this stability condition is not satisfied.

Temperature Gradients Perpendicular to the Centrifugal Body Force

An example of a case when the imposed temperature is perpendicular to the centrifugal body force is a rectangular porous domain which is rotating about a vertical axis and it is heated from above and cooled from below. For this case the centrifugal buoyancy term in equation (7) becomes $Ra_\omega T x \hat{\boldsymbol{e}}_x$ which leads to

$$\boldsymbol{q} = -\nabla p - Ra_\omega T x \hat{\boldsymbol{e}}_x \tag{9}$$

An analytical two-dimensional solution to this problem (see Figure 2) for a small aspect ratio of the domain was presented by Vadasz [12]. The solution to the non-linear set of partial differential equations was obtained through an asymptotic expansion in terms of a small parameter representing the aspect ratio of the domain.

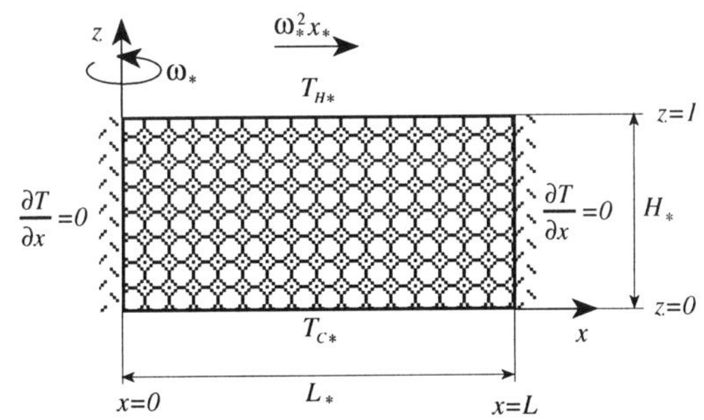

Figure 2 Rotating porous domain heated from above and cooled from below

The convection in the core region far from the domain's sidewalls was the objective of the investigation. To first-order accuracy, the heat transfer coefficient represented by the Nusselt number (Nu) was evaluated in the following form $Nu = \left[1 + (H Ra_\omega)/24 + O\left(H^2\right)\right]$, where $H = H_*/L_*$ is the aspect ratio of the domain.

A different approach was used by Vadasz [13] to solve a similar problem without the restriction of a small aspect ratio. A direct extraction and substitution of the dependent variables was found to be useful for de-coupling the non-linear **partial** differential equations, resulting in a set of independent non-linear **ordinary** differential equations which was solved analytically. To obtain an analytical solution to the non-linear convection problem we assume that the vertical component of the filtration velocity, w, and the temperature T are independent of x, i.e. $\partial w/\partial x = \partial T/\partial x = 0 \quad \forall x \in (0, L)$ and are therefore functions of z only. It is this assumption that will subsequently restrict the validity domain of the results to moderate values of Ra_ω (practically $Ra_\omega < 5$). Subject to the assumptions of two-dimensional flow $v = 0$ and $\partial(\cdot)/\partial y = 0$ and that w and T are independent of x, it became possible to decouple the governing equations and to present them in the form of the nonlinear ordinary differential equation

$$w''' - w w'' = 0 \tag{10}$$

where $(\cdot)'$ stands for $\mathrm{d}(\cdot)/\mathrm{d}z$. An interesting observation regarding equation (10) is the fact that it is identical to the Blasius equation for boundary-layer flows of pure fluids (non-porous domains) over a flat plate. To observe this one simply has to substitute $w(z) = -f(z)/2$ to obtain $2f''' + f f'' = 0$, which is the Blasius equation. Unfortunately, no further analogy to the boundary-layer flow in pure fluids exists, mainly because of the quite different boundary conditions. The solutions for the temperature T and the horizontal component of the filtration velocity u, are related to the solution of the ordinary differential equation

$$\varphi'\varphi''' - \varphi''^2 + \varphi\varphi'^2 = 0 \tag{11}$$

and are given explicitly by

$$u = x\varphi(z) \qquad \text{and} \qquad T(z) = -\frac{1}{Ra_\omega}\left[P + \varphi(z)\right] \tag{12}$$

where P is a constant defined by $P = -Ra_\omega \int_0^1 T(z)\,dz$. This relationship is a result of imposing a condition of no net flow through any vertical cross-section in the domain, namely $\int_0^1 u\,dz = 0$.

The following boundary conditions are required to the solution of equation (10) for w: $w = 0$ at $z = 0$ and $z = 1$ representing the impermeability condition at the solid boundaries and $T = 0$ at $z = 0$ and $T = 1$ at $z = 1$. Since $\partial u/\partial x = \varphi$, according to equation (12), then following the continuity equation $\varphi = -dw/dz$, and the temperature boundary conditions can be converted into conditions in terms of w by using equation (12), leading to the following complete set of boundary conditions

$$z = 0: \quad w = 0 \quad \text{and} \quad w' = P \tag{13}$$

$$z = 1: \quad w = 0 \quad \text{and} \quad w' = P + Ra_\omega \tag{14}$$

Equations (13) and (14) represent four boundary conditions, while only three are necessary to solve the third-order equation (10). The reason for the fourth boundary condition comes from the introduction of the constant P, whose value remains to be determined. This constant appears in two of the boundary conditions therefore making it impossible to obtain a third boundary condition which is independent of P. Hence, we use the two boundary conditions which are independent of P and both boundary conditions which are expressed in terms of the unknown constant P expecting the solution subject to these four conditions to determine the value of P as well. A method, similar to Blasius's method of solution, was applied to solve equation (10). Therefore, $w(z)$ was expressed as a finite power series and the objective of the solution was to determine the power series coefficients. Once the solution for $w(z)$ and P was obtained then u and T were evaluated by using $\varphi(z) = -dw/dz$ and equation (12).

Then, for presentation purposes, a stream function ψ was introduced to plot the results ($u = \partial\psi/\partial z$, $w = -\partial\psi/\partial x$). An example of the flow field represented by the stream-lines is presented in Figure 3 for $Ra_\omega = 4$ and for an aspect ratio of 3 (excluding a narrow region next to the sidewall at $x = L$). Outside this narrow region next to $x = L$ the stream-lines remain open on the right hand side and they are expected to close in the end region. Nevertheless, the stream-lines close on the left hand side, throughout the domain. The reason for this is the centrifugal acceleration,

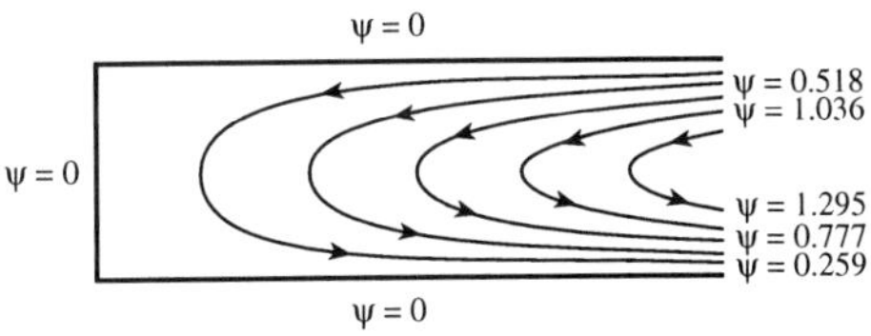

Figure 3 Graphical description of the resulting flow field; $\psi_{\min} = 0$ and $\psi_{\max} = 1.554$. The values in the figure correspond to every other stream-line

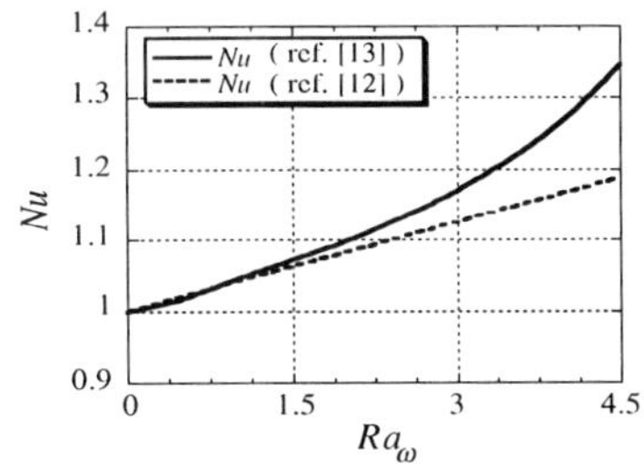

Figure 4 Graphical representation of the local Nusselt number (Nu) versus the centrifugal Rayleigh number (Ra_ω)

which causes u to vary linearly with x, thus creating (due to the continuity equation) a non-vanishing vertical component of the filtration velocity w at all values of x. The local Nusselt number Nu, representing the local vertical heat flux was evaluated as well by using its definition $Nu = \left|-\partial T/\partial z\right|_{z=0}$ and using the solution for T. A comparison between the heat flux results obtained from this solution and the results obtained by Vadasz [12] using an asymptotic method is presented in Figure 4. The two results compare well as long as Ra_ω is very small. However, for increasing values of Ra_ω the deviation from the linear relationship pertaining to the first-order asymptotic solution ($Nu = 1 + Ra_\omega/24$), according to Vadasz [12]) is evident.

Temperature Gradients Collinear with the Centrifugal Body Force

The problem of stability of free convection in a rotating porous layer when the temperature gradient is collinear with the centrifugal body force was treated by Vadasz [14, 15] for a narrow layer adjacent to the axis of rotation [14], and distant from the axis of rotation [15], respectively. The problem corresponding to the latter case is presented in Figure 5. In order to include explicitly the offset distance from the axis of rotation x_0, and to keep the coordinate system linked to the porous layer, equation (9) was presented in the form

$$\boldsymbol{q} = -\nabla p - \left[Ra_{\omega o} + Ra_\omega x\right]T\hat{\boldsymbol{e}}_x \tag{15}$$

Two centrifugal Rayleigh numbers appear in equation (15); the first one, $Ra_{\omega o} = \beta_{T*}\Delta T_c \omega_*^2 x_{0*} L_* k_* / \alpha_{eo} \nu_o$, represents the contribution of the offset distance from the rotation axis to the centrifugal buoyancy, while the second, $Ra_\omega = \beta_{T*}\Delta T_c \omega_*^2 L_*^2 k_* / \alpha_{eo} \nu_o$, represents the contribution of the horizontal location

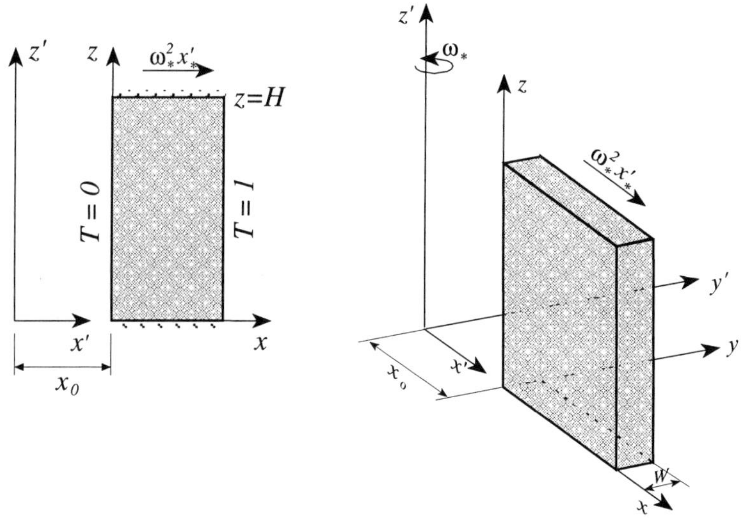

Figure 5 A rotating fluid saturated porous layer distant from the axis of rotation and subject to different temperatures at the sidewalls

within the porous layer to the centrifugal buoyancy. The ratio between the two centrifugal Rayleigh numbers is

$$\eta = \frac{Ra_\omega}{Ra_{\omega o}} = \frac{1}{x_0} \tag{16}$$

and can be introduced as a parameter in the equations by transforming equation (15) into the form

$$\boldsymbol{q} = -\nabla p - Ra_{\omega o}[1 + \eta x] T \hat{\boldsymbol{e}}_x \tag{17}$$

From equation (17) it is observed that when the porous layer is far away from the axis of rotation then $\eta << 1$ ($x_0 >> 1$) and the contribution of the term ηx is not significant, while for a layer close enough to the rotation axis $\eta >> 1$ ($x_0 << 1$) then the contribution of the first term becomes insignificant. In the first case the only controlling parameter is $Ra_{\omega o}$, while in the latter case the only controlling parameter is $Ra_\omega = \eta Ra_{\omega o}$. The flow boundary conditions are $\boldsymbol{q} \cdot \hat{\boldsymbol{e}}_n = 0$ on the boundaries, where $\hat{\boldsymbol{e}}_n$ is a unit vector normal to the boundary. These conditions stipulate that all boundaries are rigid and therefore non-permeable to the fluid flow. The thermal boundary conditions are: $T = 0$ at $x = 0$, $T = 1$ at $x = 1$ and $\nabla T \cdot \hat{\boldsymbol{e}}_n = 0$ on all other walls representing the insulation condition on these walls.

The governing equations accept a basic motionless conduction solution in the form

$$[\boldsymbol{q}_b, T_b, p_b] = \left[0,\ x,\ \left(-Ra_{\omega o}(x^2/2 + \eta x^3/3) + \text{constant}\right)\right] \tag{18}$$

The objective of the investigation was to establish the condition when the motionless solution (18) is not stable and consequently the resulting convection pattern. Therefore, a linear stability analysis was employed, representing the solution as a sum of the *basic solution* (18) and *small perturbations* in the form

$$[\boldsymbol{q},\ T,\ p] = [\boldsymbol{q}_b + \boldsymbol{q}',\ T_b + T',\ p_b + p'] \tag{19}$$

where $(\cdot)'$ stands for perturbed values. Solving the resulting linearised system for the perturbations, by assuming an expansion into normal modes in the y and z directions, i.e. $T' = A_\kappa \theta(x) \exp\left[\sigma t + i\left(\kappa_y y + \kappa_z z\right)\right]$ and using the Galerkin method to solve for $\theta(x)$ one obtains at marginal stability, i.e. for $\sigma = 0$, a homogeneous set of linear algebraic equations. This homogeneous linear system accepts a non-zero solution only for particular values of $Ra_{\omega o}$ such that its determinant vanishes. The solution of this system was evaluated up to order 7 for different values of η, representing the offset distance from the axis of rotation. However, useful information was obtained by considering the approximation at order 2. At this order the system reduces to two linear algebraic equations which lead to the characteristic values of $Ra_{\omega o}$ in the form

$$R_{0,c} = \frac{\beta\left[(1+\alpha)^2 + (4+\alpha)^2\right]}{2\alpha\left(\beta^2 - \gamma^2\right)} \pm \\ \pm \frac{\sqrt{\beta^2\left[(1+\alpha)^2 + (4+\alpha)^2\right]^2 - 4\left(\beta^2 - \gamma^2\right)(1+\alpha)^2(4+\alpha)^2}}{2\alpha\left(\beta^2 - \gamma^2\right)} \tag{20}$$

where the following notation was used

$$R_o = \frac{Ra_{\omega o}}{\pi^2}\ ;\ R = \frac{Ra_\omega}{\pi^2}\ ;\ \alpha = \frac{\kappa^2}{\pi^2}\ ;\ \beta = 1 + \frac{\eta}{2}\ ;\ \gamma^2 = \frac{256\eta^2}{81\pi^4} \tag{21}$$

and κ is the wavenumber such that $\kappa^2 = \kappa_y^2 + \kappa_z^2$ while the subscript c in equation (20) stands for identifying characteristic values. A singularity in the solution for $R_{o,c}$, corresponding to the existence of a single root for $R_{o,c}$, appears when $\beta^2 = \gamma^2$. This singularity persists at higher orders as well. Resolving for the value of η when this singularity occurs shows that it corresponds to negative η values, implying that the location of the rotation axis falls within the boundaries of the porous domain (or to the left side of the cold wall - a case of little interest due to its inherent unconditional stability). This particular case will be discussed later in this section. The critical values of the centrifugal Rayleigh number, as obtained from the solution up to order 7, are presented graphically in Figure 6(a) in terms of both $R_{o,cr}$ and R_{cr} as a function of the offset parameter η. The results presented in Figure 6 are particularly useful in order to indicate the stability criterion for all positive values of η. It can be observed from the figure that as the value of η becomes small, i.e. for a porous layer far away from the axis of rotation, the critical centrifugal Rayleigh number approaches a limiting value of $4\pi^2$. This corresponds to the critical Rayleigh number in a porous layer subject to gravity and heated from below. For high values of η it is appropriate to use the other centrifugal Rayleigh number R, instead of R_o, by introducing the relationship $R = \eta R_o$ (see equations (16) and (21)) in order to establish and present the stability criterion. It is observed from Figure 6(a) that as the value of η becomes large, i.e. for a porous layer close to the axis of rotation, the critical centrifugal Rayleigh number approaches a limiting value of $7.81\pi^2$. This corresponds to the critical Rayleigh number for the problem of a rotating layer adjacent to the axis of rotation as presented by Vadasz [14]. The stability map on the $Ra_\omega - Ra_{\omega o}$ plane is presented in Figure 6(b), showing that the plane is divided between the stable and unstable zones by the straight line $\left(Ra_{\omega,cr}/7.81\pi^2\right)+\left(Ra_{\omega o,cr}/4\pi^2\right)=1$.

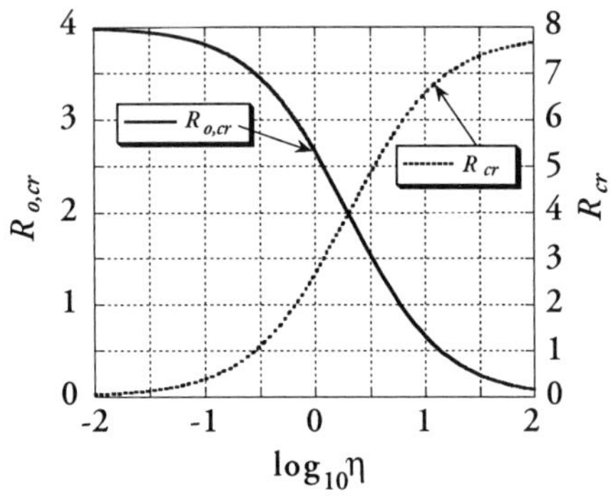

(a)

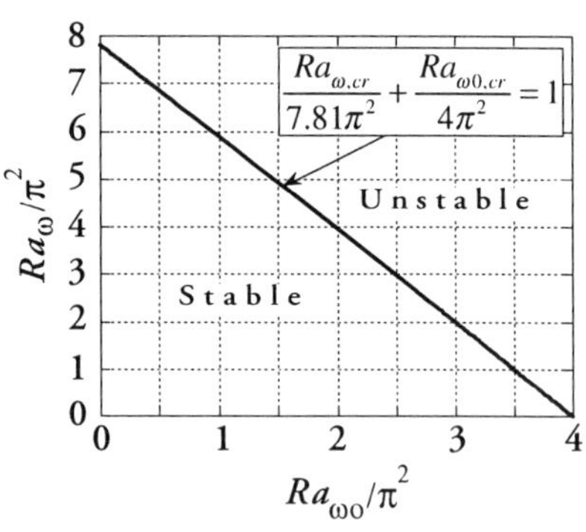

(b)

Figure 6 (a) The variation of the critical values of the centrifugal Rayleigh numbers as a function of η; (b) The stability map on the $Ra_\omega - Ra_{\omega o}$ plane showing the division of the plane into stable and unstable zones

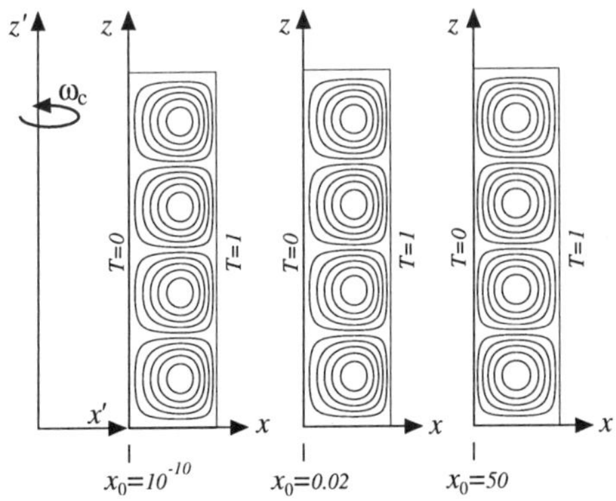

Figure 7 The convective flow field at marginal stability for three different values of x_0; 10 stream-lines equally divided between $\psi_{\min}$ and $\psi_{\max}$

The results for the convective flow field are presented graphically in Figure 7 for three different values of η corresponding to a very small offset distance from the axis of rotation, i.e. $x_0 = 10^{-10}$ $(\eta = 10^{10})$, to a large offset distance, i.e. $x_0 = 50$ $(\eta = 0.02)$, and to a moderate offset distance, i.e. $x_0 = 0.02$ $(\eta = 50)$. From the figure it can be observed that the effect of the variation of the centrifugal acceleration **within the porous layer** is definitely felt when the box is close to the axis of rotation, corresponding to an eccentric shift of the convection cells towards the sidewall at $x = 1$. However, when the layer is located far away from the axis of rotation (e.g. $x_0 = 50$) the convection cells are concentric and symmetric with respect to $x = 1/2$ as expected for a porous layer subject to gravity and heated from below (here "below" stands for the location where $x = 1$).

Although the linear stability analysis is sufficient for obtaining the stability condition of the motionless solution and the corresponding eigenfunctions describing qualitatively the convective flow, it can not provide information regarding the values of the convection amplitudes, nor regarding the average rate of heat transfer. To obtain this additional information, Vadasz and Olek [16] analysed and provided a solution to the non-linear equations by using Adomian's decomposition method to solve a system of ordinary differential equations for the evolution of the amplitudes. This system of equations was obtained by using the first three relevant Galerkin modes for the stream function and the temperature, while including the time derivative term in Darcy's equation in the form $(1/\chi)\partial \mathbf{q}/\partial t$, where $\chi = \phi Pr/Da$, and Da, Pr are the Darcy and Prandtl numbers, respectively, defined as $Da = k_o/L_*^2$ and $Pr = \nu_o/\alpha_{eo}$. Then, the following equations were obtained for the time evolution of the amplitudes X, Y, Z

$$\dot{X} = a(Y - X) \tag{22}$$

$$\dot{Y} = RX - Y - (R-1)XZ \tag{23}$$

$$\dot{Z} = 4b(XY - Z) \tag{24}$$

where a is a parameter proportional to χ, b depends on the aspect ratio, and R is a rescaled Rayleigh number. The results obtained are presented in Figure 8 in the form of projection of trajectories data points onto the $Y-X$ and $Z-X$ planes. Different transitions as the value of R varies are presented and they relate to the convective fixed point which is a stable simple node in Figure 8 (a), a stable spiral in Figures 8 (b) and (c), and loses stability via a Hopf bifurcation in Figure 8(d), where the trajectory describes a limit cycle, moving towards a chaotic solution presented in Figures 8(e) and (f). A further transition from chaos to a periodic solution was obtained at a value of R slightly above 100, which persists over a wide range of R values. This periodic solution is presented in Figures 8(g) and (h) for $R = 250$.

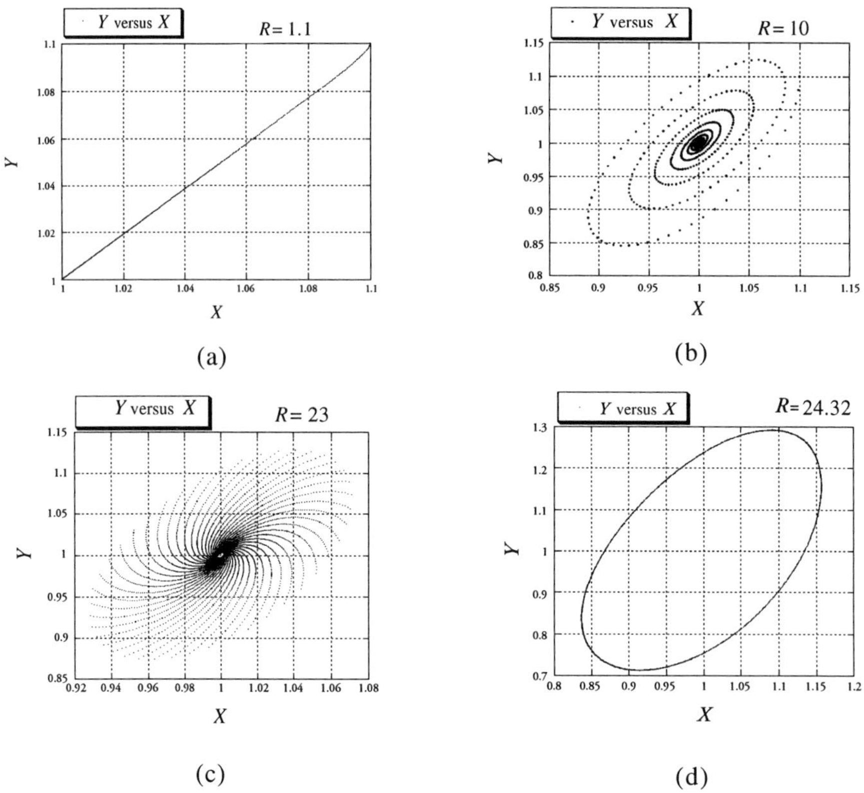

Figure 8 Different transitions in free convection in a rotating porous layer (continued next page)

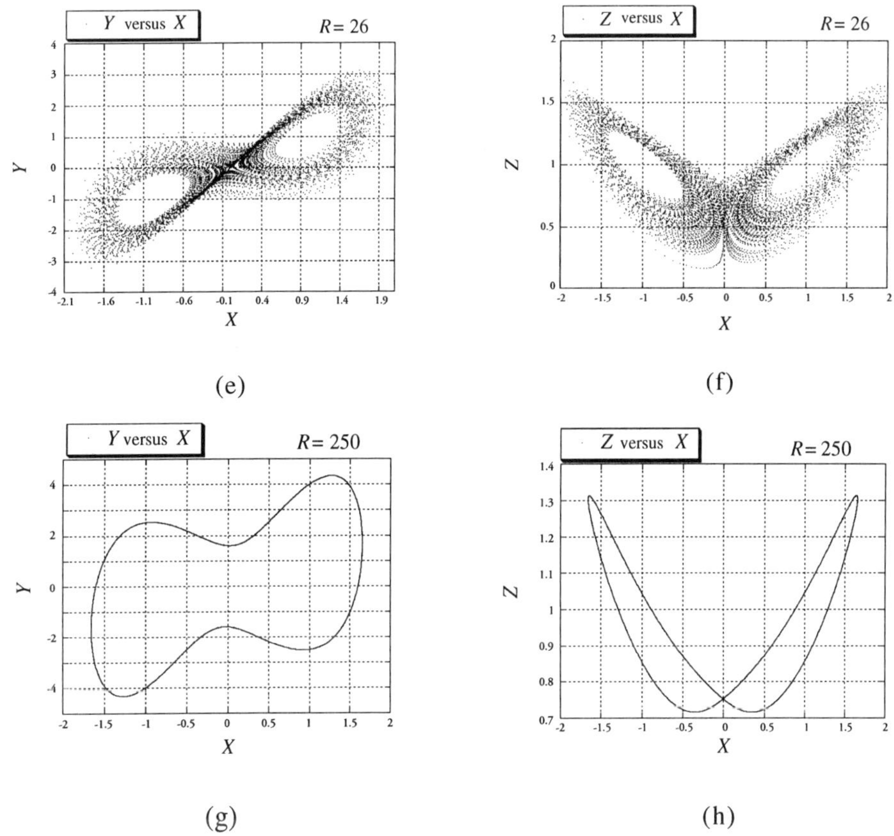

Figure 8 Different transitions in free convection in a rotating porous layer (continued from previous page)

Previously in this section (see equation (20)) a singularity in the solution was identified and associated with negative values of the offset distance from the axis of rotation. It is this resulting singularity and its consequences which were investigated by Vadasz [17] and form the objective of the following presentation. As this occurs at negative values of the offset distance from the axis of rotation it implies that the location of the rotation axis falls within the boundary of the porous domain, as presented in Figure 9. This particular axis location causes positive values of the centrifugal acceleration on the right side of the rotation axis and negative values on its left side. The rotation axis location implies that the value of x_0 is not positive. It is therefore convenient to explicitly introduce this fact in the problem formulation specifying explicitly that $x_0 = -|x_0|$. As a result, equation (17) can be expressed in the form

$$\boldsymbol{q} = -\nabla p - Ra_{\omega}\left[x - |x_0|\right]T\hat{\boldsymbol{e}}_x \tag{25}$$

The solution for this case is similar to the previous case leading to the same characteristic equation for R_c at order 2, as obtained previously in equation (20) for $R_{o,c}$ with the only difference appearing in the different definition of β and γ as follows

$$\beta = \left(\frac{1}{2} - |x_0|\right); \qquad \gamma^2 = \frac{256}{81\pi^4} \tag{26}$$

The singularity is obtained when $\beta^2 = \gamma^2$, corresponding to $\beta = \gamma$ or $\beta = -\gamma$. Since β is uniquely related to the offset distance $|x_0|$ and $\gamma = 16/9\pi^2$ is a constant, one can relate the singularity to specific values of $|x_0|$. At order 2 this corresponds to $|x_0| = 0.3199$ and $|x_0| = 0.680$. It was shown in Vadasz [17] that the second value $|x_0| = 0.680$ is the only one which has physical consequences. This value corresponds to a transition beyond which, i.e. for $|x_0| \geq 0.68$, no positive roots of R_c exist. It therefore implies an unconditional stability of the basic motionless solution for all

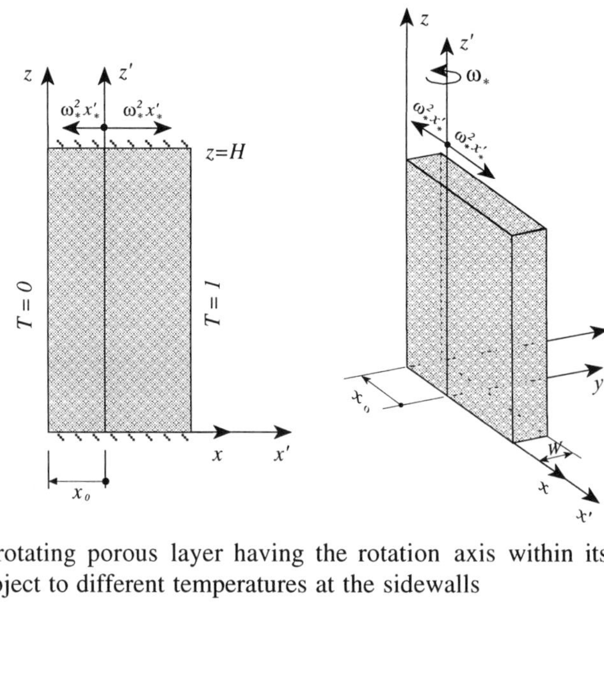

Figure 9 A rotating porous layer having the rotation axis within its boundary and subject to different temperatures at the sidewalls

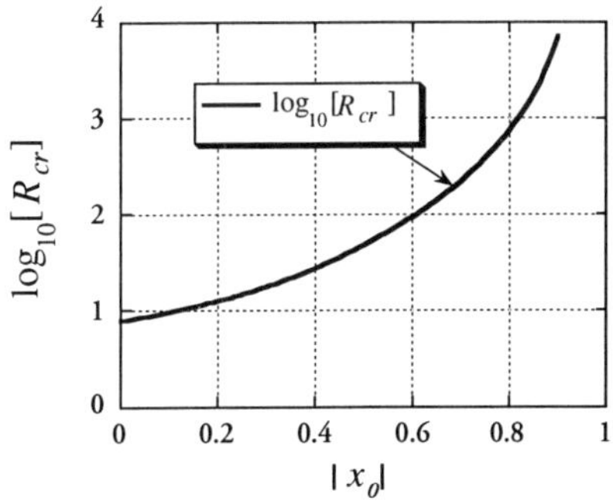

Figure 10 The variation of the critical values of the centrifugal Rayleigh number as a function of $|x_0|$

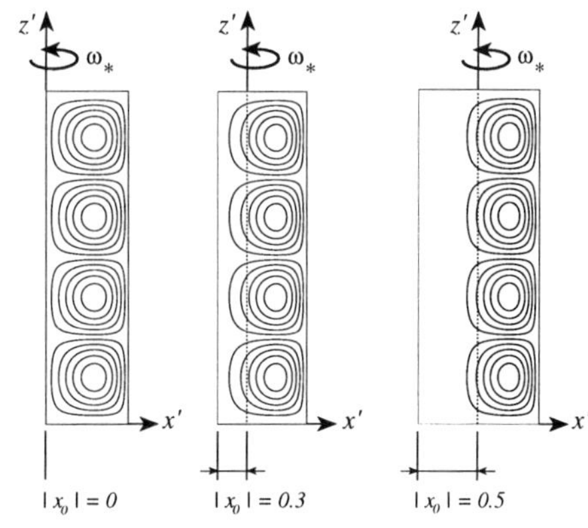

Figure 11 The convective flow field at marginal stability for three different values of $|x_0|$; 10 stream-lines equally divided between ψ_{min} and ψ_{max}

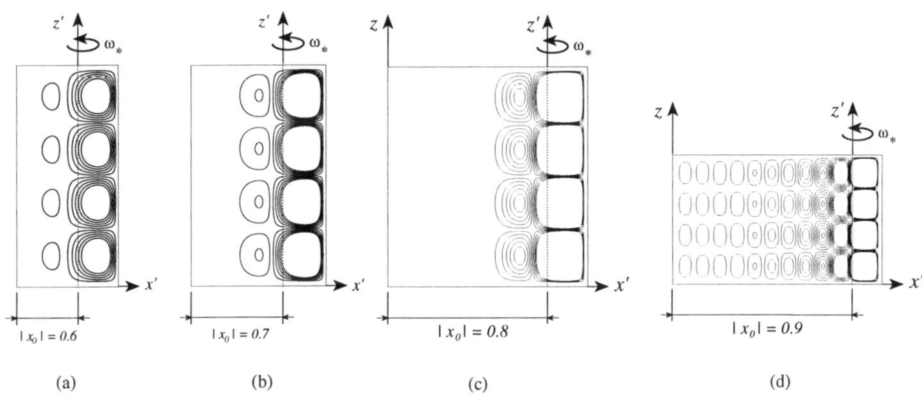

Figure 12 The convective flow field at marginal stability for four different values of $|x_0|$; 10 stream-lines equally divided between ψ_{min} and ψ_{max}

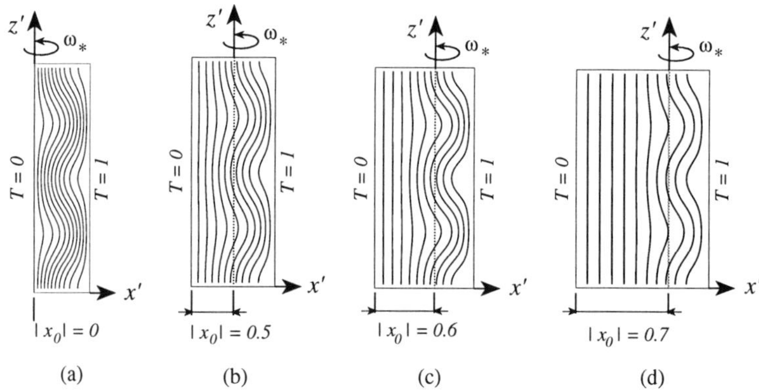

Figure 13 The convective temperature field at marginal stability for four different values of $|x_0|$; 10 isotherms equally divided between $T_{min}=0$ and $T_{max}=1$

values of R if $|x_0| \geq 0.68$. The transitional value of $|x_0|$ was investigated at higher orders showing $|x_0| \geq 0.765$ at order 3 and the value increases with increasing the order. The indications are that as the order increases, the transition value of $|x_0|$ tends towards the limiting value of 1. The result for the critical values of the centrifugal Rayleigh number expressed in terms of R_{cr} vs. $|x_0|$ are presented in Figure 10. From the figure it is clear that increasing the value of $|x_0|$ has a stabilising effect. The results for the convective flow field as obtained by Vadasz [17] are presented graphically in Figures 11 and 12 for different values of $|x_0|$. Keeping in mind that to the right of the rotation axis the centrifugal acceleration has a destabilising effect, while to its left a stabilising effect is expected, the results presented in Figure 11(b) and (c) reaffirm this expectation showing an eccentric shift of the convection cells towards the right side of the rotation axis. When the rotation axis is moved further towards the hot wall, say to $|x_0| = 0.6$ as presented in Figure 12(a), weak convection cells appear even to the left of the rotation axis. This weak convection becomes stronger as $|x_0|$ increases, as observed in Figure 12(b) for $|x_0| = 0.7$ and the formation of boundary layers associated with the primary convection cells is observed to the right of the rotation axis. These boundary layers become more significant for $|x_0| = 0.8$, as represented by sharp streamlines gradients in Figure 12(c). When $|x_0| = 0.9$, Figure 12(d) shows that the boundary layers of the primary convection are well established and the whole domain is filled with weaker secondary, tertiary and further convection cells. The results for the isotherms corresponding to values of $|x_0| = 0$, 0.5, 0.6 and 0.7 are presented in Figure 13 where the effect of moving the axis of rotation within the porous layer on the temperature is evident.

CORIOLIS EFFECT ON FREE CONVECTION DUE TO THERMAL BUOYANCY CAUSED BY CENTRIFUGAL FORCES

In the previous section, centrifugally driven free convection was discussed under conditions of small rotation rates, i.e. $Ek >> 1$, and then, as a first approximation the Coriolis effect was neglected. It is the objective of the present section to show the effect of the Coriolis acceleration on free convection even when this effect is small, i.e. $Ek >> 1$. A long rotating porous box is considered as presented in Figure 14. The possibility of internal heat generation is included but the case without heat generation, i.e. when the box is heated from above and cooled from below is dealt with separately. This case was solved by Vadasz [18] by using an asymptotic

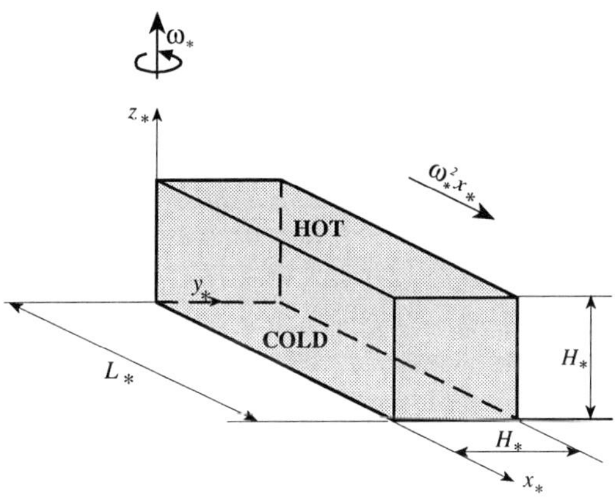

Figure 14 Three-dimensional convection in a long rotating porous box

expansion in terms of two small parameters representing the reciprocal Ekman number in porous media and the aspect ratio of the domain. Equation (5) was used with $Ra_g = 0$ and neglecting the component of the centrifugal acceleration in the y direction (a small aspect ratio). Then an expansion of the form

$$[\mathbf{q}, T, p] = \sum_{m=0}^{\infty}\sum_{n=0}^{\infty} H^m Ek^{-n}[\mathbf{q}_{mn}, T_{mn}, p_{mn}] \tag{27}$$

is introduced in the equations, where H is the aspect ratio, i.e. $H = H_*/L_*$.

To leading order, the zero powers of the aspect ratio H^m and Ek^{-n} are used, which yields

$$v_{00} = w_{00} = 0; \qquad T_{00} = z; \qquad u_{00} = -\frac{Ra_\omega x}{2}[2z-1] \tag{28}$$

This solution holds for the core region of the box. To orders 1 in Ek^{-n} and 0 in H^m the Coriolis effect is first detected in a plane perpendicular to the leading order free convection flow leading to the following analytical solution for the stream function ψ_{01} in the $y-z$ plane

$$\psi_{01} = -\frac{16Ra_\omega x}{\pi^4}\sum_{i=1}^{\infty}\sum_{j=1}^{\infty}\frac{\sin[(2i-1)\pi y]\sin[(2j-1)\pi z]}{(2i-1)(2j-1)\left[(2i-1)^2+(2j-1)^2\right]} \tag{29}$$

and the corresponding temperature solution is

$$T_{01} = -\frac{16Ra_\omega x}{\pi^5}\sum_{i=1}^{\infty}\sum_{j=1}^{\infty}\frac{\cos[(2i-1)\pi y]\sin[(2j-1)\pi z]}{(2j-1)\left[(2i-1)^2+(2j-1)^2\right]^2} \tag{30}$$

From the above solutions it was concluded that the Coriolis effect on free convection is controlled by the combined dimensionless group

$$\sigma = \frac{Ra_\omega}{Ek} = \frac{2\beta_{T*}\Delta T_c \omega_*^3 L_* H_* k_*^2}{\alpha_{eo} v_o^2 \phi} \tag{31}$$

The flow and temperature fields in the plane $y-z$, perpendicular to the leading order free convection plane as evaluated through the analytical solution are presented in Figure 15(c) in the form of stream-lines and isotherms. The single vortex in this plane is a consequence of the monotonic variation of the free convection flow field in the core region, i.e. by the sign of $\partial u_{00}/\partial z$. By extending this argument to evaluate the free convection flow and temperature field in a similar box to that shown in Figure 14 but **including internal heat generation and subject to different top and bottom boundary conditions** the following cases were considered by Vadasz [19] :

(i) a uniform rate of internal heat generation and perfectly conducting top and bottom walls, i.e. $Q=1,\ z=0: T=0$ and $z=1:\ T=0$.

(ii) a uniform rate of internal heat generation, perfectly conducting top wall and adiabatic bottom wall, i.e. $Q=1,\ z=0: \partial T/\partial z=0$ and $z=1:\ T=0$.

As a result a basic flow u_{00}, to leading order was evaluated in the form

(a) for boundary conditions - set (i)

$$u_{00} = \frac{Ra_\omega x}{2}\left[z^2 - z + \frac{1}{6}\right] \tag{32}$$

(b) for boundary conditions - set (ii)

$$u_{00} = \frac{Ra_{\omega} x}{2}\left[z^2 - \frac{1}{3}\right] \tag{33}$$

Because of the change in sign of the gradient $\partial u_{00}/\partial z$ at $z = 0.5$ in case *(i)* (see equation (32)) a double vortex secondary flow as presented in Figure 15(a) was obtained in the $y - z$ plane. Since the gradient $\partial u_{00}/\partial z$ for case *(ii)* doesn't change sign, a single vortex secondary flow is the result obtained in the $y - z$ plane for this case, as shown in Figure 15(b). A comparison of the flow direction of this vortex with

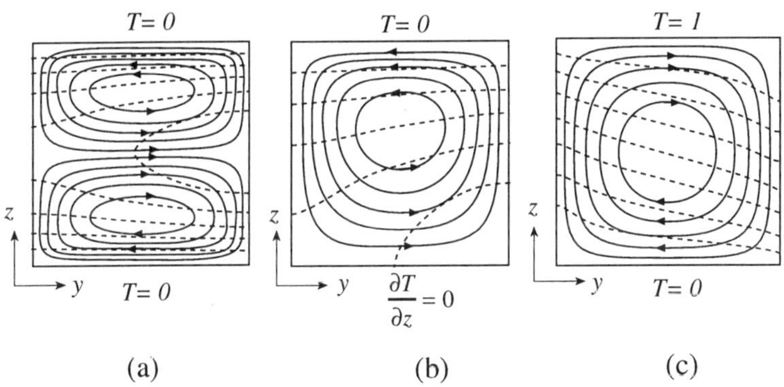

Figure 15 The flow and temperature field in the $y - z$ plane; (a) with heat generation, perfectly conducting top and bottom walls; (b) with heat generation, perfectly conducting top wall, insulated bottom wall; (c) no heat generation, heating from above

the direction of the vortex resulting from the analytical solution for the case without heat generation while heating from above (see Figure 15(c)), shows again the effect of the basic flow vertical velocity gradient $\partial u_{00}/\partial z$ on the secondary flow, i.e. a positive gradient is associated with an anti-clockwise flow and a negative gradient favours a clockwise flow. The resulting effect of these secondary flows on the temperature is presented by the dashed lines in Figure 15, representing the isotherms.

CORIOLIS EFFECT ON FREE CONVECTION DUE TO THERMAL BUOYANCY CAUSED BY GRAVITY FORCES

The problem of a rotating porous layer subject to gravity and heated from below was originally investigated by Friedrich [20] and by Patil and Vaidyanathan [5. Both studies considered a non-Darcy model which is probably subject to the limitations as given by Nield [21]. Friedrich [20] focused on the effect of Prandtl number on the convective flow resulting from a linear stability analysis as well as a non-linear numerical solution, while Patil and Vaidyanathan [5] dealt with the influence of variable viscosity on the stability condition. The latter concluded that variable viscosity has a destabilising effect while rotation has a stabilising effect. Although the non-Darcy model considered included the time derivative in the momentum equation the possibility of convection setting-in as an oscillatory instability was not explicitly investigated by Patil and Vaidyanathan [5]. It should be pointed out that for a pure fluid (non-porous domain) convection sets in as an oscillatory instability for a certain range of Prandtl number values (Chandrasekhar [22]). This possibility was explored by Friedrich [20] for a non-Darcy model and by Vadasz [23] for a Darcy model extended only to include the time derivative term. They present stability curves for both monotonic and oscillatory instability. Jou and Liaw [6, 7] investigated a similar problem of gravity driven thermal convection in a rotating porous layer subject to transient heating from below. By using a non-Darcy model, they established the stability conditions for the marginal state without considering the possibility of oscillatory convection.

An important analogy was discovered by Palm and Tyvand [8] who showed by using a Darcy model that the onset of gravity driven convection in a rotating porous layer is equivalent to the case of an anisotropic porous medium. The critical Rayleigh number was found to be

$$Ra_{g,cr} = \pi^2\left[(1+Ta)^{1/2}+1\right]^2 \tag{34}$$

where Ta is the Taylor number defined as $Ta=\left[(2\omega_* k_o)/(\phi \nu_o)\right]^2$ and the corresponding critical wave number is $\pi(1+Ta)^{1/4}$.

This result was confirmed by Vadasz [23] for a Darcy model extended to include the time derivative term, while performing a linear stability as well as a weak non-linear analysis of the problem to provide differences as well as similarities with the corresponding problem in pure fluids (non-porous domains). As such, it was found by Vadasz [23] that, in contrast to the problem in pure fluids, overstable convection in porous media at marginal stability is not limited to a particular domain of Prandtl number values (in pure fluids the necessary condition is $Pr<1$). Moreover, it was also established by Vadasz [23] that in the porous media problem the critical wave number in the plane containing the streamlines for stationary convection is not identical to the critical wave number associated with convection without rotation, and is therefore not independent of rotation, a result which is quite distinct from the corresponding pure-fluids problem. Nevertheless, it was evident that in porous media, just as in the case of pure fluids subject to rotation and heated from below, the viscosity at high rotation rates has a destabilising effect on the onset of stationary convection, i.e. the higher the viscosity the less stable is the fluid. An example of stability curves for overstable convection is presented in Figure 16 for $Ta = 5$, where κ is the wave number. The upper bound of these stability curves is represented by a stability curve corresponding to stationary convection at the same particular value of the Taylor number, while the lower bound was found to be independent of the value of Taylor number and corresponds to the stability curve for overstable convection associated with $\chi = 0$.

For overstable convection to set in at marginal stability then two conditions have to be fulfilled, *(i)* the value of Rayleigh number has to be higher than the critical Rayleigh number associated with overstable convection, and *(ii)* the critical Rayleigh number associated with overstable convection has to be greater than the critical Rayleigh number associated with stationary convection. The stability map obtained by Vadasz [23] is presented in Figure 17, which shows that the $Ta - \gamma$ $(= \chi/\pi^2)$ plane is divided by a continuous curve (almost a straight line) into two zones, one for which convection sets in as stationary, and the other where overstable convection is preferred. The dotted curve represents the case when the necessary condition *(i)* above is fulfilled but condition *(ii)* is not.

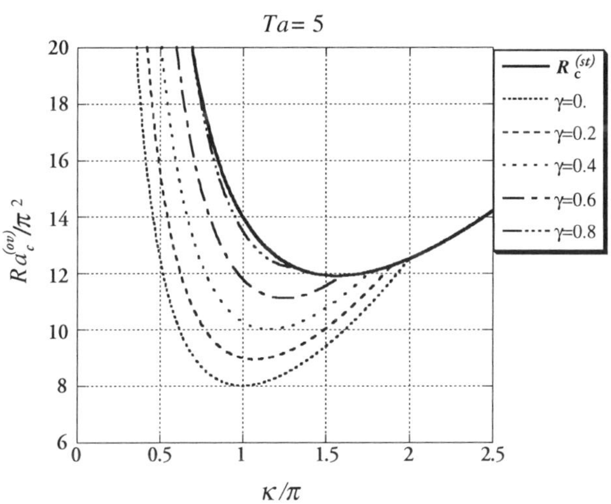

Figure 16 Stability curves for overstable gravity driven convection in a rotating porous layer heated from below ($\gamma = \chi/\pi^2$, $R = Ra_g/\pi^2$)

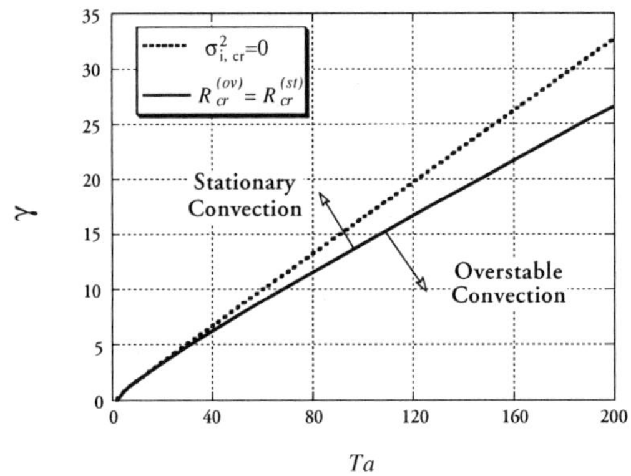

Figure 17 Stability map for gravity driven convection in a rotating porous layer heated from below ($\gamma = \chi/\pi^2$, $R = Ra_g/\pi^2$)

ONSET OF FREE CONVECTION DUE TO THERMAL BUOYANCY CAUSED BY COMBINED CENTRIFUGAL AND GRAVITY FORCES

Previous sections dealt with the onset of free convection due to thermal buoyancy caused by centrifugal or gravity body forces separately. Consequently the effect of the centrifugal body force was neglected when dealing with gravity driven convection, i.e. $Ra_\omega = 0$. Similarly, the effect of gravity was neglected when dealing with centrifugally driven convection, i.e. $Ra_g = 0$. For the latter assumption to be valid the following condition has to be satisfied: $Ra_g/Ra_\omega = g_*/\omega_*^2 L_* << 1$. It is of interest to investigate the case when $Ra_g \sim Ra_\omega$ and both centrifugal and gravity forces are of the same order of magnitude. Vadasz and Govender [24, 25] presented an investigation of cases of combined gravity and centrifugally driven free convection, while including the effect of gravity in a direction perpendicular to the centrifugal force. Figure 5 is still applicable to the present problem but with the slight modification of drawing the gravity acceleration g_* in the negative z direction. The notation remains the same and equation (17) becomes

$$\boldsymbol{q} = -\nabla p - Ra_{\omega o}[1+\eta x]T\hat{\boldsymbol{e}}_x + Ra_g T\hat{\boldsymbol{e}}_z \tag{35}$$

where $\eta = 1/x_0 = Ra_\omega/Ra_{\omega o}$ represents the reciprocal of the offset distance from the axis of rotation. The approach being the same, the solution is expressed as a sum of a basic solution and small perturbations as presented in equation (19). However, because of the presence of the gravity component in equation (35) a motionless conduction solution is not possible any more. Therefore, the basic solution far from the top and bottom walls is obtained in the form

$$\begin{aligned} &u_b = v_b = 0; \quad w_b = Ra_g(x-1/2); \qquad T_b = x \\ &p_b = (1/2)Ra_g z - Ra_{\omega o}\left[x^2/2 + \eta x^3/3\right] + \text{constant} \end{aligned} \tag{36}$$

Substituting this basic solution into the governing equations and linearising the result by neglecting terms which include products of perturbations which are small, yields a set of partial differential equations for the perturbations. Assuming an expansion into normal modes in the y and z directions in the form

$$T' = A_\kappa \theta(x)\exp\left[\sigma t + i(\kappa_y y + \kappa_z z)\right] \tag{37}$$

where κ_y and κ_z are the wave numbers in the y and z directions, respectively, i.e. $\kappa^2 = \kappa_y^2 + \kappa_z^2$, and using the Galerkin method a set of linear algebraic equations is obtained for marginal stability (i.e. for $\sigma = 0$). A particular case of interest is the configuration when the layer is placed far away from the axis of rotation, i.e. when the length of the layer L_* is much smaller than the offset distance from the rotation axis x_{0*}. Therefore for $x_0 = (x_{0*}/L_*) \to \infty$, or $\eta \to 0$, the contribution of the term ηx in equation (35) is not significant. Substitution of this limit into the set of linear equations obtained at marginal stability and solving the system at the second-order, i.e. $M = 2$, yields a quadratic equation for the characteristic values of $Ra_{\omega o}$. This equation has no real solutions for values of $\alpha = \kappa_z^2/\pi^2$ beyond a transitional value $\alpha_{tr} = (27\pi^3/16 Ra_g)^2$. This transitional value was evaluated at higher orders as well, showing that for $M = 10$ the transitional value varies very little with Ra_g beyond a certain Ra_g value around 50π. The critical values of $Ra_{\omega o}$ were evaluated for different values of R_g $(= Ra_g/\pi)$ and the corresponding two-dimensional convection solutions in terms of streamlines are presented graphically for the odd modes in Figure 18(a), showing the perturbation solutions in the $x - z$ plane as skewed convection cells when compared with the case without gravity (Figure 7). The corresponding convection solutions for the even modes are presented in Figure 18(b), where it is evident that the centrifugal effect is felt predominantly in the central region of the layer while the downwards and upwards basic gravity driven convection persists along the left and right boundaries, respectively, although not in straight lines. Beyond the transition value of α, the basic gravity driven convective flow (equation (36)) is unconditionally stable. These results were shown to have an analogy with the problem of gravity driven convection in a non-rotating, inclined porous layer (Govender and Vadasz [26]).

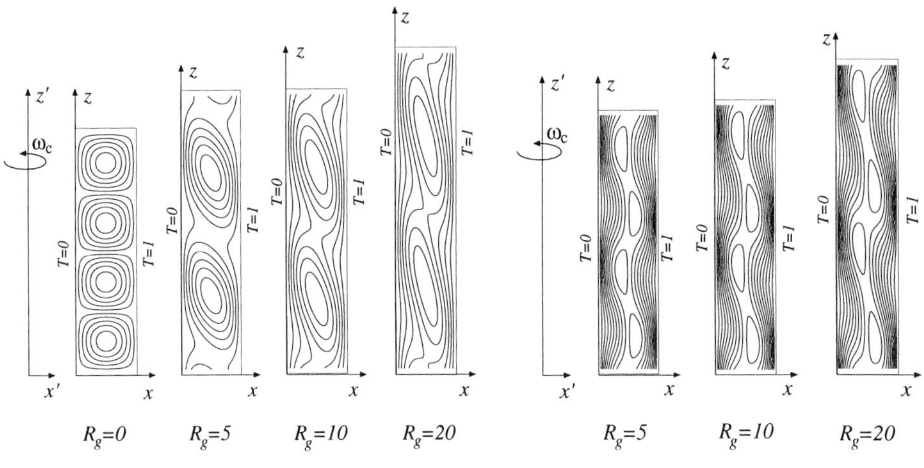

Figure 18 The convective flow field (streamlines) at marginal stability for different values of R_g $(= Ra_g/\pi)$; (a) the odd modes. (b) the even modes

The theoretical results were partially confirmed through an experimental setup using the Hele-Shaw cell analogy by Vadasz and Heerah [27], although a taller cell was recommended in order to provide a complete confirmation of the results.

When the layer is placed an arbitrary finite distance from the axis of rotation no real solutions exist for the characteristic values of $Ra_{\omega o}$ corresponding to any values of α other than $\alpha = 0$, which indicates that $\kappa_z = 0$. Therefore the presence of gravity in this case has no other role but to exclude the vertical modes of convection. The critical centrifugal Rayleigh numbers and the corresponding critical wave numbers are the same as in the corresponding case without gravity as presented in the section dealing with centrifugally driven convection subject to temperature gradients collinear with the centrifugal force. However, the eigenfunctions representing the convection pattern are essentially different as they exclude the vertical modes replacing them with a corresponding horizontal mode in the y direction. Therefore, a cellular convection in the $x - y$ plane is superimposed on the basic convection in the $x - z$ plane.

CONCLUDING REMARKS

The effect of rotation was shown to have a significant impact on free convection in porous media. The rotation may affect the flow through two distinct mechanisms, namely thermal buoyancy caused by centrifugal forces and the Coriolis force (or a combination of both). Since free convection may be driven also by the gravity force and the orientation of the buoyancy force with respect to the imposed thermal gradient has a distinctive impact on the resulting convection, a significant number of combinations of different cases is typical to the investigation of the rotation effects in non-isothermal systems. Results pertaining to some of these cases have been presented, however the lack of significant experimental confirmation is particularly noticed. It is therefore recommended to expand the effort of getting experimental confirmation of theoretical results.

REFERENCES

1. O. A. Plumb, Heat transfer during unsaturated flow in porous media, *Convective Heat and Mass Transfer in Porous Media*, ed. S. Kakaç, B. Kilkis, F.A. Kulacki & F. Arniç, Kluwer Acad. Publ., Dordrecht, pp.435-464 (1991).
2. D. A. Nield & A. Bejan, *Convection in Porous Media*, Springer Verlag, New York, (1992).
3. A. Bejan, *Convection Heat Transfer*, 2nd ed., Wiley, New York (1995).
4. N. Rudraiah, I. S. Shivakumara & R. Friedrich, The effect of rotation on linear and non-linear double-diffusive convection in a sparsely packed porous medium, *Int. J. Heat Mass Transfer*, **29**, 1301-1317 (1986).
5. P. R. Patil & G. Vaidyanathan, On setting up of convection currents in a rotating porous medium under the influence of variable viscosity, *Int. J. Engineering Science*, **21**, 123-130 (1983).
6. J. J. Jou & J.S. Liaw, Transient thermal convection in a rotating porous medium confined between two rigid boundaries, *Int. Commun. Heat Mass Transfer*, **14**, 147-153 (1987).
7. J. J. Jou & J.S. Liaw, Thermal convection in a porous medium subject to transient heating and rotation, *Int.J.Heat Mass Transfer*, **30**, 208-211 (1987).
8. E. Palm & A. Tyvand, Thermal convection in a rotating porous layer, *J. Appl. Math. Phys. (ZAMP)*, **35**, 122-123 (1984).
9. D.A. Nield, The stability of convective flows in porous media, *Convective Heat and Mass Transfer in Porous Media*, ed. S. Kakaç, B. Kilkis, F.A. Kulacki & F. Arniç, Kluwer Acad. Publ., Dordrecht, pp.79-122 (1991).
10. J. Bear, *Dynamics of Fluids in Porous Media*, Elsevier, New York, pp.131-132 (1972).
11. J. Boussinesq, *Theorie Analitique de la Chaleur*, **2**, Gauthier-Villars, Paris, p.172 (1903).
12. P. Vadasz, Natural convection in porous media induced by the centrifugal body force - the solution for small aspect ratio, *J. Energy Resources Technology*, **114**, 250-254 (1992).
13. P. Vadasz, Centrifugally generated free convection in a rotating porous box, *Int. J. Heat Mass Transfer*, **37**, 2399-2404 (1994).
14. P. Vadasz, Stability of free convection in a narrow porous layer subject to rotation, *Int. Commun. Heat Mass Transfer*, **21**, 881-890 (1994).

15. P. Vadasz, Stability of free convection in a rotating porous layer distant from the axis of rotation, *Transport in Porous Media*, **23**, 153-173 (1996).
16. P. Vadasz & S. Olek, Transitions and chaos for free convection in a rotating porous layer, submitted for publication (1997).
17. P. Vadasz, Convection and stability in a rotating porous layer with alternating direction of the centrifugal body force, *Int. J. Heat Mass Transfer*, **39**, 1639-1647 (1996).
18. P. Vadasz, Three-dimensional free convection in a long rotating porous box: analytical solution, *J. Heat Transfer,* **115**, 639-644 (1993).
19. P. Vadasz, Coriolis effect on free convection in a long rotating porous box subject to uniform heat generation, *Int. J. Heat Mass Transfer*, **38**, 2011-2018 (1995).
20. R. Friedrich, The effect of Prandtl number on the cellular convection in a rotating fluid saturated porous medium, (in German), *ZAMM*, **63**, 246-249 (1983).
21. D.A. Nield, The limitations of the Brinkman-Forchheimer equation in modeling flow in a saturated porous medium and at an interface, *Int. J. Heat and Fluid Flow*, **12**, 269-272 (1991).
22. S. Chandrasekhar, *Hydrodynamic and Hydromagnetic Stability*, Oxford Univ. Press, (1961).
23. P. Vadasz, Coriolis effect on gravity driven convection in a rotating porous layer heated from below, submitted for publication (1997).
24. P. Vadasz & S. Govender, Two-dimensional convection induced by gravity and centrifugal forces in a rotating porous layer far away from the axis of rotation, *Int. J. Rotating Machinery*, in press.
25. P. Vadasz & S. Govender, Convection induced by gravity and centrifugal forces in a rotating porous layer distant from the axis of rotation, manuscript in preparation.
26. S. Govender, & P. Vadasz, Centrifugal and gravity driven convection in rotating porous media - an analogy with the inclined porous layer, *Proceedings of the 30th National Heat Transfer Conference*, **7**, Portland, Oregon, U.S.A., pp.93-98 (1995).
27. P. Vadasz & A. Heerah, Experimental confirmation and analytical results of centrifugally driven free convection in porous media, *J. Porous Media*, in press.

NON-DARCIAN EFFECTS IN CONFINED FORCED CONVECTIVE FLOWS

K. VAFAI[1] and A. AMIRI[2]

[1]Department of Mechanical Engineering, The Ohio State University, Columbus, Ohio, U.S.A 43210
[2]Department of Mechanical Engineering, The United Arab Emirates University, Al-Ain, U.A.E.

INTRODUCTION

Transport phenomena in forced convective fluid flow through confined porous media has been of continuing interest to researchers due to the increasing pool of industrial applications. Two applications are of particular interest to industry and academia: the packed beds which are widely considered for means of thermal energy transport and storage. Second, the augmentation of heat transfer associated with high heat flux applications and cryogenic cooling have shown promising results through the use of porous insertions with high thermal conductivity. Despite that cryogen are considered ineffective energy carriers, Kuzay [1] reported that insertion of porous matrices made of silicon or silicon carbide may significantly increase a system capability to transfer heat in cryogen applications. These types of applications have polarized a substantial amount of research work in exploring the dynamic and thermal behaviors of porous media in general. Rigorous models have been reported on the simulation of the momentum and energy transport. The current review centers on the established work in the field of forced convective flows through a confined porous structure. In particular, disordered porous structures will be examined with much emphasis being placed on packed beds and porous insertions utilized for high heat flux applications. However, some of the most pertinent studies, such as external flows in the presence of a porous medium, will also be discussed on a needed basis to consolidate this review. The review begins with highlighting the local volume-averaging method for formulating the governing equations. The evolution of the momentum equation will then be discussed in some detail and the associated phenomena related to momentum transport in porous media will be highlighted. For the analysis of the macroscopic heat transfer, the local volume-averaged quantities, such as the effective thermal conductivity and the thermal dispersion effect, will be highlighted. After that the concept of local thermal equilibrium will be introduced followed by the formulation of the two-energy equation model, or simply the two-equation model, which accounts for the local temperature variation between the fluid and the solid phases. We end by examining the convective mode of heat transfer which emerges via applying the two-equation model. The review will be documented by citing the most pertinent studies for each subject.

LOCAL VOLUME AVERAGING METHOD

The complexity of a porous structure hampers a general formulation of the detailed 'microscopic' transport phenomena. Therefore, the point transport equations are commonly integrated over a representative elementary volume, which accommodates the fluid and the solid phases in a porous structure. This technique is known as the volume-averaging method and is widely considered in formulating the transport processes in porous media. Moreover, the derivations of the governing equations are quite systematic and the work by Whitaker [2] is often cited in this regard. However, this technique requires information on numerous empirical transport coefficients that are supplemented by well-established experimental findings and analytical solutions for some rather simplified cases. A more systematic approach of determining such quantities is of great challenge to future research. Since such coefficients may vary by orders of magnitude with the span of operating conditions, it is instructive to summarize the assumptions on which the current review is accomplished:

(i) The solid matrix is incompressible and free of chemical reaction.
(ii) Gravity effects are negligible.
(iii) Two-dimensional model is considered.
(iv) The thermophysical properties are assumed constant.
(v) Viscous dissipation and compressional work are negligible.
(vi) Natural convection and radiative heat transfer modes are ignored.

GOVERNING EQUATIONS

It would be appropriate here to define two characteristic parameters;

particle Reynolds number $$Re_p = u_D d_p / \nu_f \tag{1}$$

and Darcy number $$Da = K_\infty / D_h^2 \tag{2}$$

where u_D is the Darcian velocity in the axial direction, d_p is the solid particle diameter, ν_f is the fluid kinematic viscosity, K_∞ is the free stream permeability and D_h is the hydraulic diameter of the bed.

A large number of studies have been reported on fluid flow and heat transfer through confined porous media. The familiar Darcy's law was generally adopted to describe the fluid motion in porous media. However, the slug flow model may only predict fair to acceptable results at very low flow rates. The Darcy model tends to break down in the range of $1 < Re_p < 10$ and becomes invalid for $Re_p \geq 10$. The creeping non-linearity into the fluid motion is unlikely to be due to transition from laminar to turbulent, as argued by Dybbs and Edwards [3]. Instead, it is attributed to the appreciated increase in the inertial resistance, offered by the solid matrix, as compared to the friction resistance that merely constitutes Darcy's law [4]. However, there are differing views on the transitional aspects in porous media. In addition, Darcy's law does not take into consideration the presence of a rigid wall next to a porous medium. The deviations from the Darcy regime introduced by the inertial and wall viscous forces are generally referred to as the non-Darcian effects. Vafai and Tien [5] have used the local volume-averaging method to present the macroscopic governing equations associated with transport processes in porous media as follows:

$$\nabla . < \mathbf{v} > = 0 \tag{3}$$

$$\frac{\rho_f}{\varepsilon}\left[\frac{\partial < \mathbf{v} >}{\partial t} + < (\mathbf{v}.\nabla)\mathbf{v} >\right] = -\nabla < P >^f + \tilde{\mu}_f \nabla^2 < \mathbf{v} >$$

$$-\frac{\mu_f}{K} < \mathbf{v} > - \frac{\rho F \varepsilon}{\sqrt{K}}[< \mathbf{v} > . < \mathbf{v} >]\mathbf{J} \tag{4}$$

$$(\rho c)_m \frac{\partial < T >}{\partial t} + (\rho c)_f < \mathbf{v} > . \nabla < T > = \nabla . (k_m \nabla < T >) \tag{5}$$

where $(\rho c)_m = \varepsilon(\rho c)_f + (1-\varepsilon)(\rho c)_s$, $(k)_m = \varepsilon k_f + (1-\varepsilon)k_s$, $\mathbf{v}$ is the velocity vector, ε is the porosity, F is the geometric function, K is the permeability, μ_f is the fluid dynamic viscosity, $\tilde{\mu}_f$ is the effective dynamic viscosity, $\mathbf{J} = \mathbf{v}_p / |\mathbf{v}_p|$ is the unit vector oriented along the pore velocity vector $\mathbf{v}_p$, $< P >^f$ the average pressure read off a pressure gage, ρ_f is the fluid density, ρ_s is the solid density, c the specific heat at constant pressure, k_f and k_s are the thermal conductivities of the fluid and solid phase, respectively, and k_m is the porous medium effective thermal conductivity. The flow is driven by the imposed pressure difference in the axial direction and the energy equation is applicable when the local thermal equilibrium assumption is imposed. Moreover, modeling of k_m via superposition represents one of the most common forms of k_m. However, many other empirical correlations have also been reported for the modeling of k_m and an extensive review of the literature on the effective thermal conductivity is given by Kaviany [6]. However, it should be pointed out that k_m does not take into account the dispersion effect, as will be discussed later. Tien and Vafai [7] presented a statistical approach to arrive at the upper and lower bound of k_m for micro-spheres and fibrous insulation.

The effective viscosity is approximately equal to the dynamic viscosity of the fluid. The generalized form of the momentum equation, equation (4), is also known as the Forchheimer-Brinkman extended-Darcy model. The momentum equation allows for a smooth transition between fluid flow through a porous medium and the Navier-Stokes equations in a space void of porous medium as $K \to \infty$. It is worth noting that the term $(F\varepsilon)$ that appears in the inertial term as proposed by Vafai and Tien [5] is sometime represented as (F) by other investigators. The first term on the right hand side of the momentum equation represents the frictional resistance due to the presence of the porous structure (Darcy resistance), whereas the second term accounts for the inertial effects and is also referred to as the Frochheimer's term. The quadratic term resembles the additional pressure drop introduced by flow passing over a blunt body causing a wake zone behind the body. The third term on the right hand side (Brinkman term) is quite important within an interfacial region. An order of magnitude analysis of equation (4) reveals that the momentum boundary-layer in a porous medium is of the order of $\sqrt{K/\varepsilon}$ [5]. This indicates that the viscous boundary effect is very thin for most practical applications making experimental visualization very difficult. Despite its limited contribution to the equation of motion, especially for large systems, the inclusion of the viscous term may greatly affect the heat transfer calculations. In addition, Vafai and Tien [8] have shown that the convective term $<(\mathbf{v}.\nabla)\mathbf{v}>$ responsible for boundary-layer growth is significant over a length scale of the order of Ku_D/ν_f. This quantity is very small and negligible for most engineering applications. Moreover, Kaviany [9] numerically investigated the flow between parallel plates and reported that the entrance length decreases linearly with the decrease in the Darcy number. However, the investigation did not incorporate the inertia effect into the momentum equation.

The permeability of a porous structure K and the geometric function F in the momentum equation are inherently tied to the porous structure and are generally based on experimental findings such that no universal representation does exist. In addition, Beavers et al. [10] have pointed out that the size of the porous bed could have an effect on the value of F, and have reported suggested a geometric function for fibrous metals given by

$$F = 0.55\left(1 - 0.55\frac{d_p}{D_h}\right) \tag{6}$$

Sheidegger [11] listed examples of average bulk porosities and permeabilities for various porous materials, while Dullien [12] reviewed the different techniques reported on measuring the porosity. For a randomly packed bed of spheres such transport coefficients was reported by Ergun [13] and could be expressed in terms of porosity ε and particle diameter d_p as follows:

$$K = \frac{\varepsilon^3 d_p^2}{150(1-\varepsilon)^2} \tag{7}$$

$$F = \frac{1.75}{\sqrt{150\varepsilon^3}} \tag{8}$$

The Carman-Kozeny equation is also used in some capacity to predict the permeability of packed beds, however, its applicability range is quite limited. The inertial effect retards the momentum transport causing less energy to be carried away by convection compared to the case where the inertial effects are neglected. As a result, the inertial effect slows the thermal penetration depth. Moreover, the inclusion of the impermeable boundary effect reduces the velocity in the momentum boundary-layer, thus reducing the heat transfer rate. However, the impermeable wall causes redistribution of porosity near the wall, which leads to couple of important phenomena, namely porosity maldistribution and channeling effect. The experimental investigations of Roblee et al. [14] and Benenati and Brosilow [15] show a distinct porosity variation in packed beds. Their results show a high porosity region (~$0.5d_p$ from the wall), which then decays in an oscillatory fashion as moving away from the wall until an asymptotic free stream porosity value is reached at about $5d_p$ from the wall.

Although the actual porosity variation along the normal coordinate has been successfully modeled [16,17], it is a common practice to consider an exponential decaying function since the oscillatory behavior is considered to be of a secondary effect. A typical form of such an exponential decay is given by

$$\varepsilon = \varepsilon_\infty \left[1 + a\exp(by/d_p)\right] \tag{9}$$

The free stream porosity value ε_∞ is a function of the bed-to-particle D_h/d_p ratio and ranges between 0.259 and 0.43, depending on the packing arrangement. In addition, for uniform solid-sphere particles the values of $a = 1.7$ and $b = 6$ were found to yield close approximation to the experimental data. The irregular porosity distribution causes the velocity to reach its peak near the wall and not at the centerline, thus leading to a phenomenon known as the channeling effect. A number of experimental investigations, such as Schwartz and Smith [18] and Schertz and Bischoff [19], have reported on the occurrence of such a maximum velocity close to the wall, which enhances the heat transfer rate. In addition, the overall effect of the rigid wall was found to augment heat transfer due to the large velocity magnitude close to the wall.

Chandrasekhara and Vortmeyer [20] have numerically solved for the velocity field in isothermal packed beds. Reasonable agreement was reported on the comparison of the numerical results against the above mentioned experimental results [18,19]. Closed form solutions for fully developed flows through a packed bed of spheres have been obtained by Vafai [21,22]. Moreover, Vafai [22] introduced the concept of the triple momentum boundary-layer to incorporate the channeling effect in a variable porous medium. Vafai and Kim [23] have also obtained an analytical solution for fully developed flows through fibrous insertions. A blend of experimental and numerical work were reported by Vafai et al. [24], Poulikakos and Renken [25], Renken and Poulikakos [26] and Hunt and Tien [27]. Poulikakos and Renken [25] did not incorporate the porosity variation effect into the modeling of thermal conductivity. Although close agreement with experimental data was reported, such assumption is unlikely to be valid unless the individual phases possess effective thermal conductivity of the same magnitude. Vafai and Thiyagaragja [28] presented analytical solutions for three different interface cases covering a broad range of effective Prandtl numbers. Their analytical results were in excellent agreement with their numerical findings. It is worth noting that much of the reported investigations mentioned above were performed using the steady state forms of the governing equations. Amiri and Vafai [29] have examined the estimated error in dropping the inertial effects and/or the viscous effect under steady condition. The results were based on Nusselt number predictions and were presented in terms of error maps for different α_s/α_f values. Such error maps provide a simple design scheme for estimating the errors introduced due to incorporating simplified assumptions. Transient non-Darcian effects have not received much attention in the literature. To the authors' knowledge, no studies have been reported on transient, non-Darcian, forced convective flows in confined porous media. Recently, Amiri and Vafai [30] have performed a transient investigation to explore the temporal non-Darcian effect. The error maps indicated that the non-Darcian effects are significant under transient and steady state conditions.

THERMAL DISPERSION EFFECT

An important phenomenon in porous media relates to the enhanced mixing of the local fluid stream due to the torturous path offered by the complex geometry of a porous medium. This phenomenon is classified in the literature as thermal dispersion effect. Most of the existing models view the effect of thermal dispersion as a diffusive term added to the fluid stagnant thermal conductivity as follows:

$$k_{feff} = k_o + k_d \tag{10}$$

where k_o and k_d are the stagnant and the dispersion thermal conductivities, respectively. Typically, the stagnant component is the product of the phase fractions and the individual thermal conductivities of the phases. The solid phase consists merely of the stagnant component since it is stationary. The pioneering work of Yagi and his coworkers [31,32] is widely considered in this regard in addition to other more recent experimental investigations reported by Gunn and Khalid [33] and Vortmeyer [34]. The bulk of the existing experimental studies on determining dispersion conductivity were conducted using a cylindrical packed bed with air as the working fluid. Moreover, such models were found to best fit a linear function of either the particle Reynolds number Re_p or the particle Peclet number Pe_p. Li and Finlayson [35] criticized a significant number of the earlier conducted experimental work due to their failure to incorporate the entrance length effect on the measurements of dispersion conductivity. Han et al. [36] suggested such measurements to occur after an effective length L_e defined as

$$L_e \geq 0.3(d_p/\varepsilon)Pe_p \tag{11}$$

where $Pe_p = Re_p \Pr$. Yagi et al. [32] proposed the longitudinal and radial dispersion conductivity for air flowing through a packed column of spheres as follows:

$$\frac{k_d)_x}{k_f} = \delta_x Pe_p \tag{12}$$

$$\frac{k_d)_r}{k_f} = \delta_r Pe_p \tag{13}$$

where the longitudinal dispersion coefficient $\delta_x = 0.7 \sim 0.8$ and the transverse dispersion coefficient $\delta_r = 0.1 \sim 0.3$. However, Wakao and Kaguei [37] recommended a value of 0.5 and 0.1 for the longitudinal and the transverse dispersion coefficients, respectively. Levec and Carbonell [38] used water instead of air for their experimental investigation on packed bed and reported that the dispersion coefficient is a non-linear function of the Pe_p given by

$$\frac{k_d)_x}{k_f} = 0.357 Pe_p^{\ 1.256} \tag{14}$$

$$\frac{k_d)_r}{k_f} = 0.39 Pe_p^{\ 0.683} \tag{15}$$

Cheng and Hsu [39] introduced a theoretical expression for the transverse dispersion conductivity in a packed bed on the basis of the mixing length theory that may be written as

$$\frac{k_d)_r}{k_f} = \eta_\infty u Pe_p \ell \tag{16}$$

where η_∞ is the free stream dispersion coefficient and ℓ is the Van Driest wall function defined as

$$\ell / d_p = 1 - e^{-y/\omega d_p} \tag{17}$$

where ω is an empirical constant. Hunt and Tien [40] have experimentally investigated the dispersion effect in a fibrous medium and proposed an expression for the dispersion conductivity to match their experimental results which may be expressed as follows:

$$\frac{k_d)_r}{k_f} = \gamma \sqrt{Da}\, Pe_p D_h / (\varepsilon d_p) \tag{18}$$

where γ is the dispersion coefficient. What is more, Koch and Brady [41] presented a semi analytical solution for the longitudinal and transverse dispersion conductivity in a randomly arranged sphere-packed bed. Their predictions were found to be in good agreement with experimental results. There are many different techniques associated with measuring the dispersion conductivity in both longitudinal and transverse directions. Cheng et al. [42] have surveyed some of the most prominent methods in this regard. Due to the wide pool of empirical correlations, caution must be exercised in choosing an appropriate one since some of these correlations are tailored to specific problems. In addition, incorporating empirical parameters, such as η_∞ and γ, demands coupling the analytical models with experimental findings to account for such coefficients. Therefore, it is not surprising that discrepancies up to 1000% has been reported in the literature between different numerical studies and the experiments, see the review of literature by Chou et al. [43].

The following characteristics can be depicted from the modeling of dispersion conductivity: (i) dispersion conductivity is modeled as a function of the pore velocity, which vanishes near the wall region leaving the effective conductivity equivalent to the stagnant component only, (ii) the magnitude of the dispersion conductivity reaches its maximum at the peak velocity location. This causes the fluid phase temperature profile to experience a sharp gradient near the wall since the energy transport by diffusion substantially increases as compared to that by convection. The temperature gradient smoothes on moving away from the wall due to the increase in local mixing, i.e. dispersion.

A number of studies have down played the contribution of the longitudinal dispersion conductivity to the overall thermal capability of a porous medium. Gunn and Desouza [44] and Vortmeyer [34] have reported that the effect of the longitudinal dispersion is negligible for high values of Re_p. In addition, the analysis performed by Sözen and Vafai [45], using compressible working fluids, has also shown that the effects of longitudinal dispersion were less pronounced for high Reynolds number flows. This is primarily due to the fact that wall heating is imposed perpendicular to the direction of flow and the fact that the growth of the thermal boundary-layer is more dependent on the transverse thermal conductivity. All this signifies the role of the transverse dispersion conductivity. Moreover, Amiri and Vafai [29] have examined the significance of the longitudinal and transverse dispersion over a broad range of Da, Re_p and thermophysical properties. The findings, which were performed for steady state conditions, were based on the Nusselt number predictions and were presented in terms of error maps that account for the estimated error in assuming no dispersion in both coordinates, neglecting the longitudinal dispersion, and neglecting the transverse dispersion, respectively. Amiri and Vafai [30] have recently shown the implications of the longitudinal and the transverse dispersion under transient conditions. The longitudinal dispersion was almost negligible at early times and its contribution to the overall heat transfer process reached an asymptotic value as the thermal steady state condition was approached. Based on the reviewed work in the literature, the longitudinal dispersion becomes insignificant for $Pe_p > 10$.

TWO-EQUATION MODEL

Concept of Local Thermal Equilibrium (LTE)

The literature review on heat transfer in porous media indicates that the state of local thermal equilibrium **LTE** condition was heavily employed in most of the investigations without a sound justification for this assumption. The modeling of energy transport in a porous medium demands accounting for the temperature of the individual phases. This requires representation of the individual phases by separate energy equations with an additional term in each equation to assimilate the energy exchange between the two phases. The **LTE** assumption implies that the difference between the local fluid and solid temperatures is negligible throughout the system under consideration. This assumption can be exercised when the dynamic and the thermal behaviors are relatively slow, thus providing ample time for efficient heat transfer between the individual phases. In addition, it requires that the ratio of k_s / k_f be close to unity, as pointed out by Truong and Zinsmeister [46]. However, the latter condition for **LTE** is a necessary but not a sufficient condition. An alternative reading to the **LTE** assumption suggests that the temperature difference between the fluid and solid phases within the representative elementary volume (ΔT_l) is much smaller than the temperature difference for the global system under consideration (ΔT_L), i.e. $\Delta T_l << \Delta T_L$. This definition is insufficient and misleading since the local variation of the solid and fluid temperatures is an essential design parameter for many applications, especially when safety is of a prime concern. Carbonell and Whitaker [47] have stated the criteria for the time and length scales for the validity of **LTE** assumption. The **LTE** condition is in some cases a severe constraint and would most likely introduce great inaccuracies in the modeling of energy transport. This comes despite the earlier recognition for the need of using the two-equation model in simulating the energy transport in porous media [48-50].

Formulation

The proposed model by Whitaker [2] is well formulated and quite comprehensive. Detailed investigation of the transient and steady state conditions for all the individual phases and the rate of phase change can be established. The model takes into account the local temperature difference between the individual phases in addition to the effect of the local temperature gradient from the individual phases (also known as the local gradient equilibrium). However, several coefficients as well as part of the formulation itself make it difficult and awkward for simulating practical applications. Owing to the lack of information on a number of transport coefficients, Vafai and Sözen [51] reformulated the governing equations and relaxed the local gradient of phase-averaged temperatures such that simulation of practical problems could be achieved. What is more, this assumption is not expected to be as severe as the **LTE** condition [52].

The early analysis of the two-equation model in a very simplified form was introduced by Schumann [53]. The classical Schumann model is an analytical model for a one-dimensional, semi-infinite bed that is initially at a uniform temperature and is subjected to a step change in inlet fluid temperature. The analytical solution was then obtained by neglecting the diffusion terms in both phases. It predicts the mean fluid and solid temperatures as a function of axial position and time. Despite its very limited applicability, it has been used extensively for thermal analysis in thermal storage packed beds. Subsequent related simplified analytical and numerical solutions were developed to predict the transient temperature response in packed beds for various boundary conditions [54-57]. A thorough search of the literature on the two-equation models revealed the tendency toward neglecting the diffusion term in the longitudinal, transverse or both coordinates, or relying on a one-dimensional model and formulating the fluid motion in terms of Darcy's equation. An exception was the series of investigations carried out by Vafai and his coworkers. It could be argued that the one-dimensional formulation may predict fair results in the case of insulated boundary condition, however, such an assumption is quite doubtful to be extended successfully to problems with a boundary condition of constant wall temperature. This is primarily due to the molecular diffusion in the normal coordinate, which strengthens the two-dimensionality behavior in a packed bed.

The use of the two-equation model requires information on additional modes of heat transfer that will emerge to account for the energy interaction between the phases. These modes are of radiative and convective natures, however, the convective heat transfer mode will be brought up here since forced convection is the prime source of energy transport under consideration in this review. The convective heat transfer mode requires information about the solid-to-fluid heat transfer coefficient h_{sf}. A considerable amount of experimental work has been carried out for determining its value for different geometrical shapes and packing arrangements. A review of the literature on the empirical correlations reported on the solid-to-fluid heat transfer coefficient in a packed bed of spheres prior to 1965 was reported by Barker [58]. Again, a great deal of uncertainty and inconsistency were found between the reported experimental results due to the entrance length effect, see Li and Finlayson [35], and the entrance effective length, equation (11), should be taken into consideration, see Han et al. [36], when measuring h_{sf}. In addition, Wakao et al. [59] have critically examined much of the experimental work and proposed a correlation for h_{sf} which maybe expressed as follows, see Wakao and Kaguei [37],

$$h_{sf} = k_f \left[2 + 1.1 \mathrm{Pr}^{1/3} \mathrm{Re}_p^{0.6}\right] / d_p \tag{19}$$

The correlation was established by curve fitting many of the reliable experimental data on h_{sf} in a packed-sphere bed for air being the working fluid. In addition, the correlation covers Re_p values up to 8500 and was found to closely simulate the data although it slightly under-predicts the data for $Re_p < 100$. The solid-to-fluid heat transfer coefficient for gas flows has also been reported for a number of porous structures other than packed beds [60-64]. However, there may still be a need for more experimental work to establish a firmly universal h_{sf} such as that given by equation (19).

Vafai and Sözen [51] and Sözen and Vafai [65] have numerically investigated the transient forced convective flow of a gas and forced convective flow of an oscillating compressible gas in a packed bed respectively. The analysis was based on the two-equation model coupled with the solid-to-fluid heat transfer coefficient reported by

Gamson et al. [66]. The investigation has provided a great deal of insight into the thermal behavior of packed beds under transient conditions. Two distinct regimes were observed during the thermal charging process, namely the 'initial stage' and the 'later stage'. The transition from the initial stage to the later stage was depicted once the pressure distribution across the bed becomes linear. In addition, the initial stage was found to last for a very short period of time and was characterized with minimal thermal penetration depth. What is more, the authors have examined the **LTE** condition and one-dimensional approach using three non-dimensional parameters, namely Da, Re_p and α_s/α_f. The **LTE** condition was found to be less severe for low values of Re_p and Da, whereas the two-dimensional behavior was found to be sensitive to α_s/α_f.

Sözen and Vafai [67] have also examined the condensation of a vapor flow in a packed bed. The rate of condensation was reported to be significant in the initial stage and when solid materials with large heat capacity values were used. A number of interesting observations were also pointed out toward optimizing the condensation rate in a packed bed. Amiri and Vafai [29] have performed a thorough investigation on forced convective incompressible fluid flow thorough a packed bed under steady state conditions. The parametric nature of the investigation has provided detailed information on the momentum and energy transport in packed beds. The generalized form of the momentum equation was adopted and the fluid and solid energy equations were presented respectively as follows:

$$\varepsilon < \rho_f >^f c_f \frac{\partial < T_f >^f}{\partial t} + < \rho_f >^f c_f < \mathbf{v} > . \nabla < T_f >^f$$

$$= \nabla . \left[k_{feff} . \nabla < T_f >^f \right] + h_{sf} a_{sf} \left[< T_s >^s - < T_f >^f \right] \tag{20}$$

$$(1 - \varepsilon) \rho_s c_s \frac{\partial < T_s >^s}{\partial t} = \nabla . \left[k_{seff} \nabla < T_s >^s \right] - h_{sf} a_{sf} \left[< T_s >^s - < T_f >^f \right] \tag{21}$$

The specific surface area of the packed bed a_{sf} can be modeled as, see Dullien [12],

$$a_{sf} = \frac{6(1-\varepsilon)}{d_p} \tag{22}$$

The dispersion conductivity and the solid-to-fluid heat transfer coefficient were based on equations (12), (13) and (19). The **LTE** condition was presented in integrated forms and similar conclusions to those of Vafai and Sözen [52] were drawn. Moreover, the one-dimensional approximation was found to be acceptable for α_s/α_f close to unity.

Amiri et al. [68] have numerically investigated the effect of the boundary conditions on the heat transfer in a packed bed. In addition, the choice of the boundary condition for the constant wall heat flux was found to be a challenge since it appeared that no previous studies has tackled its formulation using a two-equation model. The investigation was the first attempt to provide some physical insight into the problem and has suggested a proper representation of the boundary condition. In this work, comparisons were presented with an experimental work which displayed good agreement for the reduced Nusselt number predictions and the wall temperature variation. Amiri and Vafai [30] have recently extended their investigation to cover the temporal behavior in a packed bed. The analysis explored the temporal non-Darcian effects and has also showed that the inertial forces do hinder steady state conditions. The higher flow rates were found to allow a faster and more effective exchange of heat between the fluid and the solid and thus causing a relatively shorter elapsed time to arrive at steady state condition. The investigation did not reveal any distinct stages for the thermal charging process similar to those reported by Vafai and Sözen [51] since the external pressure is fixed for modeling of incompressible flows. In addition, the larger magnitude of the solid heat capacity $(\rho c)_s$ was found to assist in damping the transient temperature distribution owing to the increase in the solid phase capability in storing energy. The temporal **LTE** condition and the one-dimensional approach were also examined. Figure 1 illustrates the history of the local temperature difference between the fluid and the solid phases. As may be seen from the figure, the local temperature difference $\Delta T_l / (T_w - T_{in})$ reaches a maximum before decaying to an asymptotic value.

In addition, the asymptotic value is found to be a strong function of k_s / k_f with the temperature difference being the smallest when k_s / k_f is close to unity. Another interesting observation from the figure is that the highest temperature difference at early time is attained by the largest k_s / k_f value as expected. The analysis does indicate that **LTE** constraint is more severe for transient problems and would lead to inaccuracies in thermal behavior predictions. The two-dimension effects were found to increase with time until an asymptotic value is reached, indicating that the flow has reached its maximum boundary-layer growth. Moreover, the investigation shows that the one-dimensional approach under transient condition can be acceptable for α_s/α_f of the order of unity.

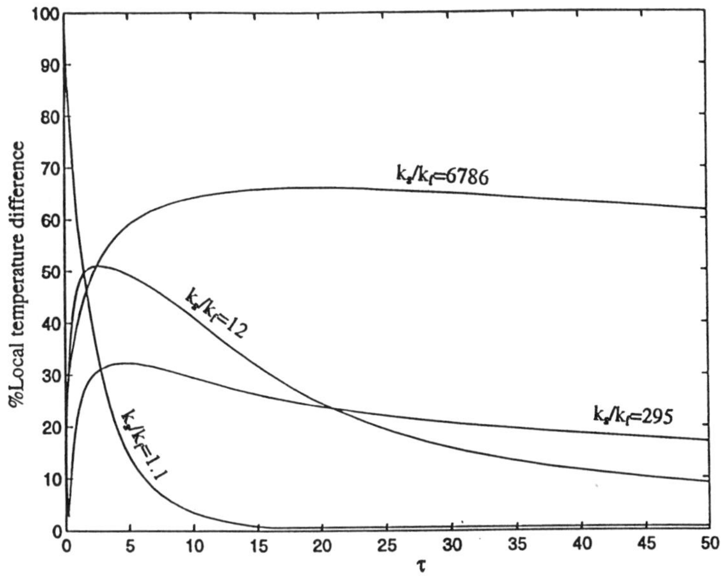

Figure 1. The temporal temperature difference between the fluid and the solid for Re_p=140 and $Da = 5.32 \times 10^{-7}$

The problem of forced convection in a channel filled with a highly conductive porous insertion is of great importance in high heat flux applications [69,70]. It was found that despite imposing high heat fluxes, the wall temperature was barely increased by few degrees while attaining a substantial enhancement in heat transfer coefficient as compared to regular channel cooling. What is more, Koh and his coworkers [71,72] have also investigated the thermal response in a bed with a highly conductive porous insertion by numerical and experimental means. However, a highly simplified numerical scheme was utilized and the momentum equation was based on Darcy's law.

CONCLUSIONS

The literature on forced convective flows through confined porous media was reviewed in addition to some related work to external flows. The pertinent transport processes and the associated phenomena were explored, which include: the non-Darcian effects, porosity maldistribution and velocity channeling and thermal dispersion effect. The essence of local volume-averaging technique was highlighted. The concept of local thermal equilibrium assumption was discussed and its validity was examined based on the pertinent investigations in this regard and an assessment of the one-dimensional approach was performed. It was observed that there is a need for more rigorous experiments for a number of porous structures including packed beds to measure the transport coefficients more accurately. At this juncture, more experimental and some analytical studies are needed to arrive at more appropriate means of identifying the transport coefficients associated with the transport equations. Such endeavor will ultimately provide us with more physical insight to transport phenomena in porous media.

References

1. T. M. Kuzay, Cryogenic Cooling of X-ray Crystals Using a Porous Matrix, *Rev. Sci. Instrum.*, **63**, 468-472 (1992).
2. S. Whitaker, Simultaneous Heat, Mass and Momentum Transfer in Porous Media: A Theory of Drying, *Advances in Heat Transfer,* **13**, 119-203 (1977).
3. A. Dybbs and R. V. Edwards, A New Look at Porous Media Fluid Mechanics–Darcy to Turbulent. In *Fundamentals of Transport Phenomena in Porous Media* (Eds. J. Bear and M. Y. Corapcioglu), pp. 199-254, Martinus Nijhof, Boston (1984).
4. D. A. Nield and A. Bejan, *Convection in Porous Media*, pp. 9, Springer-Verlag, New York (1992).
5. K. Vafai and C. L. Tien, Boundary and Inertia Effects on Flow and Heat Transfer in Porous Media, *Int. J. Heat Mass Transfer*, **108**, 195-203 (1981).
6. M. Kaviany, *Principles of Heat transfer in Porous Media*, pp. 138-141, Springer-Verlag, New York (1991).
7. C. L. Tien and K. Vafai, Statistical Bounds for the Effective Thermal Conductivity of Microsphere and Fibrous Insulation, *Prog. Astronaut. Aeronaut.*, **65**, 135-148 (1979).
8. K. Vafai and C. L. Tien, Boundary and Inertia Effects on Convective Mass Transfer in Porous Media, *Int. J. Heat Mass Transfer*, **25**, 1183-1190 (1982).
9. M. Kaviany, Laminar Flow through a Porous Channel Bounded by Isothermal Plates, *Int. J. Heat Mass Transfer*, **28**, 851-858 (1985).
10. G. S. Beavers, E. M. Sapparow and D. E. Rodenz, Influence of Bed Size on the Flow Characteristics and Porosity of Randomly Packed Bed of Spheres, *J. Appl. Mech.*, **40**, 655-660 (1973).
11. A. E. Sheidegger, *The Physics of Flow Through Porous Media* (3rd ed.), University of Toronto Press (1974).
12. F.A.L. Dullien, *Porous Media Fluid Transport and Pore Structure*, Academic Press, New York (1979).
13. S. Ergun, Fluid Flow Through Packed Columns, *Chem. Eng. Proc.*, **48**, 89-94(1952).
14. L. H. S. Roblee, R. M. Baird and J. W. Tierney, Radial Porosity Variation in Packed Beds, *AIChE J.*, **8**, 359-361 (1958).
15. R. F. Benanati and C. B. Brosilow, Void Fraction Distribution in Beds of Spheres, *AIChE J.*, **4**, 450-464 (1962).
16. H. Martin, Low Peclet Number Particle-to-Fluid Heat and Mass Transfer in Packed Beds, *Chem. Eng. Sci.*, **33**, 913-919 (1978).
17. G. E. Muller, Prediction of Radial Porosity Distribution in Randomly Packed Fixed Beds of Uniformly Sized Spheres in Cylindrical Containers, *Chem. Eng. Sci.*, **46**, 706-708 (1991).

18. C. E. Schwartz and J. M. Smith, Flow Distribution in Packed Beds, *Ind. Engng. Chem.*, **45**, 1209-1218 (1953).
19. W. M. Schertz and K. B. Bischoff, Thermal and Material Transport in Non-Isothermal Packed Beds, *AIChE J.*, **15**, 597-604 (1969).
20. B. C. Chadrasekhara and D. Vortmeyer, Flow Model for Velocity Distribution in Fixed Porous Beds Under Isothermal Conditions, *Wärme- und Stoffübertragung*, **12**, 105-111 (1979).
21. K. Vafai, Convective Flow and Heat Transfer in Variable Porosity Media, *J. Fluid Mech.*, **147**, 233-259 (1984).
22. K. Vafai, Analysis of Channeling Effect in Variable Porosity Media, *J. Energy Res. Tech.*, **108**, 131-139 (1986).
23. K. Vafai and S. J. Kim, Forced Convection in a Channel Filled with a Porous Medium: An Exact Solution, *J. Heat Transfer*, **111**, 1103-1106 (1989).
24. K. Vafai, R. L. Alkire and C. L. Tien, An Experimental Investigation of Heat Transfer in Variable Porosity Media, *J. Heat Transfer*, **107**, 642-647 (1985).
25. D. Poulikakos and K. J. Renken, Forced Convection in a Channel Filled With a Porous Medium, Including the effects of Flow Inertia, Variable Porosity, and Brinkman Friction, *J. Heat Transfer*, **109**, 880-888 (1988).
26. K. J. Renken and D. Poulikakos, Experiment and Analysis of Forced Convection Heat Transport in a Packed Bed of Spheres, *Int. J. Heat Mass Transfer*, **31**, 1399-1408 (1988).
27. M. L. Hunt and C. L. Tien, Non-Darcian Convection in Cylindrical Packed Beds, *J. Heat Transfer*, **110**, 378-384, (1988).
28. K. Vafai and R. Thiyagaraja, Analysis of Flow and heat Transfer at the Interface Region of a Porous Medium, *Int. J. Heat Mass Transfer*, **30**, 1391-1405 (1987).
29. A. Amiri and K. Vafai, Analysis of Dispersion Effects and Non-Thermal Equilibrium, Non-Darcian, Variable Porosity Incompressible Flow Through Porous Media, *Int. J. Heat Mass Transfer*, **11**, 939-954 (1994).
30. A. Amiri and K. Vafai, to be submitted for journal review.
31. S. Yagi and D. Kunii, Studies on Effective Thermal Conductivities in Packed Beds, *AIChE J.*, **3**, 373-381 (1957).
32. S.Yagi and D. Kunii and N. Wakao, Studies on Axial Thermal Conductivities in Packed Beds, *AIChE J.*, **6**, 543-546 (1960).
33. J. Gunn and M. Khalid, Thermal Dispersion and Wall Transfer in Packed Bed, *Chem. Eng. Sci.*, **30**, 261-267 (1975).
34. D. Vortmeyer, Axial Heat Dispersion in Packed Beds, *Chem. Eng. Sci.*, **30**, 999-1001 (1975).
35. C. Li and B. A. Finlayson, Heat Transfer in Packed Beds-a Reevaluation, *Chem. Eng. Sci.*, **32**, 1055-1066 (1977).
36. N. W. Han, J. Bhakta and R. G. Carbonell, Longitudinal and Lateral Dispersion in Packed Beds: Effect of Column Length and Particle Size Distribution, *AIChE J.*, **31**, 277-288 (1985).
37. N. Wakao and S. Kaguei, *Heat and Mass Transfer in Packed Beds*, Gordon and Breach Science Publishers Inc., New York (1982).

38. J. Levec and R. G. Carbonell, Longitudinal and Lateral Thermal Dispersion in Packed Beds, *AIChE J.*, **31**, 591-602 (1985).
39. P. Cheng and C. T. Hsu, Applications of Van Driest's Mixing length Theory to Transverse Thermal Dispersion in a Packed Bed With Bounding Walls, *Int. commun. Heat Mass Transfer*, **13**, 613-625 (1986).
40. M. L. Hunt and C. L. Tien, Effects of Thermal Dispersion on Forced Convection in Fibrous Media, *Int. J. Heat Mass Transfer,* **31**, 301-309, (1988).
41. D. L. Koch and F. J. Brady, Dispersion in Fixed Beds, *J. Fluid Mech.*, **154**, 399-427 (1985).
42. P. Cheng A. Chowdhury and C. T. Hsu, Forced Convection in Packed Tubes and Channels with Variable Porosity and Thermal Dispersion Effects. *Convective Heat and Mass Transfer in Porous Media*, (Eds. S. Kakaç, B. Kilkiş, F. A. Kulacki and F. Arinç), pp. 625-653, Kluwer Academic, Dordchet (1991).
43. F. C. Chou, J. H. Su and S. S. Lien, A Reevaluation of non-Darcian Forced Convection in Cylindrical Packed Tubes, *Fundamentals of Heat Transfer in Porous Media,* **HTD-Vol. 193**, 57-66, (1992).
44. J. Gunn and J. F. C. Desouza, Heat Transfer and Axial Dispersion in Packed Beds, *Chem. Eng. Sci.*, **29**, 1363-1371 (1974).
45. M. Sözen and K. Vafai, Longitudinal Heat Dispersion in Porous Beds with Real Gas Flow, *J. Thermophysics Heat Transfer*, **7**, 153-157 (1993).
46. H. V. Truong and G. E. Zinsmeister, Experimental Study of Heat Transfer in Layered Composites, *Int. J. Heat Mass Transfer*, **21**, 905-909 (1978).
47. R. G. Carbonell and S. Whitaker, Heat and Mass Transfer in Porous Media. In *Fundamentals of Transport Phenomena in Porous Media* (Eds. J. Bear and M. Y. Corapcioglu), pp. 121-198, Martinus Nijhof, Boston (1984).
48. D. Vortmeyer and R. J. Schaefer, Equivalence of One-and-Two-Phase Models for Heat Processes in Packed Beds: One Dimensional Theory, *Chem. Eng. Sci.*, **29**, 485-491 (1974).
49. E. U. Schülnder, Equivalence of One- and Two-Phase Models for Heat Transfer Processes in Packed Beds: One Dimensional Theory, *Chem. Eng. Sci.*, **30**, 449-452 (1975).
50. G. C. Glatzmaier and W. F. Ramirez, Use Of Volume Averaging For The Modeling Of Thermal Properties Of Porous Materials, *Chem. Eng. Sci.*, **43**, 3157-3169 (1988).
51. K. Vafai and M. Sözen, Analysis of Energy and Momentum Transport for Fluid Flow Through a Porous Bed, *J. Heat Transfer*, **112**, 690-699 (1990).
52. S. Whitaker, *Private Communications*, (1993).
53. T. E. W. Schumann, Heat Transfer: A liquid Flowing through a Porous Prism, *J. Franklin Inst.*, **208**, 405-416 (1929).
54. M. Riaz, Analytical Solution for Single- and Two-Phase Models of Packed-Bed Thermal Storage Systems, *J. Heat Transfer*, **99**, 489-492 (1977).
55. G. Spiga and M. Spiga, A Rigorous Solution to a Heat Transfer Two Phase Model in Porous Media and Packed Beds, *Int. J. Heat Mass Transfer*, **24**, 355-364 (1981).
56. R. J. Gross, C. E. Hickox and C. E. Hackett, Numerical Simulation of Dual-Media Thermal Energy Storage Systems, *J. Solar Energy Engng.*, **102**, 287-293 (1980).

57. D. E. Beasley and J. A. Clark, Transient Response of a Packed Bed for Thermal Energy Storage, *Int. J. Heat Mass Transfer*, **27**, 1659-1669 (1984).
58. J. J. Barker, Heat Transfer in Packed Beds, *Ind. Engng. Chem.*, **57**, 43-51 (1965).
59. N. Wakao, S. Kaguei and T. Funazkri, Effect of Fluid Dispersion Coefficients on Particle-To-Fluid Heat Transfer Coefficients in Packed Beds, *Chem. Eng. Sci.*, **34**, 325-336(1979).
60. V. M. Eroshenko and L. A. Yaskin, Heat Transfer in Forced Convection of Fluid in Porous Sintered Metals, *J. Engng. Phys.*, **30**, 1-7 (1976).
61. A. R. Balakishnan and D. C. T. Pei, Heat Transfer in Gas-Solid Packed Beds system 3. Overall Heat Transfer in Adiabatic Beds, *Ind. Engng. Chem. Process Des. Dev.*, **18**, 47-50 (1979).
62. O. M. Alifanov, A. P. Tryanin and A. L. Lozhkin, Experimental Investigation of the Method of Determining the Internal Heat Transfer Coefficient in a porous Body From the Solution of the Inverse Problem, *J. Engng. Phys.*, **52**, 340-346 (1987).
63. M. Golombok, H. Tariwala and I. C. Shirvill, Gas-Solid Heat Exchange in a Fibrous Metallic Material Measured by a Heat Regenerator Technique, *Int. J. Heat Mass Transfer*, **33**, 243-252 (1990).
64. L. B. Younis and R. Viskanta, Experimental Determination of the Volumetric Heat Transfer Coefficient Between Stream of Air and Ceramic Foam, *Int. J. Heat Mass Transfer*, **36**, 1425-1434 (1993).
65. M. Sözen and K. Vafai, Analysis of Oscillating Compressible Flow Through a Packed Bed, *Int. J. Heat Mass Transfer*, **12**, 130-136 (1993).
66. B. W. Gamson, G. Thodos and O. A. Hougen, Heat, Mass and Momentum Transfer in the Flow of gases Through Granular Solids, *AIChE J.*, **39**, 1-35 (1943).
67. M. Sözen and K. Vafai, Analysis of Non-Thermal Equilibrium Condensing Flow of a Gas Through a Packed Bed, *Int. J. Heat Mass Transfer*, **332**, 1247-1261 (1990).
68. A. Amiri, K. Vafai and T. M. Kuzay, Effects of Boundary Conditions on Non-Darcian Heat Transfer Through Porous Media and Experimental Comparisons, *Numerical Heat Transfer*, **27(A)**, 651-664 (1995).
69. F. E. Megerlin, R. W. Murphy and A. E. Bergles, Augmentation of Heat Transfer in Tubes by Use of Mesh and Brush Inserts, *J. Heat Transfer*, **96**, 145-151 (1974).
70. T. M. Kuzay, J. T. Collins, A. M. Khounsary and G. Morales, Enhanced Heat Transfer With Metal-Wool-Filled Tubes, *Proc. ASME/JSME Thermal Engng. Joint Conf.*, 145-151 (1991).
71. J. C. Y. Koh and R. Colony, Analysis of Cooling Effectiveness for Porous Material in a Coolant Passage, *J. Heat Transfer*, **96**, 324-330 (1974).
72. J. C. Y. Koh and R. L. Stevens, Enhancement of Cooling Effectiveness by Porous Materials in Coolant Passage, *J. Heat Transfer*, **97**, 309-311 (1975).

NATURAL CONVECTION IN ENCLOSURES FILLED WITH ANISOTROPIC POROUS MEDIA

P. VASSEUR and L. ROBILLARD

Ecole Polytechnique, Montréal, Québec, H3C 3A7, Canada

INTRODUCTION

Buoyancy-induced convection in a fluid-saturated porous medium is of considerable interest, due to several geophysical and engineering applications. So far, investigations have been mostly concerned with isotropic porous media. Much of this activity, both numerical and experimental, has been summarized in a recent book by Nield and Bejan [1].

Despite the fact that in several applications the porous materials are anisotropic, natural convection in such media has received relatively little attention. Anisotropy, which is generally a consequence of a preferential orientation or asymmetric geometry of the grain or fibers, is in fact encountered in numerous systems in industry and nature. Examples include fibrous materials, geological formations, oil extraction, some biological materials, and dentritic regions formed during solidification of binary alloys. Earlier studies on natural convection in saturated anisotropic porous medium are concerned mostly with layers heated from below. Several studies have also been reported concerning natural convection in anisotropic porous enclosures heated from the side. In the present chapter we focus on this last situation.

RECTANGULAR ENCLOSURES

Consider a two-dimensional enclosure of height H and horizontal length L, as shown in Figure 1. The top and bottom walls are insulated while the side walls are differentially heated. The porous medium is assumed to be both hydrodynamically and thermally anisotropic. The permeabilities along the two principal axes of the porous matrix are denoted by K_1 and K_2. The anisotropy in flow permeability of the porous medium is then characterized by the permeability ratio $K^* = K_1/K_2$, and the orientation angle θ, defined as the angle between the horizontal direction and the principal axis with permeability K_2. The principal directions of the thermal conductivities (k_1, k_2) are assumed to coincide with those of the permeability axes. Generalized Darcy's law (Bear [2]) and the Boussinesq approximation are used.

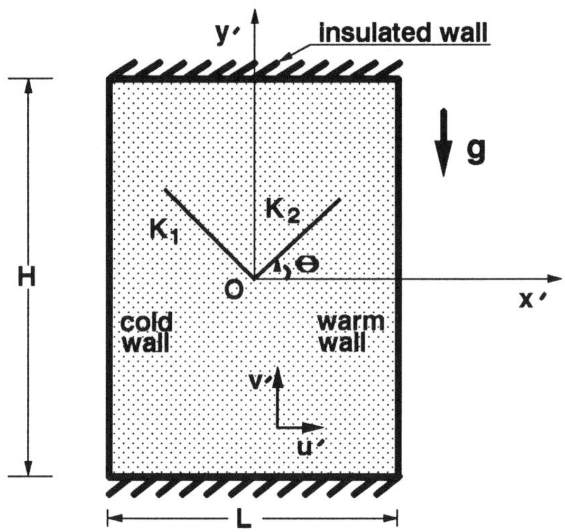

Figure 1 The physical situation and coordinate system

Then the equations that account for the conservation of mass, momentum, and energy are as follows:

$$\nabla . \overrightarrow{V'} = 0 \tag{1}$$

$$\overrightarrow{V'} = -\frac{\overline{\overline{K}}}{\mu} \left[\nabla p' + \varrho \vec{g} \beta (T' - T'_r) \right] \tag{2}$$

$$\frac{1}{(\varrho c)_f} \nabla . \left(\overline{\overline{k}} \, \nabla T' \right) = \nabla . \left(\overrightarrow{V'} T' \right) \tag{3}$$

where $\overrightarrow{V'}$ is the superficial flow velocity, p' pressure, T' temperature, μ dynamic viscosity, ρ density, β coefficient of thermal expansion, $\vec{g}$ gravitational acceleration, $(\varrho c)_f$ heat capacity of the fluid and $\overline{\overline{K}}$ and $\overline{\overline{k}}$ second-order permeability and thermal conductivity tensors, respectively. Subscript r indicates a reference state.

Eliminating the pressure in the momentum equation in the usual way, and taking L, α/L and $\Delta T'$ as respective dimensional scales for length, velocity and temperature, the governing equations may be written in nondimensional form as

$$a \, \frac{\partial^2 \psi}{\partial x^2} + b \, \frac{\partial^2 \psi}{\partial x \partial y} + c \, \frac{\partial^2 \psi}{\partial y^2} = -R \frac{\partial T}{\partial x} \tag{4}$$

$$d\,\frac{\partial^2 T}{\partial x^2} + e\,\frac{\partial^2 T}{\partial x \partial y} + f\,\frac{\partial^2 T}{\partial y^2} = u\frac{\partial T}{\partial x} + v\frac{\partial T}{\partial y} \tag{5}$$

where

$$\left.\begin{array}{ll} a = \cos^2\theta + K^*\sin^2\theta & d = \cos^2\theta + k^*\sin^2\theta \\ b = (K^*-1)\sin 2\theta & e = (1-k^*)\sin 2\theta \\ c = \sin^2\theta + K^*\cos^2\theta & f = \sin^2\theta + k^*\cos^2\theta \end{array}\right\} \tag{6}$$

In the above equations, the Rayleigh number R, defined as $R = K_1 g\beta\Delta T' L/\alpha\nu$, is based on the permeability K_1, $\alpha = k_2/(\varrho c)_f$ is the thermal diffusivity and $k^* = k_1/k_2$ is the thermal conductivity ratio.

As usual, the dimensionless stream function ψ, in equation (4), is related to the velocity components by $u = \partial\psi/\partial y$ and $v = -\partial\psi/\partial x$, so that equation (1) is automatically satisfied.

In the remainder of this section, the thermal boundary conditions applied on the cavity will be specified and various solutions to equations (4) and (5) will be discussed.

Rectangular cavity with isothermal side walls

We consider here that the right and left vertical walls of the enclosure depicted in Figure 1 are respectively heated and cooled at constant temperatures T'_h and T'_c. For this situation the side to side temperature difference $T'_h - T'_c$ is the temperature scale $\Delta T'$ and T'_c, the reference temperature T'_r.

The non-dimensional boundary conditions over the walls of the enclosure are then given by

$$\left.\begin{array}{llll} x = 1/2 & \psi = 0 & T = 1 & (a) \\ x = -1/2 & \psi = 0 & T = 0 & (b) \\ y = \pm A/2 & \psi = 0 & \dfrac{\partial T}{\partial y} = 0 & (c) \end{array}\right\} \tag{7}$$

where $A = H/L$ is the cavity aspect ratio.

Numerical integration of the coupled transport and energy equations (4) and (5), under boundary conditions (7), is readily obtained using for instance finite differences. Numerical calculations were reported by Degan and Vasseur [3]. Figures 2a-2e illustrate the effects of the permeability ratio K^* and the angle of the principal axes θ on the streamlines (left) and the isotherms (right) for a square cavity ($A = 1$). The direction and relative importance of the maximum and minimum permeabilities are depicted by the angular position and relative lengths of the perpendicular lines located between each set of flow and temperature fields. The evolution of the flow patterns and isotherms with K^* is exemplified in Figures 2a and 2b for $\theta = 45^o$, $R = 4 \times 10^2$, and $K^* = 1$ and 10^2, respectively. For $K^* = 1$, the porous medium is isotropic (independent of θ) and the flow and temperature fields of Figure 2a are similar to those reported in the literature by numerous investigators. A counterclockwise rotating cell fills up the entire cavity with hydrodynamic and thermal boundary layers along the vertical isothermal walls. Upon increasing K^* from 1 to 10^2 it is observed from Figure 2b that the resulting cell is aligned along the diagonal central region of the cavity, i.e. along the principal axis with higher permeability. Also, the numerical results indicate that the strength of the convective circulation within the cavity, as indicated by the value of the maximum stream function ψ_{max}, is considerably reduced. This follows from the fact that, for fixed values of θ and R (and consequently K_1), an increase in K^* can be interpreted as a decrease in the permeability K_2. Figures 2c and 2d illustrate the influence of the anisotropy orientation angle θ for $R = 10^4$, $K^* = 10^2$ and $\theta = 0^o$ and 90° respectively. For $\theta = 0^o$ the buoyancy-induced flow along the vertical isothermal walls is very strong, and Figure 2c shows the formation of very thin velocity boundary-layers near the upper (lower) part of the hot (cold) wall. The flow then spreads through the upper (lower) half of the cavity, and the resulting isotherms are observed to concentrate along a diagonal between the upper left and the lower right corners. As the inclination angle is turned through 90°, Figure 2d shows that the strength of the circulation is weakened, the permeability in the vertical direction being now much smaller than in the horizontal direction.

Numerical calculations were also reported by Ni and Beckermann [4] for the case when the principal directions of permeability coincide with the horizontal and vertical coordinate axes ($\theta = 0^o$). Their results indicate that, when compared to the situation where the porous medium is isotropic, a larger permeability ratio ($K^* > 1$) causes channeling of the flow along the vertical isothermal walls, a higher flow intensity in the enclosure, and, consequently a higher heat transfer rate Nu. On the other hand, as illustrated in Figure 2e, a low permeability ratio ($K^* < 1$) causes channeling of the flow along the horizontal adiabatic boundaries, a lower flow intensity in the enclosure and consequently a smaller Nu.

Boundary-layer regime. When the buoyancy-induced fluid circulation within the enclosure is strong enough, the flow has a boundary-layer structure for which an order of magnitude estimate can be derived on scaling grounds. In the boundary-layer regime, most of the fluid motion is restricted to a thin layer of thickness δ' and height H ($\delta' \ll H$). In this region, for a porous medium with isotropic thermal conductivity ($k^* = 1$), the conservation equations (4) and (5) require the following balances

$$a\frac{\psi}{\delta^2} \sim R\,\frac{1}{\delta} \tag{8}$$

$$\frac{\psi}{\delta}\,\frac{1}{A} \sim \frac{1}{\delta^2} \tag{9}$$

Equations (7), (8) and (9) imply that the scales of the vertical boundary-layer are (Degan and Vasseur [3]):

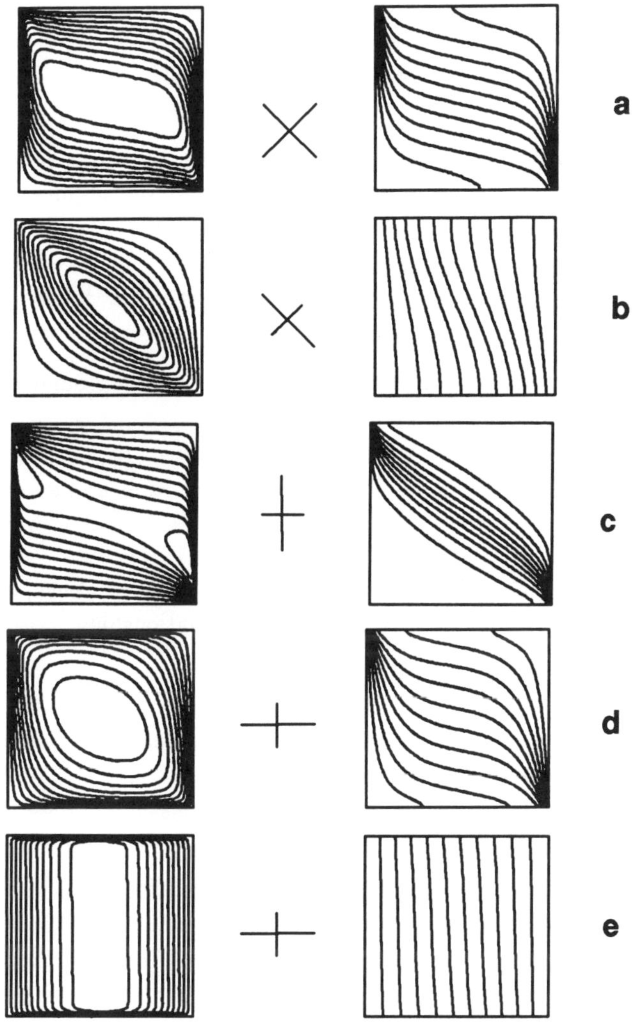

Figure 2 Streamlines and isotherms in an anisotropic porous layer ($A = 1$, $k^* = 1$) (Degan and Vasseur [3])

a) $R = 4 \times 10^2$, $K^* = 10^0$, $\theta = 45^o$, $\psi_{max} = 11.78$, $Nu = 7.86$
b) $R = 4 \times 10^2$, $K^* = 10^2$, $\theta = 45^o$, $\psi_{max} = 0.97$, $Nu = 1.09$
c) $R = 1 \times 10^4$, $K^* = 10^2$, $\theta = 0^o$, $\psi_{max} = 13.75$, $Nu = 9.97$
d) $R = 1 \times 10^4$, $K^* = 10^2$, $\theta = 90^o$, $\psi_{max} = 6.89$, $Nu = 4.34$
e) $R = 1 \times 10^0$, $K^* = 10^{-3}$, $\theta = 0^o$, $\psi_{max} = 0.12$, $Nu = 1.00$.

$$\begin{aligned} \delta &\sim A^{1/2}\, a^{1/2}\, R^{-1/2} \\ \psi &\sim A^{1/2}\, a^{-1/2}\, R^{1/2} \\ u &\sim A^{-1/2}\, a^{-1/2}\, R^{1/2} \\ v &\sim a^{-1}\, R \end{aligned} \tag{10}$$

The average Nusselt number, defined as the total heat transfer rate from one side wall to the other q', over the heat transfer rate in the pure conduction limit q'_c, has the following scale

$$Nu = \frac{q'}{q'_c} \sim \frac{kH\Delta T'/\delta'}{kH\Delta T'/L} = A^{-1/2}\, a^{-1/2}\, R^{1/2} \tag{11}$$

From equation (4) it is clear that we may use the boundary-layer hypothesis only when the conditions

$$a\,\frac{\partial^2\psi}{\partial x^2} \gg b\,\frac{\partial^2\psi}{\partial x \partial y}; \qquad a\,\frac{\partial^2\psi}{\partial x^2} \gg c\,\frac{\partial^2\psi}{\partial y^2} \tag{12}$$

are satisfied.

Upon introducing the orders of magnitude obtained above, it is found that the parameters b and c must satisfy the following relationships

$$b \ll A^{1/2}\, a^{1/2}\, R^{1/2}; \qquad c \ll A\, R \tag{13}$$

which are not too restrictive because, in order to reach the boundary-layer regime, R must be very large ($R \to \infty$).

The boundary-layer scales, equation (10), suggest the following nondimensionalization of the governing equations (4) and (5) in the boundary layer region

$$\frac{\partial^2\overline{\psi}}{\partial\overline{x}^2} = -\frac{\partial\overline{T}}{\partial\overline{x}} \tag{14}$$

$$\overline{u}\frac{\partial\overline{T}}{\partial\overline{x}} + \overline{v}\frac{\partial\overline{T}}{\partial\overline{y}} = \frac{\partial^2\overline{T}}{\partial\overline{x}^2} \tag{15}$$

where

$$\begin{aligned} &\overline{x} = x\, L/\Lambda \qquad &&\overline{y} = y/A \qquad &&\Lambda^2 = L\, H\, a/R \\ &\overline{\psi} = \psi\Lambda/H \qquad &&\overline{T} = T && \end{aligned} \tag{16}$$

Using the same notation, the boundary conditions, for the boundary-layer lining the left wall in Figure 1, are

$$\begin{aligned} &\text{i) on the wall} \qquad &&\overline{\psi} = 0 \qquad &&\overline{T} = 0 \\ &\text{ii) in the core} \qquad &&\overline{\psi} = \overline{\psi}_\infty(\overline{y}) \qquad &&\overline{T} = \overline{T}_\infty(\overline{y}) \end{aligned} \tag{17}$$

where $\overline{\psi}_{\infty}(\overline{y})$ and $\overline{T}_{\infty}(\overline{y})$ are unknown and represent the velocity and temperature distributions in the core of the cavity, sufficiently far from the boundary-layer region (Gill [5]).

The dimensionless nonlinear governing equations (14) and (15), together with boundary conditions equation (17) are exactly the same as those derived by Weber [6] while considering boundary-layer flows within an isotropic porous cavity ($K^* = 1$). Approximate analytical solutions to this problem have been reported by Weber [6], Bejan [7] and Simpkins and Blythe [8]. The solution proposed by Simpkins and Blythe [8], based on integral relations, has been adopted here, since it proved to be in good agreement with the numerical results when the chosen boundary-layer profile exhibits the correct asymptotic behaviour. Translating their results for their two-layers profile model into the present notation, it is readily found that, the total heat transfer rate between the two side walls is given by

$$Nu = 0.51 \left(\frac{R}{a\ A}\right)^{1/2} \tag{18}$$

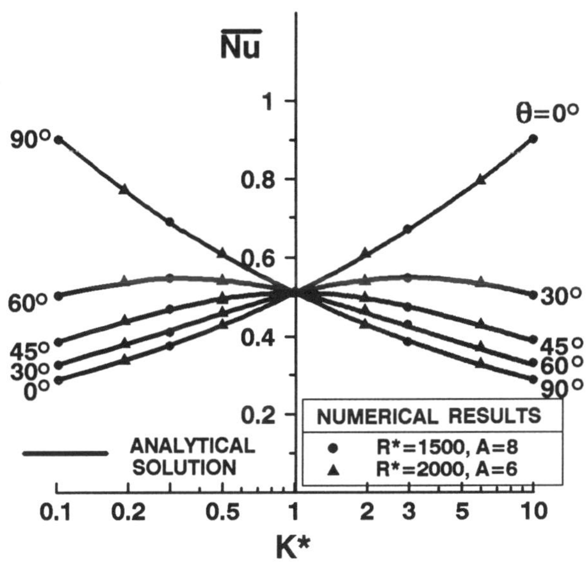

Figure 3 The effect of the parameters K^* and θ on the heat transfer rate for boundary-layer regime in an anisotropic porous layer heated isothermally from the side (Degan and Vasseur [3])

The agreement between the above results and the order of magnitude prediction, equation (10), is evident.

Figure 3 shows that the present boundary-layer solution is in good agreement with finite-difference calculations. The dependence of the groups $\overline{Nu} = Nu\,(A\sqrt{K^*}/R)^{1/2}$ on K^* and θ, anticipated from equation (18), appears to be correct in the high Rayleigh limit. As discussed by Kimura *et al.* [9] and Aboubi *et al.* [10], the modified Rayleigh number $R/\sqrt{K^*} = g\beta\Delta T' L\sqrt{K_1K_2}/(\alpha\nu)$ is more appropriate to describe the present phenomena since the two extreme permeabilities K_1 and K_2 are now part of the effects normally associated to a change in Rayleigh number. It can be easily demonstrated that if $\psi(x,y)$ and $T(x,y)$ are solutions of equations (4) and (5) at $(R/\sqrt{K^*},\ A,\ \theta,\ K^*)$, so they are at $(R/\sqrt{K^*},\ A,\ \pi/2-\theta,\ 1/K^*)$. For the particular case $\theta = 45^o$, the results obtained for a given value of K^* are equivalent to those for $1/K^*$. These points are illustrated in the graph where the symmetry obtained with respect to the vertical line $K^* = 1$ is also a consequence of the logarithmic scale for K^*.

Pseudo conduction regime. For sufficiently small values of the Rayleigh number, a perturbation method can be employed to solve the set of governing equations (4) and (5), subject to boundary conditions (7). The solution is obtained by expanding ψ and T in a power series in R as follows:

$$(\psi,T) \;=\; (\psi_o,T_o) \;+\; R(\psi_1,T_1) \;+\; R^2(\psi_2,T_2) \;+\; \ldots.. \tag{19}$$

Substituting equation (19) into equations (4) and (5) and equating terms in like powers of R yields subproblems at order 0, 1, 2, . . . in R. The zero-order problem, satisfying the boundary conditions, equation (7), yields the zero-order conduction solution $\psi_o = 0$ and $T_o = x + 1/2$. Higher-order solutions of ψ and T have been obtained by Kimura *et al.* [9] for an anisotropic porous layer having its principal axes, for both permeability and thermal conductivity aligned with the gravity vector ($\theta = 0^o$). An analytical form of the first-order solution (ψ_1,T_1) was derived by using Fourier series. However, the second-order solution (ψ_2,T_2) was obtained numerically since its was found that an analytical solution was extremely complex to obtain.

The Nusselt number is given, up to the second-order approximation, by the following expression

$$Nu \;=\; 1 \;+\; C\overline{R}^2 \tag{20}$$

where

$$C \;=\; -A^{-1}\int_{y=-A/2}^{y=A/2}\left(\frac{\partial T_2}{\partial x}\right)_{x=1/2} dy \tag{21}$$

and $\overline{R} = (A/4k^*)R$.

The integral C, which depends on the two parameters ξ and η, defined as

$$\xi \;=\; \frac{A^2}{K^*} \qquad\qquad \eta \;=\; \frac{A^2}{k^*} \tag{22}$$

has been evaluated numerically by Kimura *et al.* [9] for various combinations of ξ and η. They found that both parameters η and ξ have an equally significant impact on the overall heat transfer. Thus, for a given value of $\overline{R}$, the Nusselt number was found to be a decreasing function of K^* and an increasing function of k^*.

Rectangular cavity with uniform heat flux from the side

The porous layer illustrated in Figure 1 is now reconsidered for the case when a uniform heat flux q' is applied along both side walls. The following discussion is restricted to the particular case where the principal directions of the thermal conductivities coincide with the horizontal and vertical coordinate axes ($k_1 = k_y$ and $k_2 = k_x$) while the orientation angle θ of the anisotropy in flow permeability is arbitrary.

For this situation the characteristic temperature difference is $\Delta T' = q'L/k_2$ and it is convenient to take T'_o, the temperature at the center of the cavity, as the reference temperature T'_r. The non-dimensional boundary conditions to be considered here are given by

$$\left.\begin{array}{llll} x = \pm 1/2 & \psi = 0 & \partial T/\partial x = 1 & (a) \\ y = \pm A/2 & \psi = 0 & \partial T/\partial y = 0 & (b) \end{array}\right\} \tag{23}$$

Analytical solution to the coupled system of equations (4) and (5), subject to boundary conditions (23) is in general impossible. However, in the limit of a slender slot ($A \gg 1$) the governing equations can be considerably simplified and more tractable to solutions. In this limit, as discussed in detail by Cormack *et al.* [11], Vasseur *et al.* [12], and other authors, the flow in the central part of the cavity, far from either end can be assumed parallel, with $\psi(x,y) \simeq \psi(x)$. Also the temperature field is the sum of a linearly varying longitudinal part and an unknown transverse function so that $T = Cy + \varsigma(x)$ where C is the temperature gradient in y-direction. Substituting these approximations into the governing equations (4) and (5) and solving the resulting set of ordinary differential equations, it is found that (Degan *et al.* [13])

$$\psi = \frac{1}{C}\left[\frac{\cosh\ \alpha x}{\cosh\ \alpha/2} - 1\right] \tag{24}$$

and

$$T = Cy + \frac{1}{\alpha}\frac{\sinh\ \alpha x}{\cosh\ \alpha/2} \tag{25}$$

where $\alpha^2 = RC/a$.

The thermal boundary conditions in the y-direction, equation (23b), which cannot be satisfied exactly with the parallel flow approximation, are replaced by an equivalent energy flux condition (Bejan [14]), from which it is found that

$$\frac{1}{2k^*C^2}\ \frac{1}{\cosh^2\ \alpha/2}\left(\frac{\sinh\ \alpha}{\alpha} - 1\right) = 1 \tag{26}$$

such that the temperature gradient C can be obtained, for any combination of the controlling parameters R, K^*, k^* and θ, by numerically solving the above transcendental equation.

The heat transfer across the cavity can be expressed in terms of a Nusselt number Nu defined as

$$Nu = \frac{q'L}{k_2} \frac{1}{\overline{\Delta T}} = \frac{1}{\Delta T} = \frac{\alpha}{2} \coth \frac{\alpha}{2} \tag{27}$$

where $\overline{\Delta T}$ is the actual wall-to-wall temperature difference. According to equation (25), ΔT varies linearly in y and for convenience was evaluated arbitrarily at the position $x = 0$.

In the boundary-layer regime ($R \to \infty$), the above expression for Nu reduces to

$$Nu = \frac{\alpha}{2} \tag{28}$$

where $\alpha = k^{*-1/5}\ a^{-2/5}\ R^{2/5}$.

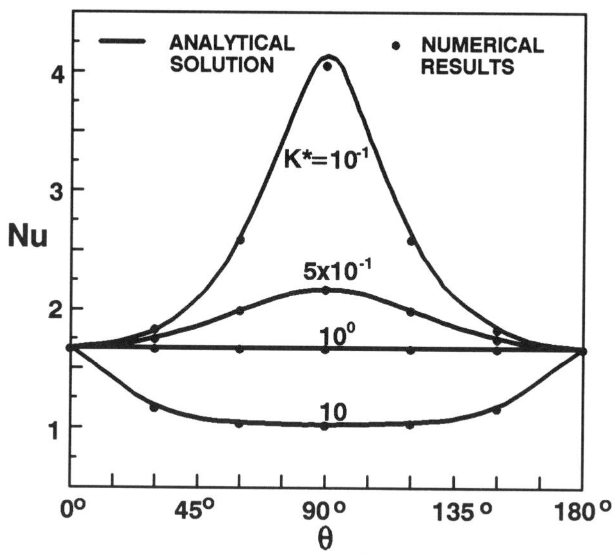

Figure 4 The effect of anisotropy orientation angle θ and permeability ratio K^* ($R = 20$, $k^* = 1$) on the Nusselt number (Degan *et al.* [13])

Figure 4 shows the effect of anisotropy orientation angle θ and permeability ratio K^* on Nu for $R = 20$ and $k^* = 1$. For an isotropic porous medium ($K^* = 1$), Nu is independent of θ, as can be expected. However, for $K^* > 1$ ($K^* < 1$) it is observed that Nu is a maximum (minimum) at $\theta = 0^o$ and 180° and a minimum (maximum) at $\theta = 90^o$. This behavior can easily be deduced from the first and second derivatives of Nu, equation (28), with respect to θ. Similar results have been reported by Zhang [15] and Degan and Vasseur [3] when considering natural convection within porous layers heated isothermally from the side. The fact that the convective heat transfer reaches a maximum (minimum) when the orientation of the principal axis with higher permeability of the anisotropic porous medium is parallel (perpendicular) to the gravity has applications in insulation techniques. Thus, the anisotropic porous medium which minimizes the loss of heat transfer through a vertical porous layer, is a medium with as small a vertical permeability as possible.

Figures 5a-5c illustrate the effect of thermal conductivity ratio k^* on the flow and temperature fields in a vertical porous layer. In Figure 5a the porous medium is thermally isotropic ($k^* = 1$) and natural convection is clearly in the boundary-layer regime. For $k^* = 10^3$, Figure 5b, the isotherms are vertical, the absence of temperature gradients in the y-direction being attributed to the fact that the thermal conductivity in that direction is much higher than that in the horizontal one. The resulting flow pattern is not of the boundary-layer type, even though the flow intensity is approximately five times higher than for $k^* = 1$. When the value of k^* is reduced to 10^{-3} ($k_x \gg k_y$), Figure 5c indicates that the isotherms are now almost horizontal in the core of the cavity. The resulting temperature pattern gives rise to a flow circulation relatively weaker than that for the isotropic situation (Figure 5a).

The effect of thermal conductivity ratio k^* on Nu has been documented by Degan *et al.* [13]. Their numerical data were found to agree well with the analytical solution, equations (24) to (27) provided that k^* is made larger than unity. It is noticed that, in the high k^* limit, Nu is asymptotically given by

$$Nu \simeq 1 + \frac{R^2}{144\, a^2\, k^*} \tag{29}$$

so that the heat transfer is almost purely conductive, in spite of the presence of an intense flow circulation (see for instance Figure 5b). On the other hand, as the value of k^* is made smaller than unity, the analytical solution predicts that Nu increases continuously with k^*. This is not in agreement with the numerical data of Ni and Beckermann [4] and Degan *et al.* [13] which indicate that, for this situation, k^* has very little effect on Nu. However, it is noticed from Figure 5c that when k^* is small the flow structure within the enclosure is not parallel anymore so that the analytical solution becomes a poor approximation.

Further numerical calculations concerning the effects of k^* on Nu were reported by Chang and Lin [16] for the case of a rectangular cavity having finite wall conductances. They pointed out that a critical value of the anisotropic thermal diffusivity ratio k^* may exist such that Nu reaches a minimum. This critical value decreases

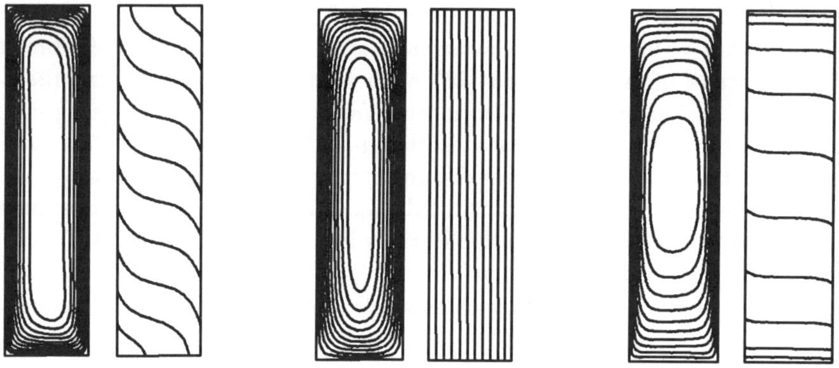

Figure 5 Streamlines and isotherms in a thermally anisotropic porous layer ($R = 100$, $A = 4$, $K^* = 1$) (Degan *et al.* [13])

a) $k^* = 10^0$, $\psi_{max} = 2.33$
b) $k^* = 10^3$, $\psi_{max} = 11.60$
c) $k^* = 10^{-3}$, $\psi_{max} = 0.90$

as the value of K^* (when $\theta = 0^o$) is made smaller.

Intuitively, it is expected that unidirectional flows, including boundary-layer type, must somehow depend on the characteristics of the anisotropic porous medium in the flow direction. This point is confirmed by the analytical results presented above which clearly indicate that, for those type of flows, the appropriate Rayleigh number is

$$R^* = \frac{R}{a} = \left(\frac{g\beta\Delta T' L}{\alpha\nu}\right) \frac{1}{P_{yy}} \tag{30}$$

where $P_{yy} = a/K_1$ is the y-component of the hydraulic resistivity tensor $\overline{\overline{K}}$ (see for instance Sinha *et al.* [17]), defined as

$$\overline{\overline{P}} = \left[\,\overline{\overline{K}}\,\right]^{-1} = \frac{1}{K_1}\begin{bmatrix} c & b/2 \\ b/2 & a \end{bmatrix} \tag{31}$$

It is thus the resistivity in the direction of the flow, or its inverse, the directional permeability (Bear [2]), that happens to be the feature of the anisotropic porous medium physically involved in unidirectional flows. In terms of this directional hydraulic resistivity the porous medium can then be treated as an isotropic one.

The tensor $\overline{\overline{P}}$ may also be expressed as

$$\overline{\overline{P}} = \frac{1}{K_1}\left\{ \begin{bmatrix} 1 & 0 \\ 0 & 1 \end{bmatrix} + (K^* - 1) \begin{bmatrix} \cos^2\theta & \sin 2\theta/2 \\ \sin 2\theta/2 & \sin^2\theta \end{bmatrix} \right\} \tag{32}$$

in which the unit matrix represents the isotropic resistivity based on the permeability K_1 and the second term the added resistivity due to the anisotropy.

Additional Effects

The behavior of the flow and heat transfer process changes substantially as the flow regime departs from the Darcy limit. The effect of boundary friction on convection in a rectangular cavity has been studied by Degan [18], using Brinkman's extension of generalized Darcy's law. Numerical calculations were reported by this author for the case of uniform heat flux. An analytical solution, for the boundary-layer regime, was developed by applying the Oseen linearization method. The resulting Nusselt number is given by

$$Nu = \frac{B}{8\ Da^*\ p} \tag{33}$$

where $Da^* = K_1/(a\ L^2)$ is the Darcy number based on the directional hydraulic resistivity, $B = (Da^*\ p^2 - 4)$ and $p = 8192\ R^{*2}\ Da^{*2}\ (B-8)/B^5$.

Figure 6 shows the effects of varying the Darcy number $Da = a\ Da^*$, i.e. the Darcy number based on permeability K_1, and anisotropy orientation angle θ on the Nusselt number. As expected, when the Darcy number is small enough ($Da < 10^{-6}$), both the numerical and analytical solutions are in agreement with Darcy's law (equation (28)), represented by a dashed line on the graph. In this limit the heat transfer rate varies considerably with the angle θ. As the permeability of the porous medium (i.e. Da) is increased, the boundary frictional resistance becomes more important and adds to the bulk resistance to slow down the convection. As a result, the heat transfer drops progressively and becomes less affected by the change in the orientation θ of the principal axes. When Da is large enough, the pure fluid situation (in the absence of inertia effects) is recovered. In Figure 6, this situation is reached for $Da \simeq 10^{-1}$, for which the Nusselt number is independent of θ. In the pure fluid limit ($Da \gg 1$), equation (33) reduces to Kimura and Bejan's [19] solution $Nu = (8192\ Ra^2)^{1/9}/8$, where $Ra = R/Da$ is the Rayleigh number for a fluid.

Double-diffusive natural convection within a porous layer, subject to horizontal thermal and compositional gradients, was studied numerically by Bennacer and Tobbal [20]. They used the Darcy-Brinkman formulation to investigate the influence

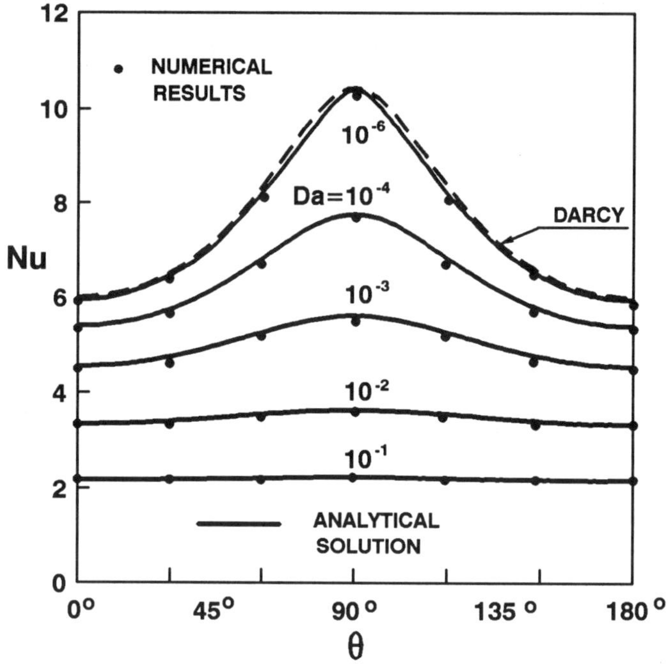

Figure 6 Effect of the inclination angle θ on Nusselt number ($R = 500$, $k^* = 1$, $K^* = 0.25$ and various values of Da), (Degan [18]).

of thermal anisotropy on heat and mass transfer. They noted that the evolution of mass transfer with k^* shows a maximum which depends on both the Darcy and the Lewis numbers. Natural convection induced by combined heat and mass transfer in an enclosure filled with anisotropic porous media saturated with a binary fluid was documented by Hoseon Yoo and Viskanta [21] and Nguyen *et al.* [22]. Their models were developed to predict transport phenomena which accompany the solidification of mixtures. The results indicate that anisotropic effects are significant on the mushy region formed during solification.

CIRCULAR ENCLOSURES

In the past, the case of a circular enclosure has been investigated by Chang and Hsiao [23]. This section focuses on horizontal circular enclosures. Two types of geometries will be considered: the annulus (Figure 7a) with uniform but different temperatures on its two boundaries and the circular cylinder (Figure 7b) with imposed non-uniform temperature on its unique boundary.

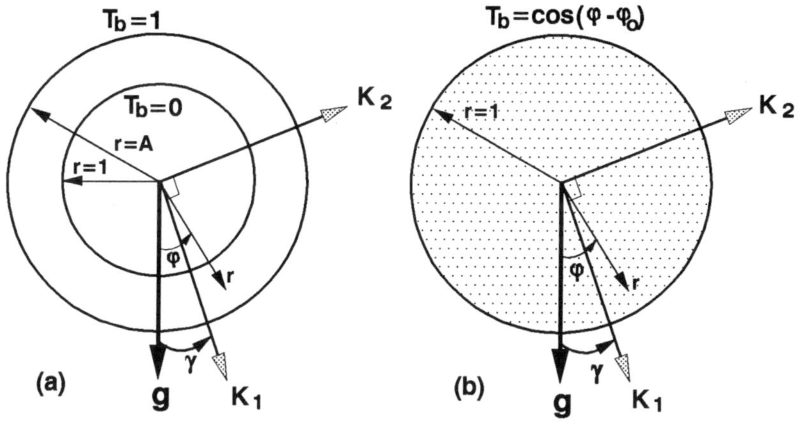

Figure 7 Geometries and coordinate systems

By refering to Figures 7a and 7b, the governing equations in cylindrical coordinates, equivalent to equations (4) and (5), are given by

$$\begin{aligned} &\frac{c}{r}\frac{\partial}{\partial r}\left(r\frac{\partial \psi}{\partial r}\right) - \frac{b}{r}\frac{\partial^2 \psi}{\partial r \partial \varphi} + \frac{a}{r^2}\frac{\partial^2 \psi}{\partial \varphi^2} \\ &+(K^*-1)\left(\frac{\cos 2\theta}{r}\frac{\partial \psi}{\partial r} + \frac{b}{r^2}\frac{\partial \psi}{\partial \varphi}\right) = -R\left[\frac{\partial T}{\partial r}\sin\varphi + \frac{1}{r}\frac{\partial T}{\partial \varphi}\cos\varphi\right] \end{aligned} \tag{34}$$

$$\frac{\partial T}{\partial t} + u\frac{\partial T}{\partial r} + \frac{v}{r}\frac{\partial T}{\partial \varphi} = \frac{1}{r}\frac{\partial}{\partial r}\left(r\frac{\partial T}{\partial r}\right) + \frac{1}{r^2}\frac{\partial^2 T}{\partial \varphi^2} \tag{35}$$

where $\theta = \varphi - \gamma$ and $k^* = 1$. The same dimensional scales as those of equations (4) and (5) have been used to make those equations dimensionless, the length scale being the inner radius of the annulus or the radius of the cylinder. The dimensionless stream function in equation (34) is related to the radial and azimuthal velocity components by $u = \partial\psi/(r\partial\varphi)$ and $v = -\partial\psi/\partial r$.

By comparison with equation (4), equation (34) contains two additional terms which arise in cylindrical coordinates when the curl of equation (2) is taken. The coefficients a, b and c on the left hand side of equation (34) are those defined in equation (6). Differences in order and sign are due to the particular position of the principal axes in Figures 7a and 7b. The angle θ involved in those coefficients is not constant but a function of the coordinate φ.

With circular boundaries, one cannot expect, for obvious reasons, simple analytical solutions such as those in rectangular cavities. Moreover, it may be worth underlining the following aspect of the present section: notwithstanding the present coordinate system, the kind of anisotropy that leads to equation (34) remains of the rectangular type; the angle between each principal axis and gravity does not change throughout the whole flow domain.

The annulus

As shown in Figure 7a, the inner and outer boundaries of the annulus are maintained at uniform temperatures T_i' and T_o', with the temperature scale $\Delta T' = T_o' - T_i' > 0$. The radius ratio is denoted by A. Thus, with the exception of the anisotropy effects, all conditions are symmetric with respect to the vertical diameter and that is why past authors when considering the isotropic case [24, 25] assumed that the solution would be symmetric with respect to the vertical diameter and solved for half the flow domain only. However, when symmetry is lost, as it is the case when the porous medium is contained between eccentric cylinders [26], or when the annulus rotates about its horizontal axis [27], or when anisotropy exists with an arbitrary angle γ, then one has to consider the whole annular domain which happens to be a doubly-connected region. In solving the stream function equation (34) for the flow field, one must consequently allow for a possible net flow $\Delta\psi$ circulating between the two boundaries. This net flow is defined as

$$\Delta\psi = \psi_1 - \psi_A = \int_1^A v\,dr \tag{36}$$

where ψ_1 and ψ_A are the values of the stream function on the inner and outer boundaries, respectively.

For two-dimensional natural convection in standard enclosures (simply-connected regions), ψ on the unique boundary may be set to any arbitrary value (usually zero) without loss of generality. For the present doubly-connected domain, an appropriate choice of boundary conditions for ψ, together with the thermal boundary conditions, is

$$\begin{array}{lll} r = 1 & T = 0 & \psi = \psi_1 \\ r = A & T = 1 & \psi = \psi_A = 0 \end{array} \tag{37}$$

The procedure to obtain ψ_1 is given in Robillard and Torrance [27] and yields the following result

$$\psi_1 = \frac{1}{(K^*+1)} \int_0^{2\pi}\int_0^A \left[R\,T\sin\varphi - \left(\frac{K^*-1}{2}\right) u \sin 2\gamma\right] dr\, d\varphi \tag{38}$$

which provides a way to evaluate ψ_1 at each time-step of a time-marching numerical method.

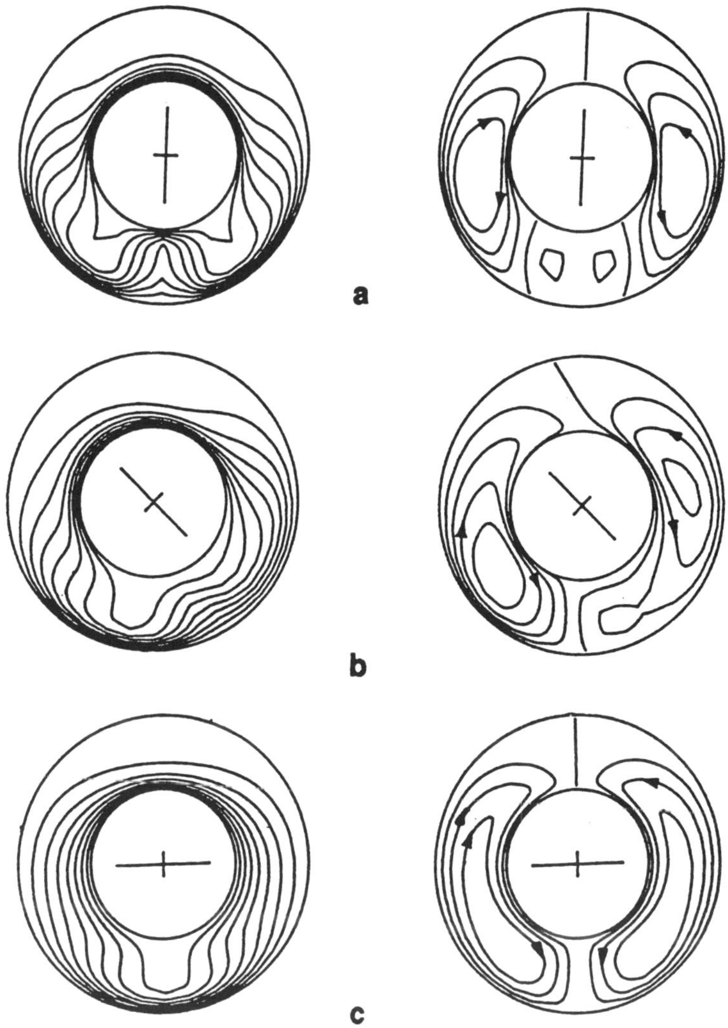

Figure 8 Effect of the angle γ on flow fields (right) and temperature fields (left) ($R = 300$, $K^* = 0.25$)

a) $\theta = 90^o$, $\psi_1 = 0$
b) $\theta = -45^o$, $\psi_1 = -2.858$
c) $\theta = 0^o$, $\psi_1 = 0$

Results obtained by solving numerically equations (34) and (35) subject to the conditions (37) and (38) are shown in Figures 9 and 10. Those figures have been obtained by Aboubi [10, 28] and are limited to a radius ratio $A = 2$.

Steady-state flow and temperature fields given by streamlines (right) and isotherms (left) are shown in Figure 8 for different values of γ, R and K^* being maintained at 300 and 0.25, respectively. At $\gamma = 0^o$ and 90°, the flow and temperature fields are symmetric with respect to the vertical diameter, as expected, and there is no net circulating flow ($\psi_1 = 0$). At $\gamma = 45^o$, symmetry is lost and a net circulating flow occurs ($\psi \neq 0$), due to the enhancement of the left convective cell which extends over a larger part of the flow domain to include the inner boundary. Figures 9a and 9b give ψ_1 and the Nusselt number Nu, functions of γ for $K^* = 0.25$. Curves are shown corresponding to $R = 200$, 300 and 400. For each curve of Figure 9a, ψ_1 starts from zero at $\gamma = 0$ and reaches a maximum before decreasing to zero at $\gamma = 90^o$. The curves are not symmetric, the maximum value being reached at a value of γ below 45°. The Nusselt number, defined as

$$Nu = \frac{lnA}{2\pi} \int_0^{2\pi} \frac{\partial T}{\partial r}\bigg|_{r=1} d\theta \tag{39}$$

is seen in Figure 9b to be a monotonically increasing function of γ from 0° to 90°.

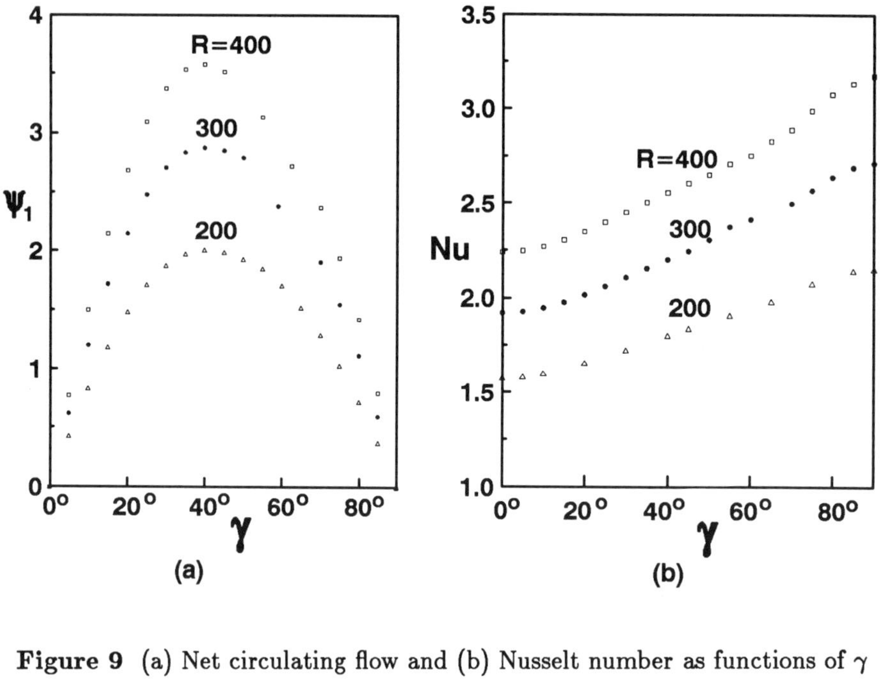

Figure 9 (a) Net circulating flow and (b) Nusselt number as functions of γ

The circular cylinder

Free convection may arise in a circular cylinder filled with a porous medium when a non-uniform temperature is imposed on the boundary [29, 30]. For instance, the cosine form $T'_b = T'_o + \Delta T' \cos(\varphi - \varphi_o)$, T'_b being the imposed boundary temperature, will produce in pure conduction a linearly stratified temperature field. When T'_o and $\Delta T'$ are taken as reference temperature and temperature scale respectively, the boundary temperature takes the dimensionless form given in Figure 7b. In particular, the heating phase angle φ_o may be chosen so that the pure conduction temperature gradient is vertical, with heating from below (adverse temperature gradient). For such a configuration, a stable pure conduction equilibrium is possible, up to a given threshold expressed in terms of a critical Rayleigh number.

Isotropic case. Storesletten and Tveitereid [29] have investigated analytically the case of an isotropic porous medium with heating from below. According to their marginal stability analysis, the incipient two-dimensional flow pattern consists of either two or three convective cells, as shown in Figure 10 from Zhao's work [31]. Either one or the other is likely to occur since both configurations have the same critical Rayleigh number, $R_c = g\beta\Delta T'_o K r'_o/(\nu\alpha) = 23.13$. They also found that two-dimensional flow patterns are limited to short pipes with $L/r'_o = 0.86$, L being the pipe length. Furthermore their nonlinear stability analysis, valid for sligthly supercritical Rayleigh numbers, indicates that both configurations remain stable beyond the threshold, as well as any linear combination of both.

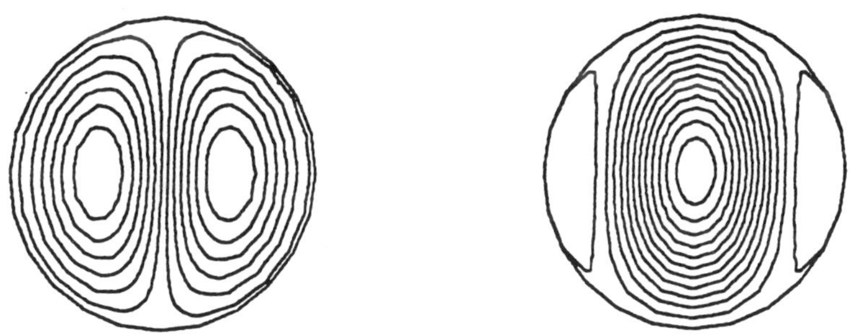

Figure 10 Flow patterns at the onset of convection (isotropic case)

One can directly deduce, by considering the dimensionless hydrodynamic circulation on the boundary

$$\Gamma_b = \int_0^{2\pi} v_b d\varphi \tag{40}$$

where v_b is the velocity on the boundary, that a flow configuration with more than one convective cell must be expected for the specific case discussed above. For an isotropic Darcy medium, the velocity in the φ direction on the boundary reduces to

$$v_b = R\ T_b \sin\varphi - \partial p/(r\partial\varphi) \tag{41}$$

With $T_b = \cos(\varphi - \varphi_o)$ being the dimensionless temperature on the boundary, equation (40) yields

$$\Gamma_b = \pi\ R\ \sin\varphi_o \tag{42}$$

Thus, for bottom heating ($\varphi_o = 0$), the hydrodynamic circulation Γ_b is zero. Apart from the particular case where v_b is zero everywhere, v_b must change sign at least twice along the closed-loop integral, equation (40), and this requirement implies the existence of at least two convective cells within the enclosure. In Figure 10, v_b reverses its direction twice and four times, respectively, for the two-cell and three-cell flow patterns.

Bifurcation diagrams of Figures 11a and 11b concern the steady-state finite amplitude convection. They are based on numerical results from Zhao's work [31]. Figure 11a is for bottom heating exclusively ($\varphi_o = 0$) and gives the maximum value of the stream function, ψ_{max}, as a function of the Rayleigh number. Reverse flows also exist with $\psi_{min} = -\psi_{max}$ and would provide branches symmetrical with respect to the abscissa. Those branches are omitted here. As shown in Figure 11a, the two types of flow configurations found at incipient convection are seen to be stable over a finite range $0 \leq R \leq 80$, in agreement with Storesletten and Tveitereid's predictions [29]. At higher Rayleigh numbers, the flow becomes of the oscillatory type. A more detailed bifurcation concerning the three-cell configuration is shown in Figure 11b, where $\psi_c = \psi_{max}$ (ψ_c being the stream function at the center) is given for different values φ_o, not too far from zero. The curves originate from $R = 0$, with no sharp transition at R_c. Again, in agreement with Storesletten and Tveitereid predictions [29], no flow configuration with two convective cells could be stabilized for $\varphi_o \neq 0$. The curves $\varphi_o > 0^o$ define an imperfect bifurcation diagram and are called *preferred branches* [32]. Other types of stable solutions with contrary motion are theoretically possible for the same heating angle φ_o and do exist in some situations [32]. Those solutions form the *isolated branches* of the imperfect bifurcation. They are obtained numerically in a transient manner, by imposing large initial perturbations on the system. As described in Zhao's work, no such solutions could be stabilized for the present problem, in conformity with Storesletten and Tveitereid's predictions [29]. In every attempt, no matter what was the initial perturbation, the flow always evolved toward the preferred branch.

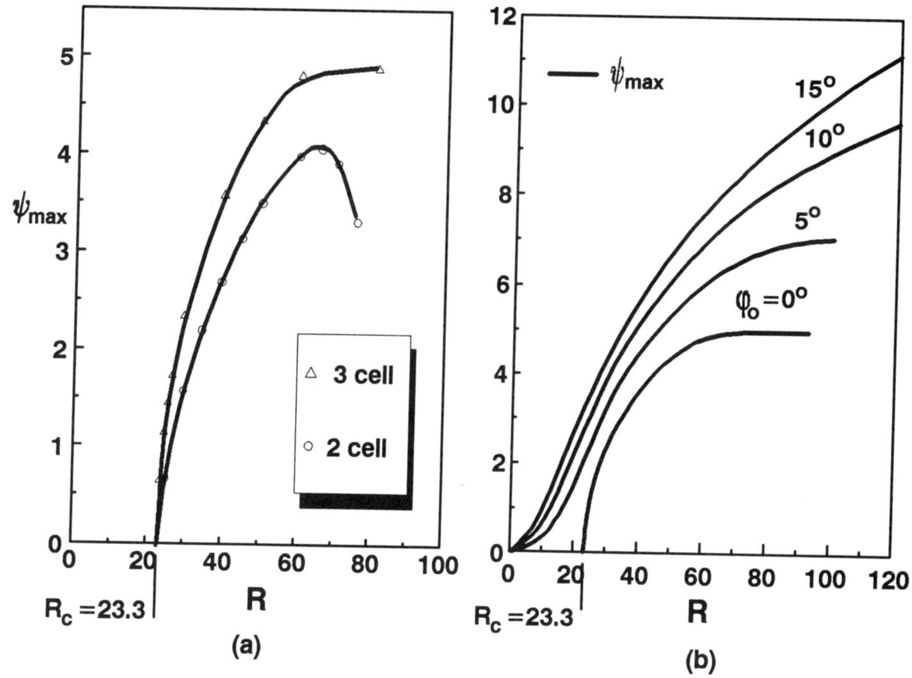

Figure 11 Bifurcation diagrams (a) two-cell and three-cell configurations and (b) effect of φ_o on the three-cell configuration

Anisotropic case. Figures 12 and 13, from Zhao's work [31], concern the anisotropic case. The linear stability analysis performed by this author relies largely on a numerical scheme, owing to the more complex situation where the threshold depends on the permeability K^* and the angle γ of the principal axes.

Assuming that the exchange of stability is satisfied, the steady-state flow and temperature fields are discretized in two N-component vectors $\vec{\Psi}$ and $\vec{T}$, with $N = (N_r - 1)N_\varphi + 1$, N_r and N_φ being the grid numbers used in the radial and azimuthal directions, respectively.

The linearized form of equations (34) and (35) can be written out as follows

$$A\,\overrightarrow{\Psi} + R\,B\,\overrightarrow{T} = 0 \tag{43}$$

$$C\,\overrightarrow{T} + D\,\overrightarrow{\Psi} = 0 \tag{44}$$

where A, B, C and D are $N \times N$ matrices. Equations (43) and (44) may then be reduced to the following canonical eigenvalue problem

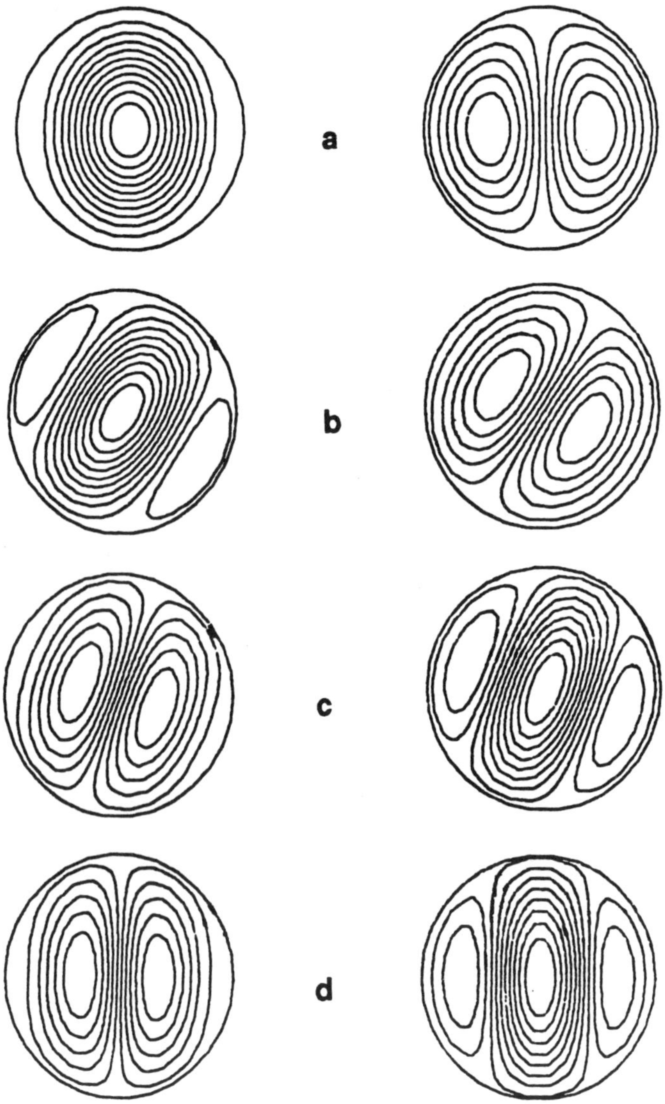

Figure 12 Incipient flow patterns at $K^* = 0.125$

a) $\gamma = 0^o$, $R_c = 46.10, 52.89$
b) $\gamma = 30^o$, $R_c = 36.77, 36.94$
c) $\gamma = 60^o$, $R_c = 23.50, 24.18$
d) $\gamma = 90^o$, $R_c = 19.88, 21.29$

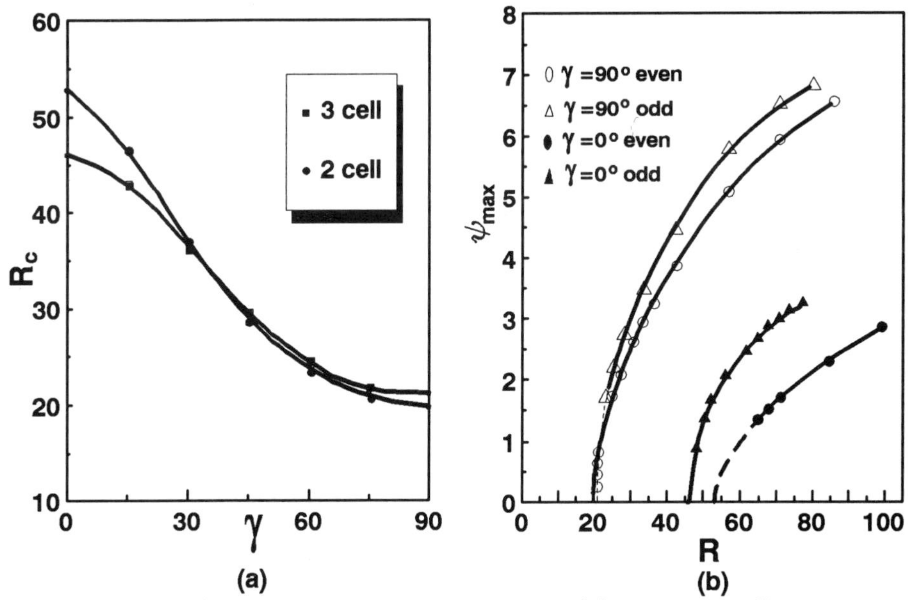

Figure 13 (a) Critical Rayleigh number and (b) bifurcation diagram

$$(\lambda I \; - \; E)\overrightarrow{\Psi} = 0 \tag{45}$$

where I is the unit matrix, $E = A^{-1}\, B\, C^{-1}\, D$ and $\lambda = 1/R$. If $\lambda_1 > \lambda_2 > \lambda_3 \ldots . > \lambda_N$ are eigenvalues of the problem, then $R_1 = 1/\lambda_1$ is the critical Rayleigh at which convective motion will occur.

The above approach to obtain the critical Rayleigh number was tested with the isotropic case. The computation lead to double roots ($R_1 = R_2 < R_3 = R_4 \ldots$) with the lowest two roots $R_1 = R_2 = 23.3$ and corresponding eigenvectors $\overrightarrow{\Psi}$ consisting of the two- and three-cell flow patterns shown in Figure 7.

In the case of an anisotropic medium, the first two roots, which again correspond to two- and three-cell flow patterns, become distinct. As stated in a paper by Aubry [33], a loss of multiplicity of eigenvalues is always associated with a breakdown in symmetry. Here, in undertaking the anisotropic medium, the centro-symmetry of the Darcy resistance has been lost.

Figure 12 shows pairs of incipient flow patterns (two-cell and three-cell) at $\gamma = 0^o$, 30°, 60° and 90°, the permeability ratio K^* being maintained at 0.125. According to the value of γ, it may consist of either two or three convective cells. The top flow field of the left column (Figure 12a) may be regarded as the limiting of the case of a three-cell configuration with vanishing secondary cells. One may also notice that the intermediate flow patterns ($0^o < \gamma < 90^o$) are no more symmetric with respect to the vertical diameter. For each γ of Figure 12, the flow pattern shown at the left is the one having the lowest threshold. Thus, with increasing R from zero, the first type of flow to occur will consist of either two or three convective cells, according to the angular position of the principal axes. That behaviour may be deduced from Figure 13a where the critical Rayleigh numbers corresponding to two-cell and three-cell flows have been plotted as functions of γ, for $K^* = 0.125$. It may be observed that those curves cross each other at a value of γ somewhere between 30° and 40°.

Finite amplitude convection gives rise to even more complex situations since a relatively large permeability in the vertical direction favours the doubling of convective cells, two-cell and three-cell flows becoming even and odd-cell flows, respectively.

When $\gamma = 0^o$ and 90°, solutions consisting of even-cell and odd-cell flow patterns may exit for given ranges of Rayleigh numbers and permeability ratios. Figure 13b shows a bifurcation diagram where ψ_{max}, the extremum value of the stream function is given as a function of R for $K^* = 0.125$ and $\gamma = 0^o$ and 90°. With increasing R, even-cell flows appears first at $\gamma = 90^o$ whereas odd cells will appear first when $\gamma = 0^o$. In each case, both flow configurations overlap for a given range of Rayleigh numbers. From all these results, it appears that the anisotropy has a strong influence on the critical value of R and on the initial flow pattern at the onset of convection. A minimum critical R can be achieved if the porous matrix is arranged such that the principal axis with the maximum permeability is in the vertical direction.

References

1. D. A. Nield and A. Bejan, *Convection in Porous Media*, Springer Verlag (1992).
2. J. Bear, *Dynamics of Fluids in Porous Media*, Elsevier, New-York (1972).
3. G. Degan and P. Vasseur, Natural Convection in a Vertical Slot Filled with an Anisotropic Porous Medium with Oblique Principal Axes, *Num. Heat Transfer*, **38**, 397-412 (1996).
4. J. Ni and C. Beckermann, Natural Convection in a Vertical Enclosure Filled with Anisotropic Porous Media, *J. Heat Transfer*, **113**, 1033-1037 (1991).
5. A. E. Gill, The Boundary Layer Regime for Convection in a Rectangular Cavity, *J. Fluid Mech.*, **25**, 515-536 (1966).
6. J. E. Weber, The Boundary-Layer Regime for Convection in a Vertical Porous Layer, *Int. J. Heat Mass Transfer*, **18**, 569-573 (1975).
7. A. Bejan, On the Boundary-Layer Regime in a Vertical Enclosure Filled with a Porous Medium, *Lett. Heat Mass Transfer*, **6**, 93-102 (1979).
8. P. G. Simpkins and P. A. Blythe, Convection in a Porous Layer, *Int. J. Heat Mass Transfer*, **23**, 881-887 (1980).
9. S. Kimura, Y. Masuda and T. K. Hayashi, Natural Convection in an Anisotropic Porous Medium Heated from the Side (Effects of Anisotropic Properties of Po-

rous Matrix), *Heat Transfer Jpn. Res.*, **22**, 139-153 (1993).
10. K. Aboubi, L. Robillard and E. Bilgen, Natural Convection in Horizontal Annulus Filled with an Anisotropic Porous Medium, *ASME/JSME Thermal Eng. Conf.*, **3**, 415-422 (1995).
11. D. E. Cormack, L. G. Leal and J. Imberger, Natural Convection in a Shallow Cavity with Differentially Heated End Walls. Part 1, Asymptotic Theory, *J. Fluid Mech.*, **65**, 209-229 (1974).
12. P. Vasseur, M. G. Satish and L. Robillard, Natural Convection in a Thin, Inclined Porous Layer Exposed to a Constant Heat Flux, *Int. J. Heat Mass Transfer*, **30**, 537-550 (1987).
13. G. Degan, P. Vasseur and E. Bilgen, Convective Heat Transfer in a Vertical Anisotropic Porous Layer, *Int. J. Heat Mass Transfer*, **38**, 1975-1987 (1995).
14. A. Bejan, The Boundary Layer Regime in a Porous Layer with Uniform Heat Flux from the Side, *Int. J. Heat Mass Transfer*, **26**, 1339-1346 (1983).
15. X. Zhang, Convective Heat Transfer in a Vertical Porous Layer with Anisotropic Permeability, *Proc. 14th Canad. Congr. of Appl. Mech.*, **2**, 579-580 (1993).
16. W. J. Chang and H. C. Lin, Natural Convection in a Finite Wall Rectangular Cavity Filled with an Anisotropic Porous Medium, *Int. J. Heat Mass Transfer*, **37**, 303-312 (1994).
17. S. K. Sinha, T. Sundarajan and V. K. Garg, A Variable Property Analysis of Alloy Solidification Using the Anisotropic Porous Medium Approach, *Int. J. Heat Mass Transfer*, **35**, 2865-2877 (1992).
18. G. Degan, Étude Numérique et Analytique de la Convection Naturelle en Milieu Poreux Anisotrope, Ph.D. Thesis, Ecole Polytechnique, University of Montreal (1997).
19. S. Kimura and A. Bejan, The Boundary-Layer Natural Convection Regime in a Rectangular Cavity with Uniform Heat Flux from the Side, *J. Heat Transfer*, **106**, 98-103 (1984).
20. R. Bennacer and A. Tobbal, Coupled Heat and Mass Transfer in Vertical Anisotropic Porous Layer, *Heat Transfer 96: 4th Int. Conf. on Adv. Comp. Meth. in Heat Transfer*, Italy, 493-502 (1996).
21. H. D. Nguyen, S. Paik and R. W. Douglas, Study of Double-Diffusive Convection in Layered Anisotropic Porous Media, *Num. Heat Transfer*, **26**, 489-505 (1994).
22. Hosen Yoo and R. Viskanta, Effect of Anisotropic Permeability on the Transport Process During Solidification of a Binary Mixture, *Int. J. Heat Mass Transfer*, **35**, 2335-2346 (1992).
23. W. J. Chang and C. F. Hsiao, Natural Convection in a Vertical Cylinder Filled with Anisotropic Porous Media, *Int. J. Heat Mass Transfer*, **36**, 3361-3367 (1993).
24. J. P. Caltagirone, Thermoconvective Instabilities in a Porous Medium Bounded by Two Concentric Cylinders, *J. Fluid Mech.*, **76**, 337-362 (1976).

25. Y. F. Rao, K. Fukuda and S. Hosegawa, Steady and Transient Analysis of Natural Convection in a Horizontal Porous Annulus with the Galerkin Method, *J. Heat Transfer*, **109**, 919-927 (1987).
26. Y. Z. Wang and H. H. Bau, Low Rayleigh Number Convection in Horizontal Eccentric Annuli, *Phys. Fluids*, **31**, 2473-2473 (1988).
27. L. Robillard and K. E. Torrance, Convective Heat Transfer Inhibition in an Annular Porous Layer Rotating at Weak Angular Velocity, *Int. J. Heat Mass Transfer*, **33**, 953-963 (1990).
28. K. Aboubi, Convection Naturelle dans un Espace Annulaire: Effets de Rotation et d'Anisotropie, Ph.D. Thesis, Ecole Polytechnique, University of Montreal (1996).
29. L. Storesletten and M. Tveitereid, Natural Convection in a Horizontal Porous Cylinder, *Int. J. Heat Mass Transfer*, **34**, 1959-1968 (1991).
30. L. Robillard, X. Zhang and M. Zhao, On the Instability of a Fluid-Saturated Porous Medium Contained in a Horizontal Circular Cylinder, *ASME Winter Annual Meeting*, **264**, 49-55, (1993).
31. M. Zhao, Natural and Mixed Convection in a Horizontal Circular Cylinder, Ph.D. Thesis, Ecole Polytechnique, University of Montreal (1996).
32. P. Ehrhard and U. Müller, Dynamic Behavior of Natural Convection in a Single-Phase Loop, *J. Fluid Mech.*, **217**, 487-515 (1990).
33. N. Aubry and R. Lima, Spatio-Temporal and Statistical Symmetries, *J. Statistical Phys.*, **81**, 793-828 (1995).

INTERNAL NATURAL, FORCED AND MIXED CONVECTION IN FLUID-SATURATED POROUS MEDIUM

B. X. WANG, J. H. DU and X. F. PENG

Thermal Engineering Department, Tsinghua University, Beijing 100084, China

INTRODUCTION

Natural, forced and mixed convection in fluid-saturated porous media occurs in a wide variety of areas in nature and engineering. Numerous applications can be found in geothermal engineering, petroleum technology, underground pollution and environmental protection, nuclear reactor engineering, grain and coal storage with natural ventilation, drying technologies, building thermal insulations, enhanced solid matrix heat exchangers, chemical reactors intensified with catalysts, etc. Natural convection induced by temperature and/or concentration gradients is dominant in many situations. Because of the high resistance of the porous media solid matrix, the flow velocity is usually very small and mixed convection may also exist in many cases. Consequently, understanding the fundamentals of transient and steady natural convection in saturated porous media is of critical importance. In addition, a reliable and practical criterion and method is needed to identify mixed convection situations.

The widespread existence and the complicated phenomena associated with transport processes in porous media have led to a continuing interest and many thorough investigations in different branches of science and engineering in past decades. Combarnous and Bories[1] and Cheng[2] provided some extensive reviews on natural convection in porous media for studies related to geothermal engineering. Bear[3] provided an excellent exposition on the fluid flow in porous media in his book and the state-of-the-art on transport in porous media can also be found in books by Kakac[4] and Kaviany[5]. This chapter summarizes our theoretical and experimental studies on internal natural, forced and mixed convection in saturated porous media in recent years. The organization of the content is: (1) The physics of natural convection and the corresponding mathematical descriptions or the governing equations are first presented for mass, momentum and energy based on the Darcy model. A numerical solution method is then described which uses the vorticity-stream function formulation and a second-order, finite-difference method(FDM). The governing equations for transient and steady-state, three-dimensional natural convection were solved numerically for tilted rectangular or annular porous closures with different boundary conditions to predict the flow field and temperature distribution in these geometries. The effects of aspect ratio and angle of inclination were compared with experimental data. (2) Developing and fully developed forced convective heat transfer in packed channels was experimentally investigated for different fluids and differently sized porous media. The importance of dispersion in forced convection in porous media is discussed and a thermal dispersion model is presented with an appropriate wall function. The factors that affect the thermal entrance length are discussed in detail. (3) The dimensionless parameters governing mixed convection in a vertical annulus filled with porous media are deduced through dimensional analysis. A criterion on $Gr^{**}Da/Re_H$ is suggested for identifying pure forced as opposed to mixed convection. Mixed convection in a vertical porous channel bounded by two parallel plates with asymmetrical heating is investigated with Brinkman's model in the momentum equation. In addition, criteria that identify the occurrence of flow reversals are deduced from an analytical solution of the Brinkman-Forchheimer-Darcy momentum equation for fully-developed flow. Finally, existing theoretical and experimental issues pertaining to internal convection in porous media are identified along wh appropriate further investigations.

NATURAL CONVECTION IN CYLINDRICAL OR RECTANGULAR ENCLOSURES

Numerical Simulation

Consider a system consisting of a fluid-saturated porous medium bounded between two concentric inclined cylinders having radii r_i and r_o, respectively. The cylinders are impermeable with the inner cylinder maintained at either constant uniform temperature, T_i, constant uniform heat flux, q_i, or axially symmetrical conditions. The outer cylinder is maintained at constant uniform temperature,T_o or constant uniform heat flux, q_o, and both end walls are thermally insulated, see Figure 1. The fluid motion can be induced by temperature differences between the inner and outer cylinders, or by internal heat generation in the medium.

The complexity of the flow and thermal transport processes between the fluid and the porous matrix material which has an irregular geometrical arrangement makes obtaining the solutions for the local velocity and temperature fields prohibitively expensive. The governing equations for the transport processes in the porous media are therefore obtained by the local averaging technique, see Whitaker[6]. The simplifications and assumptions employed in these studies can be summarized as follows:

(1) The porous media is isotropic and homogenous, and the solid matrix is made of uniform spheres. The deformation of the solid matrix due to temperature and pressure variations can be neglected.

(2) The thermophysical properties, such as the viscosity of the fluid and the thermal conductivity and the specific heat of the fluid and of the porous media, are assumed to be constant except for the density which is described by using the Boussinesq approximation.

(3) The Darcy model, which relates the volumetric flow rate with the pressure drop across the porous media in a linear relationship, is employed to describe the flow.

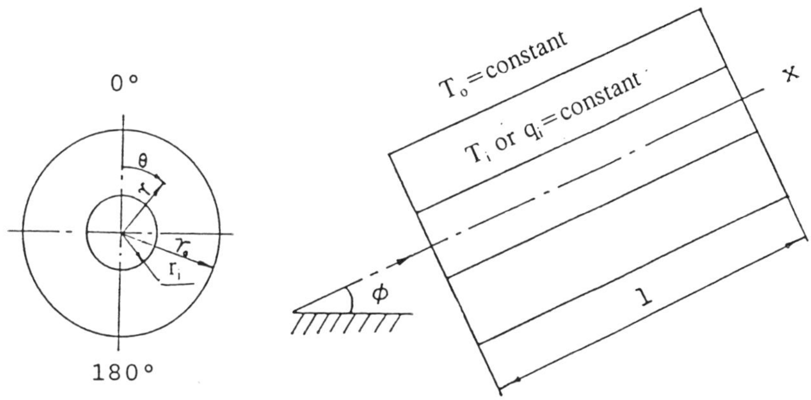

Figure 1 Analytical model

(4) The difference between the temperature of the solid matrix and fluid phase is neglected and the medium is considered to be a single continuum with one energy equation describing the average temperature $T=T_s=T_f$.
(5) Radiation in the porous medium is neglected.

With these assumptions, the volume averaged governing equations are given by Wang and Zhang[7] as:

$$\nabla \cdot \mathbf{V} = 0 \tag{1}$$

$$\frac{\rho}{\varepsilon}\frac{\partial \mathbf{V}}{\partial \tau} = -\nabla p + \mathbf{G} - \frac{\mu}{K}\mathbf{V} \tag{2}$$

$$(\rho c_p)^* \frac{\partial T}{\partial \tau} = \lambda^* \nabla^2 T - (\rho c_p)_f \nabla \cdot (\mathbf{V}\,T) \tag{3}$$

$$\rho_f = \rho_0 [\, 1 - \beta_T (T - T_0)\,] \tag{4}$$

The effective thermal conductivity of the medium,λ^*, and the effective product of the density and the specific heat $(\rho c_p)^*$, are given as

$$\lambda^* = \varepsilon\lambda + (1-\varepsilon)\lambda \tag{5}$$

$$(\rho c_p)^* = \varepsilon(\rho c_p)_f + (1-\varepsilon)(\rho c_p)_s \tag{6}$$

where $\mathbf{V}$ is the velocity vector, T is the temperature, p is the pressure, c_p is the specific heat capacity, ρ is the density, μ is the viscosity, β_T is the thermal expansion coefficient of the fluid, K is the permeability, ε is the porosity and τ is the time. The subscripts 'f' and 's' denote fluid and solid, respectively. The gravitational volumetric force vector **G**, which for the given geometry, can be expressed as

$$\mathbf{G} = [-g\beta_T \cos\theta \cos\phi,\ -g\beta_T \sin\theta \cos\phi,\ -g\beta_T \sin\phi] \begin{bmatrix} \mathbf{e}_r \\ \mathbf{e}_\theta \\ \mathbf{e}_z \end{bmatrix} \tag{7}$$

here **g** is the gravitational acceleration, $\mathbf{e}_r$, $\mathbf{e}_\theta$ and $\mathbf{e}_z$ are unit vectors in the r, θ and z directions, respectively.

As shown by Caltagirone[8], if the Darcy number, $Da=K/r_i^2$ is less than about 10^{-3}, the transient response term, $\partial\mathbf{V}/\partial\tau$, in equation (2) tends to zero within a very short time. Therefore, equation (2) becomes

$$\mathbf{V} = -\frac{K}{\mu}\,(\nabla p + \rho\,\mathbf{G}) \tag{8}$$

For two-dimensional flow, the stream function is a scalar, in order to extend this concept to the present three-dimensional investigations, the stream function ψ will be a vector with three nonvanishing components ψ_r, ψ_θ and ψ_z, which can be defined as

$$\mathbf{V} = \nabla \times \psi \tag{9}$$

Dimensionless quantities are introduced as

$$R = r/(r_o-r_i) \quad Z = z/(r_o-r_i) \quad \Psi_r = \psi_r/(l\upsilon^*) \quad \Psi_\theta = \psi_\theta/(l\upsilon^*)$$
$$\Psi_z = \psi_z/(l\upsilon^*) \quad \Theta = (T-T_0)/(T_i-T_0) \quad \bar{\tau} = \tau/[r_o^2/a^*] \tag{10}$$

where $\upsilon^*=\lambda^*/[(\rho c_p)_f(r_o-r_i)]$, which is a characteristic speed, and $a^*=\lambda^*/(\rho c_p)^*$. Then, the non-dimensional forms of equations (2) and (3) can then be expressed as derived by Wang and Zhang[7]:

$$\nabla^2\Psi_r - \left[\frac{\Psi_r}{R^2} + \frac{2}{R^2}\frac{\partial\Psi_r}{\partial\theta}\right] = -Ra\left(\frac{\sin\phi}{R}\frac{\partial\Theta}{\partial\theta} + \frac{r_i}{l}\sin\theta\cos\phi\frac{\partial\Theta}{\partial Z}\right) \tag{11}$$

$$\nabla^2\Psi_\theta - \left[\frac{\Psi_\theta}{R^2} - \frac{2}{R^2}\frac{\partial\Psi_r}{\partial\theta}\right] = Ra\left(\sin\phi\frac{\partial\Theta}{\partial R} - \frac{r_i}{l}\cos\theta\cos\phi\frac{\partial\Theta}{\partial Z}\right) \tag{12}$$

$$\nabla^2\Psi_z = Ra\left(\sin\theta\cos\phi\frac{\partial\Theta}{\partial R} + \frac{1}{R}\cos\theta\cos\phi\frac{\partial\Theta}{\partial\theta}\right) \tag{13}$$

$$\frac{\partial\Theta}{\partial\bar{\tau}} = \nabla^2\Theta - \frac{\mathbf{V}}{\upsilon^*}\cdot\nabla\Theta \tag{14}$$

where

$$\nabla^2 = \frac{\partial^2}{\partial R^2} + \frac{1}{R}\frac{\partial}{\partial R} + \frac{1}{R^2}\frac{\partial^2}{\partial \theta^2} + \frac{(r_o-r_i)^2}{l^2}\frac{\partial^2}{\partial Z^2} \tag{15}$$

and the Rayleigh number, Ra, is given as

$$Ra = Gr\ Pr = g\ \beta_T\ \rho_0\ (\ \rho c_p\ )_f\ (\ r_o-r_i\ )\ K\ (\ T_i-T_o\ )\ /\ (\ \lambda^*\mu\ ) \tag{16}$$

The dimensionless equations (11) to (14) also apply for prescribed uniform heat flux at the inner cylinder surface, q_i, with the reference temperature difference defined as

$$(\ T_i-T_o\ ) = q_i\ (\ r_o-r_i\ )\ /\ \lambda^* \tag{17}$$

and the corresponding Rayleigh number, based on the heat flux, is defined as

$$Ra = g\ \beta_T\ \rho_0\ (\ \rho c_p\ )_f\ (\ r_o-r_i\ )^2\ K\ q_i\ /\ (\ \lambda^{*2}\mu\ ) \tag{18}$$

The boundary conditions corresponding are as follows:

$$\begin{aligned}
R = R_1: &\quad \frac{\partial(R\Psi_r)}{\partial R} = \Psi_\theta = \Psi_z = 0, \quad \Theta = 1 \\
R = R_2: &\quad \frac{\partial(R\Psi)}{\partial R} = \Psi_\theta = \Psi_z = 0, \quad \Theta = 0 \\
\theta = 0,\ \pi: &\quad \frac{\partial\Psi}{\partial\theta} = \Psi_r = \Psi_z = 0 \\
Z = 0,\ A: &\quad \frac{\partial\Psi_z}{\partial Z} = \Psi_r = \Psi_\theta = 0
\end{aligned} \tag{19}$$

The respective local Nusselt numbers for the inner and outer cylinder surfaces are defined as

$$Nu_i = \ln(\frac{r_o}{r_i})\ \frac{\partial\Theta}{\partial R}\bigg|_{R=R_1} \tag{20}$$

and

$$Nu_o = (\frac{r_o}{r_i})\ \ln(\frac{r_o}{r_i})\ \frac{\partial\Theta}{\partial R}\bigg|_{R=R_2} \tag{21}$$

and the mean Nusselt numbers at each surface are given, respectively, by

$$Nu_{m,i} = \frac{1}{\pi}\int_0^1\int_0^1 Nu_i \, d\theta \, dZ \qquad (22)$$

and

$$Nu_{m,o} = \frac{1}{\pi}\int_0^1\int_0^1 Nu_o \, d\theta \, dZ \qquad (23)$$

The flow and temperature fields were determined using finite-difference equations derived from equations (11) to (14). The first and second derivatives were approximated by three-point, central-difference expressions which have second-order accuracy. The discretization equations are solved by an improved Alternating Direction Implicit (ADI) procedure and the method of Successive Over-Relaxation (SOR). The governing equations were solved iteratively. At each step, the stream function iteration was completed six times, then the latest stream function values were used to calculate the temperature distribution. The whole iteration was repeated until the following convergence criteria were satisfied at every grid point,

$$\left|\frac{\Psi^{n+1}-\Psi^{n}}{\Psi^{n+1}}\right|_{max} \le 10^{-3}, \qquad \left|\frac{\Theta^{n+6}-\Theta^{n}}{\Theta^{n+6}}\right|_{max} \le 10^{-3} \qquad (24)$$

The characteristic parameters of the porous media for solving the governing equations could be taken form data base or measured specially. Lei et al.[9] proposed a method to measure the permeability and porosity of unconsolidated porous media. The Rayleigh number was calculated based on the fluid and solid properties at the average mean temperature $(T_i+T_o)/2$. Purified water was used in the experiments and properties were measured accurately. Numerical results have been obtained for: $Ra \le 10^3$, aspect ratio, $A=l/(r_o-r_i)=1\text{-}15$, radii ratio $R_o=(r_o-r_i)/r_i$, from 1 to 10, and angle of inclination, ϕ, from 0° to 90°.

Figures 2 and 3 show the transient streamlines and isotherms at different times for vertical and horizontal porous annuli, respectively. It is seen, as time increases, the convection zone expands and develops and the temperature distribution departs from the pure conduction solution.

Figure 4 shows the transient variation of the local and mean Nusselt numbers along the inner and outer walls for a horizontal porous annulus. The local heat transfer at the inner wall is high at the beginning but decreases with increasing time. Whilst the transient response along the outer wall is low at the beginning but increases with the advance of time. The mean heat transfer at the inner wall decreases rapidly when convection begins. The time required to initiate convection and the duration of the transients decrease with increasing Ra and R_o. The mean Nusselt numbers along the inner and outer walls are identical at steady-state and depend only on the value of Ra. This is because that, in response to the constant heat flux the temperature at the inner wall rises rapidly, and as heat penetrates towards the outer side wall, by diffusion and convection of the fluid, the

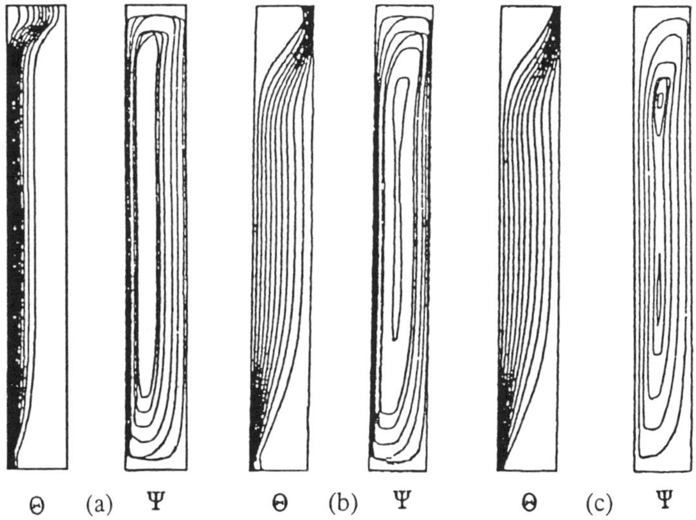

Figure 2 Transient streamlines and isotherms in a vertical annuli for Ra=200, A=7.5 and R_0=2
(a) $\bar{\tau}$=0.04, (b) $\bar{\tau}$=0.2775, (c) steady state

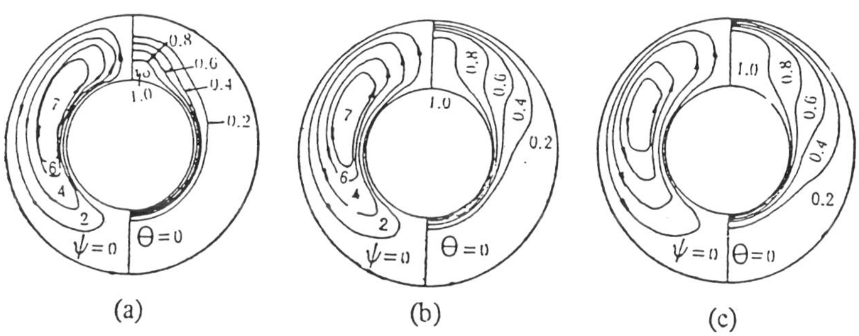

Figure 3 Transient streamlines and isotherms in a horizontal annulus for Ra=200 and R_o=2
(a) $\bar{\tau}$=0.0275, (b) $\bar{\tau}$=0.1025, (c) $\bar{\tau}$=0.2525

bulk fluid temperature increases and the heat transfer at the outer wall develops. The results for a vertical annulus can be found in Wang and Zhang[10].

The predicted effects of the inclination angle, ϕ, on the local Nusselt number indicated that, the end effect on the inner and outer local Nusselt numbers becomes more pronounced and the effect of θ decreasing as the inclination angle of the annulus

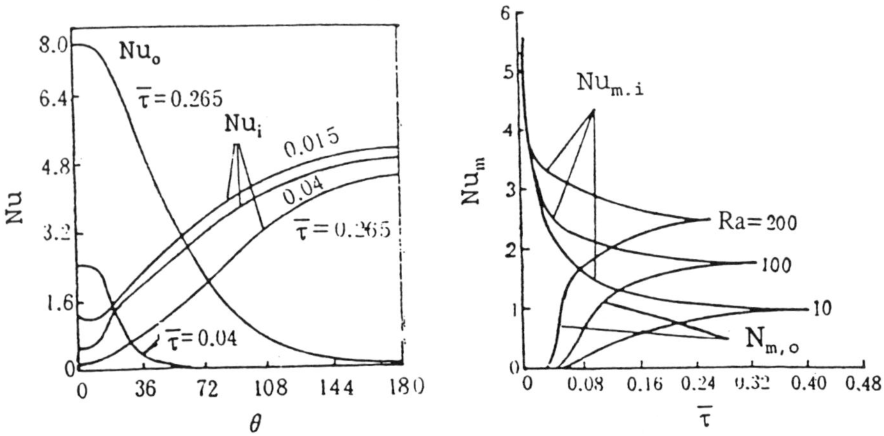

(a) $Nu=f(\theta,\bar{\tau})$ (b) $Nu_{m,i}$, $Nu_{m,o}=f(Ra,\bar{\tau})$

Figure 4 Local and mean Nusselt numbers for a horizontal annulus with $R_o=2$

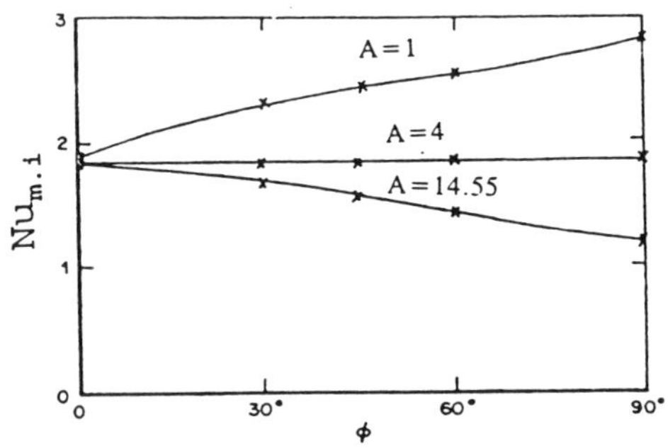

Figure 5 Effects of the aspect ratio and the inclination angle on $Nu_{m,i}$ for $Ra=100$ and $R_o=2$

increases. Further numerical studies demonstrated that the isotherms shift towards the hot wall so that the heat transfer is enhanced as Ra, A and R_o increase.

Figure 5, quoted from Wang and Zhang[7], shows the effect of A and ϕ on the mean Nusselt number along the inner wall, with $Nu_{m,i}$ decreasing for $A>4$ but increasing for $A<4$ as ϕ increases. So there exists a "critical aspect ratio", A_{cr}, for which the inclination angle ϕ has no effect on the heat transfer. Further numerical calculations show that A_{cr} decreases as R_o increases, and that the Rayleigh number has only a minor effect on the critical aspect ratio. There has been much disagreement in the literature on the effect of the angle of inclination on the heat transfer and this may be explained by these results.

Experimental Check

To verify the mathematical model and the numerical results, an experimental apparatus was designed and built, see Wang and Zhang[10]. The measured and predicted radial temperature profiles at three different locations in vertical and horizontal annuli are compared in Figure 6. The measured and predicted temperature distributions agreed well for the specified experimental conditions.

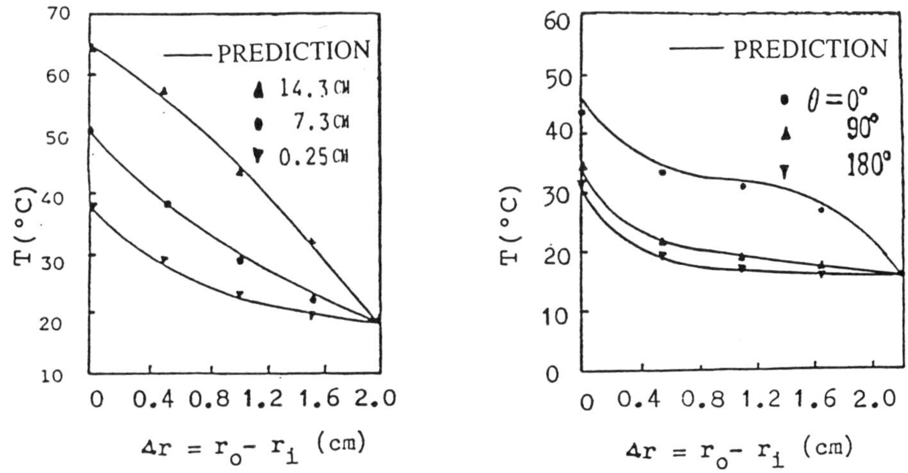

(a) Vertical annuli

A=14.55 P=169.4W/m

(b) Horizontal annuli

R_o=8.3, P=67.1W/m

Figure 6 Comparisons of predicted and experimental temperatures

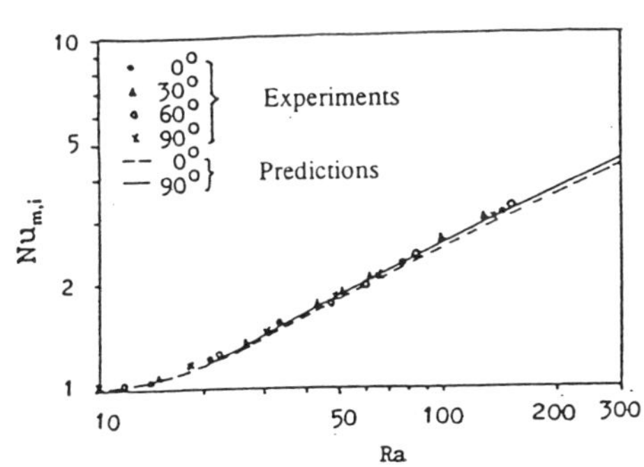

Figure 7 Mean Nusselt numbers as a function of Ra in annuli for $R_o=5$ and $A=3$

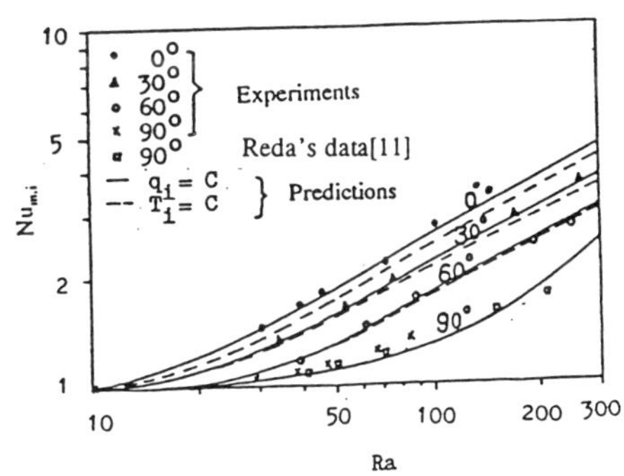

Figure 8 Effect of Ra and ϕ on $Nu_{m,i}$ for $R_o=5$ and $A=14.55$

Figure 7 shows the measured values of $Nu_{m,i}$ compared with the numerically predicted values for A=3 and R_o=5. Since A is less than A_{cr}=4, $Nu_{m,i}$ will decrease from the Nusselt number for A=2 and R_o=2 and the annulus inclination angle has little effect on the measured value of $Nu_{m,i}$ because A is near A_{cr}. As shown in Figure 8 for A=14.55 and R_o=5, increasing ϕ causes the mean Nusselt number at the inner cylinder, $Nu_{m,i}$, to rapidly decrease. The measured and predicted Nusselt numbers are consistent over the entire range of values of Ra and ϕ studied and the experimental results are confirmed.

Natural Convection with Internal Heat Sources

Natural convection in a vertical porous cylinder of radius r_o with uniform internal heat generation and side-wall cooling has also been studied numerically using the finite-difference method(FDM) by Rao and Wang[12]. Representative streamlines and isotherms for aspect ratio $A(=h/r_o)=4$ are shown in Figure 9. The Rayleigh number is taken as

$$Ra = K\beta_T g S r_o^2 / (\nu_f a^* \lambda^*) \tag{25}$$

in which ν_f is the kinematic viscosity of the fluid, $a^*=\lambda^*/(\rho c_p)^*$ is the effective thermal diffusivity and S is the uniform volumetric heat generation rate. For very low values of Ra, the streamlines are parallel in the central part of the cylinder, except near the top and bottom ends, and the value of ψ_{max} is very small, which indicates a very weak convective flow. The isotherms deviate only slightly from those for pure conduction, indicating that the convection terms and the other terms in the governing equations are negligible. Therefore, this range, i.e. Ra<50, may be defined as the 'pseudo-conduction region'.

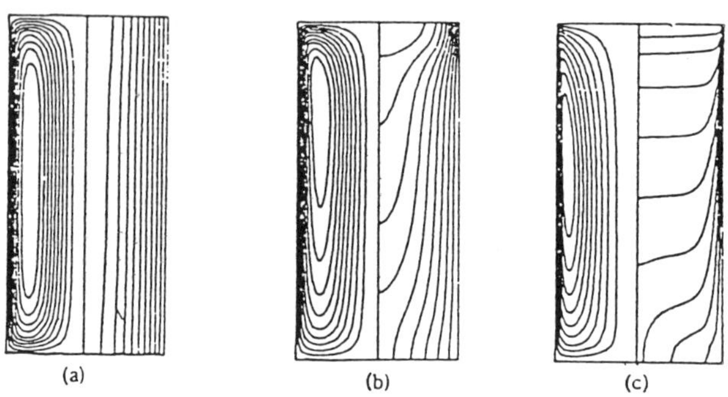

Figure 9 Streamlines and isotherms for A=4

(a) Ra=5.6, $\psi_{max}=0.088$, $\Theta_{max}=0.257$ (b) Ra=180, $\psi_{max}=2.32$, $\Theta_{max}=0.258$
(c) Ra=3.2×10^4, $\psi_{max}=40.3$, $\Theta_{max}=0.0545$

Nevertheless, the central axis of the cylinder is no longer isothermal due to the convection, and the temperature along the axis increases with height, reaching its maximum value at the center of the top of the enclosure. Thus, the temperature calculated assuming pure conduction underestimates the maximum temperature. The results show that, in the low Ra region, the variation of cylinder height does not change the flow and temperature fields near the top and bottom ends and thus the values of Ψ_{max} and Θ_{max} do not vary significantly with aspect ratio, A, as shown in Figure 10. Increasing Ra only lengthens the parallel flow region in the interior and therefore the results of the flow and heat transfer for the cylinder with a small aspect ratio can be applied to larger aspect ratio cylinders. For high values of Ra, most of the cylinder is filled by the ascending hot flow, while the descending flow along the side wall becomes very thin and forms a boundary-layer flow. Temperature stratification can be observed in the upper part of the cylinder and extends towards the lower part as Ra increases. The stratification occupies most of the cylinder, except in the bottom region, or in the vicinity of the side wall. The variation of Ψ_{max} and Θ_{max} with Ra and the cylinder height is also given in Figure 10. At high value of Ra, i.e. when convection is dominant, the circulation rate and the maximum temperature, Ψ_{max} and Θ_{max}, increase since the ascending flow path is longer for a larger cylinder height.

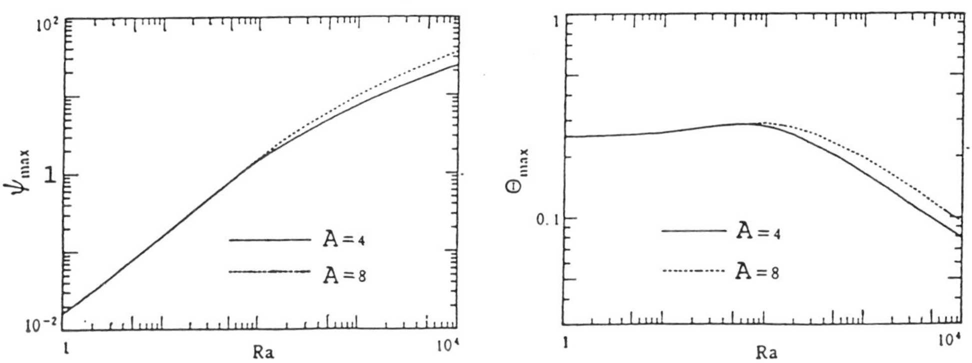

(a) ψ_{max} (b) Θ_{max}

Figure 10 Effect of A on ψ_{max} and Θ_{max}

Figure 11 shows the variation of the local Nusselt number, Nu, along the cylinder axis y as a function of Ra for different cylinder heights, y=0~h. For low values of Ra, the Nusselt number deviates from the mean value only near the ends and the deviations coincide for different aspect ratios. Further, for high values of Ra, the Nusselt numbers for A=4 and A=8 almost coincide. The flow and temperature fields are also similar for different height cylinders but for moderate Rayleigh numbers, the effects of cylinder height on the Nusselt number, Ψ_{max} and Θ_{max}, must be considered.

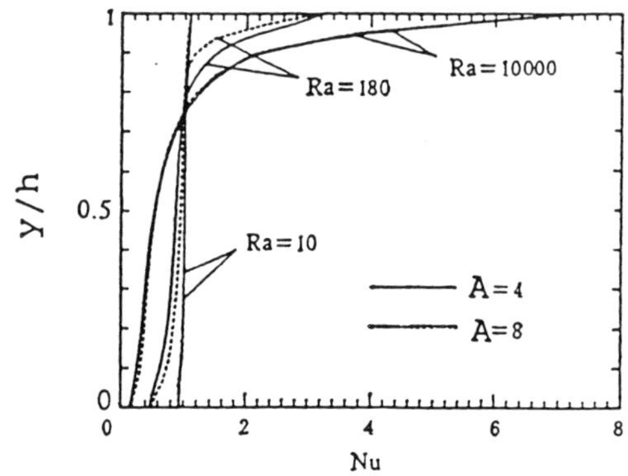

Figure 11 Local Nusselt number along the annulus wall

FORCED CONVECTION

There are far fewer experimental than theoretical investigations of forced convection flows in packed channels in the literature. Verschoor and Schuit[13] conducted experiments to investigate the forced convection of air in a cylindrical packed tube with uniform surface temperature, while Quinton and Storrow[14] performed experiments with a similar cylindrical packed tube but with uniform wall heat flux. Schroeder et al.[15] experimentally investigated the convection of water in a packed bed between parallel plates heated asymmetrically. All these experiments were limited to the fully thermally-developed region. For this reason, we conducted a series of experiments to investigate the fluid flow and convective heat transfer of water and transformer oil in a concentric annulus filled with differently sized glass beads. The main purpose of this work was to identify the forced convection and heat transfer characteristics for flow inside packed channels in the entrance region. A detailed description of the experimental installation and results were reported by Wang and Du[16].

Figure 12 illustrates the variation of the local Nusselt number along the flow direction with Re_p for water flowing through annuli of different H/d_p. The Nusselt number for the fully developed region increases with increasing Re_p, which is different from that for laminar flow in unpacked channels. This increase can be explained as the dispersion effect which is one of the main characteristics of transport phenomena in porous media. The thermal entrance effect, i.e. the development of the thermal boundary-layer in the packed annuli, is faster than that in an unpacked annular channel. However, the thermal boundary-layer development is not as fast as the hydrodynamic boundary-layer development in the packed channel. The influence of the packed bed on the momentum transfer is similar to that of turbulence, since there are successive flow disturbances due to the packed beads which force the flow through a tortuous path around the solid beads. Because of the direct moment transport induced by the packed beads and the lateral mixing of the fluid, a fully developed velocity profile is rapidly established in a very

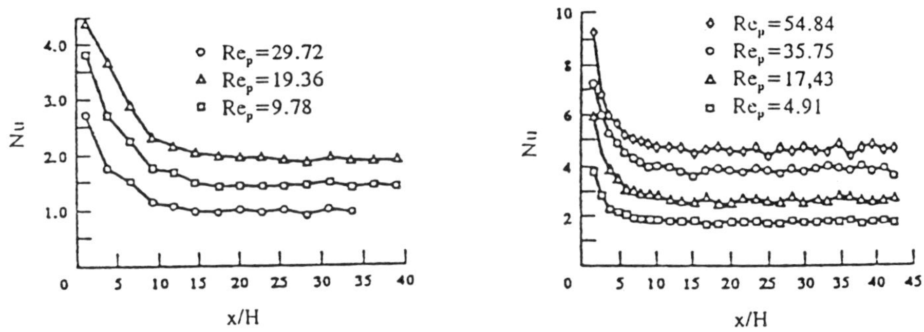

(a) $H/d_p=13.2$ (b) $H/d_p=3.7$

Figure 12 Local Nusselt number for water in the entrance region of an annulus

short distance from the channel entrance. However, the beads do not affect the heat transfer in the same way as the momentum transport and it may be assumed that conduction through the beads will significantly promote the development of the thermal boundary-layer and, hence, the heat transfer. This effect, known as thermal dispersion, acts via the diffusion of heat through the beads and the mixing of the fluid. Comparing Figures 12 (a) with (b), shows that the channelling effect causes the thermal boundary-layer to develop faster with large beads than with smaller beads.

The characteristics of the experimental results for transformer oil are very similar to those observed when using water and, hence, substantiate the above conclusions.

In the thermally fully developed region, for either constant temperature or constant heat flux at the wall, the dimensionless temperature profile remains invariant with the distance along the flow direction, that is

$$\frac{\partial}{\partial x}\left(\frac{T-T_w}{T_m-T_w}\right) = 0 \tag{26}$$

where T_w is the temperature at the wall of the channel and T_m is the mixed-mean temperature. For uniform wall flux heating, the wall temperature T_w varies linearly and with the same slope as T_m and differs from T_m by a constant amount. Therefore, for the internal flow through a packed channel with uniform wall heat flux, the thermally fully developed region can be identified by observing how far downstream the axial variation of the measured values of T_w plot as straight lines. The thermal entrance length, L_t, obtained from all the forced-convection experimental data for water and transformer oil, is shown in Figure 13. It varies with Reynolds number, which is defined using the particle diameter, Re_p, and varies also with H/d_p.

For the constant temperature wall condition, in the thermally fully developed region, the mixed-mean temperature of fluid, T_m, varies as an exponential function. Hence, we

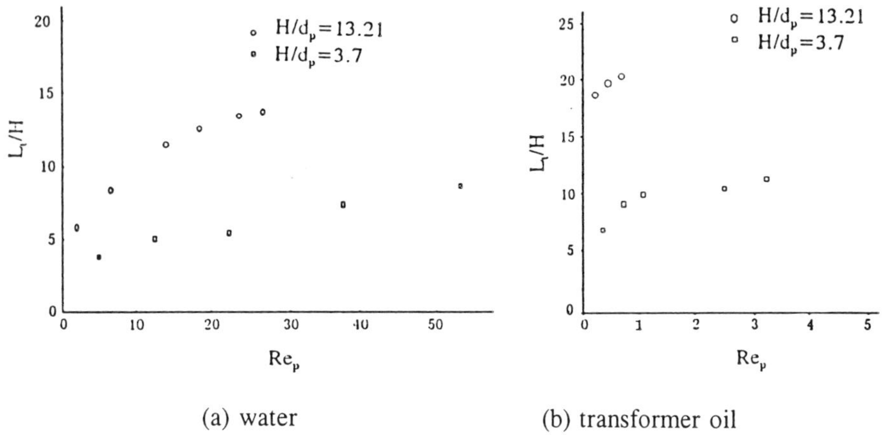

Figure 13 Thermal entrance length for flow in an annulus

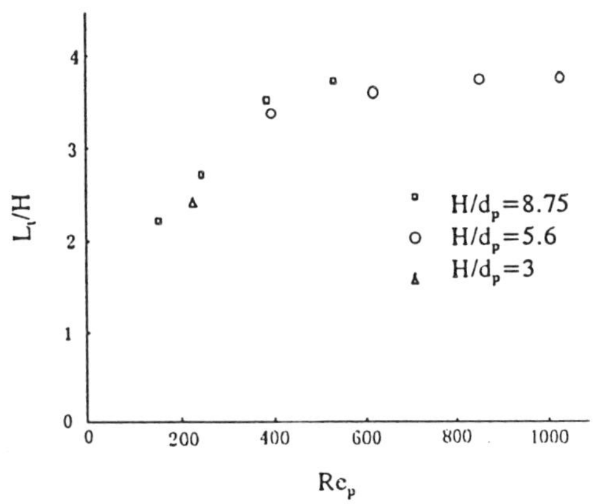

Figure 14 Thermal entrance length for air flow in an annulus

can say that the flow is thermally developed when $\ln[(T_c-T_w)/(T_i-T_w)]$ is a linear function of x/H, where T_i is the fluid inlet temperature and T_c the fluid temperature along the centerline of the passages. Figure 14 shows the variation of L_t with Re_p and H/d_p for air from the experimental results conducted by Coberly et al.[17] and Yagi et al.[18]. It is observed that the thermal entrance length increases with increasing Re_p, but the increase is much slower than that in unpacked channels and the channelling effect, H/d_p, may also influence the thermal entrance length, L_t and increasing H/d_p will cause L_t to increase. This depends on the value of the Pr number of the fluid. The data for air (Pr=0.7), in Figure 14, shows that the channelling effect on the thermal entrance length can be ignored. It is reasonable to assume that the development of the thermal boundary is also highly dependent on the ratio of the solid to fluid thermal conductivities.

As noted in the experimental results for water, dispersion becomes dominant for high Peclet numbers, and may play a reasonably significant role for low Peclet numbers. Thermal dispersion affects the thermal boundary-layer development and, hence, the convective heat transfer in packed beds both in the developing and the fully developed regions. Kwong and Smith[19] proposed the concept of effective conductivity by considering the heat diffusion in the fluid, coupled with the thermal dispersion in the porous medium. The effective conductivity is the sum of the stagnant conductivity, k_d, and the thermal dispersion conductivity, k_t, due to the fluid flow through the porous medium, is given by

$$k_e = k_d + k_t \tag{27}$$

where k_d can be evaluated by the formula proposed by Zehener[20] and k_t can be evaluated by the following empirical relation:

$$k_t / k_f = C_t \, Re_p \, Pr_f \tag{28}$$

in which C_t is determined empirically to make the prediction of the temperature profile agree with the experimental data.

In previous experiments reported in Schrieder et al.[15] and Yagi and Kunni[18], steep radial temperature gradients existed near the heated or cooled walls in the packed channel. Hsu and Cheng[21] analyzed the steep temperature gradients in forced convection in a packed channel and assumed that the local transverse thermal dispersion conductivity, k_t, is given by

$$k_t / k_f = C_t \, Re_p \, Pr_f \, l \tag{29}$$

with l being a dimensionless mixing length, normalized with respect to d_p, which was originally defined by a two-layer model in early works and more recently it has been modelled as the Van Driest wall function.

Kuo and Tien[22] also proposed a transverse thermal dispersion model, similar to equation (29), with the empirical constant C_t reduced both in the core and in the near wall regions using statistical theory.

These thermal dispersion models are chiefly based on turbulent boundary-layer theory and they cannot accurately predict the actual temperature distribution in the near wall region, especially for low values of Re_p. In addition, the actual width of the steep nonlinear temperature gradient is wider than that predicted. Although there are some similarities between dispersion and turbulence, turbulence develops from instabilities in the fluid flow whilst dispersion is a forced random flow due to the porous media. The frequency spectrum of the fluid velocity fluctuations in porous media is quite different from that in turbulence and therefore dispersion should depend on the arrangement of the packed beads. Koch and Brady[23] adopted the assemble averaging method to derive the dispersion coefficient. They showed that thermal dispersion includes mechanical thermal dispersion, which results from the stochastic fluid velocity fluctuation induced in the fluid due to the random distribution of the beads, and hold-up thermal dispersion in the non-slip boundary or zero velocity regions. Therefore, the thermal dispersion coefficient is proportional to the average fluid velocity, u, for irregular packed beds and proportional to $u^{0.1}$ for regular packed beds, that is

$$k_t / k_f = C_t^{/} \, Re_p \, Pr_f \qquad \text{for randomly packed beds}$$
$$k_t / k_f = C_t^{//} \, Re_p \, Pr_f \qquad \text{for regular packed beds} \tag{30}$$

It is well known that the bead arrangement in the near wall region is restricted by the wall. The first layer of beads contacting the wall could be relatively regular but from the second bead layer on the bead distribution gradually becomes more random.We have assumed that the thermal dispersion and the wall function are both functions of a weighted combination of the two thermal dispersion expressions, and that this weighted coefficient should be related to the regularity of the packed beads. The experimental results given by Benenati and Brosilow[24] indicate that the porosity variation in the wall region is in the form of an attenuated oscillation which disappears 4-5 bead diameters away from the wall. Their results suggest that the attenuated oscillation of the porosity may be related to the regularity of the packed beads and that the regularity of the beads could be measured by the amplitude A_m of the oscillation. Our proposal, reported in [16], is to use a weighted coefficient to describe the thermal dispersion and this may be expressed as follows, taking into account the effects of the two channel walls,

$$k_t / k_f = C_t \, (1-\phi) \, [\, f \, Re_p^{0.1} + (1-f) \, Re_p] \, Pr_f \tag{31}$$

with

$$f = \text{Max} \, [\, A_{m1}(1-A_{m2}), \, (1-A_{m1})A_{m2} \,] \tag{32}$$

where A_{m1} and A_{m2} are the oscillation amplitudes for the porosity near the two walls and C_t may be obtained by comparing the predicted distribution of the temperature by the proposed thermal dispersion model and the experiment data.

The temperature distribution predicted using equation (31) is compared with the data reported by Schroeder et al.[15] in Figure 15 and it can be observed that the model accurately predicts the actual temperature distribution near the wall surface, especially for low values of Re_p. The predicted convective heat transfer rates are compared with the experimental data for water and transformer oil in Figures 16 and 17 and again we observed that the predicted values and the corresponding experimental data coincide well in the thermally developed region, although some differences do exist in the thermal entrance region.

To identify the factors which influence the forced convection for flow inside packed channels in the entrance region, we have analyzed the heat transfer in an packed passage by using the Brinkman-Forchheimer-extended Darcy momentum equation and the energy

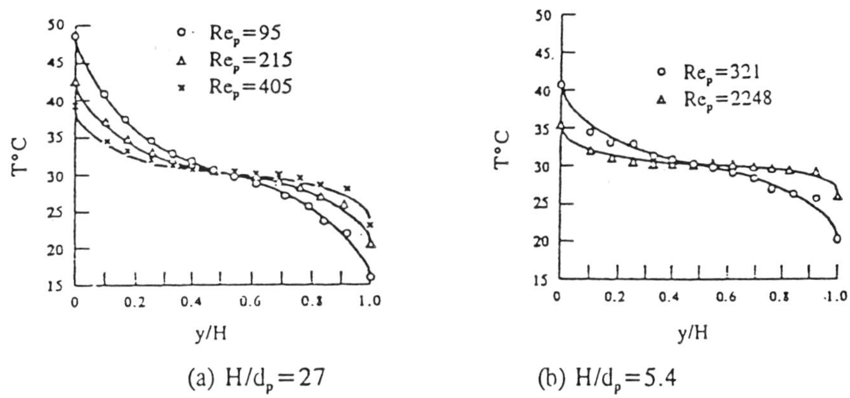

Figure 15 Comparison of temperature profiles predicted using equation (31) with Schroeder's data

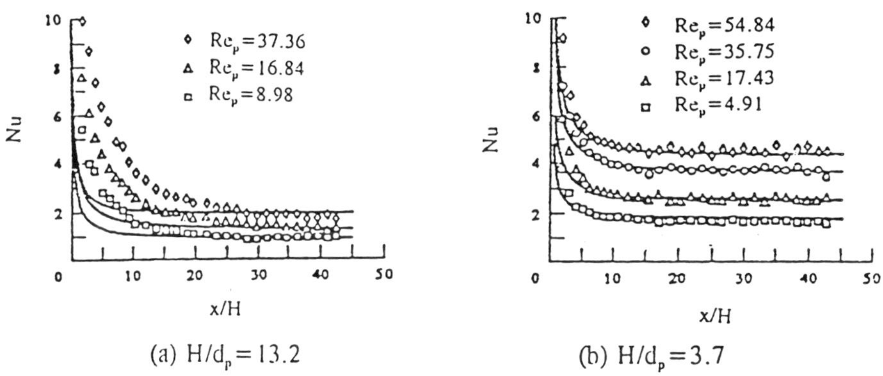

Figure 16 Comparison of predicted and measured local Nusselt number for water

equation, in which local thermal equilibrium between the fluid and solid phases was assumed. The variations of the local Nusselt number along the axis for calculations with and without the Brinkman term in momentum equation are presented in Figure 18, quoted from our report[25]. The effect of fluid viscosity is dominant only in the developing region and therefore one can reasonably assume that the reduction of the thermal dispersion near the wall with the no-slip wall boundary condition will cause Nu_x to decrease. Figure 19, also quoted from[25], illustrates the variation of the local Nusselt number along the flow direction both with and without the inertia term. It is observed that the inertia effect decreases the heat transfer coefficient in the hydrodynamically

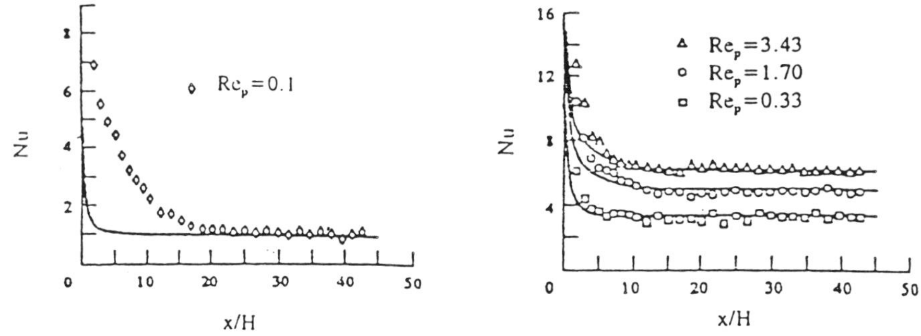

(a) $H/d_p = 13.2$ (b) $H/d_p = 3.7$

Figure 17 Comparison of predicted and measured local Nusselt number for transformer oil

developing zone but increases the heat transfer in the hydrodynamically developed region. Otherwise, the reduction of the thermal dispersion due to the wall effect reduces the heat transfer but does not change the development of the thermal boundary-layer. However, these results do not explain the discrepancy between the theoretical predictions and the experimental results and other factors, such as axial thermal dispersion and the variation of the transverse thermal dispersion along the flow direction, can also affect the heat transfer in the entrance region.

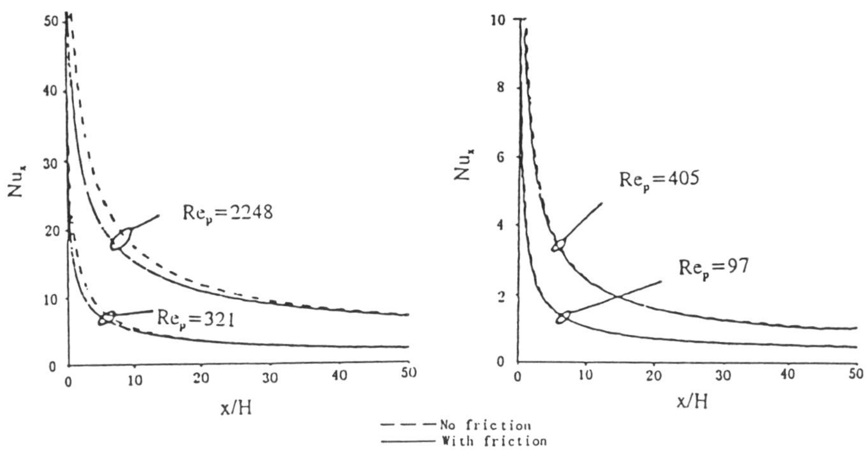

(a) $H/d_p = 5.4$ (b) $H/d_p = 27$

Figure 18 Effect of the Brinkman term on heat transfer for flow in an annulus

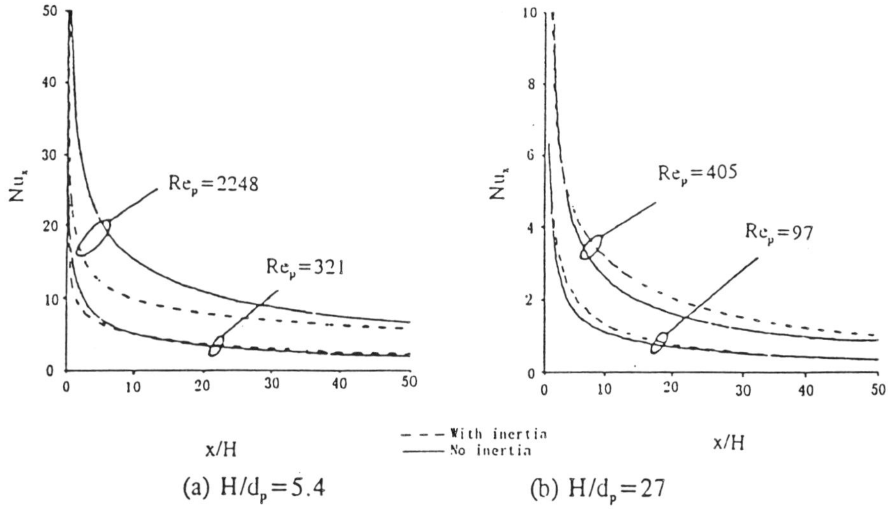

Figure 19 Effect of inertia term on heat transfer for flow in an annulus

MIXED CONVECTION

When fluid flows through packed channels, dispersion produced by the beads reduces the mean temperature difference between the fluid and the wall. For many situations, the changes in fluid properties with temperature are negligible, and uniform property analyses are generally utilized with the fluid properties evaluated at the arithmetic average of the inlet and outlet bulk temperatures. However, the buoyancy-induced flows driven by the variation of the density with temperature cannot be ignored, especially if the fluid velocities in porous media are small and natural convection weakens as the thermal boundary develops.

Mixed Convection in a Vertical Annular Porous Media

Consider a vertical porous annulus with a constant uniform heat flux, q_i, at the inner cylinder and a constant uniform heat flux, q_o, at the outer cylinder. For hydrodynamically developed regions, the governing momentum equation in the flow direction can be expressed using the extended Darcy model, i.e.

$$\frac{\partial(\varepsilon p)}{\partial x} = \mu_f \frac{1}{r}\frac{\partial}{\partial r}\left(r\frac{\partial u}{\partial r}\right) - \frac{\mu_f \varepsilon}{K}u - \rho_f \frac{F\varepsilon}{K^{1/2}}|V|u + \rho_0 \beta_T (T-T_0) g\varepsilon \qquad (33)$$

The dimensionless parameters that govern the heat transfer in mixed convection can be found by nondimensionalizing the momentum equation using following non-dimensional quantities:

$$R = r/H \quad X = x/H \quad \xi = H/d_p \quad V = v/u_0 \quad P = \frac{p}{\rho_f u_0^2} \quad \gamma = \left(\frac{\varepsilon H^2}{K}\right)^{1/2} \tag{34}$$

where H is the width of the passage and u_0 and T_0 are the fluid velocity and temperature at the passage inlet. On introducing these dimensionless variables, and noticing that $Re_H = Re_p \xi$, gives the following nondimensional equation:

$$\frac{Re_H Da}{\varepsilon}\frac{\partial(\varepsilon P)}{\partial X} = \frac{Da}{\varepsilon}\left[\frac{1}{R}\frac{\partial}{\partial R}\left(R\frac{\partial U}{\partial R}\right)\right] - U - Re_H\, Da^{1/2}\, F|V|U + \frac{Gr^* Da}{Re_H}\,\Theta \tag{35}$$

where the Darcy number, $Da = H^2/K$, U denotes the fluid velocity in the direction of fluid flow, and the modified Grashof number, Gr^*, is defined as

$$Gr^* = \frac{g\beta_T H^3 (q_i d_i + q_o d_o)}{\nu^2 k_f} \tag{36}$$

The parameter $Gr^* Da/Re_H$, in equation (35), is a measure of the relative importance of free convection in relation to forced convection. When $(Gr^* Da/Re_H) \approx 1$, free and forced convection are of the same order of magnitude and, hence, mixed convection must be considered. If $(Gr^* Da/Re_H) << 1$, the flow is primarily driven by forced convection and when $(Gr^* Da/Re_H) >> 1$, natural convection is dominant.

Another modified Grashof number Gr^{**}, which is related to mixed convection in porous media, can be defined based on the particle diameter as follow:

$$Gr^{**} = \frac{g\beta_T (q_i d_i + q_o d_o) d_p^3}{\nu^2 k_f} \tag{37}$$

The dimensionless group $Gr^* Da/Re_H$ can be rearranged as

$$\frac{Gr^* Da}{Re_H} = \frac{Gr^{**}}{Re_p}\,\frac{\varepsilon^3}{A(1-\varepsilon)^2} \tag{38}$$

in which $\epsilon^3/[A(1-\epsilon)^2]$ depends on the geometry of the packed beads and the annulus. Therefore, mixed convection in a vertical porous annulus with constant uniform heat fluxes can be characterized by two dimensionless parameters, namely Gr^{**}/Re_p and H/d_p.

Figure 20 illustrates the variation of Nu with Re_p for $Gr^{**} = 0$, 4×10^4, 8×10^4 and 12×10^4. We observed that when $Re_p \geq 60$ natural convection has little effect on the forced convection. As Re_p decreases, the effect of natural convection on forced convection becomes more important and the threshold value of Gr^{**}/Re_p for which natural convection is dominant is seen to be dependent on the value of Gr^{**}. Figure 21 shows the effect of natural convection for different value of H/d_p. It is observed that the effect of natural convection on forced convection for $H/d_p = 5$ is greater than that for $H/d_p = 27$, especially for smaller values of Re_p. Numerical results for the mixed convection in an annulus with $H/d_p=3.65$ have been compared with experimental data in Figure 22. The experiments on mixed convection were conducted with the outer wall thermally insulated and the results for $Gr^{**}=0$ indicate that natural convection need not be considered.

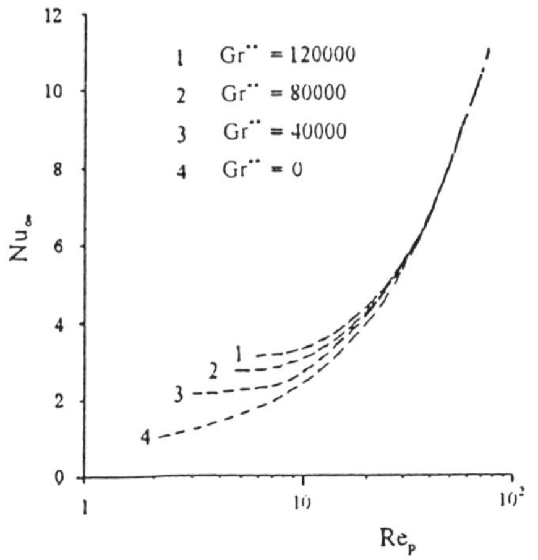

Figure 20 Fully developed Nusselt number for mixed convection for $d_i=10$cm, $d_p=5$cm and $H/d_p=10$

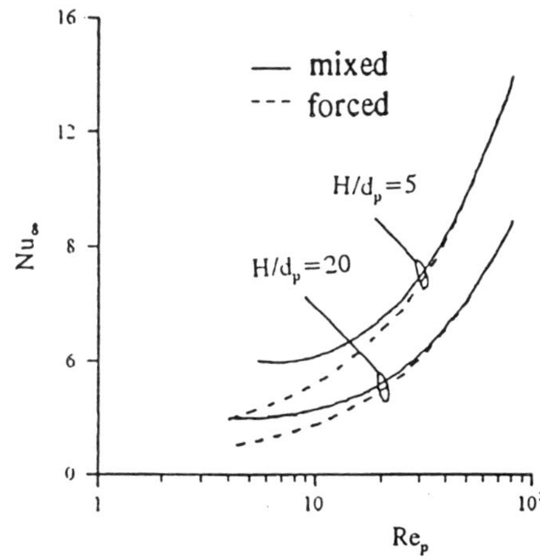

Figure 21 Fully developed Nusselt number for mixed convection for $H/d_p=10$ and 20

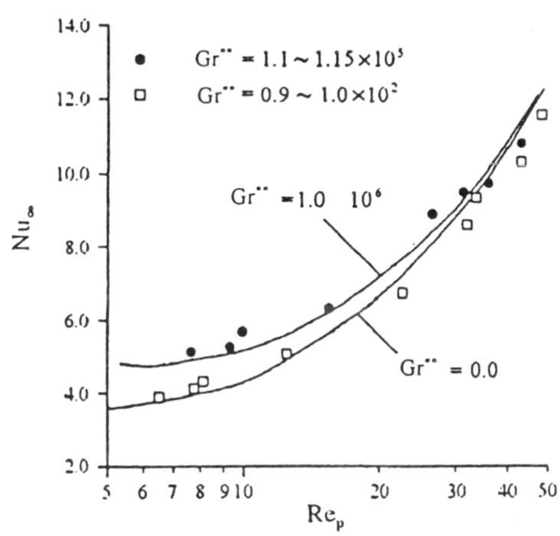

Figure 22 Comparison of predicted Nu_∞ with experimental data

Mixed Convection through a Porous Channel Bounded by Parallel Plates with Internal Heat Generation

Mixed convection through a porous channel with internal heat generation is different from that without internal heat sources in that the influence of natural convection on forced convection remains and develops in the fully developed region. Internal flows in such systems is complicated by the fact that the bulk fluid temperature generally varies with axial location. As analyzed by Bian and Wang[26], the flow pattern depends upon the relative values of the wall temperatures and the internal heat source intensity and flow reversals may occur in some regions.

The effect of internal heat sources can be investigated by considering a vertical porous channel which is bounded by two parallel plates at different constant temperatures, T_{w1} and T_{w2}, with an internal constant heat source, S. The non-dimensional Brinkman-Boussinesq-extended Darcy equations may be expressed as follows, see Bian and Wang[26]:

$$U = -\frac{\partial P}{\partial X} + Da\left(\frac{\partial^2 U}{\partial X^2} + \frac{\partial^2 U}{\partial Y^2}\right) \tag{39}$$

$$V = -\frac{\partial P}{\partial Y} + Da\left(\frac{\partial^2 U}{\partial X^2} + \frac{\partial^2 U}{\partial Y^2}\right) + \frac{Ra}{Pe}\Theta \tag{40}$$

$$U\frac{\partial \Theta}{\partial Y} + V\frac{\partial \Theta}{\partial Y} = \frac{1}{Pe}\left(\frac{\partial^2 \Theta}{\partial X^2} + \frac{\partial^2 \Theta}{\partial Y^2}\right) + \frac{Ra_s}{RaPe} \tag{41}$$

subject to the boundary conditions

$$\begin{aligned} X = 0&: \quad U = V = 0,\ \Theta = r_T \\ X = 1&: \quad U = V = 0,\ \Theta = 1 \\ Y = 0&: \quad U = 0\ ,\ V = 1,\ \Theta = 0 \end{aligned} \tag{42}$$

where the dimensionless parameters are defined as

$$Da = \frac{k}{L^2};\quad Pe = \frac{v_0 L}{\alpha_e};\quad r_T = \frac{T_2 - T_0}{T_1 - T_0};\quad Ra = \frac{Kg\beta L(T_1 - T_0)}{\alpha_e \nu};\quad Ra_s = \frac{g\beta KLS}{\alpha_e \nu K} \tag{43}$$

where r_T is the asymmetrical heating ratio, Ra is the Rayleigh number pertaining to the temperature difference while Ra_s is the Rayleigh number for the internal heat source.

The numerical results demonstrate that for symmetrical heating the fluid velocity at the centerplane of the channel will decrease in the entrance region, reach a minimum, and then increase to the limiting values, see Figure 23. The value of V_c approaches unity and the effect of the buoyancy disappears when $Ra_s = 0$. For $Ra_s \neq 0$, i.e. when there is internal heat generation, V_c is greater than unity in the fully developed region, demonstrating that natural convection is still important in this region. Figure 24 shows the fluid velocity profiles at different cross-sections and we observe that in the entrance region, the fluid rises in the vicinity of the warmer walls and falls in the cooler bulk region, see Figure 24(a). The internal heat generation will cause the position of the maximum fluid velocity to move towards the channel centerplane. The velocity profiles, given in Figures 24(b) and 24(c) show that when the buoyancy direction is opposite to

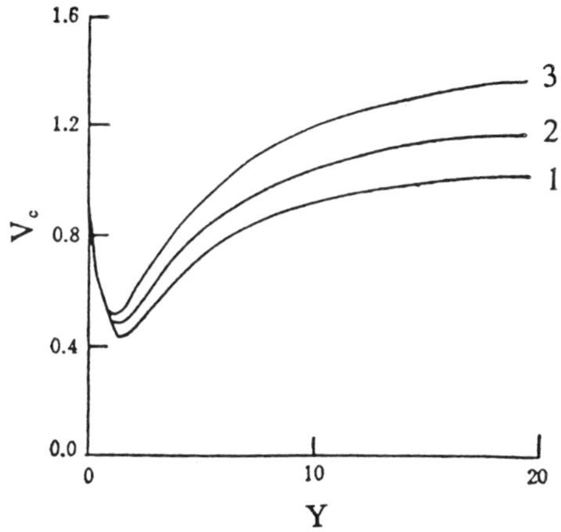

1 $Ra_s/Pe=1.5$ 2 $Ra_s/Pe=15$ 3 $Ra_s/Pe=35$

Figure 23 Velocity distribution on the centerplane for flow with internal heat sources for Ra/Pe=3

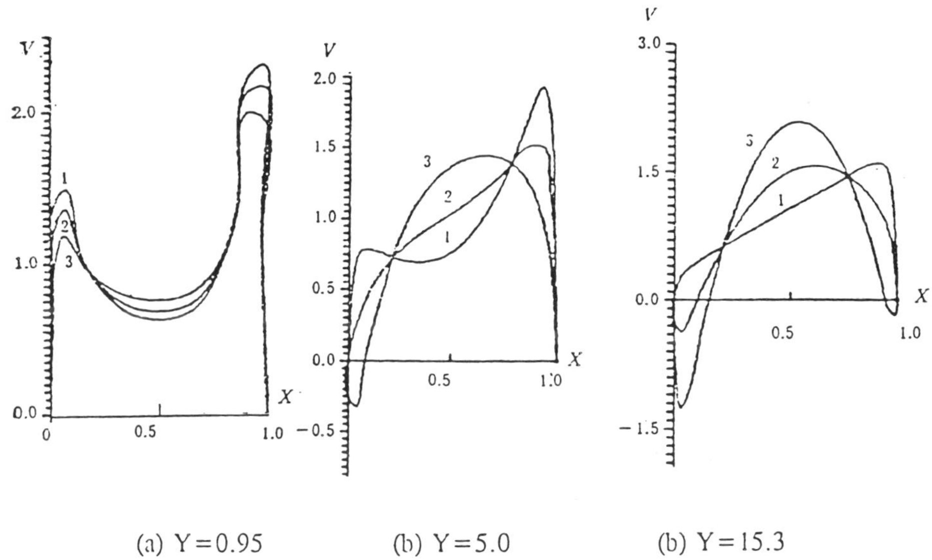

1 $Ra_s/Pe=1.5$ 2 $Ra_s/Pe=15$ 3 $Ra_s/Pe=35$

Figure 24 Velocity profiles at different locations for flow with internal heat sources that of the primary flow, and is of sufficient magnitude to overcome the pressure gradient and viscous forces, a flow reversal may occur. The condition for a reversal flow is

$$\left.\frac{\partial V}{\partial X}\right|_{X=0} < 0, \qquad \left.\frac{\partial V}{\partial X}\right|_{X=1} > 0 \tag{44}$$

On considering $Da << 1$, the analytical solution of the momentum equation in the fully developed region can be obtained, as pointed out by Bian and Wang[26]. The criteria for judging flow reversals, which depends on Ra/Pe, Ra_s and r_T, can be summarized as follows:

(i) For $r_T \neq 1$, asymmetrical heating cases, no flow reversal if $Ra_s/Pe < 12\text{-}6Ra/[Pe(1\text{-}r_T)]$; flow reversal at the cooler surface if $12\text{-}6Ra/[Pe(1\text{-}r_T)] < Ra_s/Pe < 12+6Ra/[Pe(1\text{-}r_T)]$; and flow reversal at both surfaces if $Ra_s/Pe > 12+6Ra/[Pe(1\text{-}r_T)]$.

(ii) For $r_T=1$, symmetrical heating cases, no flow reversal if $Ra_s/Pe < 12$; flow reversal at both surfaces if $Ra_s/Pe > 12$.

CONCLUSIONS

Natural and mixed convection in saturated porous media has been investigated both theoretically and experimentally for various flow geometries including horizontal, vertical and inclined annuli and flow between parallel plates. The conclusions can be made as follows:

(1) The time required to initiate convection and the duration of the transient period decreases with increasing Ra and R_o, whilst the mean Nusselt number increases.

(2) There exists a critical aspect ratio, A_{cr}, for which the inclination angle, ϕ, has no significant effect on the heat transfer. For $A < A_{cr}$, the mean Nusselt number, $Nu_{i,m}$ increases with increasing angle of inclination, while for $A > A_{cr}$, $Nu_{i,m}$ decreases as ϕ increases. The critical aspect ratio, A_{cr}, decreases as the radius ratio increases whilst the Rayleigh number has an inconsiderable effect on it.

(3) At low Rayleigh numbers (say $Ra < 50$), convection induced by internal heat generation in a vertical porous cylinder is mostly driven by the temperature distribution of pure conduction. However, the maximum temperature predicted by pure conduction underestimates the actual value due to the upward flow along the centerline. Convection at high Rayleigh number (say $Ra > 5000$) is characterized by a homogeneous upward flow in the central part of the cylinder and a very thin downward boundary-layer fluid flow on the cool wall with the effect of the curvature of the boundary being negligible.

(4) Several important mechanisms have been considered that influence the flow and heat transfer and the parameters that control these mechanisms. Thermal dispersion is an important effect in porous media and a new thermal dispersion model is proposed to effectively reproduce the experimental results.

(5) Criteria for identifying mixed convection and the occurrence of flow reversal were deduced from the governing momentum equation.

Because of the complexity and the difficulty of effectively measuring the transport processes in porous media, 'black box' empirical models are widely used. However, the intrinsic mechanisms and characteristics of the transport processes in porous media, such as preferential path effect, non-local and non-equilibrium effects, scale effects, etc., can not be explored by this method. Investigation of coupled transport phenomena in porous media is required to understand the fundamentals of the processes. Quantitative results from various models must be compared with experimental data to evaluate the models and to determine the primary and secondary mechanisms which influence the flow and heat transfer. Further work is required to establish the microscopic and mesoscopic models at the pore levels for simple artificial media, such as converging-diverging channels and periodically constricted tubes, and to incorporate these models in a macroscopic model in such a way that the model can be applied to realistic problems associated with particular applications.

Acknowledgement The research project was supported by the National Science Foundation of China and the State Education Commission of China.

REFERENCE

1. M.A. Combarnous and A. Bores, Hydrothermal Convection in Saturated Porous Media, Adv. Hydroscience **11**, 231-301 (1975).
2. P. Cheng, Heat Transfer in Geothermal System, Advances in Heat Transfer **14**,1-105 (1978).
3. J. Bear, Dynamics of fluid in porous media, American Elsevier, NY (1972).
4. S. Kakac, Convective Heat and Mass Transfer in Porous Media, Kluwer Academic, Boston (1991).
5. M. Kaviany, Principles of Heat Transfer in Porous Media, Springer-Verlag, NY (1991).
6. S. Whitaker, Simultaneous Heat, Mass and Momentum Transfer in Porous Media, Adv. Heat Transfer **13**, 119-203 (1977).
7. B.X. Wang and X. Zhang, Natural convection in liquid-saturated porous media between concentric inclined cylinders, Int. J. Heat Mass Transfer **33**, 827-833 (1990).
8. J.P. Caltagirone, Thermoconvective instabilities in a porous medium bounded by two concentric horizontal cylinders, J. Fluid Mech. **76**,337-362 (1976).
9. S.Y. Lei, S.J. Bao, W.C. Wang and B.X. Wang, Measurement of the porosity and permeability of unsolidated porous media, Chinese J. Eng. Thermophysics **13**, 408-411 (1992).
10. B.X. Wang and X. Zhang, Numerical and experimental investigations on transient and steady-state natural convection in a vertical liquid-saturated porous medium,in Heat Transfer Science & Technology 1988, ed. by B.X. Wang,pp.417-423, Hemisphere, NY (1989).
11. D.C. Reda, Natural convection experiments in stratified liquid-saturated porous medium, Trans. ASME J. Heat Transfer **108**, 660-666 (1986).
12. Y.F. Rao and B.X. Wang, Natural convection in vertical porous enclosures with internal heat generation, Int. J. Heat Mass Transfer **34**, 247-252 (1992).
13. H. Verschoor and G.C.A. Schuit, Heat transfer to fluid flowing through beds of

granular solids, Appl. Sci. Res. **A2**, 97-119 (1952).
14. J.H. Quinton and A.J. Storrow, Heat transfer to air flowing through packed tubes, Chem. Eng. Sci. **5**, 245-257 (1956).
15. K.J. Schroeder, U. Renz and K. Elegeta, Forschungsberichte des Landes Nordrhein-Westfalen No.3037 (1981).
16. B.X. Wang and J.H. Du, Forced convection heat transfer in a vertical annulus filled with porous media, Int. J. Heat Mass Transfer **36**,4207-4213 (1993).
17. C.A. Coberly and W.R. Marshall Jr.,Temperature gradients in gas streams flowing through fixed granular beds, Chem. Eng. Progress **47**, 141-150 (1951).
18. S. Yagi and D. Kunni, Studies of heat transfer near wall surface in packed beds, AIChE J. **6**, 97-104 (1960).
19. S.S. Kwong and J.M. Smith, Radial heat transfer in packed beds, Ind. Eng. Chem. **49**, 894-903 (1957).
20. P. Zehener, Waermeleitfahigkeit won Schuettung be Massigen Temperaturea, Chemie-Ingr.-Tech. **42**, 933-941 (1970).
21. C.T. Hsu and P. Cheng, Thermal dispersion in a porous media, Int. J. Heat Mass Transfer **33**, 157-1597 (1990).
22. S.M. Kuo and C.L. Tien, Transverse dispersion in packed sphere beds, Proceeding of National Heat Transfer Conference **1**, 629-634 (1989).
23. D.L. Koch and J.F. Brady, Dispersion in fixed beds, J. Fluid Mech. **154**, 321- 349 (1985).
24. R.F. Benenati and C.B. Brosilow, Void fraction distribution in beds of sphere, AIChE J. **8**, 359-361 (1962).
25. J.H. Du, B.X. Wang and X.F. Peng, Non-Darcian analysis on convective heat transfer in the entrance region for flow in packed channels, Heat Transfer Science and Technology 1996, ed. by B.X. Wang, 441-445, Higher Education Press, Beijing (1996).
26. W. Bian and B.X. Wang, Mixed convection in porous media with internal heat sources, J. Eng. Thermophyics of Chines **13**, 396-399 (1992).

MODELING MULTIPHASE FLOW AND TRANSPORT IN POROUS MEDIA

C.Y. WANG

Department of Mechanical Engineering, The Pennsylvania State University
University Park, PA 16802, USA

INTRODUCTION

Problems involving multiphase flow, heat and mass transfer in porous media arise in a number of scientific and engineering disciplines. Important technological applications include oil reservoir engineering, groundwater contamination by nonaqueous phase liquids (NAPL), Abriola [1], heat pipe technology, Faghri [2], drying processes, Whitaker [3], multiphase trickle bed reactors, Shah [4], and geothermal reservoirs, Cheng [5]. More applications were recently surveyed by Wang and Cheng [6].

A number of complex and interacting transport phenomena may take place in a multiphase, multicomponent system. In general, multiphase flows in porous media are driven by gravitational, capillary and viscous forces. Gravity causes phase migration in the direction of the gravitational field. A common example is gravity override of steam as encountered in steam injection for enhanced oil recovery and subsurface remediation, where the injected steam tends to rise to the top of a formation thus leading to a premature breakthrough of steam into production wells. Another example of the influence of gravity is thermosolutal convection in the gas phase as a result of density variations due to combined temperature and concentration gradients that are typically present in a multiphase, multicomponent system. Capillary forces play fundamental roles in controlling the phase distribution in heterogeneous porous media. Viscous forces influence relative phase motion; fluids with lower viscosity tend to migrate more rapidly due to their reduced resistance to flow. When a denser, less viscous fluid displaces another fluid of lower density and higher viscosity, the process is inherently unstable, leading to fingering of phase interfaces, Homsy [7]. Laboratory experiments have shown that unstable finger-like infiltration of an organic contaminant can significantly alter the transport and dilute characteristics of the contaminant, Güven et al. [8] and Held and Illangasekare [9].

Traditionally, complex problems of multiphase flow and transport in porous media are tackled by a multiphase approach, Abriola and Pinder [10], in which various phases are regarded as distinct fluids with individual thermodynamic and transport properties and with different flow velocities. The transport phenomena are mathematically described by the basic principles of conservation for each phase separately and by appropriate interfacial conditions between various phases. The generalized Darcy's law is employed to represent momentum conservation in each phase, with the relative permeabilities of each phase introduced to account for a decrease in the effective flow cross-section due to the presence of other fluids.

Due to the inherent nonlinearities of multiphase flow problems, exact solutions are limited to a small class of problems in one dimension and with many simplifying assumptions, see McWhorter and Sunada [11]. Solving practical problems, which usually involve multi-dimensional effects, gravity, capillarity and phase change, requires sophisticated numerical procedures and represents a challenging task as multiple coupled sets of differential equations must be solved. For this reason, there has been a large number of studies to develop robust numerical algorithms, Allen [12], or approximate models such as the unsaturated flow theory, Morel-Seytoux [13]. Another obstacle to numerical simulations of multiphase flow and transport in porous media lies in the presence of moving and irregular phase interfaces separating the single- and multi-phase subregions. The locations of these interfaces are not known *a priori* but must be determined by the coupled flows in adjacent regions. A numerical procedure for such a multi-region problem, based on the multiphase approach, needs to explicitly track the moving interfaces, thus calling for complex coordinate mapping or numerical remeshing, Ramesh and Torrance [14]. Alternatively, in the water resources literature, Panday et al. [15], the procedure of switching the primary variables and governing equations in various regions is usually adopted, which may severely affect numerical stability and convergence speed.

Recently, an alternative approach was developed to the modeling of multiphase flow and multicomponent transport in porous media, Wang and Cheng [16] and Wang and Beckermann [17]. In this so-called multiphase mixture model, the multiphase system is viewed as a chemical mixture. Multiphase flow is then described by a mass-averaged mixture velocity and a diffusive flux representing the difference between the mixture velocity and an individual phase velocity. Without making any assumptions, a set of transport equations for the multiphase mixture was derived from the classic multiphase flow model. Moreover, explicit relations describing the relative motions between the multiphase mixture and an individual phase were found, which can be utilized to determine the intrinsic characteristics of the phase dynamics (e.g. the phase velocity fields) after a solution to the model equations for the multiphase mixture is obtained. Compared to the multiphase flow formulation, the new model offers the following advantages: (i) it strongly resembles the single-phase transport theory, thus facilitating both theoretical and numerical analyses; (ii) it requires much fewer nonlinear and coupled differential equations to be solved; and (iii) it is a mixture formulation, thus eliminating the need to handle phase appearance or disappearance. Nevertheless, the multiphase mixture model still provides the same predictive capabilities as the conventional multiphase flow model because it is a mathematically equivalent but reformulated version of the multiphase flow model, i.e. without making any additional approximations.

The objectives of this chapter are to discuss the multiphase mixture model, to give examples of its applications to performing routine analyses of complex multiphase problems in porous media, and to assess the current state of the subject. The chapter is organized as follows: the next section outlines the derivation of the multiphase mixture model in sufficient detail to permit appreciation of its physical significance and mathematical rigor. Subsequently, analytical capabilities of the multiphase mixture model are explored through sample applications to unsaturated flow in the subsurface, boundary-layer two-phase flow with heat transfer, and two-phase flow with viscous coupling. Numerical capabilities of the multiphase mixture model are demonstrated in the discussion of a general numerical methodology for single-component systems with heat transfer and a numerical example of multi-component systems. Finally, the conclusions are summarized and areas for future research are indicated.

THE MULTIPHASE MIXTURE MODEL

In the multiphase mixture model, the multiple phases are regarded as constituents of a multiphase mixture. So as to distinguish this mixture from a conventional *multicomponent* mixture, Bowen [18], it is referred to as a *multiphase* mixture in this work. In this definition, phases are assumed to be distinct and separable components with nonzero interfacial areas, and their mixture represents a single fluid with smoothly varying phasic compositions. In contrast, a multicomponent mixture is defined as consisting of different chemical components. As in a classical multicomponent mixture, all physical properties of the multiphase mixture are consequences of the properties of its constituents; however, their functional forms are not assumed *a priori* but are derived strictly from the multiphase flow formulation. Hence, mathematically, the multiphase mixture model is exactly equivalent to the existing multiphase flow formulation which enjoys widespread applications in the current literature. The differences between these two models lie solely in the number and form of the governing equations, their physical meanings and amenability to numerical implementation.

To begin with, the mixture density, velocity, species concentration and enthalpy are defined, respectively, as

$$\rho = \sum_k \rho_k s_k \tag{1}$$

$$\rho \mathbf{u} = \sum_k \rho_k \mathbf{u}_k \tag{2}$$

$$\rho C^\alpha = \sum_k \rho_k s_k C_k^\alpha \tag{3}$$

$$\rho h = \sum_k \rho_k s_k h_k \tag{4}$$

where a quantity without a subscript is reserved for the multiphase mixture, while the subscript k denotes phase k. The phase saturation, s_k, denotes the volumetric fraction of the void space occupied by phase k. The superficial velocity of phase k, $\mathbf{u}_k$, is related to its intrinsic velocity by the phase volume fraction and, thus, Equation (2) implies that the mixture velocity is a mass-weighted average of the intrinsic phase velocities. The symbols C_k^α and h_k stand for the mass concentration of species α in phase k and the enthalpy of phase k, respectively.

The mixture kinematic viscosity can be defined as, see Wang and Cheng [16],

$$\nu = \left(\sum_k \frac{k_{rk}}{\nu_k} \right)^{-1} \tag{5}$$

and the mobility of each phase in the multiphase mixture is given by

$$\lambda_k = \frac{k_{rk}}{\nu_k} \nu \ ; \quad \sum_k \lambda_k = 1 \tag{6}$$

Defining a mixture pressure is not straightforward because of the differences between various phase pressures owing to the capillary forces. For the reasons that will become apparent shortly, the mixture pressure is so defined that the following differential equation holds:

$$dp = \sum_j \lambda_j dp_j = dp_k + \sum_j \lambda_j dp_{cjk} \tag{7}$$

where the capillary pressure between each pair of phases (j and k) is dependent upon the pore geometry, phase saturations and interfacial tensions, Parker [19], i.e.

$$p_{cjk} = p_j - p_k = fn(\varepsilon, \sigma_{jk}, s_1, s_2, \dots s_k, \dots) \tag{8}$$

where σ_{jk} is, in turn, a function of concentration and temperature. Taking the gradient over Equation (8) and substituting the resultant into Equation (7) yields

$$dp = dp_k + \sum_i Ca_{ik}\, ds_i + \sum_\alpha Cs_{\alpha k}\, dC^\alpha + Ct_k\, dT \tag{9}$$

where Ca_{ik}, $Cs_{\alpha k}$ and Ct_k, called the capillary, solutal-capillary and thermo-capillary factors, respectively, can be expressed as

$$Ca_{ik} = \sum_j \lambda_j \frac{\partial p_{cjk}}{\partial s_i} \tag{10}$$

$$Cs_{\alpha k} = \sum_j \lambda_j \frac{\partial p_{cjk}}{\partial \sigma_{jk}} \frac{\partial \sigma_{jk}}{\partial C^\alpha} \tag{11}$$

$$Ct_k = \sum_j \lambda_j \frac{\partial p_{cjk}}{\partial \sigma_{jk}} \frac{\partial \sigma_{jk}}{\partial T} \tag{12}$$

Integration of Equation (9) leads to the following explicit definition of the mixture pressure:

$$p = p_k + \sum_i \int_0^{s_i} Ca_{ik}\, ds_i + \sum_\alpha \int_0^{C^\alpha} Cs_{\alpha k}\, dC^\alpha + \int_0^T Ct_k\, dT \tag{13a}$$

Obviously, when the capillary forces are negligibly small, the last three terms in Equation (13a) vanish so that $p=p_k$. To illustrate how the mixture pressure is related to the phase pressures for phase saturations between zero and unity, consider a liquid-gas two-phase system in the absence of temperature and concentration gradients. In this case the mixture pressure defined in Equation (13a) is simplified to

$$p = p_l + \int_1^{s_l} (1 - \lambda_l)\left(\frac{dp_c}{ds_l}\right) ds_l = p_g - \int_1^{s_l} \lambda_l \left(\frac{dp_c}{ds_l}\right) ds_l \qquad (13b)$$

Figure 1 schematically shows a mixture pressure profile given by Equation (13b) along with the liquid and gas phase pressures in a cocurrent flow. As can be seen, in the all liquid region, the mixture pressure merges with the liquid phase pressure. Within the two-phase zone where the liquid saturation varies from unity to zero, the mixture pressure lies between the gas and liquid phase pressures. The slopes of the mixture pressure at the two endpoints, i.e. s_l=1 and 0, approach those of the liquid and gas pressures, respectively. However, the mixture pressure in the all gas region differs from the gas pressure by a constant, as evident from Equation (13b). This constant difference in pressure is of no consequence to flow behavior since flow is driven by the pressure gradient. Therefore, Equation (13b) is indeed consistent with a mixture theory. Later, we shall show how this definition can simplify the momentum conservation equation for the multiphase mixture.

Conservation of Mass

The mass conservation equation for a multiphase mixture can readily be obtained by addition of the phase conservation equations as represented by Abriola and Pinder [10]

$$\varepsilon \frac{\partial(\rho_k s_k)}{\partial t} + \nabla\cdot(\rho_k \mathbf{u}_k) = \bar{\dot{m}}_k \qquad (14)$$

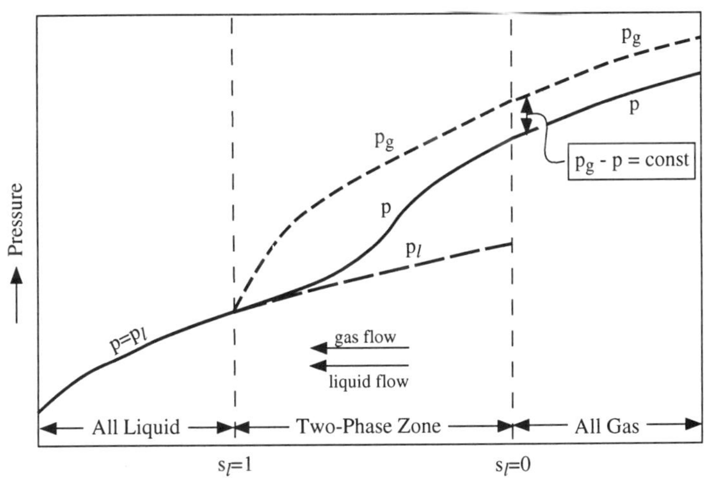

Figure 1: Schematic illustration of the mixture pressure versus liquid and gas pressures in various regions in a co-current flow

where $\bar{m}_k$ is the interfacial mass transfer rate from all other phases to phase k. Noting that the production of phase k, $\bar{m}_k$, must come at the expense of the other phases, summation of Equation (14) over all phases then results in

$$\varepsilon \frac{\partial \rho}{\partial t} + \nabla \cdot (\rho \mathbf{u}) = 0 \qquad (15)$$

It is clear that this continuity equation is just a duplicate of the corresponding equation for a single-phase mixture flow. Since the mixture density ρ is strongly variable, the multiphase mixture flow should be categorized as highly compressible.

Conservation of Momentum

The phase momentum equation in the form of the generalized Darcy's law is written as, see Abriola and Pinder [10],

$$\rho_k \mathbf{u}_k = -\frac{K k_{rk}}{\nu_k} (\nabla p_k - \rho_k \mathbf{g}) = -\frac{K}{\nu} (\lambda_k \nabla p_k - \lambda_k \rho_k \mathbf{g}) \qquad (16)$$

where the definition of the relative mobility, Equation (6), has been used. The momentum equation for the multiphase mixture is then obtained by summing the above equations for each phase. Utilizing the definitions of the mixture velocity and pressure given by Equations (2) and (7), one arrives at

$$\rho \mathbf{u} = -\frac{K}{\nu} (\nabla p - \gamma_\rho \rho \mathbf{g}) \qquad (17)$$

where γ_ρ is called the density correction factor and is defined as

$$\gamma_\rho = \frac{\sum_k \rho_k \lambda_k}{\sum_k \rho_k s_k} \qquad (18)$$

It should be noted that this correction factor is only a sole function of phase saturations and, thus, can be regarded as a property of the multiphase mixture. The physical reason that a correction factor appears in the body force term for a multiphase mixture is the relative motion between phases, so that the effective mixture density for the gravitational force contains certain dynamic properties of phases, i.e. λ_k and hence ν_k. Only when γ_ρ=1, the body force for the multiphase mixture can be represented by the mixture density defined in Equation (1). This is the case of single-phase, multicomponent mixtures.

As in a traditional mixture theory, one can define a diffusive mass flux of phase k within the multiphase mixture, such that

$$\mathbf{j}_k = \rho_k \mathbf{u}_k - \lambda_k \rho \mathbf{u} \quad ; \quad \sum_k \mathbf{j}_k = 0 \qquad (19)$$

or rearranging,

$$\rho_k \mathbf{u}_k = \mathbf{j}_k + \lambda_k \rho \mathbf{u} \qquad (20)$$

Substituting the momentum equations for phase k and for the mixture given by Equations (16) and (17) into Equation (19), the diffusive mass flux, $\mathbf{j}_k$, can be expressed as:

$$\mathbf{j}_k = K\frac{\lambda_k}{\nu}(\nabla p - \nabla p_k) + K\frac{\lambda_k}{\nu}(\rho_k - \gamma_\rho \rho)\mathbf{g} \tag{21}$$

In lieu of Equation (9) and the identity

$$\rho_k - \gamma_\rho \rho = \sum_i \lambda_i(\rho_k - \rho_i) \tag{22}$$

Equation (21) becomes

$$\mathbf{j}_k = \sum_i [-\rho_k D_{cik}\nabla s_i + \frac{\lambda_k \lambda_i K(\rho_k - \rho_i)}{\nu}\mathbf{g}] + \sum_\alpha [-\rho_k D_{s\alpha k}\nabla C^\alpha] + (-\rho_k D_{tk})\nabla T \tag{23}$$

where the capillary diffusion coefficient, D_{cik}, the solutal-capillary diffusion coefficient, $D_{s\alpha k}$, and the thermocapillary diffusion coefficient, D_{tk}, are given by

$$D_{cik} = \frac{K}{\rho_k \nu}\lambda_k \sum_j \lambda_j[-\frac{\partial p_{cjk}}{\partial s_i}] \tag{24}$$

$$D_{s\alpha k} = \frac{K}{\rho_k \nu}\lambda_k \sum_j \lambda_j[-\frac{\partial p_{cjk}}{\partial \sigma_{jk}}\frac{\partial \sigma_{jk}}{\partial C^\alpha}] \tag{25}$$

$$D_{tk} = \frac{K}{\rho_k \nu}\lambda_k \sum_j \lambda_j[-\frac{\partial p_{cjk}}{\partial \sigma_{jk}}\frac{\partial \sigma_{jk}}{\partial T}] \tag{26}$$

Physically, Equation (23) implies that the diffusive flux of phase k within the multiphase mixture results from the capillary forces due to saturation gradients, and concentration and temperature gradients through their effects on the interfacial tensions. In addition, the second term on the right hand side of Equation (23) reflects the gravitational separation due to the disparity between phase densities. Note also that the diffusion coefficients defined by Equations (24-26) depend only on phase saturations and, hence, are properties of the multiphase mixture.

Now, with the aid of Equations (20) and (23)-(26), the flow characteristics of individual phases can readily be obtained, once the flow problem of the multiphase mixture has been solved. Thus, the detailed flow aspects of each phase are not smeared out in the present model; they are still accessible wherever they are needed. This is why the model is called the *multiphase mixture model*, rather than simply a mixture model, implying that, in the latter, the intrinsic multiphase characteristics are usually lost.

Conservation of Species

Mass conservation of species α in phase k can be expressed as, see Abriola and Pinder [10],

$$\frac{\partial}{\partial t}(\varepsilon \rho_k s_k C_k^\alpha) + \nabla\cdot(\rho_k \mathbf{u}_k C_k^\alpha) = \nabla\cdot(\varepsilon \rho s_k D_k^\alpha \nabla C_k^\alpha) + \bar{J}_k^\alpha \tag{27}$$

where C_k^α stands for the mass concentration of species α in phase k, and D_k^α is a macroscopic second-order tensor accounting for molecular diffusion and/or hydrodynamic dispersion. A discussion on how to evaluate this coefficient can be found in Wang and Cheng [16]. Nonetheless, it should be noted that a great deal of research is presently being directed in this area; see, for example, Brenner and Edwards [20].

The last term $\bar{J}_k^\alpha$ in Equation (27) denotes the interphase species transfer rate caused by chemical non-equilibrium and/or phase change at the interfaces between phase k and all other phases. Recognizing that the production of species α in phase k must be accompanied by destruction of species α in other phases, Equation (27) can be summed over all phases to yield

$$\varepsilon\frac{\partial}{\partial t}\{\sum_k \rho_k s_k C_k^\alpha\} + \nabla\cdot\{\sum_k \rho_k \mathbf{u}_k C_k^\alpha\} = \nabla\cdot[\varepsilon\sum_k(\rho s_k D_k^\alpha \nabla C_k^\alpha)] \tag{28}$$

By definition of the mixture concentration given by Equation (3) and using Equation (20) to decompose the phase velocity, the phase-summed species conservation equation can be recast into

$$\varepsilon\frac{\partial}{\partial t}(\rho C^\alpha) + \nabla\cdot[(\sum_k \lambda_k C_k^\alpha)\rho\mathbf{u}] = \nabla\cdot[\varepsilon\sum_k(\rho s_k D_k^\alpha \nabla C_k^\alpha)] - \nabla\cdot\{\sum_k C_k^\alpha \mathbf{j}_k\} \tag{29}$$

To further cast Equation (29) into a form similar to the species transport equation in single-phase systems, one can define a correction factor for species advection as

$$\gamma_\alpha = \frac{\rho\sum_k \lambda_k C_k^\alpha}{\sum_k \rho_k s_k C_k^\alpha} \tag{30}$$

As a result, Equation (29) can be finally rewritten as

$$\varepsilon\frac{\partial}{\partial t}(\rho C^\alpha) + \nabla\cdot(\gamma_\alpha \rho \mathbf{u} C^\alpha) = \nabla\cdot[\varepsilon\rho D \nabla C^\alpha] + \nabla\cdot\{\varepsilon\sum_k[\rho_k s_k D_k^\alpha(\nabla C_k^\alpha - \nabla C^\alpha)]\} - \nabla\cdot(\sum_k C_k^\alpha \mathbf{j}_k) \tag{31}$$

where the effective diffusion coefficient for the multiphase mixture is defined as

$$\rho D = \sum_k \rho_k s_k D_k^\alpha \tag{32}$$

The second term on the left hand side of Equation (31) indicates that species α is advected, on the mixture level, by a modified velocity field $\gamma_\alpha \mathbf{u}$ rather than the mixture velocity field itself. This peculiar feature resembles the one related to the gravitational term in the momentum equation, Equation (17). The first two terms on the right hand side of Equation (31) combine to represent the net Fickian diffusion fluxes within various phases, while the last term on the RHS represents the diffusive flux across phases. Also, notice that the second term on the RHS of

Equation (31) vanishes in single-phase regions or when the phase compositions C_k^{α} are identical in all phases.

Conservation of Energy

Assuming local thermal equilibrium among contiguous phases, the conservation equation for energy in the multiphase mixture can be expressed as, see Cheng [5],

$$\frac{\partial}{\partial t}[(1-\varepsilon)\rho_s h_s + \varepsilon\sum_k \rho_k s_k h_k] + \nabla\cdot(\sum_k \rho_k \mathbf{u}_k h_k) = \nabla\cdot(k_{eff}\nabla T) + \dot{q} \qquad (33)$$

where k_{eff} stands for an effective thermal conductivity of the multiple flowing phases combined with the solid matrix phase, and $\dot{q}$ is a volumetric heat source.

The advective term in the energy equation, Equation (33), can be further decomposed, with the help of Equation (20), into contributions due to the bulk mixture motion and intrinsic relative phase motion. The energy equation can then be simplified to the following form in which only mixture variables are involved:

$$\frac{\partial}{\partial t}[(1-\varepsilon)\rho_s h_s + \varepsilon\rho h] + \nabla\cdot(\gamma_h \rho \mathbf{u} h) = \nabla\cdot(k_{eff}\nabla T) + \nabla\cdot[\sum_k (h_k \mathbf{j}_k)] + \dot{q} \qquad (34)$$

where the correction factor for energy advection, γ_h, is defined as

$$\gamma_h = \frac{\rho\sum_k \lambda_k h_k}{\sum_k \rho_k s_k h_k} \qquad (35)$$

and the second term on the right hand side of Equation (34) describes the energy flux due to relative phase motions, including both sensible and latent heat transport.

Supplementary Relationships

The system of conservation equations obtained above for a multiphase mixture requires constitutive equations for relative permeabilities, k_{rk}, capillary pressure functions, p_{cjk}, the effective mass diffusivity, D_k^{α}, and the thermal conductivity, k_{eff}, see Wang and Cheng [16]. A typical set of constitutive relationships for the liquid-gas system is given by

$$k_{rl} = s_l^3 \qquad (36)$$

$$k_{rg} = (1-s_l)^3 \qquad (37)$$

$$p_c = p_g - p_l = \sigma\left(\frac{\varepsilon}{K}\right)^{1/2} J(s_l) \qquad (38)$$

where

$$J(s_l) = 1.417(1-s_l) - 2.120(1-s_l)^2 + 1.263(1-s_l)^3 \qquad (39)$$

and Equation (39) was correlated by Udell [21]. The effective diffusivities within the liquid and gas phases are taken to be constant, since a reliable model for the hydrodynamic dispersion diffusivity in a two-phase system is presently lacking.

Finally, to close the mathematical system, expressions are needed to determine phase saturations s_k and compositions C_k^α within all the phases. For a wide range of practical applications, a multiphase system can be closely approximated by the assumption of local chemical equilibrium at phase interfaces. This assumption of interfacial equilibrium does not rule out the possibility of phase composition gradients on a macroscopic scale, but rather implies that within some short time scale (essentially instantaneously) contiguous phases reach a thermodynamic equilibrium. Thus, the mass concentration in the gas phase can be related to temperature and pressure through equilibrium phase diagrams, i.e.

$$C_g^\alpha = C_g^\alpha(T,p) \tag{40}$$

The concentrations of the same species in other phases can be related to the above in the gas phase through the partitioning concept:

$$C_k^\alpha = K_{kg}^\alpha\, C_g^\alpha \tag{41}$$

where K_{kg}^α is called the partition coefficient of species α between the phases k and g. In general, partition coefficients are represented by functions of phase compositions and pressures. These coefficients may be determined from Henry's law constants and solubility relations, Abriola and Pinder [10].

Once having determined the equilibrium phase compositions C_k^α, the phase saturations s_k can be calculated from the mixture concentrations which are obtained by solving Equation (31). For an *m*-phase, *n*-component system, Equation (31) thus represents *(n-1)* equations for *(m-1)* unknown phase saturations. The other phase saturation can be obtained from the constraint that the summation of all phase saturations is equal to unity. According to Gibbs phase rule, *n* is always equal to or greater than *m* when the multiphase, multicomponent system has two or more degrees of freedom (i.e. one for the temperature and others for phase saturations); therefore, Equation (31) provides a sufficient number of equations to solve for all phase saturations. To illustrate this point more clearly, Table 1 summarizes the various thermodynamic states for a two-phase, binary system, Wang and Cheng [16]. The explicit relations for determining phase saturations and concentrations corresponding to each state are given in the table.

The assumption of local thermal and chemical equilibrium does not preclude the existence of non-equilibrium conditions on a larger macroscopic scale. Macroscopic redistribution of species by both advective and diffusive transport is accommodated in Equation (31). If local equilibrium is unlikely, as in some rapid spread processes of chemical compounds, Power et al. [22], alternative means of relating the mixture concentration with the phase compositions would be required.

Model Discussion

A consistent set of model equations for the multiphase mixture has been derived from the classical multiphase flow model without making additional approximations. These model equations are now summarized in Table 2. The multiphase mixture model provides a full system of governing equations for four mixture variables, i.e. mixture velocity **u**, mixture pressure p, mixture concentration, C^{α}, and the mixture temperature, T. Several salient features of the present formulation become evident by a close inspection of Table 2 and are discussed below.

Table 1: Thermodynamic states in two-phase, binary systems

Conditions	$C < C_l^e(T,p)$	$C_l^e(T,p) < C < C_g^e(T,p)$	$C > C_g^e(T,p)$
Thermodynamic state	liquid	two-phase	gas
Liquid saturation, s_l	1	$\dfrac{\rho_g(C-C_g^e)}{\rho_l(C_l^e-C)+\rho_g(C-C_g^e)}$	0
Concentration in the liquid phase, C_l	C	$C_l^e(T,p)$	0
Concentration in the gas phase, C_g	0	$C_g^e(T,p)$	C

Table 2: Model equations in the multiphase mixture model

Conserved Quantity	Equation
Mass	$\varepsilon\dfrac{\partial\rho}{\partial t} + \nabla\cdot(\rho\mathbf{u}) = 0$
Momentum	$\rho\mathbf{u} = -\dfrac{K}{\nu}(\nabla p - \gamma_\rho\rho\mathbf{g})$
Species	$\varepsilon\dfrac{\partial}{\partial t}(\rho C^{\alpha}) + \nabla\cdot(\gamma_\alpha\rho\mathbf{u}C^{\alpha}) = \nabla\cdot[\varepsilon\rho D\nabla C^{\alpha}] - \nabla\cdot(\sum_k C_k^{\alpha}\mathbf{j}_k) + \nabla\cdot\{\varepsilon\sum_k[\rho_k s_k D_k^{\alpha}(\nabla C_k^{\alpha} - \nabla C^{\alpha})]\}$
Energy	$\dfrac{\partial}{\partial t}[(1-\varepsilon)\rho_s h_s+\varepsilon\rho h]+\nabla\cdot(\gamma_h\rho\mathbf{u}h)=\nabla\cdot(k_{eff}\nabla T) + \nabla\cdot[\sum_k(h_k\mathbf{j}_k)] + \dot{q}$

First, striking similarities can be noticed between the present model equations and those in the single-phase transport theory; the two sets differ only in that property coefficients are variable, e.g. see Figure 2, and additional source terms appear (in the mixture species and energy equations) in the case of multiphase flows. Figure 2 shows the property coefficients of the multiphase mixture model as calculated using the constitutive relations given in Equations (36)-(39). The variable properties precisely reflect the physics that multiphase flow problems are inherently nonlinear. The source terms arise from the interphase mass and heat transfer occurring internally within the multiphase mixture. In the case of single-phase flow, these source terms vanish and the variable properties simplify to those corresponding to the single-phase transport theory.

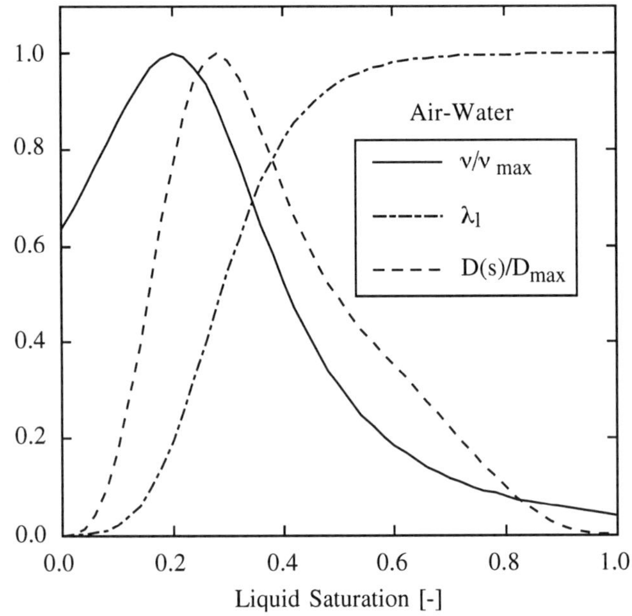

Figure 2: Characteristic curves of the effective parameters contained in the multiphase mixture model for the air-water system, including the mixture viscosity, relative mobility, and capillary diffusion coefficient

Recognizing the similar structure of the present multiphase mixture model to the single-phase transport theory, it is clear that the model equations listed in Table 2 for multiphase flows are amenable to well established single-phase CFD (Computational Fluid Dynamics) algorithms. A solution sequence can be as follows: (i) from the listed transport equations for the multiphase mixture, the mixture pressure, velocity, concentration and temperature fields are determined; (ii) from the mixture pressure, concentration and temperature at a nodal point, the thermodynamic state of this cell is identified as in Table 1 for two-phase, binary systems; (iii) phase concentrations are then calculated from equilibrium phase diagrams and the phase saturations are determined from the mixture concentration solved from the species transport equation; (iv) effective thermophysical properties of the multiphase mixture are then evaluated with the knowledge of phase saturations, and finally these properties are substituted back into the transport equations for updating. The entire procedure is repeated iteratively until convergence is achieved.

Secondly, the multiphase mixture model reduces the number of differential governing equations almost by half when compared to the existing multiphase flow model. For example, for a three-phase isothermal flow, the multiphase flow model requires a total of 12 primary variables: 2 phase saturations, 9 velocity components (for three phases in three dimensions) and any one phase pressure. In contrast, the present multiphase mixture model contains only 6 primary variables: 2 mixture concentrations or equivalently 2 phase saturations, 3 mixture velocity components and the mixture pressure. This unique feature of the new model permits efficient numerical simulations with even moderate computer resources.

Thirdly, the governing equations listed in Table 2 are valid throughout a problem domain, including single- and multi-phase regions, as the mixture variables always exist. Therefore, the present mixture formulation eliminates the need for numerical remeshing and coordinate mapping to track moving internal phase interfaces, and the use of variable or equation switching techniques. When a phase appears or disappears, the transport equations for the multiphase mixture need not to be modified, while the thermodynamic state of a multiphase system is adaptively recognized through, for example, Table 1. Therefore, no additional implementation or programming is required during phase transitions. Furthermore, it seems that the numerical difficulty of the present model does not increase in proportion with the number of phases in a system. While it is, of course, fundamentally more difficult to solve a two-phase problem than a single-phase problem because of variable properties and additional source terms involved in the former, the increase in computational burden from a two-phase problem to a three-phase problem is minor. In fact, a computer code generally for three-phase systems is identically applicable to two-phase problems without any modifications. This salient feature of the present approach allows for unification of different analysis methods existing in the literature for a wide variety of transport problems in the subsurface, including air-water two-phase flow in the unsaturated zone, single-phase groundwater flow in the saturated zone, air-water-NAPL three-phase flow in the contaminated unsaturated zone and NAPL-water two-phase flow in the contaminated saturated zone.

Most importantly, the present formulation does not smear out intrinsic characteristics of the individual phases, despite that the governing equations deal solely with the bulk behavior of the mixture. This is because the formulation embodies simple relations to determine the phase quantities from the mixture variables. These relations are summarized in Table 3. It can be seen that individual phase velocities and pressures can readily be calculated once a converged solution to the governing equations for the multiphase mixture is obtained. Moreover, this type of calculation, if needed, can be done in a post-processing manner. Hence, the computational cost required to determine the intrinsic characteristics of the individual phases is negligible. It is expected that the present model, owing to the above-discussed attributes, opens new prospects for understanding complicated multiphase flow and transport phenomena in porous media.

From an experimental point of view, both multiphase flow and multiphase mixture models require precisely the same number of effective properties to be measured. In the case of multiphase flow model, they are (*m*) relative permeabilities and (*m-1*) capillary pressures for an *m*-phase system. The effective parameters needed to be determined empirically in the current multiphase mixture model consist of (*1*) mixture kinematic viscosity, (*m-1*) relative mobilities, and (*m-1*) capillary diffusion coefficients. The both sets of effective properties are interchangeable, as given by Equations (5), (6) and (24). Figure 2 displays typical curves of the three effective parameters contained in a two-phase mixture model, namely the mixture kinematic viscosity, the liquid relative mobility (the gas mobility $\lambda_g=1-\lambda_l$) and the capillary diffusion coefficient. They were calculated from the empirical correlations for the relative permeabilities and capillary pressure function given by Equations (36)-(39).

Table 3: Relations to determine phase velocities and pressures from mixture variables

Phase Quantity	Relation
Velocity	$\rho_k \mathbf{u}_k = \mathbf{j}_k + \lambda_k \rho \mathbf{u}$ where $\mathbf{j}_k = \sum_i [-\rho_k D_{cik} \nabla s_i + \frac{\lambda_k \lambda_i K(\rho_k - \rho_i)}{\nu} \mathbf{g}] + \sum_\alpha [-\rho_k D_{s\alpha k} \nabla C^\alpha]$ $+ (-\rho_k D_{tk}) \nabla T$
Pressure	$p_k = p - (\sum_i \int_0^{s_i} Ca_{ik}\, ds_i + \sum_\alpha \int_0^{C^\alpha} Cs_{\alpha k}\, dC^\alpha + \int_0^T Ct_k\, dT)$

ANALYTICAL APPLICATIONS

Comparison to Richards' Theory

To compare the multiphase mixture model to the well-known Richards' theory for unsaturated flow in subsurface hydrology, it is instructive to substitute Equation (20) for the liquid mass flux $\rho_l \mathbf{u}_l$ into the mass conservation equation for the liquid phase, Equation (14). This results in

$$\varepsilon \frac{\partial(\rho_l s_l)}{\partial t} + \nabla \cdot (\rho \mathbf{u} \lambda_l) = - \nabla \cdot \mathbf{j}_l + \bar{\dot{m}}_l \tag{42}$$

where the diffusive mass flux $\mathbf{j}_l$ can be simplified, in the case of isothermal two-phase flow, to

$$\mathbf{j}_l = -\rho_l D_c \nabla s_l + \frac{\lambda_l \lambda_g K(\rho_l - \rho_g)}{\nu} \mathbf{g} \tag{43}$$

Substitution of the above into Equation (42) yields

$$\varepsilon \frac{\partial(\rho_l s_l)}{\partial t} + \nabla \cdot (\rho \mathbf{u} \lambda_l) = \nabla \cdot (\rho_l D_c \nabla s_l) - \nabla \cdot \left(\frac{\lambda_l \lambda_g K(\rho_l - \rho_g)}{\nu} \mathbf{g} \right) + \bar{\dot{m}}_l \tag{44}$$

This equation can be referred to as the generalized Richards' equation because it identically reduces to Richard's equation in the limit of the constant gas phase pressure or equivalently zero gas viscosity. In this case, it can be easily shown from Equations (5), (6) and (24) that

$$\lambda_l \rightarrow 0\ ;\ \lambda_g \rightarrow 1\ ;\ \frac{\lambda_l}{\nu} \rightarrow \frac{k_{rl}}{\nu_l} \quad \text{and} \quad \rho_l D_c \rightarrow -K\frac{k_{rl}}{\nu_l}\frac{dp_c}{ds_l} \tag{45}$$

Substitution of these asymptotes (45) into Equation (44) recovers Richards' equation for unsaturated flow. It thus becomes clear that Richards' approximation elegantly eliminates the advective term from Equation (44), thereby resulting in a single equation of diffusion nature for the liquid saturation. Whereas this yields great simplifications, the applicability of the unsaturated flow theory to general multiphase flow problems is severely limited as the gas phase motion remains indeterminate.

Richards' equation can also be derived from Equation (44) using the limit of $\nu_l \rightarrow \infty$. Collectively, it can be concluded that the unsaturated flow theory holds true in the limit of the viscosity ratio ν_l/ν_g approaching infinity. A numerical experiment for a boundary-layer type two-phase flow indicated that this condition would correspond to such a high threshold value, i.e. 10^7, that practically no fluid pair could possibly match, see Wang and Beckermann [23].

In the absence of Richards' approximation, Equation (44) contains both diffusion and advective terms, where the mixture velocity field must be determined from the continuity and momentum equations given by Equations (15) and (17). In view of the fact that the unsaturated flow theory introduces approximations that are difficult to justify and ascertain in general situations, the exact nature of Equation (44) may provide a new theoretical tool for an accurate understanding of nonlinear multiphase phenomena inside porous media.

Boundary-Layer, Two-Phase Flow and Transport

Two-phase flows with phase change in unbounded porous media are encountered in numerous technological applications. Examples include boiling flow along an igneous intrusion in geothermal reservoirs and condensing flow adjacent to a cold surface in heat pipes and porous insulation materials. Recently, there has been much research on the boundary-layer condensing flow because of its connection to condensation heat transfer in porous media, see Plumb [24], Majumdar and Tien [25] and Chung et al. [26]. The previous theoretical studies neglected the vapor flow in order to render the problem analytically tractable; the two-phase flow problem was thus reduced to the consideration of the liquid phase flow only. This approximate approach is essentially equivalent to the unsaturated flow theory described earlier. The order-of-magnitude analysis presented in the preceding section indicated that the unsaturated flow theory holds true for condensing flows only when the ratio of the kinematic viscosities of the liquid and vapor approaches infinity. However, in practice this ratio is very small, e.g. equal to 0.015 for a water-steam system at atmospheric pressure, thus warranting a full two-phase analysis for condensing and boiling flows.

Within the framework of the present multiphase mixture model, a set of boundary-layer equations was derived by Wang and Beckermann [27] for full two-phase flows with phase change in porous media, and a series of similarity solutions was developed in both condensing and boiling cases (Figure 3), Wang and Beckermann [23]. Figure 4a shows the liquid saturation profiles in condensing flow for different viscosity ratios ν_l/ν_g, along with the result obtained using the unsaturated flow theory of Majumdar and Tien [25]. It can be seen that the "two-phase solution" curves indeed approach the limiting case represented by the unsaturated flow theory at the value of $\nu_l/\nu_g=10^7$! Obviously, for most fluids, this condition cannot be satisfied, indicating that the unsaturated flow theory is generally questionable when applied to the condensation problem under consideration. In fact, for a water-steam system at atmospheric pressure, $\nu_l/\nu_g=0.015$, the saturation profile from the two-phase model greatly deviates from that predicted by the unsaturated flow theory. The capillary effect on condensation heat transfer is reflected by the liquid mass flux at the boundary. It was shown by Wang and Beckermann [23] that the unsaturated flow theory overpredicts this liquid mass flux by a factor of about 66, suggesting that the previous theoretical investigations based on the unsaturated flow theory, Plumb [24] and Majumdar and Tien [25], have probably exaggerated the effects of the capillary force on condensation heat transfer in porous media.

Figure 4b shows the saturation profiles for the boundary-layer boiling flow, with the liquid saturation at the wall as a curve parameter. These profiles are complex in shape and possess several points of inflection due to the fact that the effective transport properties of a two-phase mixture are highly nonlinear and exhibit maxima at certain liquid saturations (see Figure 2). The dryout heat flux corresponding to zero liquid saturation at the wall was found to be given by

$$q_{dry} = 2.716\times10^{-3}\,\frac{h_{fg}\sigma}{\nu_l}\left(\frac{\varepsilon K}{L^2}\right)^{1/4}\left[\frac{(\rho_l - \rho_v)gK}{\sigma}\right]^{1/2} \tag{46}$$

This analytical result provides a simple equation for engineering calculations.

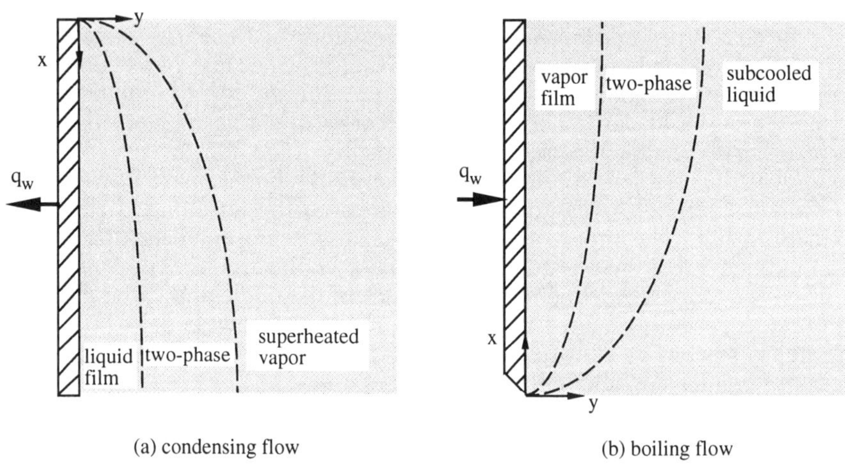

Figure 3: Schematic of boundary-layer (a) condensing and (b) boiling flows in porous media

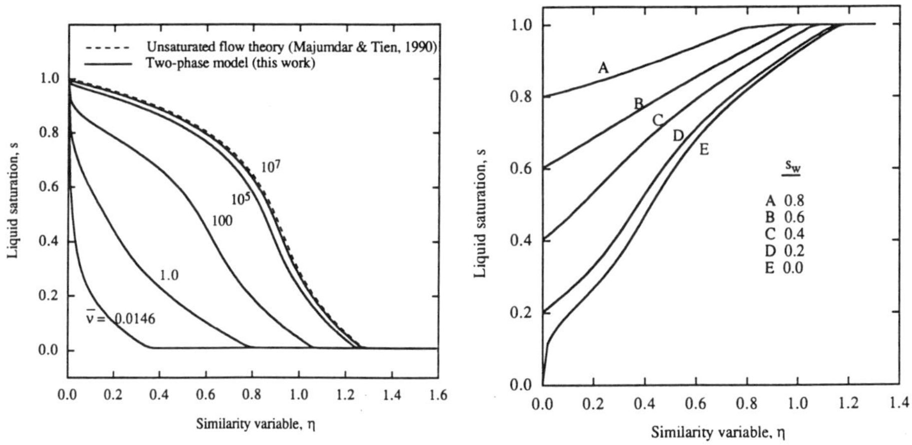

Figure 4: Saturation profiles for: (a) condensing flow, and (b) boiling flow, with the similarity variable $\eta = \frac{y}{x}\left[\frac{Kg(\rho_l - \rho_v)x}{\nu D_c}\right]^{1/2}$; $\bar{\nu}=\nu_l/\nu_g$, and s_w being the liquid saturation at the wall

Two-Phase Flow with Viscous Coupling

Most recently, the multiphase mixture model was extended to describe the viscous coupling phenomena between two immiscible fluids in porous media by Wang [28]. The viscous coupling arises from the interfacial contact between the two flowing fluids, and is usually described by the generalized Darcy's law with addition of cross-terms, i.e.

$$\mathbf{u}_1 = -K\left(\frac{k_{r11}}{\mu_1}\nabla p_1 + \frac{k_{r12}}{\mu_2}\nabla p_2\right) \tag{47}$$

$$\mathbf{u}_2 = -K\left(\frac{k_{r21}}{\mu_1}\nabla p_1 + \frac{k_{r22}}{\mu_2}\nabla p_2\right) \tag{48}$$

where subscripts 1 and 2 refer to fluids 1 and 2, respectively, and the four relative permeabilities k_{r11}, k_{r12}, k_{r21} and k_{r22} are to be determined experimentally. Wang [28] found that the mixture equations, i.e. Equations (15), (17) and (44), governing the two-phase flow without viscous coupling, remained unaltered in the case with viscous coupling, except that the property coefficients, λ_1 and D_c, were modified as

$$\lambda_1 = \frac{\frac{\rho_1 k_{r11}}{\mu_1} + \frac{\rho_1 k_{r12}}{\mu_2}}{\frac{\rho_1 k_{r11}}{\mu_1} + \frac{\rho_1 k_{r12}}{\mu_2} + \frac{\rho_2 k_{r21}}{\mu_1} + \frac{\rho_2 k_{r22}}{\mu_2}} \tag{49}$$

and

$$D_c = \frac{K\nu}{\nu_1 \nu_2} \left(k_{r11}k_{r22} - k_{r12}k_{r21}\right)\left[-\frac{dp_c}{ds_1}\right] \tag{50}$$

respectively. This result implies that two-phase flow with viscous coupling does not represent a fundamentally new problem worthy of separate studies. Mathematically, a mixture theory is possible for two-phase flow with viscous coupling because it is described by linear mechanics as evident from Equations (47) and (48).

NUMERICAL APPLICATIONS

Single-Component Systems

This subsection reduces the foregoing multiphase mixture model to a particular class of problems that involve two-phase, single-component systems. These problems often arise from heat transfer applications and are characterized by the co-existence of a nearly isothermal two-phase zone and single-phase regions with irregular and moving phase interfaces in between, see for example Figure 5. A fixed-grid numerical formulation has been developed for general problems that may simultaneously include a superheated vapor region, a two-phase zone and a subcooled liquid region in a single physical domain, Wang et al. [29,30]. This algorithm can readily be implemented into widely available single-phase CFD codes. A sample simulation concerning post-dryout heat transfer in a packed bed with subcooled liquid throughflow is presented to demonstrate the ability of this numerical model.

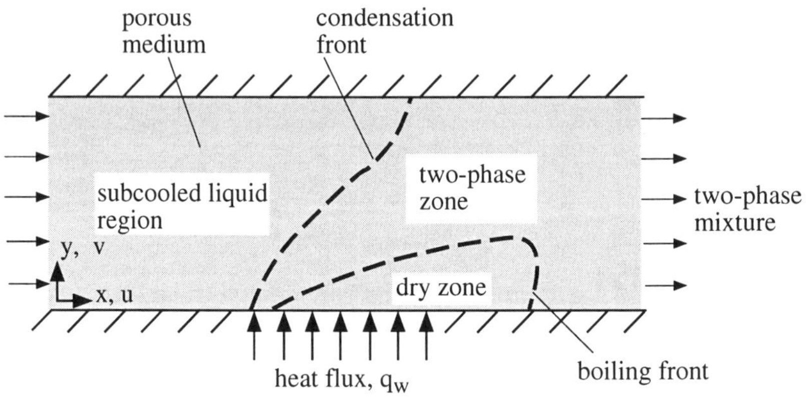

Figure 5: Boiling and post-dryout in a horizontal porous bed with liquid through-flow

Figure 6 shows numerical results representative of the post-dryout condition for a bed packed with steel beads and under the inlet conditions of u_{in}=0.25 mm/s and T_o=22°C. The bottom heat flux was set to be 350 kW/m^2, as compared to the dryout heat flux of 180 kW/m^2. Each plot displays temperature, liquid saturation, liquid and vapor velocity distributions, respectively. Two phase interfaces, namely the condensation front which separates the subcooled liquid region from the two phase zone and the boiling front which separates the two-phase zone from the superheated vapor region, are also plotted as dashed lines.

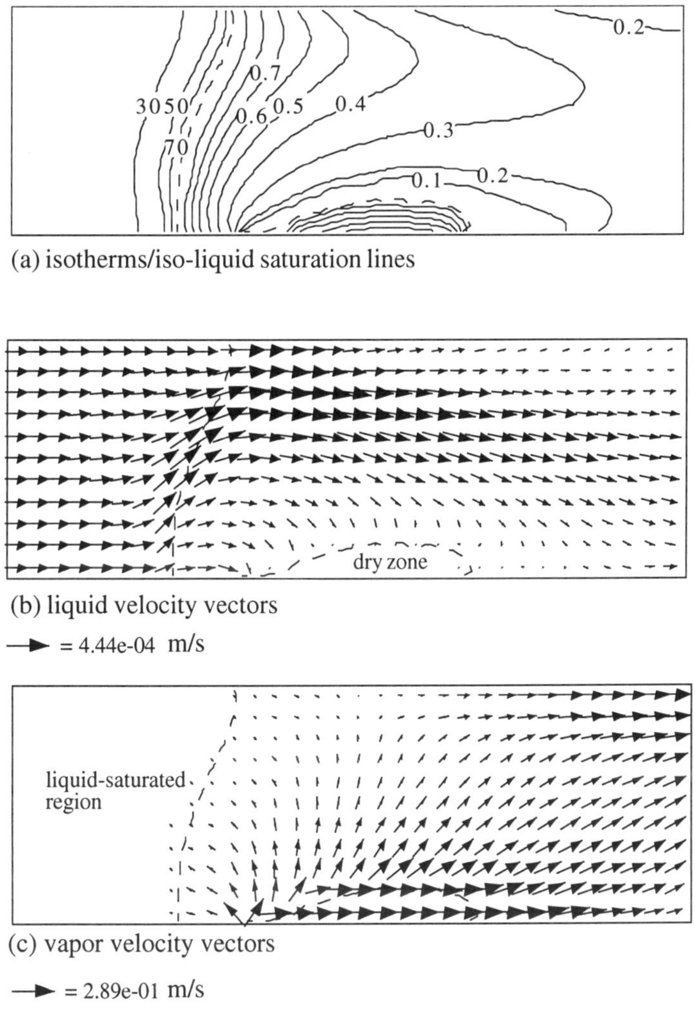

Figure 6: Numerical results for steel beads bed and with q_w=350 kW/m^2: (a) isotherms (ΔT=20^oC in the subcooled region but ΔT=100^oC in the dry zone) and liquid saturation iso-lines; (b) liquid velocity vectors; and (c) vapor velocity vectors. The dashed lines denote the phase interfaces, i.e. the condensation and boiling fronts

Figure 6a shows a dry zone, i.e. superheated vapor region, forming over the heated surface. Within this dry zone, isotherms are plotted in an interval of 100°C with the maximum temperature reaching 657°C. Such a maximum temperature is still below the melting point of steel beads and therefore the system sustains even beyond the dryout. Bounded by two dashed lines, i.e. the condensation and boiling fronts is a two-phase zone, inside which the liquid saturation distribution indicates that there are two low-liquid saturation regions, one naturally being in the vicinity of the boiling front while the other appears in the upper-right corner of the bed. The latter vapor-rich region results from the fact that the much lighter vapor migrates upward and subsequently accumulates at the adiabatic top of the bed.

Note also that the predicted thermodynamic structure appears stable, despite the gravitational instability experienced by the heavier two-phase zone overlying the lighter dry zone. The stability arises most likely because of the phase change occurring at the boiling front. The liquid velocity vectors shown in Figure 6b indicate that an appreciable portion of the liquid is drawn towards the boiling front where the liquid evaporates. Notice that in the post-dryout situation, vapor is not generated from the heated surface but on the boiling front. Part of the generated vapor ascends vertically within the two-phase zone under the influence of gravity, whereas the other part of vapor penetrates through the dry zone due to its greater permeability, Figure 6c. It can be seen that the vapor flow inside the dry zone is primarily horizontal probably because of the horizontal pressure gradient imposed externally. The superheated vapor region is seen to be contained within the bed, and is expected to break through the exit under higher heat fluxes.

Multi-Component Systems

Many groundwater contamination problems are multi-component systems containing such chemical components as water, air, and various organic compounds. Multiple phases involved in such systems include water, gas, and non-aqueous phase liquids (NAPL). It is of practical interest to numerically visualize the entire stages of groundwater contamination and remediation events, starting from the NAPL invasion into the subsurface, to the redistribution after the contaminant source is removed, and finally to the NAPL removal from the subsurface. This subsection applies the foregoing multiphase mixture model to such a computationally intensive task in order to further demonstrate the model utility.

The numerical simulation corresponds to a hypothetical two-dimensional aquifer contaminated by carbon tetrachloride, as shown in Figure 7. The aquifer initially contains air and water, while the water saturation is assumed to be uniform and at a residual value; since the water phase is practically immobile (k_{rw}=0) during the NAPL transport, it can be excluded from the system without much loss of generality. The system is also assumed to be isothermal at 25°C and at atmospheric pressure so that the energy equation needs not to be solved.

As shown in Figure 7, the NAPL liquid is introduced into the system through a small opening along the left boundary, which models a contaminant source resulting from a leaky underground tank. A constant NAPL saturation is assumed to prevail inside this region during the entire infiltration process. This first stage of the simulation consists of NAPL infiltration for 3 hours. Subsequently, the leaking contaminant source is removed and the NAPL is allowed to redistribute in the aquifer under gravitational and capillary forces for 9 hours, after which time a remediation scheme by injecting clean air from the right to left boundary is implemented. During this last stage of the simulation (i.e. remediation) the ground surface is covered to prevent the contaminant from escaping into the atmosphere. Other boundary conditions are described in Figure 7, where the lower boundary is assumed to be impermeable to both phases.

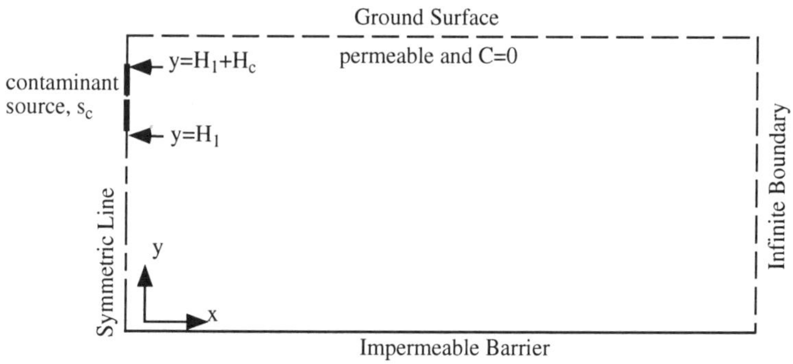

Figure 7: Description of a groundwater contamination and remediation problem

The entire process of NAPL infiltration, redistribution, and remediation is numerically predicted using a single simulator developed based on the multiphase mixture model. The solution domain consists of a fine 90×30 grid in order to accurately capture NAPL migration, Cheng and Wang [31], and to resolve many fine-scale phenomena as will be shown shortly. This mesh was also shown to be sufficient to generate accurate results. The entire simulation for all stages of infiltration, redistribution and remediation required approximately 3 CPU hours on a HP 715/100 workstation. Details of the numerical procedures were reported by Cheng and Wang [32].

It is noteworthy that this numerical study accounts for both the movement of the NAPL in the subsurface and the density-induced convection in the gas phase. The gas flow induced by the solutal buoyancy forces represents an important mechanism for the gas phase transport of chemical compounds in the vadose zone. In contrast, previous work, see Falta et al. [33], Mendoza and Frind [34] and Conant et al. [35] on vapor phase transport in the unsaturated zone assumed a stagnant NAPL, thus failing to model the NAPL infiltration process after a spill.

Figure 8 shows selected results from the numerical simulation, which include the NAPL saturation distribution (the gray scale images), the iso-concentration contours in the gas phase (the lines), and the gas phase velocity vectors (the arrows). The mass concentration of the organic vapor in the gas phase can be read from Figure 8 using the conversion relation $C_v = (1\text{-}C_g)$. Thus, the 0.99 contours shown in Figure 8 denote one percent of organic vapor present in the gas phase.

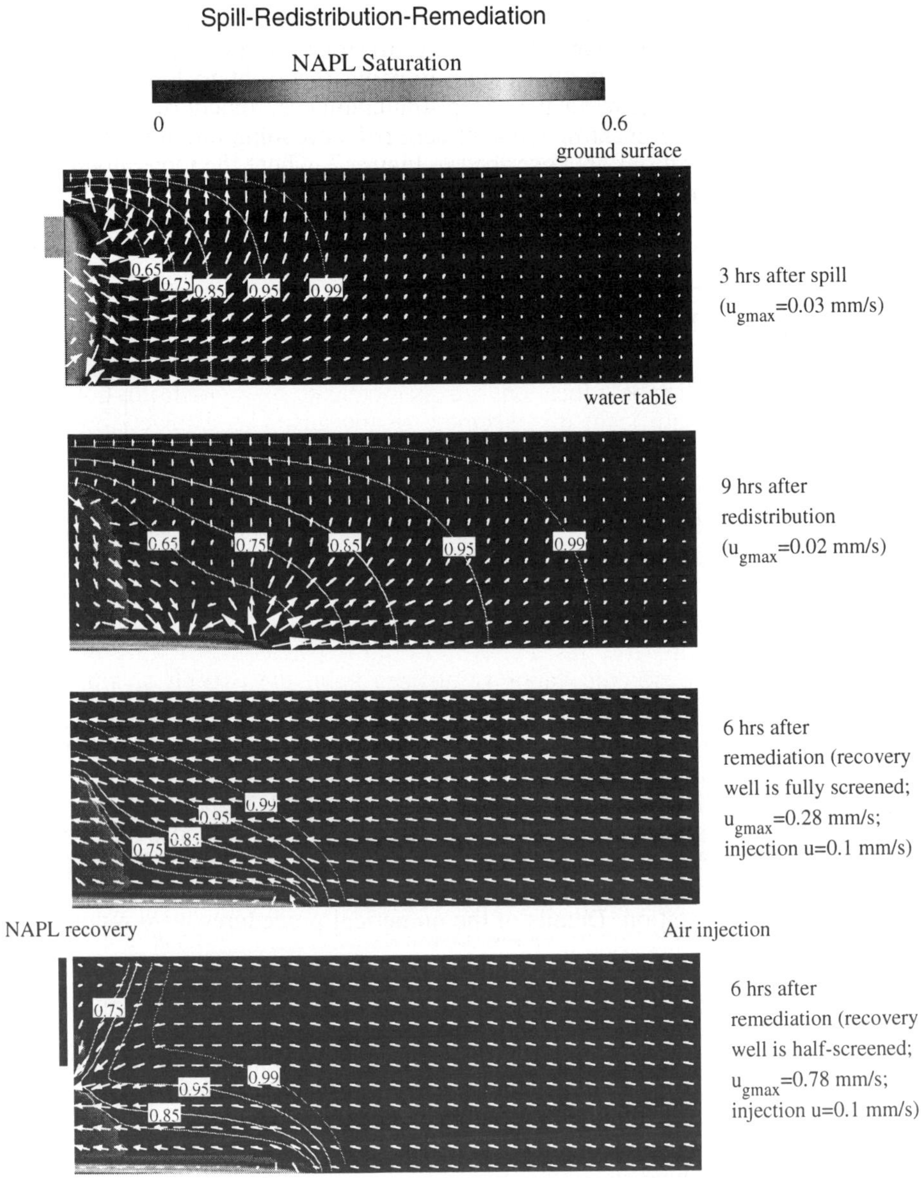

Figure 8: Composite plots of NAPL saturation, air concentration, and gas phase velocity for: (a) invasion stage; (b) redistribution stage; (c) remediation stage by injecting air from right to left boundary with fully screened wells; and (d) remediation stage by injecting air from right to left boundary with a half-screened recovery well

Figure 8a displays the composite plots at 3 hours after the spill. The NAPL is seen to reach the impermeable bottom. At the same time, the organic vapor emanating from the NAPL-gas two-phase zone has widely spread into the aquifer due to a displacement flow in the gas phase caused by the infiltrating NAPL. Previous single-phase studies, Falta et al. [33] and Mendoza and Frind [34] identified that the wide spreading of organic vapor is attributed to solutally induced convection. However, in a conjugate single- and two-phase flowing system, the displacement flow in the gas phase caused by the propagation of the phase interface overrides the density-driven flow, Cheng and Wang [32]. In Figure 8a the velocity scale has been chosen small enough to show the details of the velocity field in the pure gas region while omitting some large velocity vectors in the two-phase zone. It can be seen that the displacement stems mainly from the vicinity of the contaminant source at the left boundary as well as the leading infiltration front along the bottom impermeable boundary; both locations are characterized by rapid infiltration of the NAPL and hence significant displacement of the gas. The displaced gas exits through the permeable ground surface.

The plots displayed in Figure 8b show the NAPL redistribution process after the contaminant source is removed. At 9 hours of redistribution, the NAPL is seen to accumulate at the bottom because of the NAPL being much heavier than the gas. The maximum NAPL saturation is found therein, which may greatly endanger the underlying groundwater system. The NAPL saturation in the area adjacent to the original contaminant source is, however, decreasing so as to maintain the mass balance in the system. The gas phase in the most part of the aquifer has been contaminated by the organic vapor volatilized from the two-phase zone. Fast spreading of the contaminant is partly due to rapid infiltration of the NAPL along the bottom, and partly due to the displacement of the gas phase by the infiltrating NAPL.

Figure 8c displays results of a remediation process in which clean air is injected uniformly along the right boundary, i.e. the injection well, and the organic vapor is recovered through a production well sitting along the left boundary. Both injection and production wells were fully screened to the bottom of the aquifer. Such a remediation method is termed soil vapor extraction. The ground surface is now covered to prevent the organic vapor from escaping to the atmosphere. The air velocity blown into the aquifer is 0.36 m/h. Comparing Figures 8b and 8c, it can be seen that the two-phase zone shrinks due to the evaporation of the organic liquid into the injected clean air which is then removed. However, a substantial portion of clean air bypasses the two-phase zone where the soil pores are partially occupied by the liquid contaminant and hence offers a higher resistance to air flow; only partial air actually penetrates through the two-phase zone and becomes saturated with the organic vapor. The bypassing phenomenon is also evident from the inclined iso-concentration contours towards the outlet at the top, which apparently limits the effectiveness of the remediation system. To alleviate this problem, a second remediation system is analyzed in which the production well is only half-screened from the mid-height to the bottom, and all other conditions remain the same. The screened zone is designed to direct the injected air through the contamination zone so as to augment the recovery efficiency. Figure 8d shows that the two-phase zone indeed recedes more significantly than the first remediation system. The iso-concentration contours also demonstrate a nearly piston-like displacement. Clearly, the two designs shown in Figures 8c and 8d indicate a great need for better understanding and control of the multi-dimensional distribution of the injected fluid in the subsurface. Herein lies the practical significance of a powerful and efficient analysis tool like this multiphase mixture model.

In general, it appears that the NAPL zone suspended within the subsurface can be effectively remediated by soil vapor extraction provided the screened zones of the injection and production wells are properly designed. However, the thin NAPL lens floating over the bottom surface is very difficult to remove because of an insufficiently large area for mass transfer between the NAPL and the incoming clean air. A horizontal well located right above the bottom surface may be effective for recover this thin NAPL lens. In any case, a viable numerical simulator capable of incorporating considerations of well placement, screened zones and implementation of barriers appears extremely useful for the design of cost-effective remedial systems.

CONCLUSIONS AND OUTLOOK

A multiphase mixture model for multiphase, multicomponent transport in porous media has been reviewed along with various applications. A number of salient features of this model were revealed. In particular, the present model has significantly reduced the number of governing equations while remaining to be mathematically equivalent to the classic multiphase flow formulation. It was also found that Richards' theory for unsaturated flow is just a simplified case of the present multiphase mixture model in the limit of infinitely large liquid-to-gas viscosity ratio. Furthermore, the mixture model was shown to represent a unified theory for two-phase flow through porous media with and without viscous coupling.

A general numerical formalism for single-component, two-phase problems was discussed along with a simulation of an all-region problem including simultaneously a superheated vapor region, a two-phase zone, and a subcooled liquid region. The numerical example of multi-component systems, i.e. simulating simultaneously contamination, redistribution, and remediation events in the subsurface, demonstrated more capabilities of the multiphase mixture model. Noticeably, the simulation was carried out on an engineering workstation and required very reasonable CPU times.

Future work will apply the present multiphase mixture model to a variety of practically challenging problems, extend the present version of the multiphase mixture model to accommodate interfacial non-equilibrium effects, as well as provide thorough experimental validation.

Model Applications

For practical applications, the following areas deserve immediate attention:

- numerical modeling of three-phase systems so as to provide predictive capabilities for a wide variety of subsurface transport problems.
- modeling of multicomponent transport. Although the application of the multiphase mixture model to single-component organic liquids has been demonstrated, many groundwater pollution problems involve multicomponent

organic liquids, such as gasoline. When the gas phase is contaminated by multiple organic components with differing molecular diffusivities, the density-driven gas flow may become unstable and thus enhance the transport of contaminants. Under these circumstances, multi-diffusive convection occurs as a result of local buoyancy-driven instabilities formed by the difference in diffusivity of the various components.

- the effects of heterogeneities. Heterogeneities, for example zones of differing permeability, will contribute to contaminant spreading in a number of fundamental ways. For example, a NAPL tends to spread out horizontally when it encounters a low-permeability layer. In addition, the vapor phase transport is diffusion-dominated in low-permeability areas.
- the effects of externally-induced flow and non-isothermal conditions arising in remediation processes. For example, in the thermally-enhanced soil vapor extraction, hot air is injected into a contaminated subsurface to drive NAPLs and organic vapors into recovery wells. It is of practical interest to develop predictive capabilities for these restoration practices using the present multiphase mixture model.

Model Extension

In the present work, common assumptions invoked in multiphase flow models have been used in the model development, such as interfacial thermal and chemical equilibrium. It should be noted, however, that the multiphase mixture model is not inherently an equilibrium model; in fact, the present model has accounted for dynamic non-equilibrium because it allows for differing phase velocities. Similarly, the assumptions of interfacial thermal and chemical equilibrium can be relaxed. Under such circumstances, the model differential equations derived in this work remain valid, whereas the supplementary relationships for determining phase saturations, phase temperatures and compositions need to be revised. Under thermal and chemical non-equilibrium, the bulk phase temperature and concentration depart from their interfacial values as determined from equilibrium phase diagrams, with the magnitudes of departures measured by the local heat and mass transfer coefficients. A recent work of Peterson and Chang [36] showed that a thermal non-equilibrium, two-phase flow problem can indeed be tackled by the two-phase mixture model. More efforts are needed to develop a generalized formulation to cover the interfacial non-equilibrium cases.

Model Validation

Comprehensive predictions made possible by means of the multiphase mixture model require experimental verification. Extensive model validation would extend the utility of the multiphase mixture model eventually to field settings, e.g. for site-specific assessment of contamination and for developing cost-effective remediation schemes. To date, quantitative experimental data suitable for model validation are quite scarce and, thus, coordinated efforts between modeling and experimentation are needed to provide an ultimate understanding of the multiphase flow and transport phenomena in porous media.

REFERENCES

1. L.M. Abriola, Multiphase flow and transport models for organic chemicals: a review and assessment, *EPRI EA-5976*, California (1988).
2. A. Faghri, *Heat Pipe Science and Technology*. Taylor and Francis, Washington, D.C. (1995).
3. S. Whitaker, Simultaneous heat, mass, and momentum transfer in porous media: a theory of drying, *Advances in Heat Transfer* **13**, Academic Press, New York (1977).
4. Y.T. Shah, *Gas-Liquid-Solid Reactor Design*, McGraw-Hill, New York (1979).
6. Dhir, V.K. (1994). Boiling and two-phase flow in porous media. *Ann. Rev. Heat Transfer* **5**, 303-350.
5. P. Cheng, Heat transfer in geothermal systems, *Advances in Heat Transfer* **14**, 1-104 (1978).
6. C.Y. Wang and P. Cheng, Multiphase flow and heat transfer in porous media, *Advances in Heat Transfer* **30**, Academic Press, New York (1997).
7. G.M. Homsy, Viscous fingering in porous media, *Annu. Rev. Fluid Mech.* **19**, 271-311 (1987).
8. O. Güven, J.H. Dane, M. Oostrom and J.S. Hayworth, Physical model studies of dense solute plumes in porous media, *EPRI TR-101387*, California (1992).
9. R.J. Held and T.H. Illangasekare, Fingering of dense nonaqueous phase liquids in porous media 1. experimental investigation, *Water Resour. Res.* **31**, 1213-1222 (1995).
10. L.M. Abriola and G.F. Pinder, A multiphase approach to the modeling of porous media contamination by organic compounds 1. Equation development, *Water Resour. Res.* **21**, 11-18 (1985).
11. D.B. McWhorter and D.K. Sunada, Exact integral solutions for two-phase flow, *Water Resour. Res.* **26**, 399-414 (1990).
12. M.B. Allen III, Numerical modeling of multiphase flow in porous media, *Adv. Water Resour.* **9**, 162-187 (1985).
13. H.J. Morel-Seytoux, Two-phase flows in porous media, *Adv. in Hydroscience* **9**, 119-202 (1973).
14. P.S. Ramesh and K.E. Torrance, Numerical algorithm for problems involving boiling and natural convection in porous materials, *Numerical Heat Transfer B* **17**, 1-24 (1990).
15. S. Panday, P.A. Forsyth, R.W. Falta, Y.S. Wu and P.S. Huyakorn, Considerations for robust compositional simulations of subsurface nonaqueous phase liquid contamination and remediation, *Water Resour. Res.* **31**, 1273-1289 (1995).
16. C.Y. Wang and P. Cheng, A multiphase mixture model for multiphase, multicomponent transport in capillary porous media -- I: model development, *Int J Heat Mass Transfer* **39**, 3607-3618 (1996).
17. C.Y. Wang and C. Beckermann, A two-phase mixture model of liquid-gas flow and heat transfer in capillary porous media -- I. formulation, *Int. J. Heat Mass Transfer* **36**, 2747-2758 (1993).

18. R. Bowen, Theories of mixtures, in *Continuum Physics III* (ed. A.C. Eringer), Academic Press, New York (1976).
19. J.C. Parker, Multiphase flow and transport in porous media, *Reviews of Geophysics* **27**, 311-328 (1989).
20. H. Brenner and D.A. Edwards, *Macrotransport Processes*, Butterworth-Heinemann, Boston (1993).
21. K.S. Udell, Heat transfer in porous media considering phase change and capillarity --- the heat pipe effect, *Int. J. Heat Mass Transfer* **28**, 485-495 (1985).
22. S.E. Power, L.M. Abriola and W.J. Weber, Jr., An experimental investigation of nonaqueous phase liquid dissolution in saturated subsurface systems: transient mass transfer rates, *Water Resour. Res.* **30**, 321-332 (1994).
23. C.Y. Wang and C. Beckermann, Boundary layer analysis of buoyancy-driven two-phase flow in capillary porous media, *ASME J Heat Transfer* **117**, 1082-1087 (1995).
24. O.A. Plumb, Capillary effects of film condensation in porous media, *AIAA-84-1789* (1984).
25. A. Majumdar and C.L. Tien, Effects of surface tension on film condensation in a porous medium, *ASME J Heat Transfer* **112**, 751-757 (1990).
26. J.N. Chung, O.A. Plumb and W.C. Lee, Condensation in a porous medium bounded by a cold vertical surface, *ASME J. Heat Transfer* **114**, 1011-1018 (1992).
27. C.Y. Wang and C. Beckermann, A two-phase mixture model of liquid-gas flow and heat transfer in capillary porous media -- II. application to pressure-driven boiling flow adjacent to a vertical heated plate, *Int. J. Heat Mass Transfer* **36**, 2759-2768 (1993).
28. C.Y. Wang, An alternative description of viscous coupling in two-phase flow through porous media, *Transport in Porous Media*, in press (1997).
29. C.Y. Wang, C. Beckermann and C. Fan, Numerical study of boiling and natural convection in capillary porous media using the two-phase mixture model, *Numerical Heat Transfer A* **26**, 375-398 (1994).
30. C.Y. Wang, A fixed-grid numerical algorithm for two-phase flow and heat transfer in porous media, *Numerical Heat Transfer B Fundamentals*, in press (1997).
31. K. Rathfelder and L.M. Abriola, Impact of grid resolution and parametric representation on multiphase flow simulations, in *Computational Methods in Water Resources X* (eds. by A. Peters et al.), 983-990, Kluwer Academic Publishers, Netherlands (1994).
32. P. Cheng and C.Y. Wang, A multiphase mixture model for multiphase, multicomponent transport in capillary porous media -- II: numerical simulation of the transport of organic compounds in the subsurface, *Int. J. Heat Mass Transfer* **39**, 3619-3632 (1996).
33. R.W. Falta, I. Javandel, K. Pruess and P.A. Witherspoon, Density-driven flow of gas in the unsaturated zone due to the evaporation of volatile organic compounds, *Water Resour. Res.* **25**, 2159-2169 (1989).
34. C.A. Mendoza and E.O. Frind, Advective-dispersive transport of dense organic vapors in the unsaturated zone 2. sensitivity analysis, *Water Resour. Res.* **26**, 388-398 (1990).

35. B.H. Conant, R.W. Gillham and C.A. Mendoza, Vapor transport of trichloroethylene in the unsaturated zone: field and numerical modeling investigations, *Water Resour. Res.* **32**, 9-22 (1996).
36. G.P. Peterson and C.S. Chang, Heat transfer analysis and evaluation for two-phase flow in porous-channel heat sinks, *Numerical Heat Transfer A* **31**, 113-130 (1997).

CONVECTIVE HEAT FLOW FROM SUDDENLY HEATED SURFACES EMBEDDED IN POROUS MEDIA

R. BRADEAN[1], P.J. HEGGS[2], D.B. INGHAM[1] and I. POP[3]

[1]Department of Applied Mathematical Studies, University of Leeds, Leeds LS2 9JT, UK
[2]Department of Chemical Engineering, UMIST, Manchester M60 1QD, UK
[3]Faculty of Mathematics, University of Cluj R-3400 Cluj, CP253, Romania

INTRODUCTION

Convective heat transfer process is of fundamental importance in a variety of practical applications, such as in porous insulations, cooling of rotating electric windings, geothermal reservoirs, irrigation systems and the exploration of petroleum and gas fields. Recent books or review articles by Cheng [1], Bejan [2], Kakac *et al.* [3], Tien and Vafai [4] and Nield and Bejan [5] give some indications of the extent of the research information in this area.

The earliest studies of heat convection through porous media have investigated the steady state situation in which asymptotic analysis and numerical techniques have been extensively used. In the free convection configuration, regular perturbation solutions of the full governing equations in terms of the small Rayleigh number, Ra, have been obtained by Yamamoto [6] for an isothermal sphere and by Bradean *et al.* [7, 8] for a spatially heated and cooled, vertical and horizontal flat plate. At very large values of the Rayleigh number, the boundary-layer approximation is valid and similarity solutions of the boundary-layer equations have been obtained by Cheng and Minkowycz [9] and Ingham and Brown [10] for a vertical plate, Cheng and Chang [11] for a horizontal plate and Merkin [12] for a two-dimensional or axisymmetric body of arbitrary shape. Higher-order boundary-layer effects have been investigated by Cheng and Chang [13], Cheng and Hsu [14], Chang and Cheng [15] and Joshi and Gebhart [16] using the matched asymptotic expansions technique. In order to obtain approximate solutions, which are valid for any value of the Rayleigh number, numerical techniques are usually used, and such numerical solutions have been obtained by Bradean *et al.* [7, 8] for a vertical and a horizontal flat plate, Ingham and Pop [17] for a horizontal circular cylinder and Pop and Ingham [18] for a sphere.

In the steady state mixed convection configuration, similarity solutions of the boundary-layer equations have been obtained by Cheng [19, 20, 21] for an inclined and a horizontal flat plate and for a horizontal circular cylinder or a sphere, whereas Joshi and Gebhart [22] have analysed higher-order boundary-layer effects using the matched asymptotic expansions technique. For an isothermal horizontal circular cylinder, analytical results which are valid for small values of the Peclét number and for ratios between the Grashof and Reynolds number of order unity have been obtained by Sano [23]. Numerical results of the full equations in this configuration have been obtained by Badr and Pop [24] for both aiding and opposing flows.

In recent years, the study of heat convection through porous media has been extended to unsteady situations, and analytical and numerical results have been extensively obtained using the boundary-layer approximation. Thus, the free convection flow past an impulsively heated or cooled, vertical or horizontal flat surface has been investigated by Ingham *et al.* [25, 26], Cheng and Pop [27] and Pop and Cheng [28] using asymptotic analysis, integral methods and numerical techniques. For an impulsively heated horizontal circular cylinder, regular perturbation solutions as well as numerical results have been reported by Ingham *et al.* [29]. The effects of variable permeability have been studied by Mehta and Sood [30] using the Kármán-Pohlhausen integral method, whereas the inertial effects in forced convection have been investigated by Nakayama and Ebinuma [31] using the Forchheimer model.

Asymptotic analysis has often been used in order to obtain series solutions in terms of the small Rayleigh number, which are valid for all times. Such analytical studies have been performed by Bejan [32] for a point heat source, Purushothaman *et al.* [33] for a pulsating point heat source, Nield and White [34] for a line heat source and Sano and Okihara [35] for a sphere. The viscous effects have been studied analytically at small values of the Rayleigh number by Ganapathy [36] using the Brinkman model. For a vertical plate whose temperature is oscillating with time, Singh *et al.* [37] have obtained two asymptotic expansions in powers of either the small or large frequency parameter.

The objective of this chapter is to review some of the latest work performed on the two-dimensional, unsteady free and mixed convection from an impulsively heated surface which is embedded in a porous medium at finite values of the Rayleigh number. Singular perturbation solutions in terms of the small time are presented in some particular configurations involving differently shaped impermeable surfaces subjected to different thermal boundary conditions. These analytical solutions have been obtained by matching inner and outer expansions in order to take into account the outer flow from the thermal boundary-layer which develops adjacent to the impulsively heated surface at small times. Numerical solutions, which are valid for all times and for any value of the Rayleigh number, are also presented in these configurations. The growth of the thermal boundary-layer at small times is taken into account in the formulation of the problem by making appropriate transformations in the direction normal to the surface.

GOVERNING EQUATIONS

Consider a two-dimensional model of unsteady heat convection through a fluid saturated porous medium. Assuming that the convective fluid and the solid structure are in local thermal equilibrium, the properties of the porous medium are constant, the Boussinesq approximation is valid and Darcy's law is applicable, the governing equations in terms of the temperature, T^*, and stream function, ψ^*, can be written as, see Nield and Bejan [5],

$$\frac{\partial^2 \psi^*}{\partial x^{*2}} + \frac{\partial^2 \psi^*}{\partial y^{*2}} = \frac{K\beta}{\nu}\left(g_y \frac{\partial T^*}{\partial x^*} - g_x \frac{\partial T^*}{\partial y^*}\right) \tag{1}$$

$$\sigma \frac{\partial T^*}{\partial t^*} + \frac{\partial \psi^*}{\partial y^*}\frac{\partial T^*}{\partial x^*} - \frac{\partial \psi^*}{\partial x^*}\frac{\partial T^*}{\partial y^*} = \alpha\left(\frac{\partial^2 T^*}{\partial x^{*2}} + \frac{\partial^2 T^*}{\partial y^{*2}}\right) \tag{2}$$

where (x^*, y^*) are arbitrary Cartesian coordinates, t^* is the time, ψ^* is the stream function which is defined so as to satisfy the continuity equation, i.e.

$$u^* = \frac{\partial \psi^*}{\partial y^*} \quad , \quad v^* = -\frac{\partial \psi^*}{\partial x^*} \tag{3}$$

u^* and v^* are the components of the fluid velocity in the x^* and y^* directions, respectively, g_x and g_y are the components of the acceleration due to gravity in the x^* and y^* directions, respectively, ν, K, σ and α are the kinematic viscosity of the fluid, permeability, heat capacity ratio and thermal diffusivity of the porous medium, respectively, and β is the coefficient of thermal expansion.

However, there are configurations of two-dimensional convective flows through a porous medium, such as the convection adjacent to a heated horizontal circular cylinder, for which Cartesian coordinates are not convenient. In terms of polar coordinates (r^*, θ), with $\theta = 0$ in the direction of the gravitational force, the governing equations (1) and (2) can be written as, see Nield and Bejan [5],

$$\frac{\partial^2 \psi^*}{\partial r^{*2}} + \frac{1}{r^*}\frac{\partial \psi^*}{\partial r^*} + \frac{1}{r^{*2}}\frac{\partial^2 \psi^*}{\partial \theta^2} = \frac{Kg\beta}{\nu}\left(\frac{\partial T^*}{\partial r^*}\sin\theta + \frac{\partial T^*}{\partial \theta}\frac{\cos\theta}{r^*}\right) \tag{4}$$

$$\sigma \frac{\partial T^*}{\partial t^*} + \frac{1}{r^*}\left(\frac{\partial \psi^*}{\partial r^*}\frac{\partial T^*}{\partial \theta} - \frac{\partial \psi^*}{\partial \theta}\frac{\partial T^*}{\partial r^*}\right) = \alpha\left(\frac{\partial^2 T^*}{\partial r^{*2}} + \frac{1}{r^*}\frac{\partial T^*}{\partial r^*} + \frac{1}{r^{*2}}\frac{\partial^2 T^*}{\partial \theta^2}\right) \tag{5}$$

where the stream function ψ^* is now defined by

$$u^* = -\frac{1}{r^*}\frac{\partial \psi^*}{\partial \theta} \quad , \quad v^* = \frac{\partial \psi^*}{\partial r^*} \tag{6}$$

so as to satisfy the continuity equation and u^* and v^* are the components of the fluid velocity in the r^* and θ directions, respectively.

FREE CONVECTION FROM A VERTICAL SURFACE

Consider the two-dimensional unsteady, free convection fluid flow due to an impulsively heated and cooled, in space, infinite vertical surface which is embedded in an isothermal fluid saturated porous medium. It is assumed that the surface, which is initially at the same temperature, T_∞, as the surrounding porous medium, is given at time $t^* = 0^+$ a sudden sinusoidal temperature distribution along the surface, namely,

$$T^*(x^*, y^*, 0) = T_\infty \qquad -\infty < x^* < \infty\ ,\ 0 \le y^* < \infty \tag{7}$$

$$T^*(x^*, 0, t^*) = T_a \sin\left(\frac{x^*}{L}\right) + T_\infty \qquad -\infty < x^* < \infty\ ,\ t^* > 0 \tag{8}$$

where the Cartesian coordinates (x^*, y^*) have been chosen along (upward) and normal (horizontal) to the surface, respectively, and T_a is the amplitude and $2\pi L$ the period of the wall temperature variations.

In general, the non-dimensional variables are defined as follows

$$x = \frac{x^*}{q}\ ,\ y = \frac{y^*}{q}\ , \tag{9}$$

$$u = \frac{u^*}{U_c}\ ,\ v = \frac{v^*}{U_c}\ ,\ T = \frac{T^* - T_{ref}}{\Delta T}\ ,\ \psi = \frac{\psi^*}{U_c q}\ ,\ t = \frac{t^* U_c}{\sigma q} \tag{10}$$

where q is a characteristic length, U_c is a characteristic speed, T_{ref} is a reference temperature and ΔT is a characteristic temperature variation. In the free convection configuration, a characteristic speed can be defined as

$$U_f = \frac{Kg\beta\Delta T}{\nu} \tag{11}$$

and in terms of the non-dimensional variables defined by equations (9)–(11), where $q = L$, $T_{ref} = T_\infty$, $\Delta T = T_a$ and $U_c = U_f$, the governing equations (1) and (2) can be written as

$$\frac{\partial^2\psi}{\partial x^2} + \frac{\partial^2\psi}{\partial y^2} = \frac{\partial T}{\partial y} \tag{12}$$

$$\frac{\partial T}{\partial t} + \frac{\partial\psi}{\partial y}\frac{\partial T}{\partial x} - \frac{\partial\psi}{\partial x}\frac{\partial T}{\partial y} = \frac{1}{Ra}\left(\frac{\partial^2 T}{\partial x^2} + \frac{\partial^2 T}{\partial y^2}\right) \tag{13}$$

where

$$Ra = \frac{Kg\beta\Delta T q}{\nu\alpha} \tag{14}$$

is the Rayleigh number.

At small times the heat flows from the impermeable surface into the porous medium and a thermal boundary-layer is formed adjacent to the surface. Convection is not important at this early stage and the heat is mainly conducted normal to the surface, and this suggests that the thickness of the thermal boundary-layer is proportional to $(t/Ra)^{1/2}$. Therefore, the following scalings

$$\psi = \frac{\tilde{\psi}}{Ra^{1/2}} \quad , \quad y = \frac{\tilde{y}}{Ra^{1/2}} \tag{15}$$

are first introduced to obtain the governing equations (12-13) in the form

$$\frac{1}{Ra}\frac{\partial^2 \tilde{\psi}}{\partial x^2} + \frac{\partial^2 \tilde{\psi}}{\partial \tilde{y}^2} = \frac{\partial T}{\partial \tilde{y}} \tag{16}$$

$$\frac{\partial T}{\partial t} + \frac{\partial \tilde{\psi}}{\partial \tilde{y}}\frac{\partial T}{\partial x} - \frac{\partial \tilde{\psi}}{\partial x}\frac{\partial T}{\partial \tilde{y}} = \frac{1}{Ra}\frac{\partial^2 T}{\partial x^2} + \frac{\partial^2 T}{\partial \tilde{y}^2} \tag{17}$$

and to take account of the growth of the thermal boundary-layer at small times, the further transformations

$$\tilde{\psi} = 2t^{1/2} f(x,\xi,t) \quad , \quad \tilde{y} = 2t^{1/2}\xi \tag{18}$$

are made into equations (16) and (17), which become

$$\frac{4t}{Ra}\frac{\partial^2 f}{\partial x^2} + \frac{\partial^2 f}{\partial \xi^2} = \frac{\partial T}{\partial \xi} \tag{19}$$

$$\frac{\partial T}{\partial t} + \frac{\partial f}{\partial \xi}\frac{\partial T}{\partial x} - \frac{\partial f}{\partial x}\frac{\partial T}{\partial \xi} = \frac{1}{Ra}\frac{\partial^2 T}{\partial x^2} + \frac{1}{4t}\left(\frac{\partial^2 T}{\partial \xi^2} + 2\xi\frac{\partial T}{\partial \xi}\right) \tag{20}$$

Thus, at small times, i.e. $0 < t \leq t_p$ where t_p is a time to be specified, the problem reduces to solving equations (19) and (20) in the domain

$$D_1 = \{(x,\xi) \in \mathbf{R}^2 \ : \ 0 \leq x \leq 2\pi \ , \ 0 \leq \xi < \infty\} \tag{21}$$

subject to the initial conditions

$$T_0 = \mathrm{erfc}(\xi)\sin(x) \tag{22}$$

$$f_0 = \left[\xi \,\mathrm{erfc}(\xi) - \frac{1}{\sqrt{\pi}}(e^{-\xi^2} - 1)\right]\sin(x) \tag{23}$$

and the boundary conditions

$$T(x,0,t) = \sin(x)\ , \ f(x,0,t) = 0 \qquad 0 \leq x \leq 2\pi \tag{24}$$

$$T(0,\xi,t) = T(2\pi,\xi,t)\ , \ f(0,\xi,t) = f(2\pi,\xi,t) \qquad 0 \leq \xi < \infty \tag{25}$$

$$T(x,\xi,t) \to 0\ , \ \frac{\partial f}{\partial \xi}(x,\xi,t) \to 0 \qquad \text{as } \xi \to \infty\ , \ 0 \leq x \leq 2\pi \tag{26}$$

for any $t \in (0, t_p)$. Note that the initial conditions (22) and (23) have been obtained by solving equations (19) and (20) subject to the boundary conditions (24)–(26) at $t = 0$.

Since at larger values of time, the thermal boundary-layer will only occur adjacent to the surface when Ra is large, it is appropriate to return to equations (16) and (17), and they are solved for $t > t_p$ in the domain

$$\widetilde{D}_1 = \{(x, \widetilde{y}) \in \mathbf{R}^2 \ : \ 0 \leq x \leq 2\pi \ , \ 0 \leq \widetilde{y} < \infty\} \tag{27}$$

The initial conditions for equations (16) and (17) are given by the solution of equations (19) and (20) obtained at $t = t_p$, and the boundary conditions are the following:

$$T(x, 0, t) = \sin(x) \ , \ \widetilde{\psi}(x, 0, t) = 0 \qquad 0 \leq x \leq 2\pi \tag{28}$$

$$T(0, \widetilde{y}, t) = T(2\pi, \widetilde{y}, t) \ , \ \widetilde{\psi}(0, \widetilde{y}, t) = \widetilde{\psi}(2\pi, \widetilde{y}, t) \qquad 0 \leq \widetilde{y} < \infty \tag{29}$$

$$T(x, \widetilde{y}, t) \to 0 \ , \ \frac{\partial \widetilde{\psi}}{\partial \widetilde{y}}(x, \widetilde{y}, t) \to 0 \qquad \text{as } \widetilde{y} \to \infty \ , \ 0 \leq x \leq 2\pi \tag{30}$$

for any $t \in (t_p, \infty)$.

Analytical Results

A series expansion solution in terms of the small time has been obtained by Bradean *et al.* [38] using the method of matched asymptotic expansions. The method consists of matching locally valid expansions inside and outside the thermal boundary-layer, in order to take into account the outer flow from the thermal boundary-layer. The composite solution, which is valid in the entire solution domain, was found to be

$$T = T_0 + T_1 t + \mathbf{O}(t^2) \tag{31}$$

$$f = f_0 + f_1 t + \mathbf{O}(t^2) \tag{32}$$

where

$$T_0 = \mathrm{erfc}(\xi)\sin(x) \tag{33}$$

$$T_1 = \frac{2}{Ra}\left(\xi^2\,\mathrm{erfc}(\xi) - \frac{1}{\sqrt{\pi}}\xi\,\mathrm{e}^{-\xi^2}\right)\sin(x) + g_1(\xi)\sin(2x) \tag{34}$$

$$f_0 = \left[\xi\,\mathrm{erfc}(\xi) - \frac{1}{\sqrt{\pi}}(\mathrm{e}^{-\xi^2} - \mathrm{e}^{-y})\right]\sin(x) \tag{35}$$

$$f_1 = \frac{2}{Ra}g_2(\xi)\sin(x) + g_3(\xi)\sin(2x) \tag{36}$$

and

$$g_1(\xi) = \frac{3\pi - 4}{12\pi}\left[(2\xi^2 + 1)\,\mathrm{erfc}(\xi) - \frac{2}{\sqrt{\pi}}\xi\,\mathrm{e}^{-\xi^2}\right]$$
$$+\frac{1}{4}(2\xi^2 - 1)\,\mathrm{erfc}^2(\xi) - \frac{3}{2\sqrt{\pi}}\xi\,\mathrm{e}^{-\xi^2}\,\mathrm{erfc}(\xi) + \frac{1}{\pi}\mathrm{e}^{-2\xi^2} - \frac{2}{3\pi}\mathrm{e}^{-\xi^2} \tag{37}$$

$$g_2(\xi) = \frac{1}{6}(4\xi^3 + 3\xi)\,\mathrm{erfc}(\xi) - \frac{1}{6\sqrt{\pi}}(4\xi^2 + 1)\,\mathrm{e}^{-\xi^2} + \frac{1}{\sqrt{\pi}}\xi^2 + \frac{1}{6\sqrt{\pi}}\mathrm{e}^{-y} \tag{38}$$

$$\begin{aligned} g_3(\xi) = {} & \frac{1}{12}(2\xi^3 - 3\xi)\,\mathrm{erfc}^2(\xi) - \frac{1}{12\sqrt{\pi}}(4\xi^2 - 11)\,\mathrm{e}^{-\xi^2}\,\mathrm{erfc}(\xi) \\ & - \frac{4}{3\sqrt{2\pi}}\mathrm{erfc}(\sqrt{2}\xi) + \frac{1}{3}\left[\frac{3\pi - 4}{12\pi}(2\xi^3 + 3\xi) + \frac{1}{\sqrt{\pi}}\right]\mathrm{erfc}(\xi) \\ & - \frac{3\pi - 4}{18\pi}\left[\frac{1}{\sqrt{\pi}}\xi^2 - \frac{3}{3\pi - 4}\xi + \frac{1}{\sqrt{\pi}}\right]\mathrm{e}^{-\xi^2} - \frac{(39 - 24\sqrt{2})\pi + 8}{36\pi\sqrt{\pi}}\mathrm{e}^{-2y} \end{aligned} \tag{39}$$

From this solution, analytical expressions for the mean fluid velocity along the surface, $\widetilde{Mv}$, and the mean Nusselt number, $\widetilde{Nu}$, are obtained in the form

$$\widetilde{Mv} = \int_0^{\pi} \frac{\partial\tilde{\psi}}{\partial\tilde{y}}(x, 0)\,\mathrm{d}x = 2 - \frac{4}{\sqrt{\pi Ra}}t^{1/2} + \frac{2}{Ra}t - \frac{4}{3Ra\sqrt{\pi Ra}}t^{3/2} + \mathbf{O}(t^2) \tag{40}$$

$$\widetilde{Nu} = -\int_0^{\pi} \frac{\partial T}{\partial\tilde{y}}(x, 0)\,\mathrm{d}x = \frac{1}{2t^{1/2}}\left[\frac{4}{\sqrt{\pi}} + \frac{4}{Ra\sqrt{\pi}}t + \mathbf{O}(t^2)\right] \tag{41}$$

Numerical Results

In order to obtain an approximate solution which is valid for all times, and for any value of the Rayleigh number, the problem has been solved numerically by Bradean *et al.* [39] using finite differences. Since the solution domain is infinite, the problem was reformulated for numerical purposes, both at small and large times, using a scaling in the direction normal to the surface, such that the solution domain becomes finite. The Crank-Nicolson implicit method was then used to discretise the energy equation, whereas central differences were used in the momentum equation. The resulting non-linear system of algebraic equations obtained after the discretisation of the governing equations was solved iteratively at each time step using the Successive Over Relaxation method.

The mean Nusselt number $\widetilde{Nu}$ along the surface, calculated analytically and numerically, is presented in Figure 1 for different values of the Rayleigh number. The analytical and finite-difference results are observed to be in very good agreement up to a time which increases as Ra increases, and this is because the dominant term in the analytical expansion of the solution was obtained from the boundary-layer equations. However, it is observed that for values of the Rayleigh number greater than about 100 the solution appears to be in the boundary-layer regime.

For small values of the Rayleigh number, the flow patterns are in the form of a row of counter rotating cells, and the steady state solution obtained by Bradean *et al.* [7] is approached in a short period of time. For large values of the Rayleigh number and at small times, the cellular flow separates from the surface at the point of collision of two boundary layers and this gives rise to two recirculating flow regions,

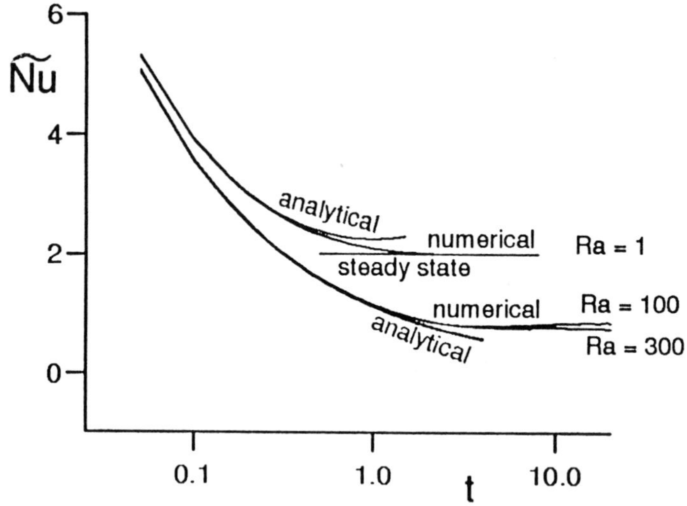

Figure 1 The mean Nusselt number, $\widetilde{Nu}$, calculated analytically and numerically at small times for $Ra = 1$, 100 and 300 and the steady state value for $Ra = 1$ as obtained by Bradean *et al.* [7]

see Figure 2 which presents the streamline patterns for $Ra = 100$ at $t = 6$ and 12. The separation process was deduced by Bradean *et al.* [39] from the time evolution of the vertical component of the fluid velocity and the horizontal temperature gradient. From Figure 2, the unsteady solution at small times appears to develop towards the steady state patterns obtained by Bradean *et al.* [7], who used the following additional symmetrical conditions in the x-direction

$$T(x,y) = 0 \quad , \quad \psi(x,y) = 0 \qquad x = k\pi \ , \ k \text{ integer} \ , \ 0 \le y < \infty \tag{42}$$

but at larger values of time small oscillations of the solution were observed in the recirculating regions. As the size of the recirculating flow regions increases in time, the amplitude of the oscillations increases until there is a complete breakdown of symmetry in the x-direction when the oscillations extend into the main cellular flow. It was found that the point of collision of two adjacent boundary layers which flow along the surface oscillates about the plane $x = \pi$, and the solution becomes periodic in time. When the Rayleigh number is sufficiently large and at very large times, only one recirculating region is developed at the time when the upper or lower cell crosses the plane $x = \pi$.

Figure 3 illustrates the time variation of the mean Nusselt number calculated for $Ra = 100$ using different grid systems and the additional symmetrical conditions (42), as well as the steady state value as obtained by Bradean *et al.* [7]. The numerical results presented for small and large times in Figure 3, show that the steady state numerical solution obtained by Bradean *et al.* [7] is not stable. The additional symmetrical conditions (42) considered in the steady state situation are only valid up to a time which decreases as the Rayleigh number increases, whereas when using conditions (42) the unsteady numerical solution at large times is in very good agreement with the steady state numerical results.

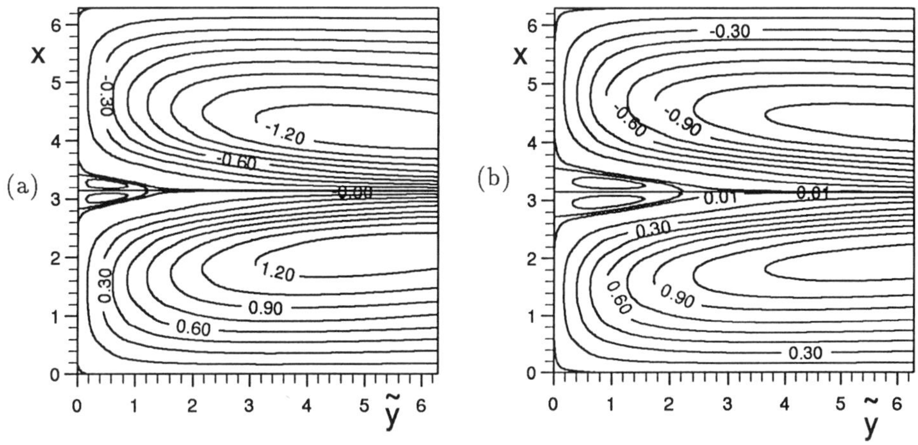

Figure 2 Streamlines for $Ra = 100$ at (a) $t = 6$ and (b) $t = 12$

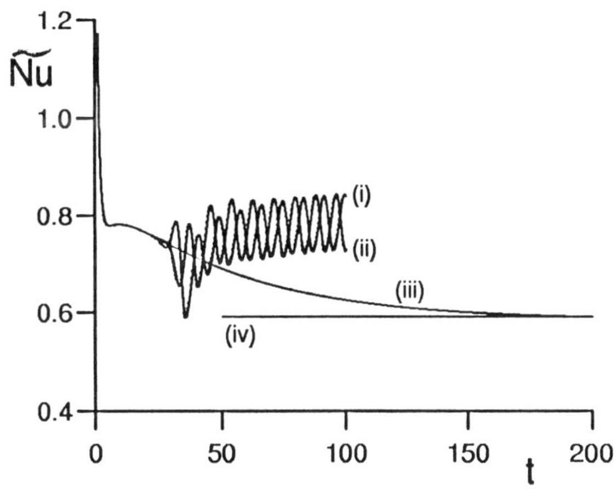

Figure 3 The mean Nusselt number, $\widetilde{Nu}$, for $Ra = 100$ calculated using (i), (ii) different grid systems, (iii) the additional symmetrical conditions (42) and (iv) the steady state as obtained by Bradean *et al.* [7]

FREE CONVECTION FROM A HORIZONTAL SURFACE

Consider the two-dimensional free convection generated by an impulsively heated and cooled, in space, horizontal surface which is embedded in a porous medium. The Cartesian coordinates (x^*, y^*) are chosen along and normal (upward) to the horizontal surface, respectively. Initially, i.e. at time $t^* = 0$, both the surface and the porous medium are at the same temperature T_∞, but at $t^* = 0^+$ the surface is suddenly heated and cooled sinusoidally along its length, namely

$$T^*(x^*, y^*, 0) = T_\infty \qquad -\infty < x^* < \infty \, , \; 0 \leq y^* < \infty \tag{43}$$

$$T^*(x^*, 0, t^*) = T_a \cos\left(\frac{x^*}{L}\right) + T_\infty \qquad -\infty < x^* < \infty \, , \; t^* > 0 \tag{44}$$

where T_a is the amplitude and $2\pi L$ the period of the surface temperature variation. The governing equations (1) and (2) can now be written in non-dimensional form as follows

$$\frac{\partial^2 \psi}{\partial x^2} + \frac{\partial^2 \psi}{\partial y^2} = -\frac{\partial T}{\partial x} \tag{45}$$

$$\frac{\partial T}{\partial t} + \frac{\partial \psi}{\partial y}\frac{\partial T}{\partial x} - \frac{\partial \psi}{\partial x}\frac{\partial T}{\partial y} = \frac{1}{Ra}\left(\frac{\partial^2 T}{\partial x^2} + \frac{\partial^2 T}{\partial y^2}\right) \tag{46}$$

where the non-dimensional variables are defined by equations (9)–(11), where $q = L$, $T_{ref} = T_\infty$, $\Delta T = T_a$ and $U_c = U_f$. Thus, the problem reduces to solving equations (45) and (46) in the domain

$$D_2 = \{(x, y) \in \mathbf{R}^2 \; : \; -\pi < x < \pi \, , \; 0 \leq y < \infty\} \tag{47}$$

subject to the initial conditions

$$T(x, y, 0) = 0 \, , \; \psi(x, y, 0) = 0 \qquad -\pi \leq x \leq \pi \, , \; 0 \leq y < \infty \tag{48}$$

and the boundary conditions

$$T(x, 0, t) = \cos(x) \, , \; \psi(x, 0, t) = 0 \qquad -\pi \leq x \leq \pi \tag{49}$$

$$T(-\pi, y, t) = T(\pi, y, t) \, , \; \psi(-\pi, y, t) = \psi(\pi, y, t) \qquad 0 \leq y < \infty \tag{50}$$

$$T(x, y, t) \to 0 \, , \; \frac{\partial \psi}{\partial y}(x, y, t) \to 0 \qquad \text{as } y \to \infty \, , \; -\pi \leq x \leq \pi \tag{51}$$

for any $t \in (0, \infty)$. In order to take into account the growth of the thermal boundary-layer at small times, Bradean *et al.* [40] reformulated the problem by introducing the transformed variables

$$\psi = \frac{\tilde{\psi}}{Ra} \quad , \quad y = \frac{\tilde{y}}{Ra^{1/2}} \tag{52}$$

and then

$$\tilde{\psi} = 4t \, f(x, \xi, t) \quad , \quad \tilde{y} = 2t^{1/2} \, \xi \tag{53}$$

in a similar way to that described in the configuration in which the surface is vertical.

Analytical Results

The matched asymptotic expansions technique has been used by Bradean *et al.* [40] to obtain an approximate analytical solution which is valid at small times. The composite expansions of the temperature, stream function, mean fluid velocity, $\widetilde{Mv}$, and mean Nusselt number, $\widetilde{Nu}$, which are valid in the entire solution domain, were obtained in the form

$$T = T_0 + T_1 t^{1/2} + T_2 t + T_3 t^{3/2} + \mathbf{O}(t^2) \tag{54}$$

$$f = f_0 + f_1 t^{1/2} + f_2 t + f_3 t^{3/2} + \mathbf{O}(t^2) \tag{55}$$

where

$$T_0 = \mathrm{erfc}(\xi)\cos(x) \tag{56}$$

$$T_1 = 0 \tag{57}$$

$$T_2 = \frac{2}{Ra}\left[\xi^2\,\mathrm{erfc}(\xi) - \frac{1}{\sqrt{\pi}}\xi\,\mathrm{e}^{-\xi^2}\right]\cos(x) \tag{58}$$

$$T_3 = \frac{4}{Ra^{1/2}}[g_4(\xi) + g_5(\xi)\cos(2x)] \tag{59}$$

$$f_0 = \left[\left(\frac{1}{2}\xi^2 + \frac{1}{4}\right)\mathrm{erfc}(\xi) - \frac{1}{2\sqrt{\pi}}\xi\,\mathrm{e}^{-\xi^2} - \frac{1}{4}\mathrm{e}^{-y}\right]\sin(x) \tag{60}$$

$$f_1 = 0 \tag{61}$$

$$f_2 = \frac{2}{Ra}\left[\left(\frac{1}{6}\xi^4 + \frac{1}{4}\xi^2\right)\mathrm{erfc}(\xi) - \frac{1}{6\sqrt{\pi}}(\xi^3 + \xi)\,\mathrm{e}^{-\xi^2}\right]\sin(x) \tag{62}$$

$$f_3 = g_6(\xi)\sin(2x) \tag{63}$$

$$g_4(\xi) = -\left(\frac{1}{3}\xi^3 + \frac{1}{4}\xi\right)\mathrm{erfc}^2(\xi) + \frac{1}{12\sqrt{\pi}}\left(7\xi^2 + \frac{5}{2}\right)\mathrm{e}^{-\xi^2}\,\mathrm{erfc}(\xi)$$

$$+\frac{13}{48}\left(\xi^3 + \frac{3}{2}\xi\right)\mathrm{erfc}(\xi) - \frac{1}{4\pi}\xi\,\mathrm{e}^{-2\xi^2} + \frac{1}{16\sqrt{\pi}}\mathrm{e}^{-\xi^2} - \frac{13}{48\sqrt{\pi}}(\xi^2 + 1)\,\mathrm{e}^{-\xi^2} \tag{64}$$

and $g_5(\xi)$ is the solution of the equation

$$g_5''(\xi) + 2\xi\, g_5'(\xi) - 6\, g_5(\xi) = \xi\,\mathrm{erfc}^2(\xi)$$

$$+\frac{1}{\sqrt{\pi}}\left(\xi^2 - \frac{1}{2}\right)\mathrm{e}^{-\xi^2}\,\mathrm{erfc}(\xi) - \frac{1}{\pi}\xi\,\mathrm{e}^{-2\xi^2} - \frac{1}{2\sqrt{\pi}}\mathrm{e}^{-\xi^2} \tag{65}$$

subject to homogeneous boundary conditions, namely

$$g_5(\xi) = 0 \qquad \text{at } \xi = 0 \tag{66}$$

$$g_5(\xi) \to 0 \qquad \text{as } \xi \to \infty \tag{67}$$

$g_6(\xi)$ is obtained by matching with the fourth-order outer flow and

$$\widetilde{Mv} = \int_0^\pi \frac{\partial \widetilde{\psi}}{\partial \widetilde{y}}(x,0)\,\mathrm{d}x = -\frac{4}{\sqrt{\pi}}t^{1/2} + \frac{2}{\sqrt{Ra}}t - \frac{4}{3Ra\sqrt{\pi}}t^{3/2} + \mathbf{O}(t^{5/2}) \qquad (68)$$

$$\widetilde{Nu} = -\int_0^{\pi/2} \frac{\partial T}{\partial \widetilde{y}}(x,0)\,\mathrm{d}x$$

$$= \frac{1}{\sqrt{\pi}}t^{-1/2} + \frac{1}{Ra\sqrt{\pi}}t^{1/2} + \frac{1}{\sqrt{Ra}}\left(\frac{2}{3} - \frac{5\pi}{32}\right)t + \mathbf{O}(t^{3/2}) \qquad (69)$$

Numerical Results

The finite-difference technique described in the vertical configuration has also been used by Bradean *et al.* [39] to obtain numerical results for the present problem. Figure 4 presents the mean Nusselt number, $\widetilde{Nu}$, along the surface at small times and shows very good agreement between the analytical and numerical results up to a time which increases as the Rayleigh number increases. The values of the mean Nusselt number calculated at small times indicate that the boundary-layer regime is approached at $Ra \approx 100/\pi$. It should be noted that the boundary-layer scalings with the Rayleigh number at small times and the steady state do not coincide and this was also obtained by Pop *et al.* [41].

The streamline and isotherm patterns for $Ra = 100/\pi$ and at different times are plotted in Figure 5, and show that a cellular flow forms adjacent to the horizontal surface, with each cell developed between the cold and hot regions of the surface.

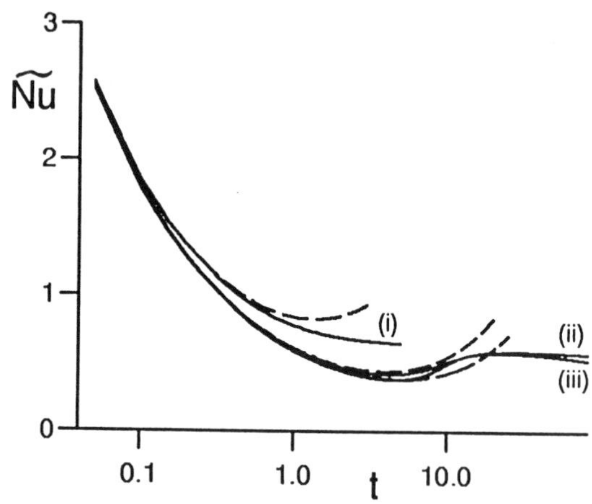

Figure 4 The mean Nusselt number, $\widetilde{Nu}$, along the surface, calculated – – – analytically and ——— numerically as a function of time for (i) $Ra = 10/\pi$, (ii) $Ra = 100/\pi$ and (iii) $Ra = 200/\pi$

Initially, both the heat and fluid flow penetrate to larger distances into the porous medium as time increases. However, the height of the cellular flow penetration reaches a maximum value at a large time, which increases as the Rayleigh number increases, and then slowly decreases towards the steady state value obtained by Bradean *et al.* [8]. As *Ra* increases, the maximum height of the cellular flow increases, whereas the height of the steady state cellular flow slightly decreases. Unlike the vertical configuration, the flow does not separate from the horizontal surface at the point of collision of the two adjacent boundary layers.

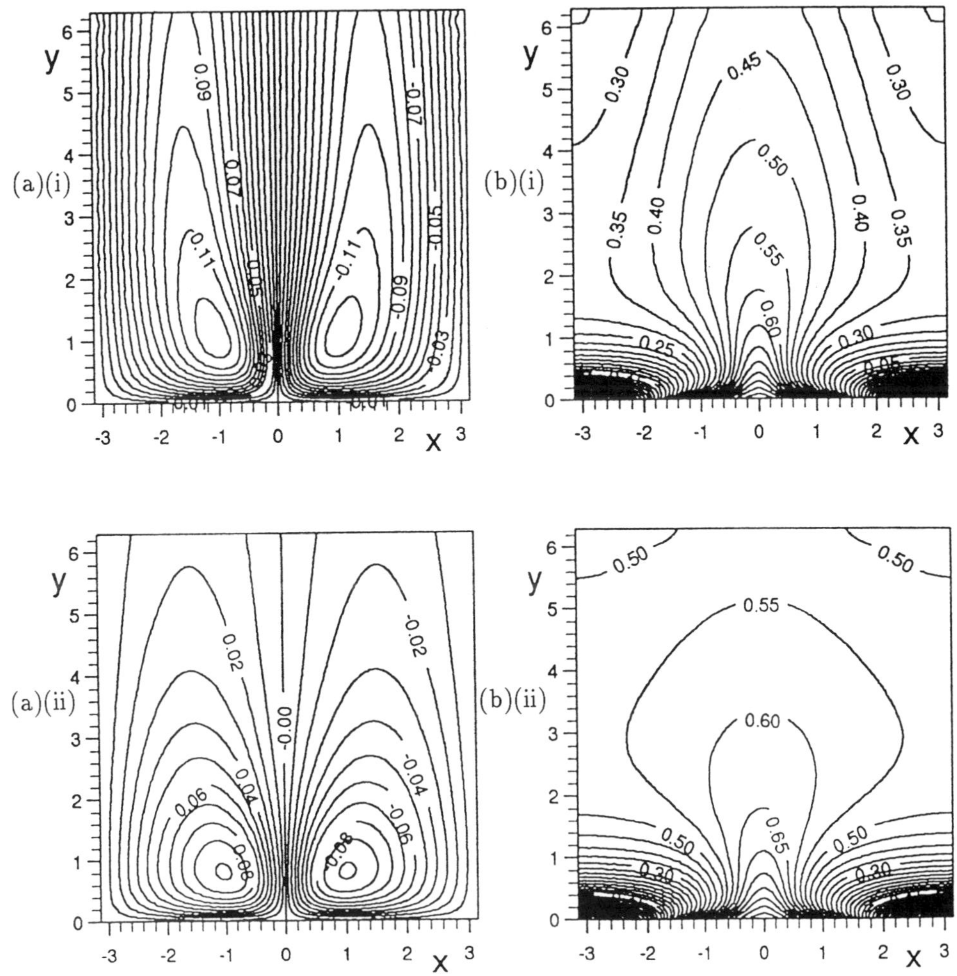

Figure 5 (a) Streamlines and (b) isotherms for $Ra = 100/\pi$ at (i) $t = 300$ and (ii) $t = 2000$

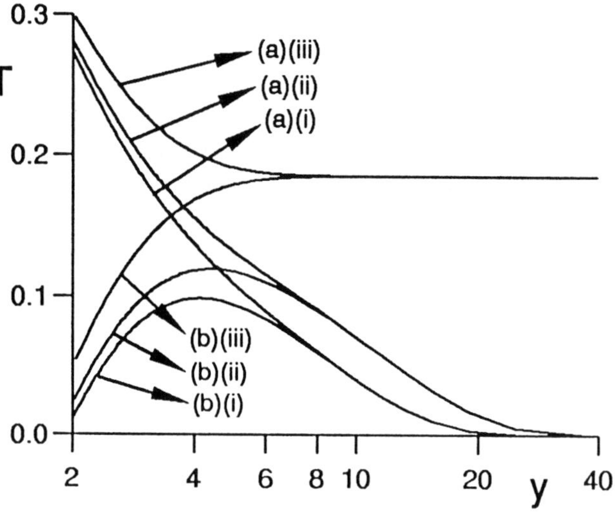

Figure 6 The temperature distribution vertically above the surface at the locations (a) $x = 0$ and (b) $x = \pi$ calculated for $Ra = 10/\pi$ at (i) $t = 100$, (ii) $t = 200$ and (iii) the steady state as obtained by Bradean *et al.* [8]

Since the hot fluid rises and the cold fluid is trapped near the surface, the temperature of one cell becomes hotter as time, or the Rayleigh number, increases and at large times above the cellular flow the heat is transferred by conduction to infinity. The heat penetration into the porous medium can be described from Figure 6, which presents the temperature distribution vertically above the surface at $x = 0$ and π for $Ra = 10/\pi$ at different large times, as well as the steady state solution as obtained by Bradean *et al.* [8]. At large times, the temperature above the cellular flow does not depend on x and decreases from a positive value to zero at infinity. As time increases, the solution very slowly develops towards the steady state solution, for which the temperature at infinite distances from the surface is a non-zero constant which depends on the Rayleigh number. However, a steady state solution which is of similar shape to that obtained at $t = 200$ and which satisfies $T \to 0$ as $y \to \infty$ is not possible. This is because the heat conduction equation has to be satisfied above the cellular flow and the solution is a constant temperature which depends on the Rayleigh number. Thus, the heat penetrates to infinite distances from the surface and the steady state solution is approached at larger times as the distance from the surface increases.

FREE CONVECTION FROM A HORIZONTAL CIRCULAR CYLINDER

Consider an infinitely long horizontal circular cylinder of radius a which is embedded in a fluid saturated porous medium which is at rest. Initially, the cylinder and the porous medium are maintained at a constant temperature, T_∞, but at time $t^* = 0^+$ the impermeable surface is impulsively heated to the constant temperature T_w ($> T_\infty$) and maintained at this value thereafter. It is considered that the resulting flow is two-dimensional and polar coordinates (r^*, θ) are chosen such that $r^* = 0$ is the axis of the cylinder and $\theta = 0$ is the direction of the gravitational force. Then, the non-dimensional governing equations (4) and (5) can be written as

$$\nabla^2 \psi = \frac{\partial T}{\partial r}\sin(\theta) + \frac{\partial T}{\partial \theta}\frac{\cos(\theta)}{r} \tag{70}$$

$$\frac{\partial T}{\partial t} + \frac{1}{r}\left(\frac{\partial \psi}{\partial r}\frac{\partial T}{\partial \theta} - \frac{\partial \psi}{\partial \theta}\frac{\partial T}{\partial r}\right) = \frac{1}{Ra}\nabla^2 T \tag{71}$$

where the non-dimensional variables are defined by

$$r = \frac{r^*}{q} \tag{72}$$

and equations (10) and (11), where $q = a$, $T_{ref} = T_\infty$, $\Delta T = T_w - T_\infty$ and $U_c = U_f$. Thus, the problem reduces to solving equations (70) and (71) in the domain

$$D_3 = \{(\theta, r) \in \mathbf{R}^2 \ , \ 0 \le \theta \le 2\pi \ , \ 1 \le r < \infty\} \tag{73}$$

subject to the initial conditions

$$T(\theta, r, 0) = 0 \ , \ \psi(\theta, r, 0) = 0 \qquad 0 \le \theta \le 2\pi \ , \ 1 \le r < \infty \tag{74}$$

and the following boundary conditions

$$T(\theta, 1, t) = 1 \ , \ \psi(\theta, 1, t) = 0 \qquad 0 \le \theta \le 2\pi \tag{75}$$

$$T(0, r, t) = T(2\pi, r, t) \ , \ \psi(0, r, t) = \psi(2\pi, r, t) \qquad 1 \le r < \infty \tag{76}$$

$$T(\theta, r, t) \to 0 \ , \ \frac{\partial \psi}{\partial r}(\theta, r, t) \to 0 \qquad \text{as } r \to \infty \ , \ 0 \le \theta \le 2\pi \tag{77}$$

for any $t \in (0, \infty)$. For this problem, small and large time formulations, similar to those described for a vertical plate, were presented by Bradean *et al.* [42] using the transformations

$$\psi = \frac{\widetilde{\psi}}{Ra^{1/2}} \quad , \quad R = r - 1 = \frac{\widetilde{R}}{Ra^{1/2}} \tag{78}$$

and then

$$\widetilde{\psi} = 2t^{1/2} \, f(\theta, \xi, t) \qquad , \qquad \widetilde{R} = 2t^{1/2} \, \xi \tag{79}$$

Analytical Results

By matching inner and outer expansions, a singular perturbation solution in terms of the small time has been derived by Pop *et al.* [43], who obtained the composite expansions for the temperature and stream function as follows

$$T = T_0 + T_1 t^{1/2} + T_2 t + \mathbf{O}(t^3/2) \tag{80}$$

$$f = f_0 + f_1 t^{1/2} + \mathbf{O}(t) \tag{81}$$

where

$$T_0 = \mathrm{erfc}(\xi) \tag{82}$$

$$T_1 = -\frac{1}{\sqrt{Ra}}\xi\ \mathrm{erfc}(\xi) \tag{83}$$

$$T_2 = \frac{1}{Ra}\xi^2\ \mathrm{erfc}(\xi) + \frac{1}{2Ra\sqrt{\pi}}\xi\ \mathrm{e}^{-\xi^2} - g_7(\xi)\cos(\theta) \tag{84}$$

$$f_0 = \left[\xi\ \mathrm{erfc}(\xi) - \frac{1}{\sqrt{\pi}}\mathrm{e}^{-\xi^2} + \frac{1}{r\sqrt{\pi}}\right]\sin(\theta) \tag{85}$$

$$f_1 = -\frac{1}{\sqrt{Ra}}\left[\left(\frac{3}{2}\xi^2 + \frac{1}{4}\right)\mathrm{erfc}(\xi) - \frac{3}{2\sqrt{\pi}}\xi\ \mathrm{e}^{-\xi^2} - \frac{1}{4r}\right]\sin(\theta) \tag{86}$$

and

$$g_7(\xi) = \frac{1}{4}(1+2\xi^2)\,\mathrm{erfc}^2(\xi) + \frac{2}{\sqrt{\pi}}\xi\ \mathrm{e}^{-\xi^2}\,\mathrm{erfc}(\xi) - \frac{2}{\pi}\mathrm{e}^{-2\xi^2}$$
$$-\frac{3\pi-8}{12\pi}\left[(1+2\xi^2)\,\mathrm{erfc}(\xi) - \frac{2}{\sqrt{\pi}}\xi\ \mathrm{e}^{-\xi^2}\right] \tag{87}$$

The mean fluid velocity, $\widetilde{Mv}$, and the mean Nusselt number, $\widetilde{Nu}$, have been calculated from this solution and are given by

$$\widetilde{Mv} = \int_0^{\pi} \frac{\partial\widetilde{\psi}}{\partial\widetilde{R}}(\theta,0)\,\mathrm{d}\theta = 2 + \mathbf{O}(t) \tag{88}$$

$$\widetilde{Nu} = -\frac{1}{\pi}\int_0^{\pi}\frac{\partial T}{\partial\widetilde{R}}(\theta,0)\,\mathrm{d}\theta = \frac{1}{\sqrt{\pi}}t^{-1/2} + \frac{1}{2\sqrt{Ra}} - \frac{1}{4Ra\sqrt{\pi}}t^{1/2} + \mathbf{O}(t) \tag{89}$$

Singular perturbation solutions in terms of the small time have also been obtained by Pop *et al.* [41] for a horizontal circular cylinder with a sudden increase in the heat flux, and by Tyvand [44] for an impulsively heated horizontal cylinder of arbitrary shape.

Numerical Results

A numerical study of the free convection from an impulsively heated horizontal circular cylinder has been performed by Bradean *et al.* [42] in a similar way to that described for a vertical plate. The mean Nusselt number, $\widetilde{Nu}$, along the surface of the circular cylinder as a function of time for different values of the Rayleigh number is plotted in Figure 7 and shows that the analytical and numerical results are in very

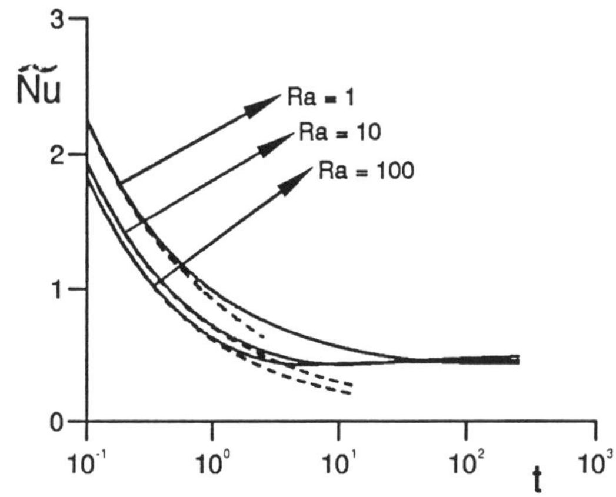

Figure 7 The mean Nusselt number, $\widetilde{Nu}$, as a function of time calculated – – – analytically and ——— numerically for $Ra = 1, 10$ and 100

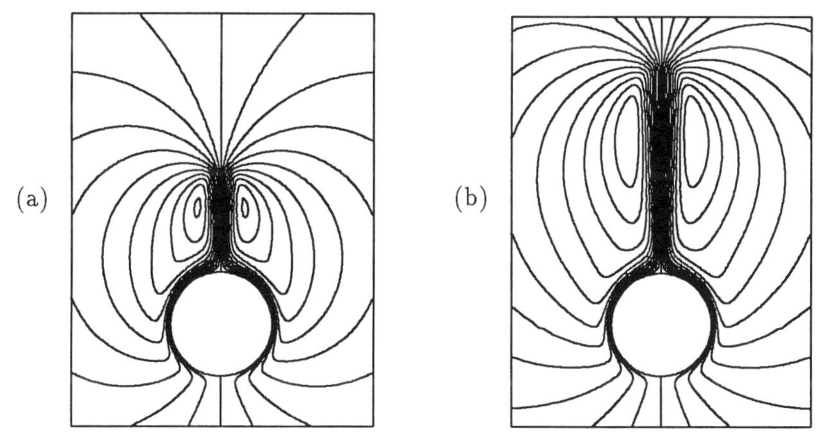

Figure 8 Streamlines for $Ra = 300$ at (a) $t = 8$ and (b) $t = 14$

good agreement up to a time which increases as the Rayleigh number increases. The numerical results obtained as Ra increases indicate that the solution is in the boundary-layer regime for values of the Rayleigh number greater than about 100. The minimum obtained in the variation of the mean Nusselt number with time can be associated with a maximum in the boundary-layer thickness on the upper half of the cylinder just before convection becomes significant.

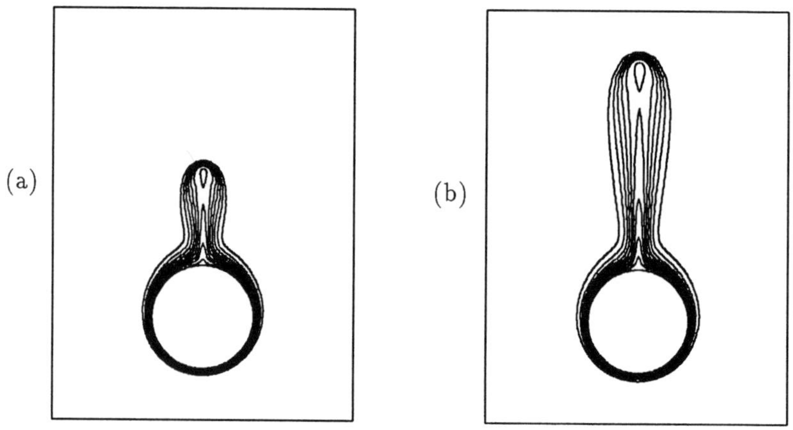

Figure 9 Isotherms for $Ra = 300$ at (a) $t = 8$ and (b) $t = 14$

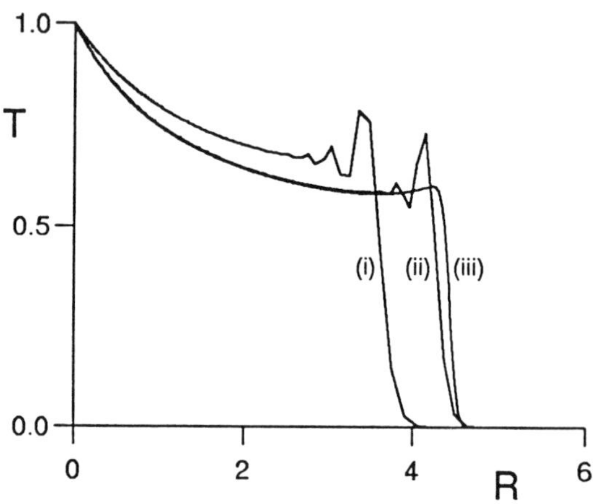

Figure 10 The temperature distribution vertically above the cylinder for $Ra = 300$ at $t = 15$ obtained using the progressively refined discretisations (i), (ii) and (iii)

Figures 8 and 9 present the streamlines and isotherms, respectively, for $Ra = 300$ at $t = 8$ and 14, and show that the flow consists of two counter rotating cellular patterns which develop near the sides of the cylinder, whose centre of rotation occurs further from the surface of the cylinder as time increases. The collision of free convection boundary layers at the top of the cylinder occurs without separation vortices, which is contrary to the non-porous situation, see Wang *et al.* [45]. After the initial conductive stage, the convective effects become increasingly more dominant and a plume region develops above the top of the cylinder. As time increases, the heat and flow penetrate to larger distances vertically above the cylinder, but only to a finite distance vertically below the cylinder. The steady state solution obtained by Ingham and Pop [17] is approached near the surface of the cylinder at larger values of time as the Rayleigh number decreases.

For an impulsively heated sphere, Yan *et al.* [46] found that the temperature distribution vertically above the sphere exhibits spatial temperature oscillations, which were attributed to a physical instability rather than to numerical errors. The temperature distribution vertically above the cylinder for $Ra = 300$ at $t = 15$ is plotted in Figure 10 using different grid systems and shows that these oscillations are probably not real as they were only obtained when using large mesh sizes. When using a sufficiently refined mesh, the spatial temperature oscillations vertically above the cylinder disappear and the temperature distribution exhibits only one local maximum and one local minimum near the top of the plume region. Thus, a single hot cell of fluid forms vertically above the cylinder and it rapidly moves away from the cylinder as time increases, see Figure 8.

MIXED CONVECTION FROM A HORIZONTAL CIRCULAR CYLINDER

An infinitely long horizontal circular cylinder of radius a, which is embedded in a fluid saturated porous medium, is initially at the same temperature T_∞ as the surrounding porous medium. It is assumed that there is an upward, or downward, fluid flow past the cylinder with a uniform constant velocity U_∞ at large distances from the cylinder. We choose polar coordinates (r^*, θ) such that $r^* = 0$ is the axis of the cylinder and $\theta = 0$ is the direction of the gravitational force and the initial flow past the cylinder can be written as

$$\psi^* = \pm U_\infty \left(r^* - \frac{a^2}{r^*} \right) \sin(\theta) \tag{90}$$

at $t^* = 0$, $a \leq r^* < \infty$ and $0 \leq \theta \leq 2\pi$, where the signs + and − correspond to upward and downward flows, respectively. At time $t^* = 0^+$ the surface of the cylinder is impulsively heated to a higher constant temperature T_w than that of the surrounding porous medium and, therefore, two-dimensional mixed convection is generated adjacent to the surface of the cylinder. The non-dimensional governing equations (4) and (5) become

$$\nabla^2\psi = \frac{Ra}{Pe}\left[\frac{\partial T}{\partial r}\sin(\theta) + \frac{\partial T}{\partial \theta}\frac{\cos(\theta)}{r}\right] \tag{91}$$

$$\frac{\partial T}{\partial t} + \frac{1}{r}\left(\frac{\partial \psi}{\partial r}\frac{\partial T}{\partial \theta} - \frac{\partial \psi}{\partial \theta}\frac{\partial T}{\partial r}\right) = \frac{1}{Pe}\nabla^2 T \tag{92}$$

where

$$Pe = \frac{U_\infty a}{\alpha} \tag{93}$$

is the Peclét number and the non-dimensional variables are defined by equations (72) and (10), where $q = a$, $T_{ref} = T_\infty$, $\Delta T = T_w - T_\infty$ and $U_c = U_\infty$.

It is now convenient to introduce the transformed stream function χ defined by

$$\psi = \chi \pm \left(r - \frac{1}{r}\right)\sin(\theta) \tag{94}$$

so that $\chi = 0$ at $t = 0$, $1 \le r < \infty$ and $0 \le \theta \le 2\pi$, and the governing equations (91) and (92) become

$$\nabla^2\chi = \frac{Ra}{Pe}\left[\frac{\partial T}{\partial r}\sin(\theta) + \frac{\partial T}{\partial \theta}\frac{\cos(\theta)}{r}\right] \tag{95}$$

$$\frac{\partial T}{\partial t} + \frac{1}{r}\left(\frac{\partial \chi}{\partial r}\frac{\partial T}{\partial \theta} - \frac{\partial \chi}{\partial \theta}\frac{\partial T}{\partial r}\right) \pm E(T) = \frac{1}{Pe}\nabla^2 T \tag{96}$$

where the operator E is defined by

$$E = \frac{\sin(\theta)}{r}\left(1 + \frac{1}{r^2}\right)\frac{\partial}{\partial \theta} - \cos(\theta)\left(1 - \frac{1}{r^2}\right)\frac{\partial}{\partial r} \tag{97}$$

Thus, the problem reduces to solving equations (95) and (96) in the domain D_3, given by expression (73), subject to the initial and boundary conditions

$$T(\theta, r, 0) = 0\ ,\ \chi(\theta, r, 0) = 0 \qquad 0 \le \theta \le 2\pi\ ,\ 1 \le r < \infty \tag{98}$$

$$T(\theta, 1, t) = 1\ ,\ \chi(\theta, 1, t) = 0 \qquad 0 \le \theta \le 2\pi \tag{99}$$

$$T(0, r, t) = T(2\pi, r, t)\ ,\ \chi(0, r, t) = \chi(2\pi, r, t) \qquad 1 \le r < \infty \tag{100}$$

$$T(\theta, r, t) \to 0\ ,\ \frac{\partial \chi}{\partial r}(\theta, r, t) \to 0 \qquad \text{as } r \to \infty\ ,\ 0 \le \theta \le 2\pi \tag{101}$$

for any $t \in (0, \infty)$. Proceeding in a similar way to that described in the free convection configuration, the problem can be reformulated at small and large times using the transformations

$$\chi = \frac{Ra}{Pe^{3/2}}\widetilde{\chi}\quad ,\quad R = r - 1 = \frac{\widetilde{R}}{Pe^{1/2}} \tag{102}$$

and then

$$\widetilde{\chi} = 2t^{1/2}\, f(\theta, \xi, t)\quad ,\quad \widetilde{R} = 2t^{1/2}\,\xi \tag{103}$$

Analytical Results

In this mixed convection configuration, an analytical solution, which is valid at small times, has been obtained using the matched asymptotic expansions technique in a similar way to that described by Pop *et al.* [42]. The composite expansions of the temperature and stream function were found to be given by expressions (80) and (81) where T_0 and f_0 are given by expressions (82) and (85), respectively, and

$$T_1 = -\frac{1}{\sqrt{Pe}}\xi \operatorname{erfc}(\xi) \tag{104}$$

$$T_2 = \frac{1}{Pe}\xi^2 \operatorname{erfc}(\xi) + \frac{1}{2Pe\sqrt{\pi}}\xi \, e^{-\xi^2} - \frac{Ra}{Pe} g_8(\xi) \cos(\theta) \tag{105}$$

$$f_1 = -\frac{1}{\sqrt{Pe}}\left[\left(\frac{3}{2}\xi^2 + \frac{1}{4}\right) \operatorname{erfc}(\xi) - \frac{3}{2\sqrt{\pi}}\xi \, e^{-\xi^2} - \frac{1}{4r}\right] \sin(\theta) \tag{106}$$

where

$$g_8(\xi) = g_7(\xi) \pm \frac{2Pe}{Ra\sqrt{\pi}}\xi \, e^{-\xi^2} \tag{107}$$

and $g_7(\xi)$ is given by expression (87). The fluid velocity along the surface, $\widetilde{v}$, the mean fluid velocity, $\widetilde{Mv}$, and the mean Nusselt number, $\widetilde{Nu}$, have been calculated as follows

$$\widetilde{v}(\theta,0) = \frac{\partial \widetilde{\psi}}{\partial \widetilde{R}}(\theta,0) = \left(1 \pm \frac{2Pe}{Ra}\right) \sin(\theta) + \mathbf{O}(t) \tag{108}$$

$$\widetilde{Mv} = \frac{1}{\pi}\int_0^\pi \frac{\partial \widetilde{\psi}}{\partial \widetilde{R}}(\theta,0)\, d\theta = \frac{2}{\pi}\left(1 \pm \frac{2Pe}{Ra}\right) + \mathbf{O}(t) \tag{109}$$

$$\widetilde{Nu} = -\frac{1}{\pi}\int_0^\pi \frac{\partial T}{\partial \widetilde{R}}(\theta,0)\, d\theta = \frac{1}{\sqrt{\pi}}t^{-1/2} + \frac{1}{2\sqrt{Pe}} - \frac{1}{4Pe\sqrt{\pi}}t^{1/2} + \mathbf{O}(t) \tag{110}$$

where the transformed stream function $\widetilde{\psi}$ is defined by

$$\psi = \frac{Ra}{Pe^{3/2}}\widetilde{\psi} \tag{111}$$

In the configuration of an initial downward fluid flow past the cylinder, expression (108) shows that the direction of the fluid velocity along the surface at small times depends on the values of the Rayleigh and Peclét numbers. Thus, depending on whether $Ra > 2Pe$ or $Ra < 2Pe$, the direction of the fluid velocity along the surface at small times is opposite or in the same direction as that of the initial fluid flow past the cylinder. In order to determine the direction of the fluid velocity along the surface at small times in the situation when $Ra = 2Pe$, then higher-order approximations have to be obtained in the composite expansion solution.

Numerical Results

The numerical solution for both upward (aiding) and downward (opposing) external flows has been obtained using a finite-difference technique, similar to those described by Bradean *et al.* [39, 42] in the free convection configuration.

Aiding flow. In the situation in which the direction of the initial fluid flow past the cylinder is vertically upward, the Nusselt number is plotted for $Ra = 140$ and $Pe = 70$ as a function of time in Figure 11. The analytical and numerical results are found to be in very good agreement up to a non-dimensional time which decreases as the Rayleigh number increases. This is contrary to the free convection configuration since the non-dimensional time is defined in a different way. The numerical results obtained for different values of the Rayleigh and Peclét numbers have shown that the minimum in the time variation of the mean Nusselt number occurs only in the situation in which the free convection is significant in comparison with the forced convection.

The heat penetration into the porous medium was found to be similar to that in the free convection configuration, with a plume region being developed vertically above the cylinder for any value of Ra and Pe. However, the flow patterns qualitatively depend on the ratio between the Rayleigh and Peclét numbers. Figure 12, which presents the streamlines for $Ra = 140$ and $Pe = 7$ at different times, shows that for values of the Rayleigh number sufficiently larger than those of the Peclét number, two vortices develop near the sides of the cylinder. These vortices are then swept away from the cylinder by the external flow.

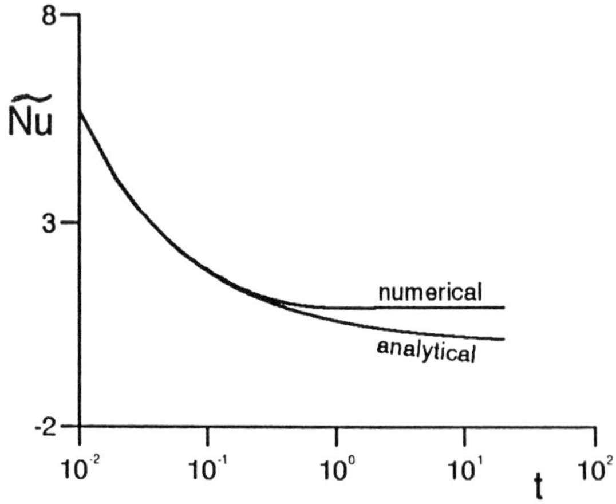

Figure 11 The mean Nusselt number, $\widetilde{Nu}$, along the surface calculated analytically and numerically for $Ra = 140$ and $Pe = 70$

(a) (b)

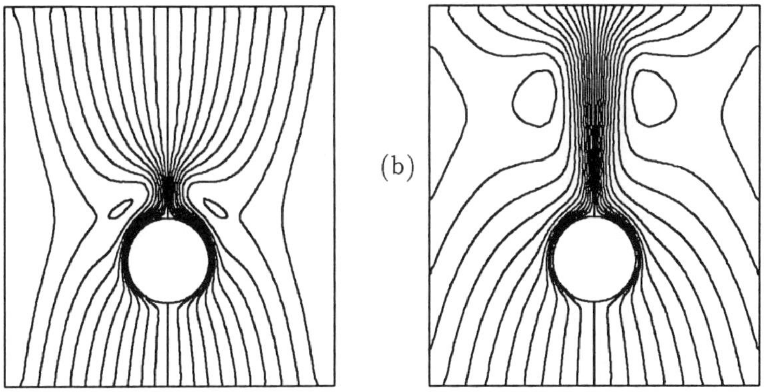

Figure 12 Streamlines for $Ra = 140$ and $Pe = 7$ at (a) $t = 0.25$ and (b) $t = 0.75$

As time increases, the unsteady numerical solution near the surface of the cylinder approaches an asymptotic value which is in reasonable agreement with the steady state numerical solution obtained by Badr and Pop [24], who enforced a zero non-dimensional temperature at a very large distance from the cylinder. However, the unsteady numerical results obtained at very large times indicate that the heat penetrates upward with almost constant speed, and therefore, the zero non-dimensional temperature boundary condition at very large distances from the cylinder is not appropriate in the steady state situation.

Opposing flow. When the direction of the initial fluid flow past the cylinder is vertically downward, the numerical results obtained for different ratios between the Rayleigh and Peclét numbers have shown that both the streamline and isotherm patterns qualitatively depend on this ratio. Thus, in the situation in which forced convection is dominant, namely $Ra \ll Pe$, the flow patterns are very slightly perturbed from those of the external flow past the cylinder. The heat is convected to larger distances vertically below the cylinder as time increases but the upward heat penetration is finite.

In the free convection dominated regime, namely $Ra \gg Pe$, two counter rotating cellular patterns develop near the sides of the cylinder. The heat transfer is almost entirely confined to the cellular patterns and the evolution of the isotherm and streamline cellular patterns is qualitatively similar to that described in the free convection configuration. However, the external flow delays the heat and cellular flow penetration vertically upward into the porous medium.

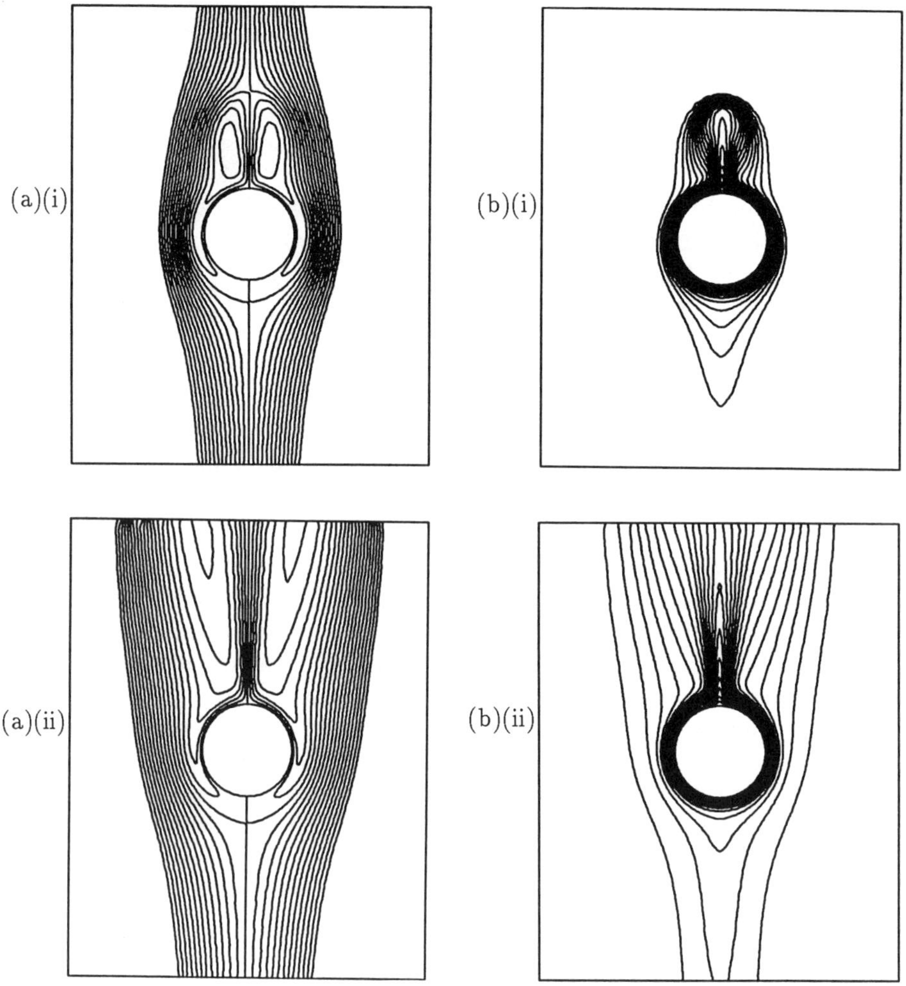

Figure 13 (a) Streamlines and (b) isotherms for $Ra = 280$ and $Pe = 70$ at (i) $t = 6$ and (ii) $t = 40$

Finally, when neither free nor forced convection is dominant, namely $Ra \sim Pe$, Figure 13 presents the evolution of the streamline and isotherm patterns for $Ra = 280$ and $Pe = 70$. It is found that two counter rotating cellular patterns also develop near the sides of the cylinder, and that these cellular patterns penetrate further upward into the porous medium as time increases, but only to a finite distance below the cylinder. The heat transfer is not confined only to the cellular patterns and the heat which escapes into the mainstream of the external flow is convected to larger distances vertically below the cylinder as time increases. Therefore, the heat penetrates into the porous medium infinitely above and below the cylinder.

CONCLUSIONS

Recent theoretical work performed on the two-dimensional heat convection from suddenly heated surfaces embedded in a porous medium at finite values of the Rayleigh number have been reviewed in this chapter. In all the particular configurations presented, an analytical solution in terms of the small time can be obtained using the matched asymptotic expansions technique. The small and large time formulations predict the growth of the thermal boundary-layer at small times and large times and give confidence in the accuracy of the numerical solutions.

The analytical and numerical results showed that the free convection from heated and cooled, in space, flat surfaces is qualitatively different in the horizontal and vertical configurations. Thus, the flow only separates from the vertical surface, whereas the heat penetrates to infinite distances from the surface in the horizontal configuration. In the free convection from an impulsively heated horizontal circular cylinder, an important new phenomena, namely that the temperature distribution becomes non-monotonic and a cell of hot fluid develops vertically above the cylinder, is observed. In the mixed convection configuration, vortices near the surface of the cylinder can develop in both situations of aiding and opposing external flows.

In all the free convection configurations reviewed in this chapter, the mean Nusselt number for large values of the Rayleigh number exhibits a minimum value in time, which is associated with a maximum thickness of the thermal boundary-layer at small times. This effect has different causes, namely the non-uniform temperature distribution along the flat surface, and the curvature of the surface of the isothermally heated horizontal circular cylinder. However, in the mixed convection configuration in which the forced convection is dominant, the values of the mean Nusselt number are monotonic in time.

REFERENCES

1. P. Cheng, Heat transfer in geothermal systems, *Adv. Heat Transfer* **14**, 1-105 (1978).
2. A. Bejan, Convective heat transfer in porous media. Handbook of Single-Phase Convective Heat Transfer (eds. S. Kakac, R. K. Shah and W. Aung), Chapter 16, Wiley, New-York (1987).
3. S. Kakac, B. Kilkis, F. A. Kulacki and F. Arinc, Convective Heat and Mass Transfer in Porous Media, Kluwer Academic Publishers, Dordrecht (1991).
4. C. L. Tien and K. Vafai, Convective and radiative heat transfer in porous media, *Adv. Appl. Mech.* **27**, 225-281 (1990).
5. D. A. Nield and A. Bejan, Convection in Porous Media, Springer, New-York (1992).
6. K. Yamamoto, Natural convection about a heated sphere in a porous medium, *J. Phys. Soc. Japan* **37**, 1164-1166 (1974).

7. R. Bradean, D. B. Ingham, P. J. Heggs and I. Pop, Free convection fluid flow due to a periodically heated and cooled vertical plate embedded in a porous media, *Int. J. Heat Mass Transfer* **39**, 2545-2557 (1996).

8. R. Bradean, D. B. Ingham, P. J. Heggs and I. Pop, Buoyancy-induced flow adjacent to a periodically heated and cooled horizontal surface in porous media, *Int. J. Heat Mass Transfer* **39**, 615-630 (1996).

9. P. Cheng and W. J. Minkowycz, Free convection about a vertical flat plate embedded in a porous medium with application to heat transfer from a dike, *J. Geophys. Res.* **82**, 2040-2044 (1977).

10. D. B. Ingham and S. N. Brown, Flow past a suddenly heated vertical plate in a porous medium, *Proc. R. Soc. Lond.* **A403**, 51-80 (1982).

11. P. Cheng and I. D. Chang, Buoyancy induced flows in a saturated porous medium adjacent to impermeable horizontal surfaces, *Int. J. Heat Mass Transfer* **19**, 1267-1272 (1976).

12. J. H. Merkin, Free convection boundary layers on axisymmetric and two dimensional bodies of arbitrary shape in a saturated porous medium, *Int. J. Heat Mass Transfer* **22**, 1461-1462 (1979).

13. P. Cheng and I. D. Chang, Convection in a porous medium as a singular perturbation problem, *Lett. Heat Mass Transfer* **6**, 253-258 (1979).

14. P. Cheng and C. T. Hsu, Higher-order approximations for Darcian free convective flow about a semi-infinite vertical flat plate, *ASME J. Heat Transfer* **106**, 143-151 (1984).

15. I. D. Chang and P. Cheng, Matched asymptotic expansions for free convection about an impermeable horizontal surface in a porous medium, *Int. J. Heat Mass Transfer* **26**, 163-173 (1983).

16. V. Joshi and B. Gebhart, Vertical natural convection flows in porous media: calculations of improved accuracy, *Int. J. Heat Transfer* **27**, 69-75 (1984).

17. D. B. Ingham and I. Pop, Natural convection about a heated horizontal cylinder in a porous medium, *J. Fluid Mech.* **184**, 157-181 (1987).

18. I. Pop and D. B. Ingham, Natural convection about a heated sphere in a porous medium, Proc. 9th Int. Heat Transfer Conf. **2**, 567-572 (1990).

19. P. Cheng, Combined free and forced boundary layer flows about inclined surfaces in a porous medium, *Int. J. Heat Mass Transfer* **20**, 807-814 (1977).

20. P. Cheng, Similarity solutions for mixed convection from horizontal impermeable surfaces in saturated porous media, *Int. J. Heat Mass Transfer* **20**, 893-898 (1977).

21. P. Cheng, Mixed convection about a horizontal cylinder and a sphere in a fluid saturated porous medium, *Int. J. Heat Mass Transfer* **25**, 1245-1247 (1982).

22. Y. Joshi and B. Gebhart, Mixed convection in porous media adjacent to a vertical uniform heat flux surface, *Int. J. Heat Mass Transfer* **28**, 1783-1786 (1985).

23. T. Sano, Mixed convection around a horizontal circular cylinder immersed in a Darcy flow, *J. Engng. Math.* **27**, 221-232 (1993).

24. H. Badr and I. Pop, Combined convection from an isothermal horizontal rod buried in a porous medium, *Int. J. Heat Mass Transfer* **31**, 2527-2541 (1988).

25. D. B. Ingham, J. H. Merkin and I. Pop, Flow past a suddenly cooled vertical flat surface in a saturated porous medium, *Int. J. Heat Mass Transfer* **25**, 1916-1919 (1982).

26. D. B. Ingham, J. H. Merkin and I. Pop, Flow past a suddenly cooled horizontal flat surface in a saturated porous medium, *Acta Mech.* **56**, 205-217 (1985).

27. P. Cheng and I. Pop, Transient free convection about a vertical flat plate embedded in a porous medium, *Int. J. Engng. Sci.* **22**, 253-264 (1984).

28. I. Pop and P. Cheng, The growth of a thermal layer in a porous medium adjacent to a suddenly heated semi-infinite horizontal surface, *Int. J. Heat Mass Transfer* **26**, 1574-1576 (1983).

29. D. B. Ingham, J. H. Merkin and I. Pop, The collision of free convection boundary layers on a horizontal cylinder embedded in a porous medium, *Q. Jl. Mech. Appl. Math.* **36**, 313-335 (1983).

30. K. N. Mehta and S. Sood, Transient free convective flow about a non- isothermal vertical flat plate immersed in a saturated inhomogeneous porous medium, *Int. Comm. Heat Mass Transfer* **19**, 687-699 (1992).

31. A. Nakayama and C. D. Ebinuma, Transient non-Darcy forced convective heat transfer from a flat plate embedded in a fluid saturated porous medium, *Int. J. Heat Fluid Flow* **11**, 249-263 (1990).

32. A. Bejan, Natural convection in an infinite porous medium with a concentrated heat source, *J. Fluid Mech.* **89**, 97-107 (1978).

33. R. Purushothaman, R. Ganapathy and P. S. Hiremath, Free convection in an infinite porous medium due to a pulsating point heat source, *ZAMM Z. angew. Math. Mech.* **70**, 41-47 (1990).

34. D. A. Nield and S. P. White, Natural convection in an infinite porous medium produced by a line heat source, Mathematics and models in Engineering Science (ed. McNabb *et al.*) Dept. Sci. Indust. Res., Wellington, New Zealand, 121-128 (1982).

35. T. Sano and R. Okihara, Natural convection around a sphere immersed in a porous medium at small Rayleigh numbers, *Fluid Dyn. Res.* **13**, 39-44 (1994).

36. R. Ganapathy, Thermal convection in an infinite porous medium due to a source in sphere, *Fluid Dyn. Res.* **9**, 223-234 (1992).

37. P. Singh, J. K. Misra and K. A. Narayan, A mathematical analysis of unsteady flow and heat transfer in a porous medium, *Int. J. Engng. Sci.* **24**, 277-287 (1986).

38. R. Bradean, D. B. Ingham, P. J. Heggs and I. Pop, The penetration of unsteady free convection from a heated and cooled vertical flat surface in a porous media. Accepted for publication in the Proceedings of the International Symposium on Advances in Computational Heat Transfer, Cesme, Turkey (1997).

39. R. Bradean, D. B. Ingham, P. J. Heggs and I. Pop, The unsteady penetration of free convection flows caused by heating and cooling flat surfaces in a porous media, *Int. J. Heat Mass Transfer* **40**, 665-687 (1997).

40. R. Bradean, D. B. Ingham, P. J. Heggs and I. Pop, Unsteady free convection from a horizontal surface embedded in a porous media, Proc. 2nd European Thermal-Sciences and 14th UIT National Heat Transfer Conference (eds. G. P. Celata, P. Di Marco and A. Mariani), Rome, Italy, 329-335 (1996).

41. I. Pop, D. B. Ingham and R. Bradean, Transient free convection about a horizontal circular cylinder in a porous medium with constant surface flux heating, *Acta Mech.* **119**, 79-91 (1996).

42. R. Bradean, D. B. Ingham, P. J. Heggs and I. Pop, Unsteady free convection adjacent to an impulsively heated horizontal circular cylinder in porous media. Accepted for publication in *Num. Heat Transfer* (1997).

43. I. Pop, D. B. Ingham and P. Cheng, Transient free convection about a horizontal circular cylinder in a porous medium, *Fluid Dyn. Res.* **12**, 295-305 (1993).

44. P. A. Tyvand, First-order transient free convection about a horizontal cylinder embedded in a porous medium, *Fluid Dyn. Res.* **15**, 277-294 (1995).

45. P. Wang, R. Kahawita and D. L. Nguyen, Transient laminar natural convection from horizontal cylinders, *Int. J. Heat Mass Transfer* **34**, 1429-1442 (1991).

46. B. Yan, I. Pop and D. B. Ingham, A numerical study of unsteady free convection from a sphere in a porous medium, *Int. J. Heat Mass Transfer* **40**, 893-903 (1997).